PRENTICE-HALL INTERNATIONAL SERIES IN ENGINEERING

BAZOVSKY *Reliability Theory and Practice*
BLACKWELL AND KOTZEBUE *Semiconductor-Diode Parametric Amplifiers*
DRESHER *Games of Strategy: Theory and Applications*
EDMUNDSON, ED. *Proceedings of the National Symposium on Machine Translation*
KOLK *Modern Flight Dynamics*
LONG *Mechanics of Solids and Fluids*
ROHSENOW AND CHOI *Heat, Mass, and Momentum Transfer*
STANTON *Numerical Methods for Science and Engineering*

PRENTICE-HALL, INC.
PRENTICE-HALL INTERNATIONAL, INC., UNITED KINGDOM AND EIRE
PRENTICE-HALL OF CANADA, LTD., CANADA

PRENTICE-HALL SERIES IN ENGINEERING
OF THE PHYSICAL SCIENCES

K. R. Wadleigh and J. B. Reswick, editors

LONG *Mechanics of Solids and Fluids*
ROHSENOW AND CHOI *Heat, Mass, and Momentum Transfer*

HEAT, MASS
AND
MOMENTUM TRANSFER

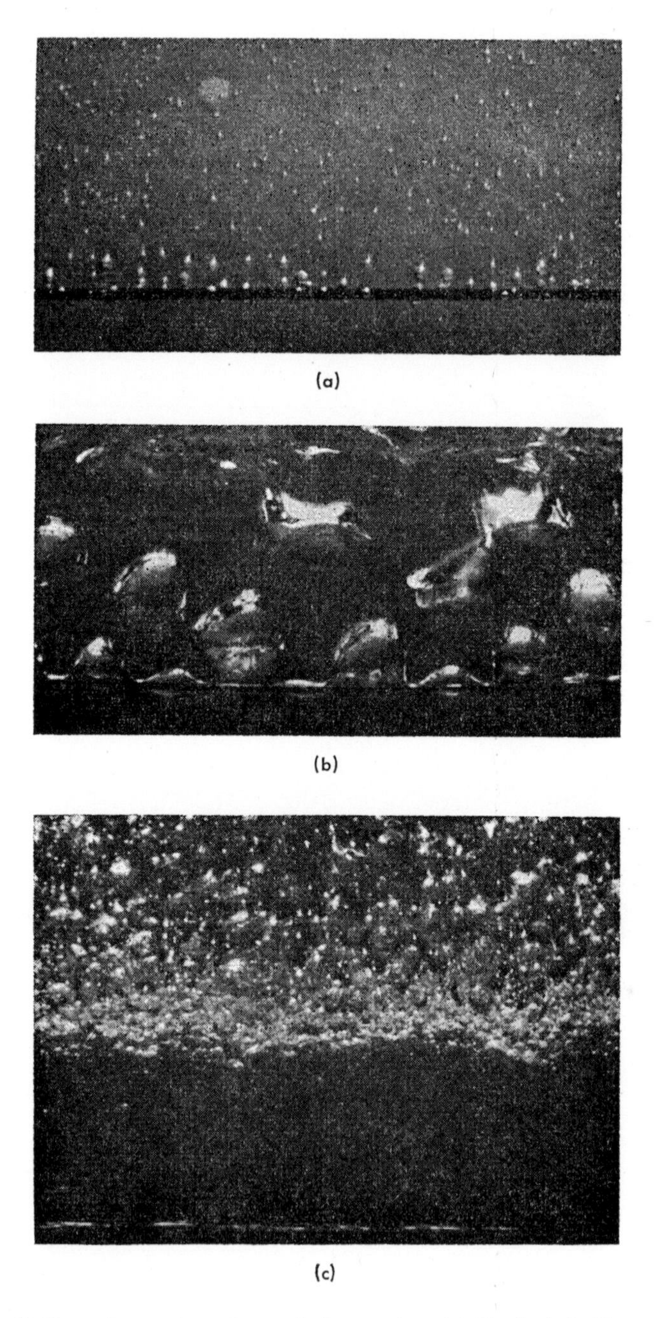

Boiling of n-pentane on a platinum wire. (a) Nucleate boiling; (b) film boiling; (c) boiling in presence of radial electrostatic field.

HEAT, MASS,
AND
MOMENTUM TRANSFER

Warren M. Rohsenow

PROFESSOR OF MECHANICAL ENGINEERING
MASSACHUSETTS INSTITUTE OF TECHNOLOGY

Harry Y. Choi

ASSOCIATE PROFESSOR OF MECHANICAL ENGINEERING
TUFTS UNIVERSITY

Prentice-Hall, Inc.

ENGLEWOOD CLIFFS, NEW JERSEY

Library of Congress Catalog Card No. 61–15516
Printed in the United States of America
48418–C

Current printing (last digit):
13 12 11 10 9 8 7 6 5

TO

TOWNELEY SMITH ROHSENOW

SZU-WEN WU CHOI

Preface

In recent years our knowledge of heat and mass transfer processes has been considerably extended. Our aim is to provide a comprehensive and fundamental treatment of the various transfer processes, and also to explore selectively the state of the art of experimental knowledge in the major areas of application. The subject matter is developed in the sequence dictated by the logic of analysis, and experimental results are integrated with the analysis to emphasize its validity and to provide a rational basis for design. In many areas, analysis lags well behind the accumulation of experimental results.

The organization of the material provides parallel treatments of momentum transfer (Chapters 2, 3, and 4), heat transfer (Chapters 5, 6, 7, and 8), and mass transfer (Chapters 14, 15, and 16). Analogous electricity transfer processes appear whenever appropriate. The parts of heat transfer that are concerned with boiling, condensation, radiation, high speed flow, rarefied gases, and dissociating gases—all of which have no convenient counterparts in mass transfer—are placed in Part II together with material on heat exchanger design. These chapters (9–13) are organized so that any topic can be taken up after Chapter 8. Aids in the treatment of both analysis and experiment—dimensional analysis, finite difference calculations, and analog solutions—are grouped together in Part IV.

Kinetic theory explanations of transport properties and coupled processes (irreversible thermodynamics) applied to heat conduction in anisotropic materials and to thermoelectricity are given introductory treatments in Part IV.

This arrangement of material provides maximum flexibility. For example, the analytical treatment emphasizes one- and two-dimensional cases but at the end of Chapters 3 and 6 three-dimensional equations are derived for momentum and heat transfer. This could be rearranged to begin with the three-dimensional formulation as we do in Chapter 15 for mass transfer. One person may prefer to read momentum, heat, and mass transfer separately as outlined in the book, but another may prefer a simultaneous attack—making Chapters 2, 5, and 14, a unit; Chapters 3, 6, 7, and 15, a second unit; and Chapters 4, 8, and 16, still a third unit. The material on dimensional analysis, finite differences, and analogs (Chapters 17 and 18) may be read colaterally with other chapters according to individual preference. The subject of conduction in anisotropic materials has been placed in Chapter 19 since it gives a logical introduction to the subject of coupled phenomena and irreversible thermodynamics.

This book contains many carefully selected problems arranged in sequence with the text material. These problems are an integral part of the text; in fact, many of them extend the subject matter of the text, and provide the engineering applications of the text material.

While we suggest the study of momentum, heat, and mass transfer as one subject, it should be emphasized that the similarity of these processes is limited. The physics of the three processes is quite different, being similar only in those special cases where the mathematical formulation is identical. For example, when frictional effects in the energy equation are not negligible the similarity disappears because there are no analogous terms in the momentum equations and in the mass transfer equations. Nevertheless, there are many significant cases in which the mathematical formulations are identical so that ease in teaching and learning along with a certain economy of time favor the study of the three processes together.

Departing from tradition, we have acknowledged only two modes of heat transfer—conduction and radiation. Convection is treated simply as fluid motion (Chapter 5). Also, in order to emphasize the similarity in the rate equations, we have defined positive shear stress in terms of the negative of the velocity gradient (Chapter 1). Although contrary to the conventional definition in the literature of fluid mechanics, it permits the suggestion that shear stress can be thought of as rate of momentum transfer in the direction of decreasing potential (velocity). The final equation in terms of velocities (Navier-Stokes Equation) is unchanged. The conversion factors $g_0 = 32.16\ \text{lb}_m\ \text{ft/lb}_f\ \text{sec}^2$ and $J = 778\ \text{ft–lb}_f/\text{Btu}$ have been omitted except in Chapter 9.

The form of the material presented evolved from teaching the subject at M.I.T. over the past fifteen years and at Tufts University for the past four years. The idea of presenting a unified treatment of heat, mass, and momentum transfer evolved quite naturally and was discussed for some time among teachers of heat transfer. A most significant work on the unified treatment is the book by Bosworth[1] which extends the treatment to include electricity flow, chemical reaction, magnetic displacement, and elastic systems as transfer processes. The ASEE Committee on Evaluation of Engineering Education, active in the mid-1950's, emphasized the desirability of a unified treatment of these processes.[2] Recently, books by Bird, et al,[3] Foust, et al,[4] and Lykov and Mikhaylov[5] have emphasized again these suggestions.

This book is addressed to students, engineers, and applied scientists. It has been used for some time as a text in a one-semester, senior-level course and, with some augmenting, has formed the basis of a graduate-level course. It is assumed that the reader has studied mathematics through the first course in ordinary differential equations and has studied one semester of thermodynamics. Although it is not absolutely necessary, he would find it helpful to have studied one semester of fluid mechanics.

We are indebted to many people who have influenced this work either knowingly or unknowingly. After a number of years, it is difficult to know exactly which teachers and colleagues were instrumental in implanting specific ideas in one's mind. Even though they cannot be identified, the effects of these associations are real. We are grateful for close associations with Professors J. H. Keenan, W. H. McAdams, H. C. Hottel, J. Kaye, L. Trefethen, J. A. Clark, P. Griffith, R. J. Nickerson, L. C. Hoagland, G. A. Brown, and S. W. Gouse. In particular, we are indebted to Professors Nickerson and Hoagland for their very careful reviewings and their many suggestions for changes in the manuscript. We are particularly grateful to Nicki for her patient interpretation of our scribblings, necessary to convert them into a manuscript.

W. R.
H. C.

---

[1] Bosworth, R. C. L., *Transport Processes In Applied Chemistry*, Wiley, New York, 1956.

[2] "Report on Engineering Sciences," *ASEE Jour. of Engineering Education*, **49**, 1, Oct. 1958.

[3] Bird, R. B., W. E. Stewart, and E. N. Lightfoot, *Transport Phenomena*, Wiley, New York, 1961.

[4] Foust, A. S., *et al*, *Principles of Unit Operations*, Part II, Wiley, New York, 1960.

[5] Lykov, A. V., and Y. A. Mikhaylov, *Theory of Energy and Mass Transfer*, Prentice-Hall, Englewood Cliffs, N. J., 1961.

Table of Contents

LIST OF PRINCIPAL SYMBOLS

A	area
A	thermal accommodation coefficient, Eq. (11.40)
b	width
C	capacity rate, wc
C	electrical capacitance
C_D	drag coefficient, Eq. (4.46)
C_f	friction coefficient, Eq. (3.30)
c	speed of light = 2.997902×10^{10} cm/sec
c_i	mass concentration of component i
c_i^*	mass fraction of component i, c_i/ρ
c_p	specific heat at constant pressure
c_v	specific heat at constant volume
D, d	diameter
D_b	bubble diameter
D_e	equivalent diameter, defined by Eq. (4.14)
D	mass diffusivity, Eq. (1.3)
$\mathfrak{D}$	diffusivity, a general symbol
E	internal energy
e	internal energy per unit mass
E	ratio of eddy diffusivities, ϵ_h/ϵ_m
E_D	ratio of eddy diffusivities, ϵ_D/ϵ_m
$\mathcal{E}$	surface efficiency, Eq. (12.4)
e	base of natural logarithm
e	roughness of inner surface of pipe, ft.
e	electrical potential
e	radiant energy flux or emissive power
e_b	black body emissive power
e_λ	monochromatic emissive power
e_v	radiant energy density
$e_{v\lambda}$	monochromatic radiant energy density
F	degrees Fahrenheit
F	factor, defined by Eq. (12.17)
F	specular reflection coefficient, Eq. (11.37)
F_{12}, $\bar{F}_{12}$	geometric shape factor between black bodies
$\mathfrak{F}$	geometric shape-emissivity factor between grey surfaces

F_D	drag force
f	friction factor, defined by Eq. (4.3)
f	function of η, defined by Eq. (3.22)
G	mass flow rate, ($G = w/S = \rho V$)
g	gravitational acceleration
g_0	constant, 32.17 $\text{lb}_m\text{ft}/\text{lb}_f\ \text{sec}^2$, $4.17 \times 10^8\ \text{lb}_m\text{ft}/\text{lb}_f\ \text{hr}^2$
h	heat transfer coefficient
h_x	local heat transfer coefficient
h	Planck's constant $= 6.62377 \times 10^{-27}$ erg-sec
h	transport property, defined by Eq. (19.21), Chap. 19 only
h_D	mass transfer coefficient, defined by Eq. (14.11)
h_e	effective heat transfer coefficient, defined by Eq. (11.21)
h_{ei}	effective heat transfer coefficient based on enthalpies, Eq. (11.28a)
h_{fg}	latent heat of condensation or evaporation
H	head, Eq. (4.25)
I	intensity of radiation, Eq. (13.9)
I	enthalpy
i	enthalpy per unit mass
i	electrical current
j_H, j_D	defined by Eq. (16.18)
J	mechanical equivalent of heat, 778 ft lb/Btu
J	flux, Eq. (19.5)
K	degrees Kelvin
K_c	contraction coefficient, Eq. (4.27)
K_e	expansion coefficient, Eq. (4.28)
k	thermal conductivity defined by Eq. (1.2)
L, l	length
L	transport coefficient, Eq. (19.5), Chap. 19 only
L	radiant mean beam length, Chap. 13
M	molecular weight
M	Mach number, V/V_s
M	$(\Delta x)^2/\alpha\Delta t$, Chap. 18 only
m	mass
m	mass of a molecule
N, n	number in general
N	$h(\Delta x)/k$, Chap. 18 only
N_a	heat transfer area number, NTU, Eq. (12.18c)
n	molecular density
n_i	molal concentration of component i

n_i^*	mole fraction of component i, n_i/n
N_i/A	molal flux of component i
P	property per unit volume, a general symbol
P, p	pressure
P	wetted perimeter
p_i	partial pressure of component i
p/A	property flux, a general symbol
Q	volume flow rate
q	rate of heat transfer
q/A, q''	heat flux
R	degrees Rankine
R, R_T, R_D	resistance (electrical, thermal, and mass diffusional)
R	perfect gas constant
$\mathcal{R}$	universal gas constant
r	radius
r	temperature recovery factor, defined by Eq. (11.5)
r_i	enthalpy recovery factor, defined by Eq. (11.28b)
r	radiance, Eq. (13.29)
S	molecular speed ratio, $\sqrt{\gamma/2}\, M$
S	cross-sectional area
S	entropy
s	entropy per unit mass
T	temperature
T_b	bulk fluid temperature
T_0, T_w	wall temperature
T_{aw}	adiabatic wall temperature
T_s	stagnation temperature
T_∞	free stream temperature
T'	fluctuation temperature
t	time
U	internal energy
U	over-all heat transfer coefficient, Eq. (6.21)
V	volume
v	specific volume
V	atomic volume, Chap. 14
V	velocity
V_s	speed of sound
v_x, v_y, v_z	velocity components; also bulk velocity components of a mixture, defined by Eq. (14.3)

$\mathcal{V}_x, \mathcal{V}_y, \mathcal{V}_z$	molal bulk velocity components of a mixture, defined by Eq. (14.7)
v_{ix}, v_{iy}, v_{iz}	statistical mean velocity of component i with respect to stationary coordinate axes
v_x', v_y', v_z'	fluctuation velocity components
v^+	$v/\sqrt{\tau_0/\rho}$
W	thermodynamic probability
W_i	heat source or rate of heat generation per unit volume
w	flow rate
w_i/A	mass flux of component i relative to bulk velocity of mixture
X	a potential gradient or "driving force," Eq. (19.5)
x, y, z	coordinate axes
y^+	$(y/\nu)\sqrt{\tau_0/\rho}$
Z	collision frequency
α	coefficient, defined by Eq. (4.23)
α	thermal diffusivity, $k/\rho c_p$
α	absorptivity
α_λ	monochromatic absorptivity
β	coefficient, defined by Eq. (4.26)
β	thermal coefficient of volume expansion
Γ	mass rate of flow of condensate per unit width of wetted wall
Γ	$hP(\Delta x)^2/Sk$
γ	ratio of specific heats, c_p/c_v
$\delta, \delta_T, \delta_D$	boundary layer thickness (hydrodynamic, thermal, and concentration)
δ	liquid film thickness in condensation
δ^*	displacement thickness of boundary layer, Eq. (2.3)
ϵ	exchanger heat transfer effectiveness, defined by Eq. (12.18b)
ϵ	emissivity
ϵ_λ	monochromatic emissivity
ϵ_ϕ	directional emissivity
$\epsilon_h, \epsilon_m, \epsilon_D$	eddy diffusivity (thermal, momentum, and mass)
η	efficiency
η	boundary layer variable, defined by Eq. (3.23)
θ	dimensionless temperature ratio
θ	angle in cylindrical or spherical coordinates
θ	momentum thickness of boundary layer, Eq. (2.5)
κ	Boltzmann constant $= 1.38026 \times 10^{-16}$ erg/K

λ	wavelength of electromagnetic radiation
λ	molecular mean free path
μ	absolute viscosity, defined by Eq. (1.1)
μ	electrochemical potential, Sec. 19.3 footnote, Chap. 19 only
ν	kinematic viscosity μ/ρ
ν	frequency of electromagnetic radiation
ρ	mass density
ρ	reflectivity, Chap. 13
σ	Stefan-Boltzmann constant $= 0.1713 \times 10^{-8}$ Btu/ft^2 hr R^4
σ	surface tension
σ	molecular collision cross section
σ_x, σ_y, σ_z	normal stress components
τ	transmissivity, Chap. 13
τ	reciprocal temperature T^{-1}, Chap. 19 only
τ, τ_{yx}, etc.	shear stress components
Φ	rate of entropy production per unit volume, Eq. (19.10)
ϕ	angle in spherical coordinates
ϕ	potential function in ideal fluid flow, defined by Eq. (18.32)
ψ	stream function in ideal fluid flow, defined by Eq. (3.20)
ω	solid angle
ω	specific humidity

CHAPTER 1

Introduction

The study of transfer phenomena, which include transfer of momentum, energy, mass, electricity, etc., has in recent years evolved from a rather loose collection of theories and empirical information in diverse branches of engineering into a basic subject of engineering science. This has resulted from the recognition of transfer processes as a unified discipline of fundamental importance and as a topical extension of the concepts and laws of mechanics, thermodynamics, and fluid mechanics.

1.1 Transfer Phenomena

The transfer process typified by a diffusion process may be defined as the *tendency toward equilibrium*—a process which tends to establish equilibrium. For example, in a solid body with a nonuniform temperature distribution, energy is transferred to tend to establish a uniform temperature distribution. Heat is defined as *energy transferred by virtue of a temperature difference or gradient* and is vectorial in the sense that it is said, arbitrarily, to be transferred in the direction of decreasing temperature, a negative temperature gradient. In the science of thermodynamics, the important magnitude is the *quantity* of heat transferred during a process. In the area of knowledge known as "heat transfer," attention is directed toward the relationship between *rate* of heat transfer, temperature distribution, geometry, and other properties of the materials involved.

When a small amount of perfume vapor is sprayed into a room of air, the mass transfer process causes the perfume vapor to diffuse throughout the room until its concentration is uniform—an equilibrium condition.

In an electrically conducting material with a nonuniform electrical potential (voltage) distribution, electric charge will flow until a uniform potential distribution exists.

Our main interest is in heat and mass transfer, but we shall study them within the general framework of transfer processes. In all transfer processes we are chiefly concerned with rates at which changes in properties of a system occur. In the flow of a viscous fluid, the frictional phenomenon or viscous stresses may be related to the rate of change of momentum of a system. Likewise, heat conduction may be related to the rate of change of internal energy of a system, and mass diffusion in any one of its various forms may be related to the rate of change of composition of a mixture due to transfer of one of the component species.

Many important applications involve simultaneous occurrence of several transfer processes. We cite as an example the cooling of a surface by the introduction of fluid through pores in the surface (mass transfer cooling, Chap. 15), in which simultaneous transfers of momentum, energy, and mass occur.

Under certain conditions of multiple transfer, the individual fluxes may interfere with one another and give rise to interesting cross-phenomena. The best-known example is the thermoelectric phenomenon in which an electric current is set up in a circuit by a temperature potential gradient or, conversely, heat flow results from the application of an electrical potential gradient (Chap. 19).

These transfer processes, such as the transfer of energy, chemical component, and electricity, which occur relative to a mass of material, must be distinguished from the transport of a property by motion of the mass of material. For example, a fluid flowing through a heat exchanger is raised in temperature by heat transfer from the hot walls. Because of the fluid flow, energy is also transported by the flow of mass into and out of the exchanger. Similarly, any of the properties previously mentioned may be transported by mass movement. Superimposed on this are the transfer processes.

1.2 Rate Equations

Macroscopic theories of transfer processes are based on the phenomenological approach in which the basic transfer relations are postulated from experience without reference to the details of mechanisms of transfer. By definition, a transfer process consists of net flow of a property under the influence of a driving force. The rate of transfer is the *flux* and the intensity of the driving force is the *potential gradient*. The basis of the phenomenological method is the assumption that a linear dependence exists between the flux and the potential gradient. The transfer process (or flux) takes place in the direction of decreasing potential.

The phenomenological rate equations for transfer of momentum, energy, and mass are presented next.

(*a*) *Newton's Equation of Viscosity*

Consider a fluid confined between two parallel plates, the upper one being at rest and the lower one being set in motion with a velocity V. If the plates are sufficiently close, the motion of fluid will be laminar.

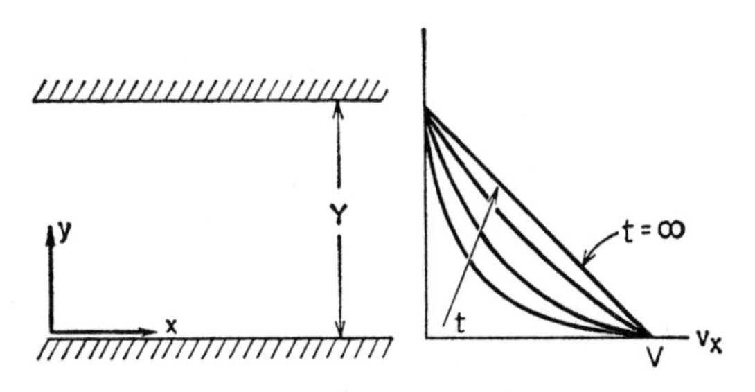

Figure 1.1

Imagine the fluid as being subdivided into infinitesimally thin layers parallel to the plates. From experience, we know the layer adjacent to the stationary plate will be at rest and the layer adjacent to the moving plate will have velocity V. As a consequence of internal friction in the fluid, each layer exerts a forward drag upon the layer immediately above it and, in time, a steady velocity profile is established in the fluid as shown in Fig. 1.1. When the steady state is reached, a force F must be exerted on the lower plate to maintain the motion, and an equal and opposite force must be exerted on the stationary plate. If A is the area of the plate on which the shear force F acts, shear stress is given by $\tau_{yx} = F/A$, where the first subscript gives the direction of the inward drawn normal to the surface and the second the direction of the force. Experiments indicate the existence of direct proportionality between τ_{yx} and the velocity gradient. In the steady state

$$\tau_{yx} = \mu \frac{V}{Y} \tag{1.1a}$$

or, more generally, at any position in the fluid during the transient,

$$\tau_{yx} = -\mu \frac{\partial v_x}{\partial y} \tag{1.1b}$$

This empirical relation is known as *Newton's equation of viscosity*, and fluids which obey this equation are called *Newtonian fluids*.

τ_{yx} may be interpreted alternately in terms of a momentum transfer. Owing to the velocity gradient, a fluid layer closer to the moving plate has more x-direction momentum than layers farther away. The force applied on the moving plate is thus "transmitted" as x-direction momentum from one fluid "layer" to the next.

Some insight into the mechanism of viscous action in a gas is afforded by the elementary kinetic theory of gases. Imagine a plane at $y = y_1$ which is being crossed by molecules from above and below. The molecules in the x-direction have an average velocity component equal to the velocity of mass motion v_x, but in the y-direction their average velocity component is zero. Hence, molecules cross the plane in both directions with equal frequency, but those crossing from below to above possess more x-direction momentum than those crossing from above to below. As a consequence, there is a net transfer of x-direction momentum in the upward y-direction, and kinetic analysis shows the magnitude of transfer per unit area is equal to the shear stress τ_{yx}.

The negative sign is introduced in Eq. (1.1b) so that momentum transfers in a positive direction along a coordinate axis in a natural downhill direction of decreasing potential. This follows the generally accepted convention for heat and mass transfer.

The dimensions of μ from Eq. (1.1) are

$$\frac{FT}{L^2} \quad \text{or} \quad \frac{M}{LT}$$

where symbols F, T, L, and M represent the primary dimensions of force, time, length, and mass. Thus, in engineering units, μ may be expressed in lb_f-hr/ft^2 or lb_m/ft-hr.

It has been found useful to define another quantity known as *kinematic viscosity* and denoted by ν:

$$\nu \equiv \frac{\mu}{\rho}$$

where ρ is the fluid density. It is readily seen that ν has the dimensions L^2/T or ft^2/hr. Magnitudes of μ are tabulated in Appendix E.

We shall confine our attention to Newtonian fluids, which include all gases and most liquids. It should be understood, however, that many industrially important fluids, such as molten plastics, are non-Newtonian and the study of transfer processes in these media is complicated by lack of precise knowledge of the functional forms of the rate equations.

(*b*) *Fourier's Equation of Heat Conduction*

Consider a homogeneous solid slab, initially at T_0, the upper plane of which is maintained at temperature T_0 and the lower plane suddenly changed to a temperature T_1. It will be assumed that the temperature difference is not so great as to cause any significant change in other properties of the solid. In time, a steady temperature profile becomes established in the solid as shown in Fig. 1.2.

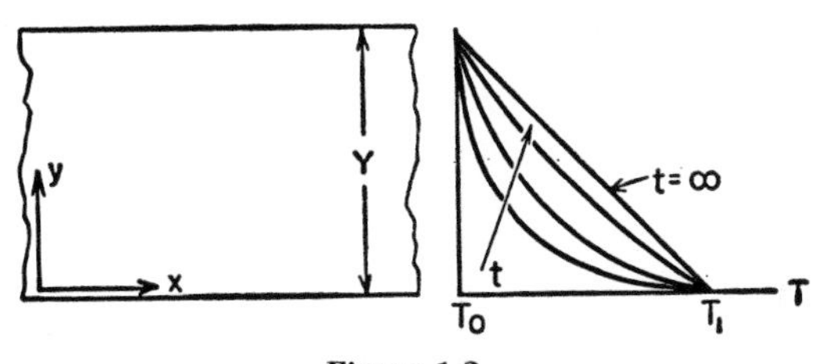

Figure 1.2

Experiments indicate the existence of direct proportionality between the temperature gradient and the amount of energy transferred across a unit area of the solid in the y-direction. The direction of transfer must be from the region of higher temperature to the region of lower temperature in accordance with the Second Law of thermodynamics. The constant of proportionality is called *thermal conductivity* and is given the symbol k. In the steady state

$$\left(\frac{q}{A}\right)_y = k\frac{T_1 - T_0}{Y} \tag{1.2a}$$

or, more generally, at any position during the transient,

$$\left(\frac{q}{A}\right)_y = -k\frac{\partial T}{\partial y} \tag{1.2b}$$

This empirical relation is known as *Fourier's equation of heat conduction.*

The units of k follow from its definition by Eq. (1.2) and are conventionally expressed as Btu/(hr ft^2) (F/ft) or Btu/(hr ft F). The thermal conductivity of insulating materials, however, is usually given as Btu/(hr ft^2) (F/in).

A quantity known as *thermal diffusivity* and denoted by α occurs frequently in heat transfer analyses. Its definition follows:

$$\alpha \equiv \frac{k}{\rho c}$$

where ρ is the density and c is the specific heat. It has the dimensions L^2/T or ft^2/hr.

(*c*) *Fick's Equation of Diffusion*

Consider two parallel plates with initially dry air between them, the bottom plate being covered with gauze and the top plate coated with a substance such as silica gel which absorbs essentially all water vapor which it contacts. Suddenly the gauze is wetted with water so that the partial density of the water vapor at the wet surface is maintained at c_{w_0} lb water vapor per ft^3. Then diffusion takes place until a steady state is reached. At the silica gel surface c_w is assumed to be zero (Fig. 1.3).

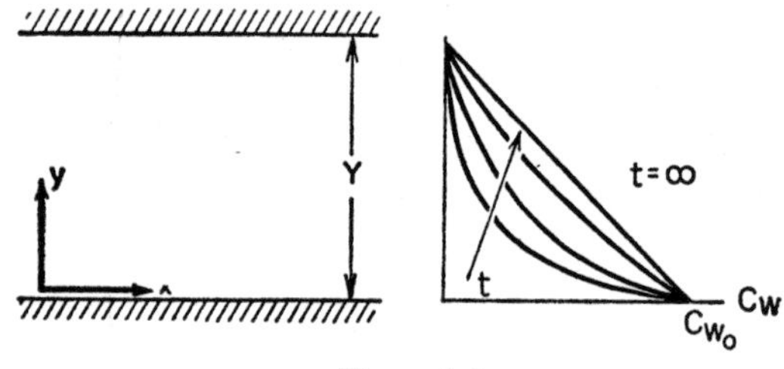

Figure 1.3

Experimental evidence indicates the existence of a direct proportionality between the diffusion rate of the water vapor and the concentration gradient. In the steady state,

$$\left(\frac{w_w}{A}\right)_y = D\frac{c_{w_0}}{Y} \tag{1.3a}$$

or, in general, at any position during the transient

$$\left(\frac{w_w}{A}\right)_y = -D\frac{\partial c_w}{\partial y} \tag{1.3b}$$

where D is called the *diffusivity* or *coefficient of diffusion* and has the units L^2/T such as ft^2/hr. Equation (1.3) is Fick's equation.

The preceding equations, (1.1), (1.2), and (1.3), are not in fact laws of nature, but rather definitions of μ, k, and D which are found, experimentally, to be properties of the material, e.g., functions of two independent properties for pure substances.

(*d*) *Diffusivities*

Among the transfer coefficients defined by the rate equations, we note that the kinematic viscosity ν, the thermal diffusivity α, and the diffusion coefficient D all have the same dimensions L^2/T. Clearly, a dimensionless number can be formed from the ratio of any two of these quantities.

Prandtl Number:

$$\text{Pr} \equiv \frac{\nu}{\alpha} = \frac{c\mu}{k}$$

It is a significant parameter in the study of systems undergoing simultaneous energy and momentum transfer, as in problems involving convection. Physically, it expresses relative speeds at which momentum and energy are propagated through the system. For most cases Pr is approximately unity, but for other fluids it varies over a wide range. Liquid metals have very low Pr, whereas viscous fluids like oil have Pr of the order of a thousand.

Lewis Number:

$$\text{Le} \equiv \frac{\alpha}{D} = \frac{k}{\rho c D}$$

It is an important parameter of systems subject to simultaneous energy and mass transfer, as for example, a wet-bulb thermometer. Its magnitude expresses relative rates of propagation of energy and mass within a system.

Schmidt Number:

$$\text{Sc} = \frac{\nu}{D} = \frac{\mu}{\rho D}$$

It is a significant parameter of isothermal systems undergoing simultaneous momentum and mass transfer processes. It is approximately unity in gases, but is large for liquids. In the case of a gaseous system, it is possible to obtain expressions for Prandtl, Lewis, and Schmidt numbers on the basis of the kinetic theory.

1.3 Conservation Laws for One-Dimensional Cases

It was stated earlier that in transfer processes we are chiefly concerned with finding out the rates at which some property of a system changes. In most instances, we are interested in changes due to fluxes of that property crossing a control surface; and in some special cases we are able to postulate from experience that the property is conserved. In the following paragraphs, we shall formulate a mathematical statement of the conservation laws for certain one-dimensional cases.

Consider the three systems represented by Figs. 1.1, 1.2, and 1.3. Figure 1.4 represents all of these cases and shows the distribution of properties v_x, T, and c_w at a particular instant of time.

For the case in which the space between $y = 0$ and L contains a fluid and the wall at $x = 0$ is suddenly set into motion, the net "momentum flux" (associated with shear stress) into element $A\ \Delta y$ (shown crosshatched) is

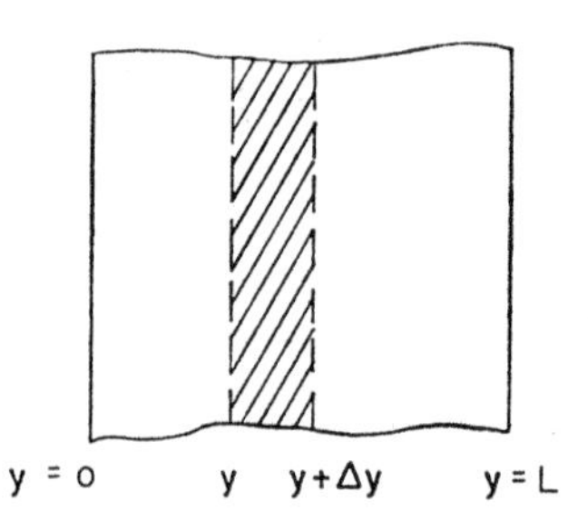

$$A(\tau_{yx} - \tau_{y+\Delta y,x}) \quad \text{or} \quad -\left(\frac{\partial \tau_{yx}}{\partial y} A\ \Delta y\right)$$

By the principle of conservation of momentum, this should equal the rate of increase of momentum of the fluid in the element or $\partial(A\ \Delta y \rho v_x)/\partial t$. Equating these and canceling $A\ \Delta y$ because it is constant, there results for constant ρ,

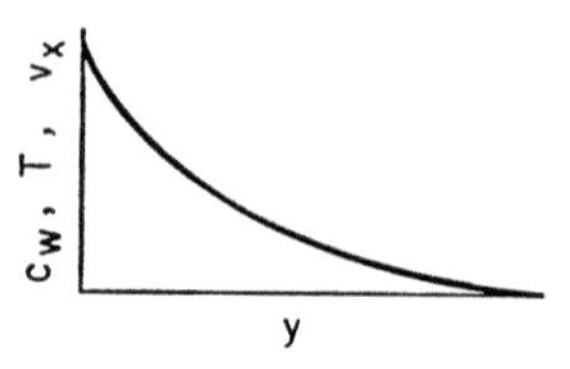

Figure 1.4

$$-\frac{\partial \tau_{yx}}{\partial y} = \rho \frac{\partial v_x}{\partial t} \tag{1.4a}$$

or with Eq. (1.1b)

$$\frac{\partial}{\partial y}\left(\mu \frac{\partial v_x}{\partial y}\right) = \rho \frac{\partial v_x}{\partial t} \tag{1.4b}$$

which is the differential equation to be solved to determine the velocity distribution for any particular boundary and initial conditions.

Similarly, if the space between $y = 0$ and L is a homogeneous solid body experiencing a transient in temperature distribution, at any instant of time the net heat flux into the element $A\ \Delta y$ is

$$q_y - q_{y+\Delta y} \quad \text{or} \quad -\left(\frac{\partial q_y}{\partial y}\right) \Delta y$$

By energy conservation this net heat flux must equal the rate of rise of internal energy, $\partial(A\ \Delta y \rho e)/\partial t$. Equating these, cancelling Δy, and realizing A is a constant, this energy conservation equation for ρ constant becomes

$$-\frac{\partial}{\partial y}\left(\frac{q_y}{A}\right) = \rho \frac{\partial e}{\partial t} \tag{1.5a}$$

For a solid, de may be expressed as $c\ dT$. Then, with Eq. (1.2b) this becomes

$$\frac{\partial}{\partial y}\left(k \frac{\partial T}{\partial y}\right) = \rho c \frac{\partial T}{\partial t} \tag{1.5b}$$

which, for a particular set of boundary and initial conditions, determines the transient temperature distributions for this case.

The conservation equation for mass transfer follows a similar development. If the space between $y = 0$ and L is air with water vapor in dilute concentration diffusing from one wall to the other, the net rate of mass of water-vapor diffusing into the element $A\ \Delta y$ is

$$w_{w(y)} - w_{w(y+\Delta y)} = -\frac{\partial w_w}{\partial y}\,\Delta y$$

Then, by conservation of mass this should equal the rate of accumulation of water vapor in the element or $\partial(A\ \Delta y c_w)/\partial t$. Then

$$-\frac{\partial}{\partial y}\left(\frac{w_w}{A}\right) = \frac{\partial c_w}{\partial t} \tag{1.6a}$$

or with Eq. (1.3b)

$$\frac{\partial}{\partial y}\left(D\,\frac{\partial c_w}{\partial y}\right) = \frac{\partial c_w}{\partial t} \tag{1.6b}$$

which with boundary and initial conditions determines the transient concentration distribution.

For special cases such as these there is a striking similarity among these various equations. To emphasize this similarity, we now rewrite them for the cases in which ρ and c are constant. Equations (1.1b), (1.2b), and (1.3b) may be written as follows:

$$\begin{aligned}
\tau_{yx} &= -\nu\,\frac{\partial(\rho v_x)}{\partial y}\\
\frac{q}{A} &= -\alpha\,\frac{\partial(\rho e)}{\partial y}\\
\frac{w_w}{A} &= -D\,\frac{\partial c_w}{\partial y}
\end{aligned} \tag{1.7}$$

and Eqs. (1.4), (1.5), and (1.6) as follows:

$$\begin{aligned}
\frac{\partial}{\partial y}\left(\nu\,\frac{\partial(\rho v_x)}{\partial y}\right) &= \frac{\partial(\rho v_x)}{\partial t}\\
\frac{\partial}{\partial y}\left(\alpha\,\frac{\partial(\rho e)}{\partial y}\right) &= \frac{\partial(\rho e)}{\partial t}\\
\frac{\partial}{\partial y}\left(D\,\frac{\partial c_w}{\partial y}\right) &= \frac{\partial c_w}{\partial t}
\end{aligned} \tag{1.8}$$

These equations are of the following form:

Rate Equation:

$$\frac{p_x}{A} = -\mathfrak{D}\,\frac{\partial P}{\partial y} \tag{1.9}$$

Conservation Equation:

$$\frac{\partial}{\partial y}\left(\mathfrak{D}\,\frac{\partial P}{\partial y}\right) = \frac{\partial P}{\partial t} \qquad (1.10)$$

where P is some property per unit volume, (p_x/A) is interpreted as the flux of property P, and $\mathfrak{D}$ is called the *diffusivity*. In the case of momentum transfer, P is ρv_x, the x-direction momentum per unit volume of fluid; in energy transfer P is ρe, the internal energy per unit volume of the body; and in mass transfer P is c_w, the mass of the transferred material per unit volume. The following table summarizes the comparison of these equations.

Transfer process	Transferred property, P	Flux, p/A	Diffusivity, $\mathfrak{D}$
Mass.	c_w	w_a/A	D
Energy.	ρe	q/A	α
Momentum.	ρv_x	τ	ν

This general type of comparison shows the desirability of interpreting τ as a flux of momentum as suggested in Sec. 1.2. Further, it is possible to include other transfer processes in this over-all structure. For example, the transient flow of electricity in a cable having appreciable capacitance-to-ground is expressed by equations of the form of (1.9) and (1.10).

This similarity of mathematical form plus similarity of boundary conditions defines a group of processes which are said to be analogous. Hence, mass transfer processes or electricity transfer processes, for example, may be used to solve analogous heat transfer problems.

It must be emphasized that this one-to-one correspondence, although valid for certain of the simple cases, cannot be extended to all cases. The conservation equations and rate equations applicable to the general processes in three dimensions and including external force fields, viscous effects in energy transfer, chemical reactions, coupling phenomena, etc., are not analogous in form. Some of the more general equations are presented in Chaps. 3, 7, and 15. Our purpose here is merely to point out an area of equivalence among the familiar laws of conservation, with regard to both transfer mechanism and mathematical formulation.

1.4 Continuity Equations

Relations expressing the over-all continuity of matter in flowing systems will be useful in simplifying conservation equations for transfer processes in nonstationary systems. Consider a flow of a single-phase,

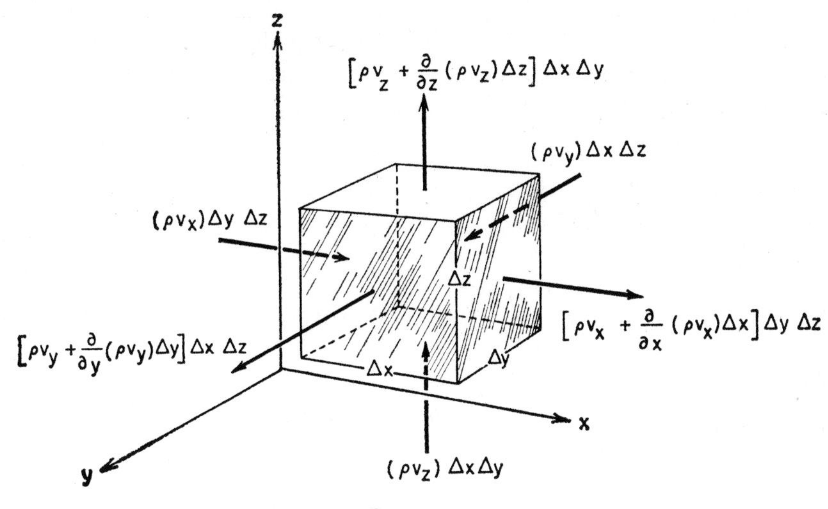

Figure 1.5

single-component fluid with velocity V having v_x, v_y, and v_z components. The x-direction flow rates entering and leaving a control volume Δx, Δy, Δz fixed in space are shown in Fig. 1.5. The net effluxes of mass in the three component directions are then:

x-direction:

$$\frac{\partial(\rho v_x)}{\partial x}\,\Delta x\;\Delta y\;\Delta z$$

y-direction:

$$\frac{\partial(\rho v_y)}{\partial y}\,\Delta y\;\Delta z\;\Delta x$$

z-direction·

$$\frac{\partial(\rho v_z)}{\partial z}\,\Delta z\;\Delta x\;\Delta y$$

The sum of these will equal the rate of change (decrease) of mass within the control volume, $-(\partial\rho/\partial t)\;\Delta x\;\Delta y\;\Delta z$, or dividing by $\Delta x\;\Delta y\;\Delta z$,

$$-\frac{\partial\rho}{\partial t} = \frac{\partial(\rho v_x)}{\partial x} + \frac{\partial(\rho v_y)}{\partial y} + \frac{\partial(\rho v_z)}{\partial z}$$

or

$$\frac{\partial\rho}{\partial t} + \text{div }(\rho\mathbf{V}) = 0 \tag{1.11a}$$

This is known as the *continuity equation* or *conservation of mass*. In Sec. 15.5 it is shown that this equation applies equally well for multi-component diffusing systems provided each component velocity is the mass-average velocity, Eq. (15.18). Equation (1.11a) may be rearranged as follows:

$$\frac{\partial \rho}{\partial t} + v_x \frac{\partial \rho}{\partial x} + v_y \frac{\partial \rho}{\partial y} + v_z \frac{\partial \rho}{\partial z} + \rho\left(\frac{\partial v_x}{\partial x} + \frac{\partial v_y}{\partial y} + \frac{\partial v_z}{\partial z}\right) = 0$$

or

$$\frac{D\rho}{Dt} + \rho \operatorname{div} \mathbf{V} = 0 \tag{1.11b}$$

where

$$\frac{D}{Dt} \equiv \frac{\partial}{\partial t} + v_x \frac{\partial}{\partial x} + v_y \frac{\partial}{\partial y} + v_z \frac{\partial}{\partial z}$$

is the substantial derivative. The first term in the substantial derivative $(\partial/\partial t)$ expresses the time rate of change at a point in the fluid, and the remaining terms account for changes associated with the motion of the fluid. In other words, the substantial derivative of density, $D\rho/Dt$, expresses the change of density in a system following the motion and is that change seen by an observer moving at velocity $\mathbf{V}$. Equation (1.11b) is therefore the continuity equation written from the viewpoint of an observer moving along with the fluid whereas Eq. (1.11a) is the same equation written from the viewpoint of a stationary observer.

If the density is constant in time and space, Eq. (1.11) reduces to

$$\frac{\partial v_x}{\partial x} + \frac{\partial v_y}{\partial y} + \frac{\partial v_z}{\partial z} = 0$$

or

$$\operatorname{div} (\mathbf{V}) = 0 \tag{1.12}$$

In the study of flow through circular ducts, it is generally more convenient to work in cylindrical coordinates (r, θ, z). The transformation of equations from Cartesian to cylindrical coordinates is rather lengthy but straightforward.

In cylindrical coordinates, Eqs. (1.11) and (1.12) become respectively

$$-\frac{\partial \rho}{\partial t} = \frac{1}{r}\frac{\partial}{\partial r}(\rho r v_r) + \frac{1}{r}\frac{\partial}{\partial \theta}(\rho v_\theta) + \frac{\partial}{\partial z}(\rho v_z) \tag{1.13}$$

and for constant density,

$$\frac{\partial v_r}{\partial r} + \frac{v_r}{r} + \frac{1}{r}\frac{\partial v_\theta}{\partial \theta} + \frac{\partial v_z}{\partial z} = 0 \tag{1.14}$$

1.5 Kinetic-Theory Explanation of Transport Phenomena

An alternative to the phenomenological approach to transfer processes is that based on the molecular theories of matter. The molecular approach has wide appeal because it gives a more complete description of the transfer mechanism. However, mathematical complexity limits its application to the simplest systems, and validity of its results rests to a large extent on a judicious choice of the molecular model *for a particular system*. The microscopic theory of transfer phenomena is the kinetic theory:

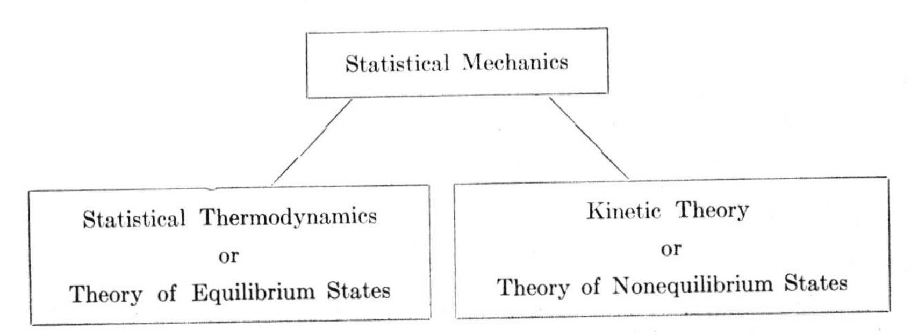

Statistical thermodynamics yields information on the equilibrium properties such as energy and entropy of a system, whereas the kinetic theory is useful for evaluating the nonequilibrium properties or transfer coefficients such as viscosity, thermal conductivity, and diffusivity of mass.

The discussion that follows considers only the kinetic theory of the gaseous state, partly because useful results can be obtained from elementary considerations and partly because the kinetic theories of liquids and theories of the solid state are not at comparable stages of development.

The kinetic study of transfer phenomena can be taken up at several levels of sophistication. A rather crude approach assumes molecules can be represented as hard, elastic spheres (billiard balls) which, at any given instant, have a Maxwellian distribution of velocities. This simple model is quite adequate for a qualitative description of the transfer mechanisms.

Kinetic theory pictures viscosity as a momentum transport in the following way: Imagine a gas flowing between parallel plates aligned in the x-direction. Because a velocity gradient exists perpendicular to the plates, molecules from a higher-velocity region move randomly into a lower-velocity region and vice versa. The shear stress at any plane is parallel to the plates, and the coefficient of viscosity, Eq. (1.1), is interpreted in terms of the net interchange of x-direction momentum across the plane resulting from this random molecular motion.

In a stationary gas (except for molecular motion) with temperature decreasing in one direction, the random motion of the molecules brings molecules of higher kinetic energy (higher temperature) across a perpendicular plane into a region where molecules have lower kinetic energy. The net interchange of energy across this plane is interpreted as the thermal conductivity.

In a similar way the diffusivity D can be interpreted as net exchange of molecules of gas a in a stagnant gas b from a region of higher to one of lower concentration.

A significant parameter that governs the mechanism of transfer in gases is the free path, defined as the distance travelled by a molecule between two successive collisions. A molecule describing a free path of length λ is in effect transporting momentum, energy, or mass through the distance λ. If the gaseous system is in equilibrium, each such transfer is exactly balanced by an equal and opposite transfer, so that the net transfer is zero. However, if the system is not in equilibrium, a net transfer takes place. The mechanism of transfer can be identified on a basis of the relative lengths of the mean free path λ and a characteristic dimension of the system L.

If λ is negligibly small compared to L, the transfer takes place by a series of random collisions. It explains why diffusion is a relatively slow process, although molecules travel at velocities of several hundred feet per second. The calculation of transfer coefficients for this case reduces essentially to a problem of dynamics of collisions; however, in view of the enormous number of particles involved (a cubic centimeter of gas at ambient state contains about 3×10^{19} molecules) and the random behavior of each particle, statistical methods must be used. Values of transfer coefficients calculated by this method are of the right orders of magnitude.

If λ is large compared to L, as in a high vacuum, random collision becomes negligible as a mechanism of transfer. Molecules travel through a rarefied system in straight lines essentially of pencil-like rays. Discussion may be extended to include radiant transfer itself, in which properties are carried from a source to a sink by electromagnetic waves. It is quite apparent that λ no longer governs the transfer mechanism; instead, the most significant parameter of transfer will be a quantity which measures how effectively properties are exchanged when the molecules or waves impinge upon a surface. In radiant energy transfer, this quantity is known as the *surface absorptivity*.

Rigorous application of the kinetic theory, even for the simplest monatomic gases, is possible only when λ is either large compared with L or negligibly small compared with L. When λ is of the same order of magnitude as L, analysis becomes extremely complicated. This material is discussed in greater detail in Chapters 11 and 20.

1.6 Units

Generally, all quantities will be expressed in the English system of engineering units—Btu, °F, ft, hr, lb_f, and lb_m. Departures from this practice will be clearly noted. In this system of units with F in lb_f and m in lb_m, Newton's second law equation is

$$F = \frac{m}{g_0} a \tag{1.15}$$

where, if acceleration a is in ft/sec², then $g_0 = 32.16\ lb_m\ ft/lb_f\ sec^2$ or if a is in ft/hr², then $g_0 = 4.16 \times 10^8\ lb_m\ ft/lb_f\ hr^2$.

A unit of mass commonly used in fluid mechanics work is the *slug*, defined as 32.16 lb_m.

The mass density ρ in lb_m/ft^3 is related to the weight density γ in lb_f/ft^3 by Eq. (1.15) with $a = g$, the acceleration of gravity. Then

$$\gamma = \frac{g}{g_0} \rho \tag{1.16}$$

The units of specific heat c are Btu/lb_m F.

Air-conditioning and refrigeration engineers are accustomed to expressing k in Eq. (1.2) in terms of temperature gradient in F/in. Then the units of thermal conductivity k are Btu/hr ft² (F/in). In order to maintain uniformity of units, most heat transfer engineers express temperature gradient in F/ft and k in Btu/hr ft² (F/ft) or Btu/hr ft F.

The coefficient of viscosity, or simply the viscosity μ, relates the shear stress in lb_f/ft^2 on a plane parallel with the flow to the velocity gradient in (ft/hr)/ft, thereby $\tau = -\mu(\partial v_x/\partial y)$. Then μ_F is in $lb_f\ hr/ft^2$. Quite often in heat transfer problems, g_0 and μ_F are found appearing as a product; therefore, it has been found convenient to use a viscosity μ_M which equals $g_0\mu_F$. The units of μ_M are then lb_m/ft hr. Another common unit of viscosity is the *centipoise* which is defined as $\frac{1}{100}$ gm/sec cm. The following are conversion factors for viscosity:

viscosity in centipoise × 0.000672 = viscosity in lb_m/sec ft

viscosity in centipoise × 0.0000209 = viscosity in $lb_f\ sec/ft^2$

viscosity in centipoise × 2.42 = viscosity in lb_m/hr ft

viscosity in centipoise × 3.60 = viscosity in kg/hr m

Other conversion factors are found in Appendix E.

PROBLEMS

1.1 Two parallel plates are $\frac{1}{4}$ in. apart. The lower plate is stationary and the upper plate is in motion with a velocity of 10 ft/sec. If a force of 0.05 lb_f/ft^2 is needed to maintain the upper plate in motion, find μ of the fluid contained between the plates. If density of fluid is 50 lb_m/ft^3, find the kinematic viscosity ν

1.2 Water at 60 F flows over a flat surface. If the velocity profile at a point $x = x_1$ is given by $v_x = 3y - y^3$, find the shear stress at the wall at that point. (Properties of water at 60 F: $\rho = 62.4\ lb_m/ft^3$, $c_p = 1\ Btu/lb_m\ F$, $k = 0.34$ Btu/hr-ft^2 F/ft, $\mu = 2.71\ lb_m$/hr-ft) ($|v_x| =$ ft/sec: $|y| =$ ft).

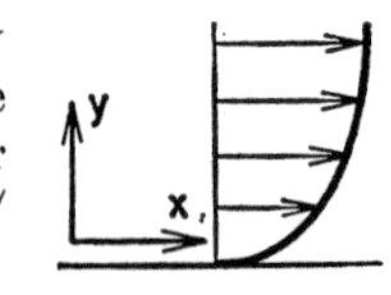

1.3 In the same system described in Prob. 1.2, the temperature profile at $x = x_1$ is given by

$$T = 6 \sin (\pi/2)y, \qquad 0 \leq y \leq 1.$$

Find the heat flux through the wall at that point. Calculate the Prandtl number and interpret its significance. ($|T| =$ °F; $|y| =$ ft.)

1.4 The temperature drop through a 1-in. thick slab of asbestos is 1000 F. If the heat flux through the slab is 1050 Btu/hr ft^2, find its thermal conductivity in Btu/hr ft^2 F/ft. What is the thermal conductivity in the cgs system?

1.5 Express the rate equation of mass diffusion in molal form.

1.6 The sketch shows gases a and b in a chamber separated by an impermeable partition.

(a) Sketch the history of the concentration variation in the chamber from time $t = 0$, when the partition is removed, to time $t = t_\infty$, when equilibrium is restored. Temperature and pressure uniform.

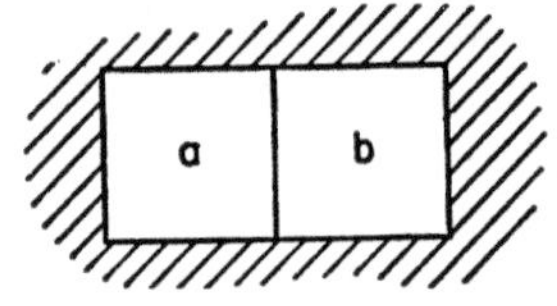

(b) The transfer process is irreversible, and therefore entropy must increase during the diffusion process. Find the entropy change, assuming the chamber is insulated from surroundings.

1.7 Dry air at 60 F and 14.7 psia flows over a very porous wet plate. If the concentration profile at a point $x = x_1$ is given by

$$c_w = 5e^{-2y}, \qquad 0 \leq y \leq 1,$$

calculate the rate at which water vapor evaporates from the plate. Assume the plate temperature is maintained constant at 60 F and $D = 0.9\ ft^2/hr$. Calculate the Schmidt number, using as an approximation the value of kinematic viscosity for air. ($|c_w| = 1\ lb/ft^3$; $|y| =$ ft.)

1.8 Write down the rate equation and the conservation equation in one dimension for the flow of electricity.

1.9 The sketch shows a simple electrical resistance circuit. Give schematic sketches of physically analogous fluid, thermal, and mass diffusion circuits. Pick your examples from among familiar systems.

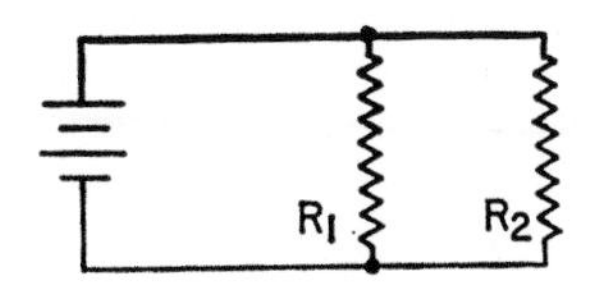

1.10 Given the following expressions for the velocity distribution in a two-dimensional incompressible flow,

$$v_x = \sin yt + \ln xy; \qquad v_y = \cos 2xt - y/x$$

is continuity of flow satisfied?

PART ONE **Momentum Transfer**

CHAPTER 2

Introduction to Momentum Transfer

In the formulation of Newton's equation of viscosity, we described fluid flowing in parallel layers or laminae between parallel plates and the establishment of a velocity gradient due to the action of viscosity. The resulting shear stress in the fluid was interpreted in terms of a momentum transfer induced by the velocity gradient. It should be clearly understood at this point that although fluid layers are conceived as sliding over each other, viscosity in fluid is not caused by "sliding friction" between molecular layers, but by a momentum exchange.

We shall confine our discussion of momentum transfer to fluid systems, and in this sense, the next three chapters may well be called an outline of viscous flow theory. Momentum transfer in solids is comprised in the extensive area of mechanics of solid bodies. We shall simply note in passing that the rate of momentum transfer in solids generally is several orders of magnitude larger than that in fluids, and is infinite in an ideal, rigid solid. Before taking up the analysis of momentum transfer, we propose first to explain the phenomenon of turbulence and the concept of a boundary layer.

2.1 Turbulence

In addition to the laminar type of flow already described, a distinct irregular flow is frequently observed in nature. The flow is termed *turbulent* and is extremely complicated. However, certain basic postulates about turbulent flow may be formulated from experience.

Careful measurements, for instance with a hot-wire anemometer, reveal that the velocity at a fixed point in space fluctuates randomly about some mean value, called the *temporal mean value* (see Fig. 2.1). At a given instant, the velocity is also seen to fluctuate randomly in space around a mean value, which may be called the *spatial mean value*. In the study of transfer processes in turbulent flow, we are usually interested in the temporal mean value, and unless otherwise stated, temporal mean value will be implied in subsequent discussions.

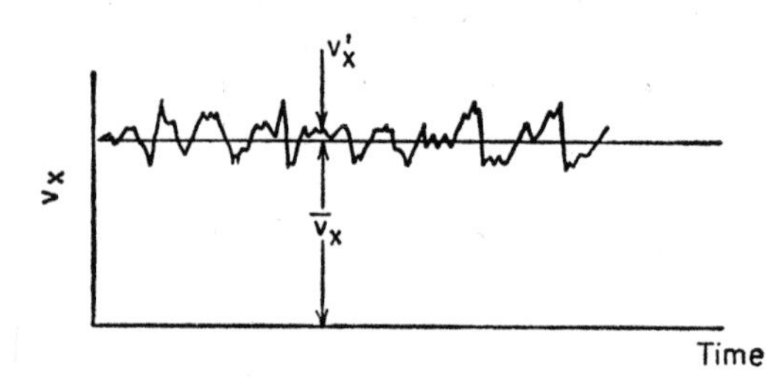

Fig. 2.1. Instantaneous velocity variation in turbulent flow.

Turbulent motion may now be defined in a precise mathematical language by postulating that it is comprised of a randomly fluctuating motion superimposed on a mean motion. Instantaneous velocity is then given by

$$
\begin{aligned}
v_x &= \bar{v}_x + v_x' \\
v_y &= \bar{v}_y + v_y' \\
v_z &= \bar{v}_z + v_z'
\end{aligned}
\tag{2.1}
$$

Here

$$\bar{v}_x \equiv \frac{1}{\Delta t}\int_t^{t+\Delta t} v_x\,dt$$

and similar definitions apply to $\bar{v}_y$ and $\bar{v}_z$. The time increment Δt is taken sufficiently large so that the mean quantities will be independent of time, but not so large that slow periodic motion not associated with turbulence is averaged out. A time increment of the order of seconds is generally suitable. Turbulent flow is defined as steady flow if the barred quantities are constant with time, and the primed quantities averaged over the period of observation are zero.

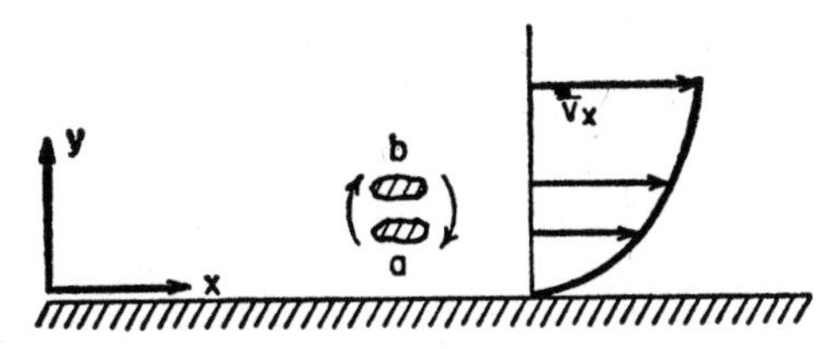

Fig. 2.2. Exchange of eddy packets in turbulent flow.

Consider a turbulent flow between parallel plates (Fig. 2.2). Assume there is no mass motion in the y-direction, $\bar{v}_y = 0$, but $v_y' \neq 0$ and in fact can be quite large. A packet of fluid a with x-direction momentum $m\bar{v}_{xa}$ moves with velocity v_y' from level y_a to level y_b. To satisfy continuity, an equal mass of fluid b possessing x-direction momentum $m\bar{v}_{xb}$ must move from y_b to y_a. The net transfer of x-direction momentum in the downward y-direction due to the fluctuating motion is then $m(\bar{v}_{xb} - \bar{v}_{xa})$.

A striking similarity may be seen between the macroscopic eddy mechanism described above and the molecular transfer mechanism described earlier. Since discrete packets of fluid act as "carriers" in place of molecules, the transfer scale is considerably magnified, which can explain, for instance, the large turbulent shear stress. It is not surprising to find efforts to develop theories of turbulence based upon mixing-length concepts, the role of a mixing length in eddy transfer being roughly equivalent to that of a mean free path in diffusion. A more sophisticated and undoubtedly the correct approach to turbulence is a statistical one, but the theories have not yet been advanced to the stage where useful engineering answers can be obtained.

Turbulent flow prevails in the majority of engineering systems. In the study of transfer processes, especially, turbulence is of the utmost importance because of the high transfer rates possible. As a result, there is a vast amount of literature on this subject, but because a basic and general theory of turbulence and turbulent flow is not yet available, most of this information is of a semiempirical nature.

One of the earliest systematic investigations of turbulent flow was conducted by Reynolds (1) in 1883, when he suggested the parameter VD/ν as the criterion for the onset of turbulence in round pipes, where V is the average fluid velocity, D is the pipe diameter, and ν is the kinematic viscosity of the fluid. In consistent sets of units, the parameter is dimensionless and is called the *Reynolds number* (Re).

The magnitude of Re at which transition from laminar to turbulent flow occurs is called the *critical* or the *transition* Reynolds number. The complex nature of transition is evidenced by the fact that the transition Reynolds number varies with different systems—even for a given system, it varies over a wide range of values depending on such external factors as surface roughness and initial disturbances in the fluid. In usual engineering applications of pipe flow, $(\text{Re})_{tr} \approx 2000$, but values as high as 10,000 have been attained under controlled conditions. Calculation of transfer rates in turbulent systems will be taken up in later chapters.

Example 2.1: Compute the Reynolds number for water flow at $V = 10$ ft/sec at 70 F in a 1-in. diameter pipe.

At 70 F

$$\mu = 2.36 \text{ lb/hr ft}$$

$$\rho = 62.4 \text{ lb/ft}^3$$

$$\text{Re} = \frac{\rho VD}{\mu} = \frac{(62.4)(10)(3600)(\frac{1}{12})}{2.36} = 79{,}200$$

which is assuredly turbulent flow.

2.2 Boundary Layer

In fluid mechanics, as in other sciences, many idealizations are made in order to simplify analysis. One such assumption with far-reaching implications is the postulate that a fluid may be considered "ideal," that is, nonviscous and incompressible. Since shear stress is associated with viscosity, an ideal fluid would slip on a solid surface. However, observations on real fluids in continuum flow ($\lambda \ll L$) show that a fluid layer adjacent to a wall adheres to it, with zero relative velocity between the fluid and the wall.

Consequently, although it may seem reasonable to assume that fluids of low viscosity such as air and water would behave next to a wall like an ideal fluid, the assumption is erroneous. In transfer processes, it is usually the phenomena at the boundary between a fluid and a wall which are of the greatest import, and it is precisely here that the ideal-fluid analysis fails.

The dilemma is neatly resolved by the boundary-layer theory postulated by Prandtl in 1904. Before stating the theory, let us examine a simple example of flow of water over a flat plate at zero angle of incidence.

At any point x downstream from the leading edge of the plate, careful observations show velocity increases from zero at the wall to very near V_∞ at a very short distance δ from the plate. Also δ is seen to increase in thickness in the downstream direction. Figure 2.3 shows this, with the y-coordinate considerably magnified in relation to the x-coordinate. In the case of fluids of low kinematic viscosity such as water, δ is exceedingly thin. Since the shear stress is given by $\tau_{yx} = -\mu(\partial v_x/\partial y)$, the shear stress or, equivalently, the momentum transfer at the wall can be appreciable. Outside the thin layer, however, the velocity gradient is negligible, which means the shear stress is very small. Consequently, fluid in this outer region can be treated as ideal.

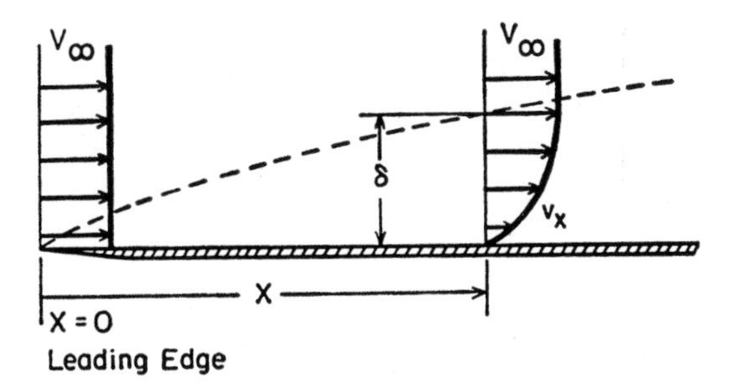

Fig. 2.3. Laminar boundary layer on flat plate.

Prandtl (2) suggested that the field of flow can be divided into two regions: a thin layer next to the wall, which he called the *boundary layer*, where the shear stress is significant, and the region outside this layer, where the ideal-fluid analysis is applicable. We shall see in Chaps. 3, 7, and 15 that the boundary-layer theory not only serves as a useful physical concept, but also introduces fundamental simplifications in the mathematical analysis of the various transfer processes.

(*a*) *Boundary-Layer Thickness*

The definition of a boundary-layer thickness δ is somewhat arbitrary since velocity in the boundary layer approaches the free-stream velocity V_∞ asymptotically. Practically, we define δ as the distance from the wall where

$$v_x = 0.99V_\infty \tag{2.2}$$

The following alternate definitions in terms of the volume flow or the momentum loss in the boundary layer are also used.

Let Q be the volume flow in the boundary layer. From Fig. 2.4

$$Q = \int_0^\delta v_x\,dy = V_\infty(\delta - \delta^*)$$

or

$$\delta^* = \int_0^\delta \left(1 - \frac{v_x}{V_\infty}\right) dy \tag{2.3}$$

δ^* is known as the *displacement thickness* of boundary layer. It may be thought of as the distance by which an equivalent ideal-fluid stream must be displaced from the wall to give identical volume flow. For laminar flow over a flat plate, δ^* is approximately one-third of δ (see Sec. 3.3).

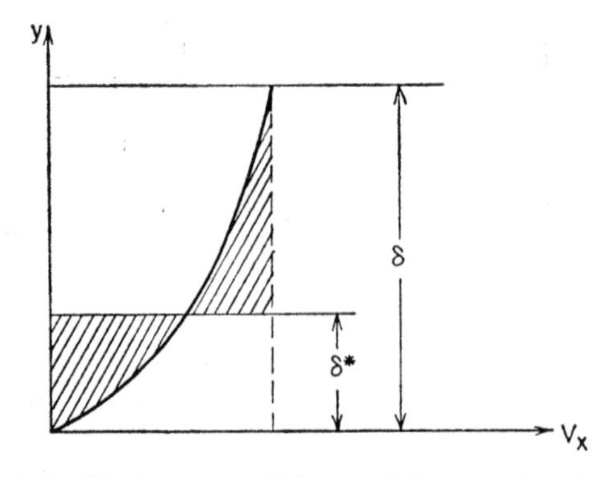

Fig. 2.4. Displacement thickness of the boundary layer.

Momentum loss in the boundary layer due to viscous shear, can be expressed as

$$\int_0^\infty \rho v_x (V_\infty - v_x)\, dy$$

Momentum thickness θ is defined in terms of the momentum loss by

$$\int_0^\delta \rho v_x (V_\infty - v_x)\, dy = \rho V_\infty^2 \theta$$

or

$$\theta = \int_0^\delta \frac{v_x}{V_\infty}\left(1 - \frac{v_x}{V_\infty}\right) dy \tag{2.4}$$

For laminar flow over a flat plate, θ is approximately one-seventh of δ (see Sec. 3.3).

(b) Boundary-Layer Transition

In our example of flow of water over a flat plate, if the plate were sufficiently long or if V_∞ were sufficiently large, flow in the boundary layer would become turbulent. The onset of turbulence is indicated by a sudden thickening of the boundary layer and a sudden increase in the wall shear stress. The transition point is given by a length Reynolds number defined by

$$\text{Re}_x \equiv V_\infty x/\nu$$

where V_∞ is the free-stream velocity, x is the distance from the leading edge, and ν is the kinematic viscosity of fluid. Figure 2.5 shows oscillograms obtained by Schubauer and Scramstad (3) showing growth of fluctuations in the boundary layer as the flow passes through the transition region.

The flow in the boundary layer near the leading edge is laminar, regardless of the level of turbulence existing in the approaching free stream. The

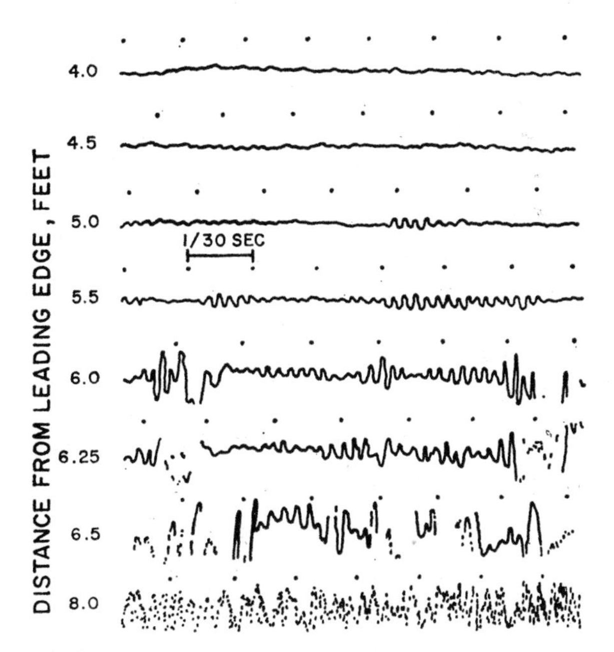

Fig. 2.5. Oscillograms of fluctuations in the boundary layer transition on a flat plate. Distance from surface = 0.025 in., $\mathbf{V}_{\infty} = 80$ ft/sec, time interval between dots = 1/30 sec (3).

location of the transition to turbulence, x_{tr} or $(\text{Re}_x)_{tr}$, depends greatly on the turbulence in the free stream. In streams with very low turbulence level, $(\text{Re}_x)_{tr}$ has been observed as high as 10^6, but in normally encountered engineering apparatus, 3.2×10^5 is considered more representative of average conditions. Actually, the location of transition is not well defined since a transition zone occupies a finite length of plate as shown in Fig. 2.6.

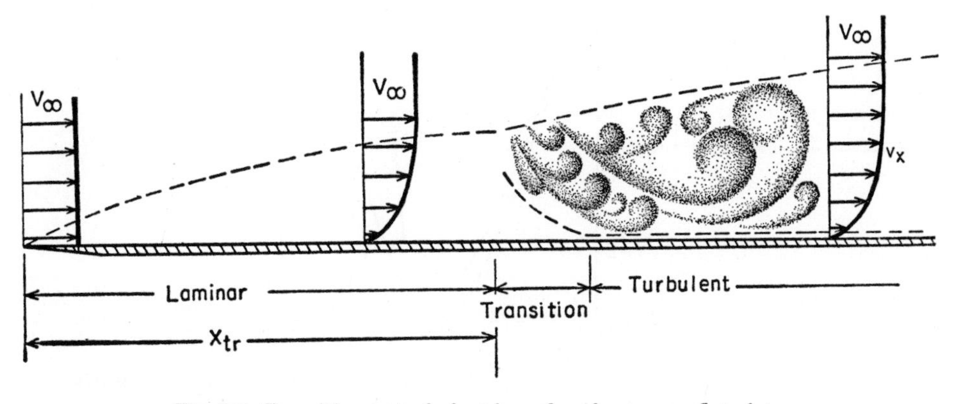

Fig. 2.6. Transition to turbulent boundary layer on a flat plate.

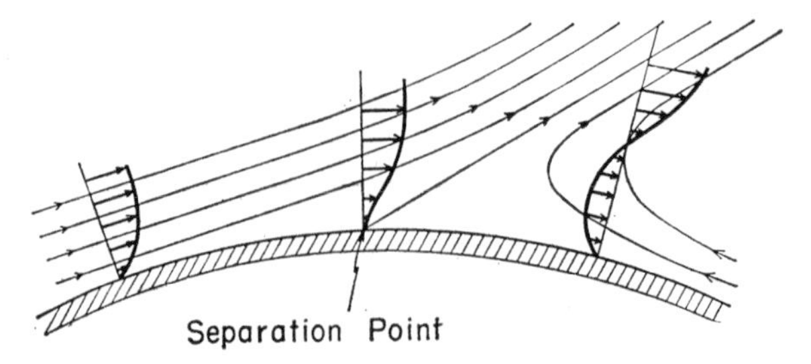

Fig. 2.7. Separation in boundary layer on cylinder.

For the completely turbulent boundary layer, the magnitude of the turbulent fluctuations gradually approaches zero at the wall.

(*c*) *Flow Separation*

The phenomenon of flow separation is often observed when the static pressure is increasing in the direction of flow. In flow over bluff bodies such as cylinders and spheres, pressure decreases from 0 to 90° and increases from 90 to 180°. The rising pressure on the downstream half can distort the velocity distribution in the boundary layer as shown in Figs. 2.7 and 2.8. The location where the profile contains an inflection point is called the *separation point*, and beyond that point a reversal of flow occurs near the wall.

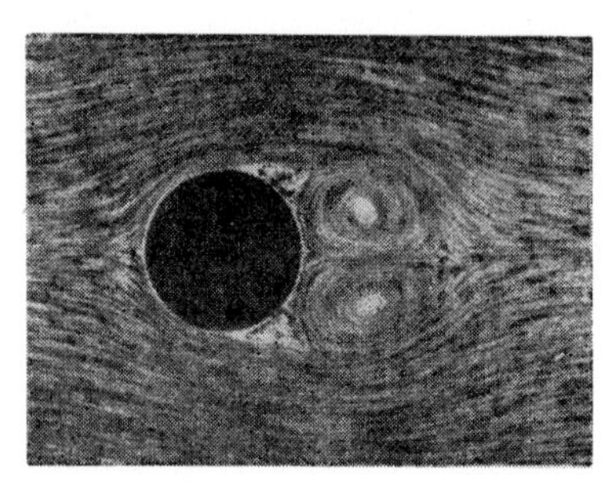

Fig. 2.8. Picture of flow across horizontal cylinder with boundary layer separation. [Prandtl and Tietjens (4)].

Figure 2.9 shows the pressure distribution around a cylinder for flow of an ideal fluid (curve *A*); for a laminar boundary layer with separation,

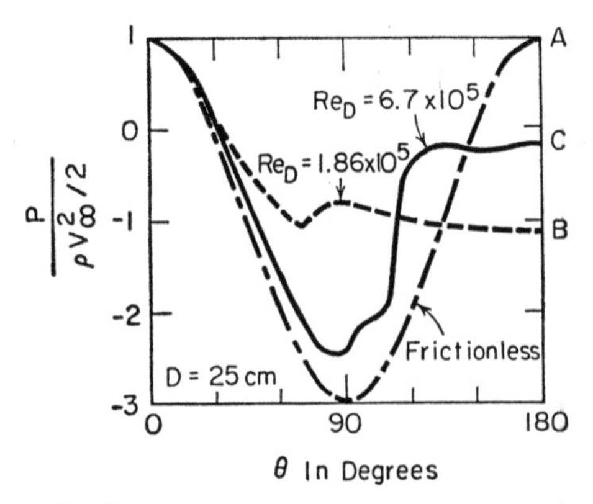

Fig. 2.9. Pressure distribution around cylinder with crossflow [Flachsbart (5)].

$Re_D = 1.86 \times 10^5$ (curve B); and for a laminar boundary layer which changes to turbulent before separation occurs, $Re_D = 6.7 \times 10^5$ (curve C). It is seen that curves B and C diverge widely from curve A.

An extremely useful assumption in connection with the boundary-layer theory, as we shall see in Chap. 3, is that since the layer is very thin for fluids of low viscosity, the pressure outside the boundary layer is transmitted unchanged through the layer to the wall. This means the pressure distribution inside a boundary layer may be calculated by an ideal-fluid analysis applicable to the external flow field. When there is flow separation at the wall, this assumption is obviously not valid, as Fig. 2.9 shows.

Example 2.2: The velocity distribution in a boundary layer over a plate is given as

$$\frac{v_x}{V_\infty} = \frac{3}{2}\left(\frac{y}{\delta}\right) - \frac{1}{2}\left(\frac{y}{\delta}\right)^3$$

Find the displacement thickness and the momentum thickness of the boundary layer in terms of δ.

Solution: From Eq. (2.3),

$$\delta^* = \int_0^\delta \left(1 - \frac{v_x}{V_\infty}\right) dy$$

$$= \delta \int_0^1 \left[1 - \frac{3}{2}\left(\frac{y}{\delta}\right) + \frac{1}{2}\left(\frac{y}{\delta}\right)^3\right] d\left(\frac{y}{\delta}\right)$$

$$= \delta \int_0^1 \left(1 - \frac{3}{2}\eta + \frac{1}{2}\eta^3\right) d\eta$$

where $\eta \equiv y/\delta$. Integration gives

$$\frac{\delta^*}{\delta} = \frac{3}{8}$$

From Eq. (2.5)

$$\theta = \int_0^\delta \frac{v_x}{V_\infty}\left(1 - \frac{v_x}{V_\infty}\right) dy$$

$$= \delta \int_0^1 \left(\frac{3}{2}\eta - \frac{1}{2}\eta^3\right)\left(1 - \frac{3}{2}\eta + \frac{1}{2}\eta^3\right) d\eta$$

Integration gives

$$\frac{\theta}{\delta} = \frac{39}{280}$$

REFERENCES

1. Reynolds, O., *Trans. Roy. Soc. (London)*, **174**:3, 935 (1883).
2. Prandtl, L., *Proc. 3rd Intern. Math. Congr.*, Heidelberg, 1904; reprinted in *Nat. Advisory Comm. Aeronaut., Tech. Mem.* 452 (1928).
3. Schubauer, G. B., and H. K. Skramstad, *Nat. Bur. Standards (U.S.) Research Paper*, 1772. Also *Nat. Advisory Comm. Aeronaut.*, Rept. 909.
4. Prandtl, L., and O. G. Tietjens, *Applied Hydro- and Aeromechanics*, Engineering Societies Monographs, McGraw-Hill, 1934.
5. Flachsbart, O., *Reports of the Aerodynamic Versuchanstalt, Goettingen*, Fourth Series, **134** (1932).

PROBLEMS

2.1 Find the Reynolds number for a spherical dust particle 0.015 in. in diameter settling in stagnant air at 60 F. Its velocity is estimated to be 3 in/min.

2.2 What must be the flow rate of light oil, sp. gr. 0.90, at 80 F in a 4-in. diameter pipe to give a Reynolds number of 1200? If the Reynolds number is to remain the same, what must be the flow rate of water at 80 F?

2.3 (a) A plate 5 ft wide and 40 ft long is towed through still water at 10 ft/sec. If $(\mathrm{Re}_x)_{\mathrm{tr}}$ is estimated to be 3×10^5, find the distance from the leading edge at which the boundary layer becomes turbulent. $T = 60$ F.

(b) If the velocity profile beyond the transition point is given by $v_x/V_\infty = (y/\delta)^{1/7}$ find δ^* and θ in terms of δ.

2.4 Answer true or false to the following questions.

(a) In a fluid of very low viscosity, the drag on a body may be found to a good approximation by the ideal fluid analysis.

(b) Flow separation at a solid surface is caused by an adverse pressure gradient.

(c) At the separation point, the pressure is a minimum.

(d) The purpose of streamlining is to minimize drag due to skin friction.

(e) In flow over a circular cylinder, if the Reynolds number is very small $(\mathrm{Re} < 1)$, the pressure distribution around the cylinder approximates that of an ideal frictionless fluid.

CHAPTER 3

Momentum Transfer in Laminar Flow

We shall first derive the differential equations of momentum for a few special cases such as flow between two flat plates, flow in circular tubes, and flow over a single flat plate. The same equations may be obtained by direct reduction of the general, three-dimensional, momentum conservation equations (Navier-Stokes equations) presented in Sec. 3.7.

The momentum equations are solved by starting with the simplest examples of fully developed flows in channels. The solutions give the velocity distributions; the velocity gradients at the wall in turn give the shear stresses. The momentum equations of the boundary layer are solved next, and for even the simplest geometry of a flat plate, lengthy numerical integration is necessary. The boundary-layer equations are solved alternatively by the approximate integral method of Kármán and Pohlhausen. The method involves writing the conservation of momentum in integral form; it is a powerful analytical tool in that it avoids much of the laborious mathematics and gives results that agree remarkably well with the exact solutions. In a similar way, the boundary-layer equations are solved for the inlet regions of channels.

Finally, the transient motion of a plate in an infinite fluid medium is studied; analogous problems in heat conduction and mass diffusion will be taken up in Secs. 6.15 and 15.3.

3.1 Flow between Parallel Plates

Consider the flow of a fluid between parallel plates, as shown in Fig. 3.1. The velocity at entrance is uniform, and as the flow progresses, a boundary layer forms and grows at either wall. Farther downstream where the velocity profile no longer changes along the flow path, the flow is said to be fully developed.

Now focus attention on a control volume fixed in space—of size Δx by Δy and of unity depth. Within the developing boundary layer there are cross velocities v_y. We write a conservation of momentum equation by identifying the significant momentum terms. Consider the x-direction momentum quantities. As discussed in Chap. 1, the force due to shear stress along the x-direction may be interpreted as a momentum transfer quantity; likewise the force due to pressure drop along the flow path may be interpreted as a momentum quantity. The arrows representing the shear stress τ are shown here in their assumed positive direction. The net forces (or equivalent momentum transfer) acting on the fluid within the control volume are

$$-\frac{\partial p}{\partial x}\,\Delta x\,\Delta y - \frac{\partial \tau}{\partial y}\,\Delta y\,\Delta x$$

The x-direction momentum flux due to motion in the x-direction is $(\rho v_x\,\Delta y)v_x$, and the x-direction momentum flux due to motion in the y-

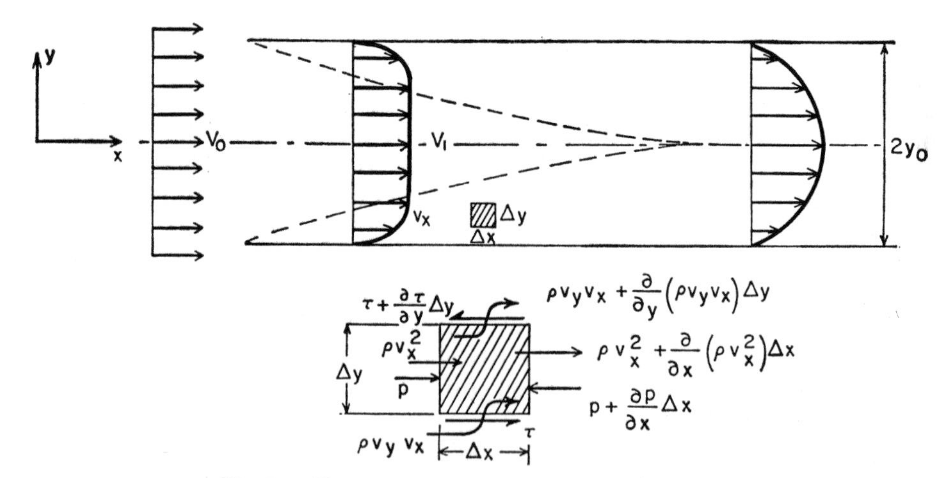

Fig. 3.1. Flow in entrance region between flat plates.

direction is $(\rho v_y \, \Delta x)v_x$. The net efflux of x-direction momentum is

$$\frac{\partial}{\partial x}(\rho v_x^2) \, \Delta x \, \Delta y + \frac{\partial}{\partial y}(\rho v_y v_x) \, \Delta y \, \Delta x$$

In the unsteady state the rate of increase of momentum within the control volume is

$$\frac{\partial}{\partial t}(\rho v_x) \, \Delta x \, \Delta y$$

Equating the forces and the momentum fluxes crossing the boundary to the change in momentum within the control volume and dividing by the volume $(\Delta x \, \Delta y \cdot 1)$ leads to

$$-\frac{\partial (\rho v_x)}{\partial t} = \frac{\partial p}{\partial x} + \frac{\partial \tau}{\partial y} + \frac{\partial (\rho v_x^2)}{\partial x} + \frac{\partial (\rho v_y v_x)}{\partial y} \tag{3.1}$$

which is the conservation of momentum equation for the case considered. For constant density and viscosity, with Eq. (1.7) and Eq. (1.12),

$$-\rho \frac{\partial v_x}{\partial t} = \frac{\partial p}{\partial x} - \mu \frac{\partial^2 v_x}{\partial y^2} + \rho v_x \frac{\partial v_x}{\partial x} + \rho v_y \frac{\partial v_x}{\partial y} \tag{3.2}$$

or

$$\frac{Dv_x}{Dt} = -\frac{1}{\rho}\frac{\partial p}{\partial x} + \nu \frac{\partial^2 v_x}{\partial y^2}$$

In the steady state $\dfrac{\partial v_x}{\partial t} = 0$; then

$$v_x \frac{\partial v_x}{\partial x} + v_y \frac{\partial v_x}{\partial y} = -\frac{1}{\rho}\frac{\partial p}{\partial x} + \nu \frac{\partial^2 v_x}{\partial y^2} \tag{3.3}$$

A momentum equation in the y-direction, derived by exactly the same procedure as above, reduces to $\partial p/\partial y = 0$, or p is independent of y over any particular cross section. Downstream beyond the point where the boundary layers have essentially joined, the flow is fully developed and $v_y = 0$. From continuity, $\partial v_x/\partial x = 0$, or v_x is independent of x.

Introducing all of these simplifications into Eq. (3.3), we get

$$\frac{dp}{dx} = \mu \frac{d^2 v_x}{dy^2} \tag{3.4}$$

We note that in the fully developed region, p is independent of y and v_x is independent of x. This would be possible only if each side of Eq. (3.4) were equal to a constant. In the following examples, Eq. (3.4) is integrated for two typical sets of boundary conditions.

Example 3.1: Fully developed flow between parallel plates—plane Poiseuille flow.

Take $y = 0$ at the centerline (Fig. 3.1) and integrate Eq. (3.4) between $v_x = 0$ at $y = y_0$ and v_x at any position y in the stream. Then

$$v_x = -\frac{1}{2\mu}\frac{dp}{dx}(y_0^2 - y^2) \tag{3.5}$$

The average velocity, V, is then

$$V = \frac{1}{y_0}\int_0^{y_0} v_x\, dy = -\frac{1}{3\mu}\frac{dp}{dx}y_0^2 \tag{3.6}$$

or

$$\frac{v_x}{V} = \frac{3}{2}\left[1 - \left(\frac{y}{y_0}\right)^2\right] \tag{3.7}$$

a parabolic velocity distribution.

Example 3.2: Lower plate is stationary and upper plate moves with a velocity V_1—Couette flow.

Take $y = 0$ at the lower plate with $v_x = 0$, and $y = l$ at the upper plate with $v_x = V_1$. Then integration of Eq. (3.4) results in

$$\frac{v_x}{V_1} = \frac{y}{l} - \frac{l^2}{2\mu V_1}\frac{dp}{dx}\frac{y}{l}\left(1 - \frac{y}{l}\right) \tag{3.8}$$

The solution is shown graphically in Fig. 3.2.

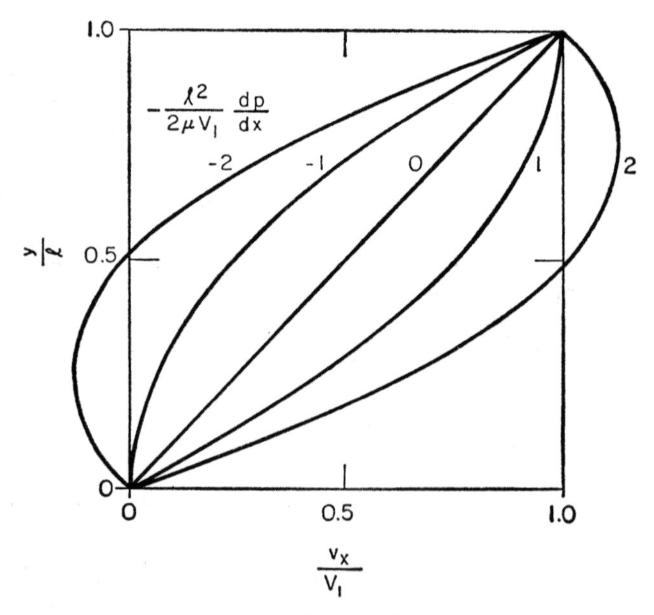

Fig. 3.2. Velocity distribution in Couette flow with a pressure gradient.

The velocity profile given by Eq. (3.8) is a superposition of a linear profile, corresponding to shear motion induced by the moving plate, and a parabolic profile corresponding to plane Poiseuille flow between stationary parallel plates. For a given fluid and plate width, final profile shape depends on the pressure gradient dP/dx, which may be positive or negative, and the plate velocity V_1. If dp/dx is strongly positive, flow may reverse at the lower plate, and if dp/dx is strongly negative, local fluid velocity may exceed the upper plate velocity V_1.

3.2 Flow in Circular Tubes

Imagine the sketch in Fig. 3.1 to represent flow in the inlet region of a circular tube of radius r_0. Consider a ring-shaped control volume of size $2\pi r\ \Delta r\ \Delta x$, as shown in Fig. 3.3. Then, in a manner similar to the previous development, but noting that area changes in the r-direction:

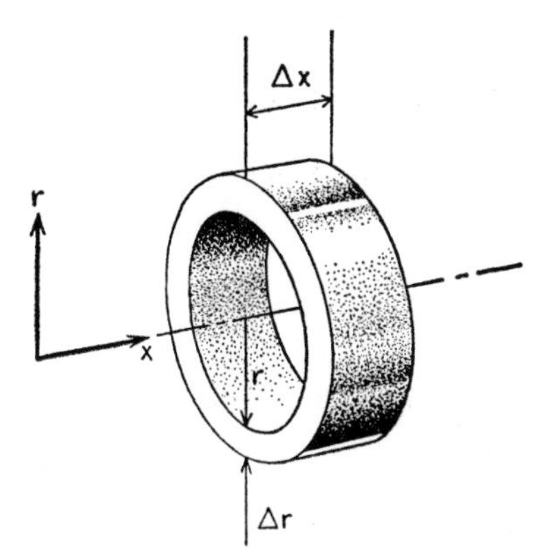

Fig. 3.3. Control surface in cylindrical coordinates.

Net x-direction forces (equivalent momentum transfer):

$$-\frac{\partial p}{\partial x}\ \Delta x(2\pi r\ \Delta r)\ -\ \frac{\partial}{\partial r}(\tau 2\pi r\ \Delta x)\ \Delta r$$

Net x-direction momentum efflux associated with fluid motion in x-and r-directions:

$$\frac{\partial(\rho v_x^2)}{\partial x}\ \Delta x(2\pi r\ \Delta r)\ +\ \frac{\partial}{\partial r}(\rho v_r v_x 2\pi r\ \Delta x)\ \Delta r$$

Net change in x-direction momentum of fluid inside control volume:

$$\frac{\partial}{\partial t}(\rho v_x)2\pi r\ \Delta r\ \Delta x$$

Summing all of these momentum fluxes and forces leads to

$$-\frac{\partial(\rho v_x)}{\partial t} = \frac{\partial p}{\partial x} + \frac{1}{r}\frac{\partial(r\tau)}{\partial r} + \frac{\partial(\rho v_x^2)}{\partial x} + \frac{1}{r}\frac{\partial(\rho r v_x v_r)}{\partial r} \tag{3.9}$$

For constant density and viscosity with Eqs. (1.1b) and (1.14)

$$-\rho\frac{\partial v_x}{\partial t} = \frac{\partial p}{\partial x} - \frac{\mu}{r}\frac{\partial}{\partial r}\left(r\frac{\partial v_x}{\partial r}\right) + \rho v_x\frac{\partial v_x}{\partial x} + \rho v_r\frac{\partial v_x}{\partial r} \tag{3.10}$$

In the steady state $\partial v_x/\partial t = 0$; then

$$v_x \frac{\partial v_x}{\partial x} + v_r \frac{\partial v_x}{\partial r} = -\frac{1}{\rho}\frac{\partial p}{\partial x} + \frac{\nu}{r}\frac{\partial}{\partial r}\left(r \frac{\partial v_x}{\partial r}\right) \tag{3.11}$$

In the fully developed flow region $v_r = 0$. By a similar reasoning as in the case of parallel plates, we can show that Eq. (3.11) reduces to

$$\frac{dp}{dx} = \mu\left[\frac{d^2 v_x}{dr^2} + \frac{1}{r}\frac{dv_x}{dr}\right] \tag{3.12}$$

where dp/dx is a constant.

Example 3.3: For fully developed flow in tubes with $v_x = 0$ at $r = r_0$, and $v_x =$ finite at $r = 0$, Eq. (3.12) integrates to

$$v_x = -\frac{1}{4\mu}\left(\frac{dp}{dx}\right)(r_0^2 - r^2) \tag{3.13}$$

The velocity distribution is parabolic as in the case of fully developed flow between parallel plates. The average velocity is

$$V = \frac{1}{\pi r_0^2}\int_0^{r_0} v_x(2\pi r)\, dr = -\frac{r_0^2}{8\mu}\frac{dp}{dx} \tag{3.14}$$

Mass flow rate is given by

$$w = \pi r_0^2 \rho V = -\frac{\pi r_0^4 \rho}{8\mu}\frac{dp}{dx} \tag{3.15}$$

or

$$\frac{dp}{dx} = -\frac{8\mu V}{r_0^2} \tag{3.16}$$

and from Eqs. (3.13) and (3.14)

$$\frac{v_x}{V} = 2\left[1 - \left(\frac{r}{r_0}\right)^2\right] \tag{3.17}$$

At the centerline $r = 0$; then, $v_c/V = 2$. Hence, for laminar flow the pressure drop dp/dx is proportional to the mean velocity V, and the centerline velocity is twice the average velocity. This type of flow is called *Hagen-Poiseuille flow*.

3.3 Flow over a Flat Plate—Exact Solution

The sketch in Fig. 3.4 represents the developing boundary layer on a flat plate parallel with the stream.

If V_∞ is assumed constant, pressure everywhere outside the boundary layer is uniform; therefore within the boundary layer the pressure gradient

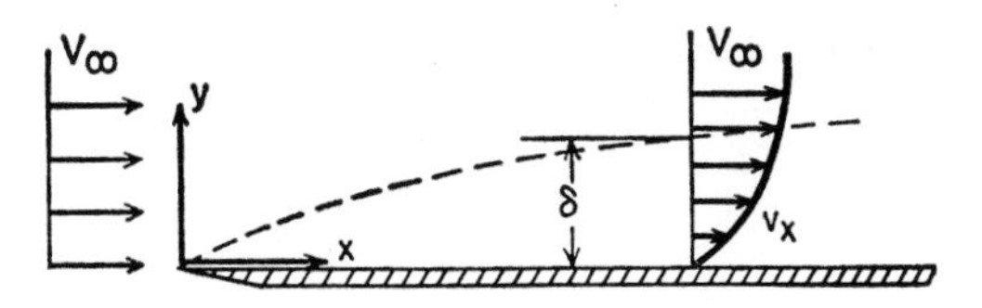

Fig. 3.4. Hydrodynamic boundary layer on flat plate.

$(\partial p/\partial x)$ is zero, and Eqs. (3.1), (3.2), and (3.3) apply. For the steady-state case with uniform density, Eq. (3.3) becomes

$$v_x \frac{\partial v_x}{\partial x} + v_y \frac{\partial v_x}{\partial y} = \nu \frac{\partial^2 v_x}{\partial y^2} \tag{3.18}$$

which is the *boundary-layer equation* for a flat plate with zero pressure gradient.

We now proceed to solve Eq. (3.18) in conjunction with the continuity equation (1.12) for the following boundary conditions:

$$\begin{aligned} y &= 0, \qquad v_x = v_y = 0 \\ y &= \infty, \qquad v_x = V_\infty \end{aligned} \tag{3.19}$$

To simplify Eq. (3.18), define a stream function ψ (7) by

$$v_x \equiv \frac{\partial \psi}{\partial y} \quad \text{and} \quad v_y \equiv -\frac{\partial \psi}{\partial x} \tag{3.20}$$

This satisfies the continuity equation (1.12) since

$$\frac{\partial^2 \psi}{\partial x\, \partial y} = \frac{\partial^2 \psi}{\partial y\, \partial x}$$

Substitute Eq. (3.20) into (3.18) to obtain

$$\frac{\partial \psi}{\partial y}\frac{\partial^2 \psi}{\partial x\, \partial y} - \frac{\partial \psi}{\partial x}\frac{\partial^2 \psi}{\partial y^2} = \nu \frac{\partial^3 \psi}{\partial y^3} \tag{3.21}$$

Blasius (2) showed that Eq. (3.21) could be solved by expressing the stream function as follows:

$$\psi \equiv \sqrt{V_\infty \nu x}\, f(\eta) \tag{3.22}$$

where

$$\eta \equiv \frac{y}{\sqrt{x}} \sqrt{\frac{V_\infty}{\nu}} \tag{3.23}$$

The derivation of this choice for ψ is left to a more advanced treatment of this subject.

From Eqs. (3.20), (3.22), and (3.23)

$$v_x = \frac{\partial \psi}{\partial y} = \frac{\partial \psi}{\partial \eta}\frac{\partial \eta}{\partial y} = V_\infty \frac{df}{d\eta} \tag{3.24}$$

$$v_y = -\frac{\partial \psi}{\partial x} = \frac{1}{2}\sqrt{\frac{\nu V_\infty}{x}}\left(\eta \frac{df}{d\eta} - f\right) \tag{3.25}$$

Then Eqs. (3.18) or (3.21) become

$$2\frac{d^3 f}{d\eta^3} + f\frac{d^2 f}{d\eta^2} = 0 \tag{3.26}$$

The procedure outlined above with ψ and η defined by Eqs. (3.22) and (3.23) reduces the nonlinear, second-order, partial differential equation (3.18) or (3.21) to a nonlinear, third-order, ordinary differential equation (3.26), with the boundary condition (3.19) of the following equivalent form:

$$\begin{aligned} &\eta = 0, \qquad f = 0, \qquad \frac{df}{d\eta} = 0 \\ &\eta = \infty, \qquad \frac{df}{d\eta} = 1 \end{aligned} \tag{3.27}$$

Equation (3.26) may be solved numerically by expressing $f(\eta)$ in a power series which, for the boundary conditions (3.27), becomes

$$f = \frac{\alpha\eta^2}{2!} - \frac{1}{2}\frac{\alpha^2\eta^5}{5!} + \frac{11}{4}\frac{\alpha^3\eta^8}{8!} - \frac{375}{8}\frac{\alpha^4\eta^{11}}{11!} + \cdots \tag{3.28}$$

where $\alpha \equiv 0.332$. Then, Eqs. (3.24) and (3.25) give expressions for v_x and v_y. Equation (3.26) was also solved (2 and 3) by the numerical method of

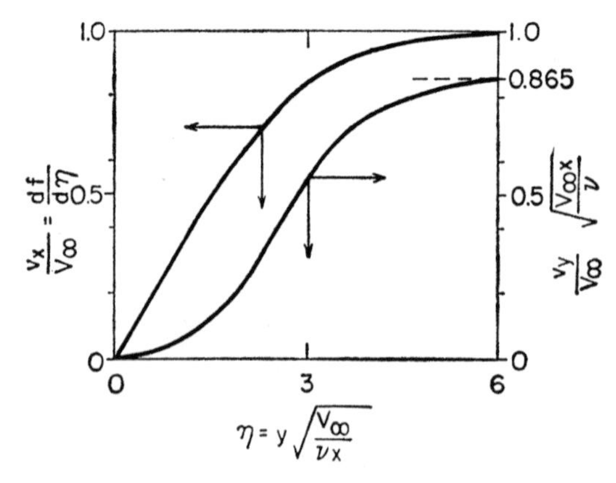

Fig. 3.5. Velocity distribution in boundary layer on flat plate.

finite differences, discussed in Chap. 18. The results of Howarth (3) are shown in Fig. 3.5. Note an anomaly in the solution—that v_y does not go to zero at large values of y; as $\eta \to \infty$

$$v_y \to 0.865 V_\infty \sqrt{\frac{\nu}{x V_\infty}}$$

Also, the position at which $v_x/V_\infty = 0.99$ is located approximately at $\eta = 5.0$ which establishes an arbitrary boundary-layer thickness δ as

$$\frac{\delta}{x} \cong \frac{5.0}{\sqrt{V_\infty x/\nu}} \tag{3.29}$$

In terms of displacement thickness δ^*, defined by Eq. (2.3) and momentum thickness θ, given by Eq. (2.5),

$$\frac{\delta^*}{x} = \frac{1.73}{\sqrt{V_\infty x/\nu}} \quad \text{or} \quad \frac{\delta^*}{\delta} \approx \frac{1}{3}$$

$$\frac{\theta}{x} = \frac{0.66x}{\sqrt{V_\infty x/\nu}} \quad \text{or} \quad \frac{\theta}{\delta} \approx \frac{1}{7}$$

The local skin friction coefficient on the plate is defined† and calculated by

$$C_{f,x} \equiv \frac{(\tau_0)_x}{\rho V_\infty^2/2} = \frac{\mu(\partial v_x/\partial y)_{y=0}}{\rho V_\infty^2/2} = \frac{2}{\sqrt{V_\infty x/\nu}}\left(\frac{\partial(v_x/V_\infty)}{\partial\eta}\right)_{\eta=0} \tag{3.30}$$

From Fig. 3.5 or Eq. (3.28), the gradient at $\eta = 0$ is

$$\left(\frac{\partial(v_x/V_\infty)}{\partial\eta}\right)_{\eta=0} = 0.332 \tag{3.31}$$

Then

$$C_{f,x} = \frac{0.664}{\sqrt{V_\infty x/\nu}} \tag{3.32}$$

and the average value of $C_{f,x}$ is

$$C_f = \frac{1}{L}\int_0^L C_{f,x}\,dx = \frac{1.328}{\sqrt{V_\infty L/\nu}} \tag{3.33}$$

Here

$$C_f \equiv \frac{(\tau_0)_{\text{avg}}}{\rho V_\infty^2/2} \tag{3.34}$$

where τ_0 is the shear force acting *on* the plate per unit of exposed area.

† In Eq. (3.30), $(\tau_0)_x$ is the sheer stress acting *on* the wall at location x $(= -\tau_{yx})$.

Example 3.4: Air at 80 F flows over a flat plate with a velocity of 40 ft/sec. At a point 3 in. from the leading edge, estimate (a) the thickness of the boundary layer, and (b) the rate of growth of the boundary layer.

At 80 F, the air properties are

$$\rho = 0.0735 \text{ lb}_m/\text{ft}^3 \quad \text{and} \quad \nu = 1.69 \times 10^{-4} \text{ ft}^2/\text{sec.}$$

The Reynolds number is

$$\text{Re}_x = \frac{V_\infty x}{\nu} = \frac{(40)\left(\frac{3}{12}\right)}{1.69 \times 10^{-4}} = 59{,}100$$

a laminar flow.

(a) From Eq. (3.29),

$$\frac{\delta}{x} = \frac{5}{\sqrt{59{,}100}} = 0.0206$$

or $\delta = 0.062$ in. thick at the 3 in. location.

(b) The rate of growth is given by differentiating Eq. (3.29):

$$\frac{d\delta}{dx} = \frac{5}{2}\sqrt{\frac{\nu}{V_\infty x}} = 0.01 \text{ in./in.}$$

3.4 Flow over a Flat Plate—Approximate Integral Method

The solution of the differential equation (3.18) yields both velocity distribution and friction coefficient. An alternative solution which provides results only for friction coefficient involves writing the momentum equation not for a small element but for total momentum quantities integrated over the entire boundary layer (Fig. 3.6).

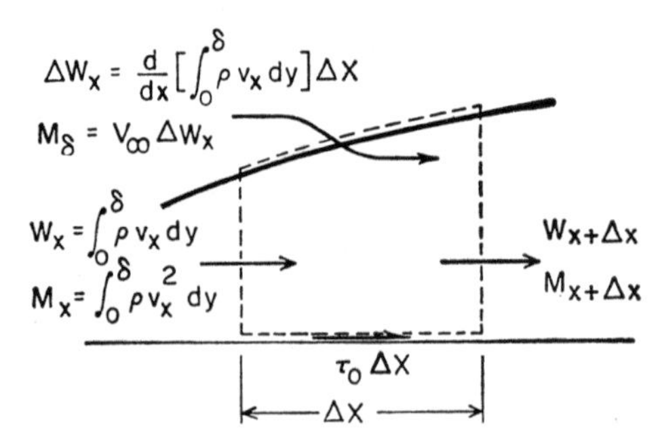

Fig. 3.6. Integrated mass and momentum quantities in boundary layer on a flat plate.

For the control surface, shown dotted, the momentum conservation is

$$(M_{x+\Delta x} - M_x) - M_\delta = (\tau_{yx})_0 \, \Delta x \tag{3.35}$$

Then with the flow and momentum expression shown in the sketch in Fig. 3.6, this becomes

$$\frac{d}{dx}\left[\int_0^\delta \rho v_x^2\, dy\right]\Delta x - V_\infty \frac{d}{dx}\left[\int_0^\delta \rho v_x\, dy\right]\Delta x = -\mu\left(\frac{\partial v_x}{\partial y}\right)_{y=0}\Delta x \tag{3.36}$$

or, since V_∞ and ρ are constants,

$$\frac{d}{dx}\left[\int_0^\delta (V_\infty - v_x)v_x\, dy\right] = \nu\left(\frac{\partial v_x}{\partial y}\right)_{y=0} \tag{3.37}$$

Equation (3.37) is known as the *von Kármán integral relation* for zero pressure gradient.

At this point, it is well to note that this equation applies equally well to a turbulent boundary layer (Sec. 4.8). We now proceed with its solution for a laminar boundary layer.

Equation (3.37) is solved by assuming a reasonable shape for the velocity distribution. Even with assumed distributions which depart significantly from the actual velocity distribution, the result for the friction coefficient is quite close to the correct one.

To illustrate the nature of the solution assume, quite arbitrarily, that the velocity distribution has the following form:

$$v_x = ay + by^3$$

Then, with $v_x = 0$ at $y = 0$, and $v_x = V_\infty$ and $\partial v_x/\partial y = 0$ at $y = \delta$, constants a and b may be evaluated with the following result:

$$\frac{v_x}{V_\infty} = \frac{3}{2}\left(\frac{y}{\delta}\right) - \frac{1}{2}\left(\frac{y}{\delta}\right)^3 \tag{3.38}$$

Substitute this into Eq. (3.37) and perform the operations indicated. Then Eq. (3.37) reduces to

$$\delta d\delta = \frac{140}{13}\frac{\nu}{V_\infty}\, dx \tag{3.39}$$

which integrates directly with $\delta = 0$ at $x = 0$ to

$$\frac{\delta}{x} = 4.64\sqrt{\frac{\nu}{xV_\infty}} \tag{3.40}$$

Evaluate $(\partial v_x/\partial y)_{y=0}$ from Eq. (3.38) and substitute into Eq. (3.30) to obtain

$$C_{f,x} = \frac{0.646}{\sqrt{V_\infty x/\nu}} \tag{3.41}$$

Equations (3.40) and (3.41) compare quite favorably, within about 3 per cent, with the exact solution results, Eqs. (3.29) and (3.32).

TABLE 3.1 RESULTS OF INTEGRAL-METHOD SOLUTION FOR LAMINAR BOUNDARY LAYER ON A FLAT PLATE

Velocity distribution $\frac{v_x}{V_\infty}$	$\frac{\delta}{x}\sqrt{\mathrm{Re}_x}$	$C_{f,x}\sqrt{\mathrm{Re}_x}$
Exact solution:	5.0	0.664
$\frac{y}{\delta}$	3.46	0.578
$\frac{3}{2}\frac{y}{\delta} - \frac{1}{2}\left(\frac{y}{\delta}\right)^3$ VON KARMAN	4.64	0.646
$\sin\frac{\pi}{2}\frac{y}{\delta}$	4.80	0.654

This example illustrates quite forcefully the great potential usefulness of the integral-method solution for many problems, particularly those for which exact solutions cannot be obtained readily.

A rather close approximation of the result can be obtained for the preceding case even when the rather absurd assumption of a linear velocity distribution is used. Table 3.1 shows the results obtained for various assumed velocity distributions compared with the results of the exact solution.

It should be noted that the integral equation (3.37) can be obtained by integrating the differential equation (3.18) directly.

$$\int_0^\delta \left(v_x \frac{\partial v_x}{\partial x} + v_y \frac{\partial v_x}{\partial y}\right) dy = \nu \int_0^\delta \frac{\partial}{\partial y}\left(\frac{\partial v_x}{\partial y}\right) dy \tag{3.42}$$

or

$$\int_0^\delta v_x \frac{\partial v_x}{\partial x}\, dy + v_x v_y \Big|_0^\delta - \int_0^\delta v_x \frac{\partial v_y}{\partial y}\, dy = \nu \int_0^\delta d\left(\frac{\partial v_x}{\partial y}\right) \tag{3.43}$$

From continuity, Eq. (1.12),

$$\left(\frac{\partial v_y}{\partial y}\right) = -\left(\frac{\partial v_x}{\partial x}\right) \quad \text{and} \quad (v_y)_\delta = -\int_0^\delta \frac{\partial v_x}{\partial x}\, dy$$

Also $(v_x)_\delta = V_\infty$. Then, Eq. (3.43) becomes

$$\int_0^\delta 2v_x \frac{\partial v_x}{\partial x}\, dy - V_\infty \int_0^\delta \frac{\partial v_x}{\partial x}\, dy = -\nu \left(\frac{\partial v_x}{\partial y}\right)_{y=0}$$

or since $2v_x\, \partial v_x/\partial x = \partial(v_x^2)/\partial x$, this becomes

$$\int_0^\delta \frac{\partial}{\partial x}[V_\infty - v_x]v_x\, dy = \nu \left(\frac{\partial v_x}{\partial y}\right)_{y=0} \tag{3.44}$$

Since the order of integration and differentiation is immaterial here, Eq. (3.44) reduces directly to Eq. (3.37).

3.5 Flow in Inlet between Parallel Plates and Circular Tubes

Schlichting (4) solved the problem of flow in the inlet regions by series solution of the differential equation, Eq. (3.3), with the results for the velocity distribution shown in Fig. 3.7(b). We now present the integral-method solution accomplished by Sparrow (5).

Because of the developing boundary layers, the velocity, V_1, in the core fluid outside the boundary layer increases along the length, Fig. 3.1. This is the region of potential flow—negligible frictional effects. Then the change in core velocity is related to the pressure change in the core by the following momentum equation for an ideal fluid:

$$V_1 \frac{dV_1}{dx} = -\frac{1}{\rho}\frac{dp}{dx} \tag{3.45}$$

which follows from Eq. (3.2) when $\mu = 0$, $v_y = 0$, and $\partial/\partial t = 0$. Within the boundary layer, the momentum conservation equation, Eq. (3.36), must be modified in this case by addition of a pressure force term:

$$-\frac{d}{dx}\left[\int_0^\delta \rho v_x^2 dy\right] + V_1 \frac{d}{dx}\left[\int_0^\delta \rho v_x \, dy\right] - \int_0^\delta \frac{dp}{dx}\, dy = \mu \left(\frac{\partial v_x}{\partial y}\right)_{y=0} \tag{3.46}$$

We now assume that the pressure in the boundary layer at any x is the same as the pressure in the core at that position. This allows us to substitute Eq. (3.45) into Eq. (3.46) for dp/dx. By adding and subtracting a term

$$\frac{dV_1}{dx}\int_0^\delta \rho v_x \, dy$$

Eq. (3.46) may be contracted to

$$\frac{d}{dx}\left[\int_0^\delta (V_1 - v_x)v_x \, dy\right] + \frac{dV_1}{dx}\int_0^\delta (V_1 - v_x)\, dy = \nu \left(\frac{\partial v_x}{\partial y}\right)_{y=0} \tag{3.47}$$

which is the integral equation for this case.

The continuity equation may be written as follows

$$\int_0^\delta \rho v_x \, dy + \rho V_1(y_0 - \delta) = \rho y_0 V_0$$

or

$$\int_0^\delta (v_x - V_1)\, dy + y_0(V_1 - V_0) = 0 \tag{3.48}$$

where y_0 is the half-width of the plate spacing, and where V_0 is the uniform inlet velocity and equals the average velocity at any position x.

We now assume a velocity distribution

$$v_x = my + ny^2$$

which for boundary conditions $v_x = 0$ at $y = 0$, and $v_x = V_1$ and $\partial v_x/\partial y = 0$ at $y = \delta$ becomes

$$\frac{v_x}{V_1} = 2\left(\frac{y}{\delta}\right) - \left(\frac{y}{\delta}\right)^2 \tag{3.49}$$

Substitute Eq. (3.49) into Eqs. (3.47) and (3.48) and rearrange to obtain the following equations:

$$dx^* = \frac{3}{10}\,\frac{(V_1^* - 1)}{(V_1^*)^2}\,(9V_1^* - 7)\,dV_1^* \tag{3.50}$$

$$\delta^* = 3\left(1 - \frac{1}{V_1^*}\right) \tag{3.51}$$

where

$$x^* = \frac{16(x/D_e)}{\mathrm{Re}_{D_e}}$$

$$V_1^* = \frac{V_1}{V_0} \tag{3.52}$$

$$\delta^* = \frac{\delta}{y_0}$$

$$D_e = 4y_0$$

Equation (3.50) is readily integrated starting with $V_1^* = 1.0$ at $x^* = 0$. This results in a single curve of V_1^* vs. x^*. Then from Eq. (3.51) δ^* may be calculated at any V_1^* and, hence, at any x^*, Fig. 3.7(a). These results agree reasonably well with those of Schlichting (4), who solved the differential equation by series solution. The latter results are shown in Fig. 3.7(b).

For the case of flow in the inlet region of circular tubes, Langhaar (6) solved Eq. (3.11) with the results shown in Fig. 3.8.

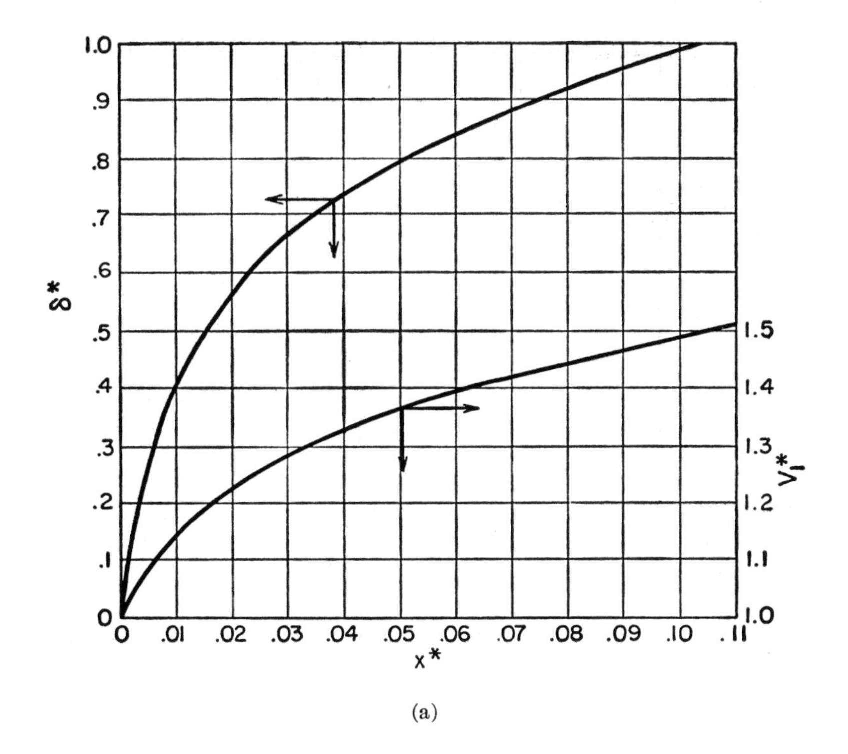

(a)

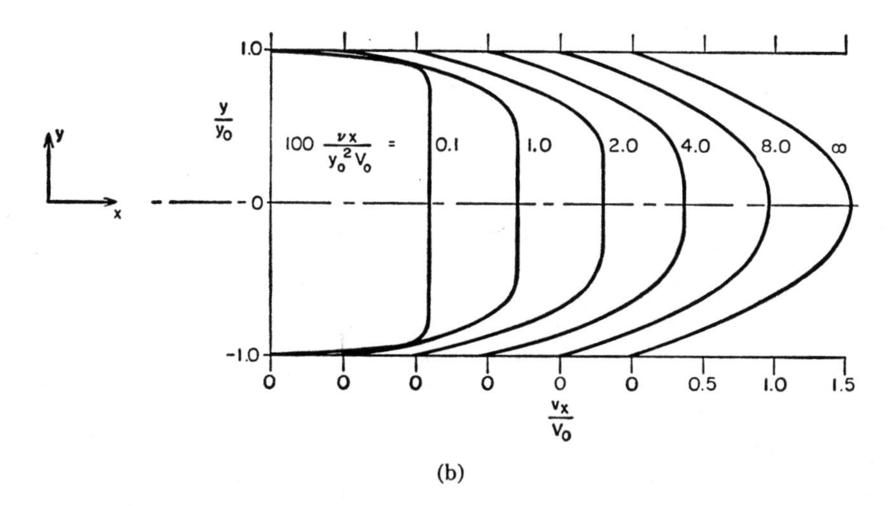

(b)

Fig. 3.7. Velocity distribution for laminar flow in inlet section between parallel plates. (a) Results of integral solution [Sparrow (5)]. (b) Results of series solution of differential equation [Schlichting (4)].

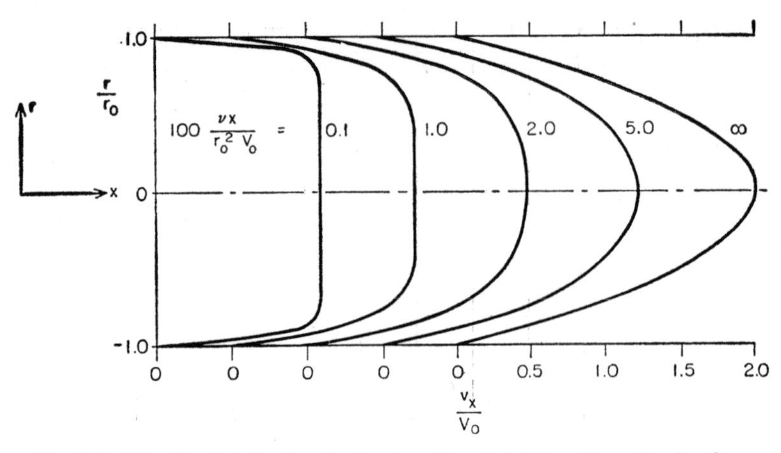

Fig. 3.8. Velocity distribution for laminar flow in inlet section of a circular pipe [Langhaar (6)].

3.6 Transient Motion of Plate in Infinite Fluid

Consider a very large flat plate bounding a fluid of semi-infinite extent. Initially, both are at rest; then, suddenly, the plate is set into motion with velocity V_0 as shown in Fig. 3.9. The velocity distribution in the fluid varies

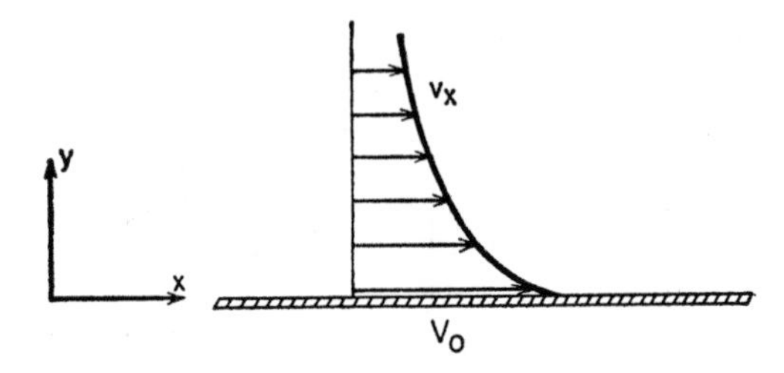

Fig. 3.9. Transient motion of large plate in fluid of infinite extent.

with time. For this one-dimensional transient problem, Eq. (3.2) reduces to

$$\frac{\partial v_x}{\partial t} = \nu \frac{\partial^2 v_x}{\partial y^2} \tag{3.53}$$

since $v_y = 0$ and $\partial v_x/\partial x = 0$ from continuity.

For this case, the initial and boundary conditions are as follows:

$$\begin{aligned} t &= 0, \quad v_x = 0 \\ t &> 0, \quad v_x = V_0 \quad \text{at } y = 0 \end{aligned} \tag{3.54}$$

Also, $$t \geq 0, \quad v_x = 0 \quad \text{at } y = \infty$$

The solution of Eq. (3.53) with Eq. (3.54) is found in Appendix C with $P_i = V_0$, $P = (V_0 - v_x)$, and $\mathfrak{D} = \nu$ to be as follows:

$$\frac{v_x}{V_0} = 1 - \operatorname{erf} \frac{y}{2\sqrt{\nu t}} \tag{3.55}$$

which expresses the velocity as a function of y and t.

3.7 General Momentum Conservation Equations—Navier-Stokes Equations

In Chap. 1, we presented a one-dimensional statement of the momentum conservation law, our purpose then being to emphasize the area of similarity among the conservation laws of various transferable properties. We considered only viscous shear forces (or equivalent momentum transfer), expressed by the simple one-dimensional rate equation (1.1). So far in this chapter, we have derived momentum balances for a number of elementary systems. We now write a more general form of the momentum conservation equations in three dimensions.

In general, forces acting on a fluid system may be classified as body forces proportional to the volume or mass of the system, such as gravity, and surface forces proportional to the area of surface on which they act, such as pressure and viscous forces.

The presence of viscous forces in a fluid system gives rise to three-dimensional stress-strain relations analogous to the well-known stress-strain relations (Hooke's law) for an elastic solid. However, whereas in an elastic solid stresses are porportional to strain, in fluids stresses are em-

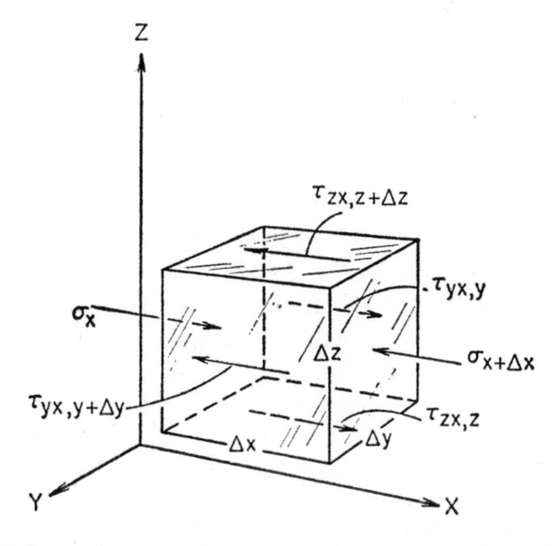

Fig. 3.10. Volume element with arrows indicating the positive direction of x-direction stresses.

pirically found to be proportional to the rate of strain, expressible in terms of velocity gradients. Physically, this means a fluid offers no resistance to change of shape, but resists time rate of change of shape. Newton's equation of viscosity given in Chap. 1 is a stress-strain relation for fluid in one-dimension, and, therefore, is accurate only so far as the assumption that τ depends solely on velocity gradient normal to the flow is valid. The general three-dimensional stress-strain relations for a viscous fluid are given below. We shall accept them as empirical formulations, although precise expressions for gases can be derived from the kinetic theory. Figure 3.10 shows the assumed positive direction of x-direction shear stresses; similar sets exist in the other two directions.*

$$\sigma_x = p - 2\mu \frac{\partial v_x}{\partial x} + \frac{2}{3}\mu \left(\frac{\partial v_x}{\partial x} + \frac{\partial v_y}{\partial y} + \frac{\partial v_z}{\partial z}\right)$$

$$\sigma_y = p - 2\mu \frac{\partial v_y}{\partial y} + \frac{2}{3}\mu \left(\frac{\partial v_x}{\partial x} + \frac{\partial v_y}{\partial y} + \frac{\partial v_z}{\partial z}\right)$$

$$\sigma_z = p - 2\mu \frac{\partial v_z}{\partial z} + \frac{2}{3}\mu \left(\frac{\partial v_x}{\partial x} + \frac{\partial v_y}{\partial y} + \frac{\partial v_z}{\partial z}\right) \tag{3.56}$$

$$\tau_{xy} = -\mu \left(\frac{\partial v_x}{\partial y} + \frac{\partial v_y}{\partial x}\right) = \tau_{yx}$$

$$\tau_{yz} = -\mu \left(\frac{\partial v_y}{\partial z} + \frac{\partial v_z}{\partial y}\right) = \tau_{zy}$$

$$\tau_{zx} = -\mu \left(\frac{\partial v_z}{\partial x} + \frac{\partial v_x}{\partial z}\right) = \tau_{xz}$$

In these equations μ is a constant of proportionality called, as before, *coefficient of viscosity*; its definition is more general than that given by Eq. (1.1). σ's are normal stresses, composed of nonviscous normal stresses or pressures p, and viscous normal stresses (remaining terms containing μ). τ's are viscous shear stresses, with the double subscripts having the same meaning as before. We note in the general case that τ_{xy}, for instance, depends not only on $\partial v_x/\partial y$, but also on $\partial v_y/\partial x$. In many problems of physical significance such as boundary-layer problems, $\partial v_y/\partial x \ll \partial v_x/\partial y$. Consequently, the simple rate equation (1.1) can be used with acceptable accuracy.

* Footnote on sign convention: It should be noted that the sign convention adopted in Eq. (3.56) is consistent with that of Eq. (1.1b), and is opposite to that found in many fluid-mechanics and heat-transfer books. This convention requires that the positive directions of shear and normal stresses be as shown in Fig. 3.10. The Navier-Stokes equations (3.59) will have the correct signs, as they must.

The momentum conservation equation given in Chap. 1 for the one-dimensional case can be written as

$$\frac{\partial}{\partial t}(\rho v_x) = -\frac{\partial}{\partial x}(\rho v_x^2) - \frac{\partial \tau_{yx}}{\partial y} \tag{3.57}$$

Generalization of the momentum conservation equation in three dimensions, in which body and normal forces are considered in addition to the shear forces, follows:

$$\begin{aligned}
\frac{\partial}{\partial t}(\rho v_x) &= -\frac{\partial}{\partial x}(\rho v_x^2) - \frac{\partial}{\partial y}(\rho v_x v_y) - \frac{\partial}{\partial z}(\rho v_x v_z) \\
&\qquad - \frac{\partial \sigma_x}{\partial x} - \frac{\partial \tau_{yx}}{\partial y} - \frac{\partial \tau_{zx}}{\partial z} + X \\
\frac{\partial}{\partial t}(\rho v_y) &= -\frac{\partial}{\partial x}(\rho v_x v_y) - \frac{\partial}{\partial y}(\rho v_y^2) - \frac{\partial}{\partial z}(\rho v_z v_y) \\
&\qquad - \frac{\partial \tau_{xy}}{\partial x} - \frac{\partial \sigma_y}{\partial y} - \frac{\partial \tau_{zy}}{\partial z} + Y \qquad (3.58) \\
\frac{\partial}{\partial t}(\rho v_z) &= -\frac{\partial}{\partial x}(\rho v_x v_z) - \frac{\partial}{\partial y}(\rho v_y v_z) - \frac{\partial}{\partial z}(\rho v_z^2) \\
&\qquad - \frac{\partial \tau_{xz}}{\partial x} - \frac{\partial \tau_{yz}}{\partial y} - \frac{\partial \sigma_z}{\partial z} + Z
\end{aligned}$$

where X, Y, and Z are body forces per unit volume in the coordinate directions.

Equation (3.58) states that the rate of change of momentum within a differential volume element fixed in space equals the net momentum influx due to fluid motion (first three terms on right), net surface forces (middle three terms on right), and body forces (last term on right).

Substitution of Eq. (3.56) into Eq. (3.58) gives:

$$\begin{aligned}
\frac{\partial}{\partial t}(\rho v_x) &= -\frac{\partial}{\partial x}(\rho v_x^2) - \frac{\partial}{\partial y}(\rho v_x v_y) - \frac{\partial}{\partial z}(\rho v_x v_z) - \frac{\partial p}{\partial x} \\
&\quad + \frac{\partial}{\partial x}\left\{\mu\left[2\frac{\partial v_x}{\partial x} + \frac{2}{3}\left(\frac{\partial v_x}{\partial x} + \frac{\partial v_y}{\partial y} + \frac{\partial v_z}{\partial z}\right)\right]\right\} + \frac{\partial}{\partial y}\left[\mu\left(\frac{\partial v_x}{\partial y} + \frac{\partial v_y}{\partial x}\right)\right] \\
&\quad + \frac{\partial}{\partial z}\left[\mu\left(\frac{\partial v_z}{\partial x} + \frac{\partial v_x}{\partial z}\right)\right] + X \qquad (3.59)
\end{aligned}$$

Similar expressions for y- and z-directions can be readily written down.

Equation (3.59) is known as the *Navier-Stokes equation.* For constant ρ (incompressible fluid) and μ (approximately constant T), Eq. (3.59) combined with the continuity equation for an incompressible fluid, Eq. (1.12), simplifies to:

$$\rho\left(\frac{\partial v_x}{\partial t} + v_x \frac{\partial v_x}{\partial x} + v_y \frac{\partial v_x}{\partial y} + v_z \frac{\partial v_x}{\partial z}\right) = -\frac{\partial p}{\partial x} + \mu\left(\frac{\partial^2 v_x}{\partial x^2} + \frac{\partial^2 v_x}{\partial y^2} + \frac{\partial^2 v_x}{\partial z^2}\right) + X \tag{3.60a}$$

$$\rho\left(\frac{\partial v_y}{\partial t} + v_x \frac{\partial v_y}{\partial x} + v_y \frac{\partial v_y}{\partial y} + v_z \frac{\partial v_y}{\partial z}\right) = -\frac{\partial p}{\partial y} + \mu\left(\frac{\partial^2 v_y}{\partial x^2} + \frac{\partial^2 v_y}{\partial y^2} + \frac{\partial^2 v_y}{\partial z^2}\right) + Y \tag{3.60b}$$

$$\rho\left(\frac{\partial v_z}{\partial t} + v_x \frac{\partial v_z}{\partial x} + v_y \frac{\partial v_z}{\partial y} + v_z \frac{\partial v_z}{\partial z}\right) = -\frac{\partial p}{\partial z} + \mu\left(\frac{\partial^2 v_z}{\partial x^2} + \frac{\partial^2 v_z}{\partial y^2} + \frac{\partial^2 v_z}{\partial z^2}\right) + Z \tag{3.60c}$$

or in vector notation:

$$\rho \frac{D\mathbf{V}}{Dt} = -\nabla p + \mu \nabla^2\mathbf{V} + \mathbf{F}$$

In cylindrical coordinates (r, θ, z), Eq. (3.60) transforms into the following set of equations:

$$\rho\left(\frac{\partial v_r}{\partial t} + v_r \frac{\partial v_r}{\partial r} + \frac{v_\theta}{r}\frac{\partial v_r}{\partial \theta} - \frac{v_\theta^2}{r} + v_z \frac{\partial v_r}{\partial z}\right)$$

$$= -\frac{\partial p}{\partial r} + \mu\left(\frac{\partial^2 v_r}{\partial r^2} + \frac{1}{r}\frac{\partial v_r}{\partial r} - \frac{v_r}{r^2} + \frac{1}{r^2}\frac{\partial^2 v_r}{\partial \theta^2} - \frac{2}{r^2}\frac{\partial v_\theta}{\partial \theta} + \frac{\partial^2 v_r}{\partial z^2}\right) + F_r \tag{3.61a}$$

$$\rho\left(\frac{\partial v_\theta}{\partial t} + v_r \frac{\partial v_\theta}{\partial r} + \frac{v_\theta}{r}\frac{\partial v_\theta}{\partial \theta} + \frac{v_r v_\theta}{r} + v_z \frac{\partial v_\theta}{\partial z}\right)$$

$$= -\frac{1}{r}\frac{\partial p}{\partial \theta} + \mu\left(\frac{\partial^2 v_\theta}{\partial r^2} + \frac{1}{r}\frac{\partial v_\theta}{\partial r} - \frac{v_\theta}{r^2} + \frac{1}{r^2}\frac{d^2 v_\theta}{d\theta^2} + \frac{2}{r^2}\frac{\partial v_r}{\partial \theta} + \frac{\partial^2 v_\theta}{\partial z^2}\right) + F_\theta \tag{3.61b}$$

$$\rho\left(\frac{\partial v_z}{\partial t} + v_r \frac{\partial v_z}{\partial r} + \frac{v_\theta}{r}\frac{\partial v_z}{\partial \theta} + v_z \frac{\partial v_z}{\partial z}\right)$$

$$= -\frac{\partial p}{\partial z} + \mu\left(\frac{\partial^2 v_z}{\partial r^2} + \frac{1}{r}\frac{\partial v_z}{\partial r} + \frac{1}{r^2}\frac{\partial^2 v_z}{\partial \theta^2} + \frac{\partial^2 v_z}{\partial z^2}\right) + F_z \tag{3.61c}$$

The Navier-Stokes equations are a general statement of Newton's second law: Mass times acceleration of a system equals the sum of all external forces acing on the system. This may be shown by writing Eq. (3.59) in its alternate form:

$$\rho \frac{Dv_x}{Dt} = -\frac{\partial p}{\partial x} + \frac{\partial}{\partial x}\left\{\mu\left[2\frac{\partial v_x}{\partial x} - \frac{2}{3}\left(\frac{\partial v_x}{\partial x} + \frac{\partial v_y}{\partial y} + \frac{\partial v_z}{\partial z}\right)\right]\right\}$$

$$+ \frac{\partial}{\partial y}\left[\mu\left(\frac{\partial v_x}{\partial y} + \frac{dv_y}{\partial x}\right)\right] + \frac{\partial}{\partial z}\left[\mu\left(\frac{\partial v_z}{\partial x} + \frac{\partial v_x}{\partial z}\right)\right] + X \quad (3.62)$$

Similar expressions can be written down for the y- and z-directions. The term on the left is mass times acceleration, where acceleration is expressed by the substantial- rather than the partial-time derivative of velocity to account for the motion of system. The terms on the right all represent forces: nonviscous surface forces or pressures (first term), viscous surface forces (all terms containing μ), and body forces (last term).

The Navier-Stokes equations are valid for both laminar and turbulent flows, but can be practically applied only to the former. Even then, exact solutions can be found only in very limited cases—mainly because the equations are nonlinear.

In earlier sections we discussed several exact solutions. We shall now show how the Navier-Stokes equations are reduced to the simple forms considered previously.

Case 1: Plane Poiseuille Flow

In Eqs. (3.60) we note that since flow is

(a) two-dimensional

$$v_z = 0, \quad \frac{\partial}{\partial z} = 0$$

(b) steady

$$\frac{\partial}{\partial t} = 0$$

(c) fully developed

$$v_y = 0$$

We shall assume no body forces are present. From the continuity equation, $\partial v_x/\partial x = 0$, or v_x is independent of x. Equation (3.60b) reduces to $\partial p/\partial y = 0$, or p is independent of y. Finally, Eq. (3.60a) simplifies to Eq. (3.4):

$$\frac{dp}{dx} = \mu \frac{\partial^2 v_x}{\partial y^2}$$

Case 2: Hagen-Poiseuille Flow

In Eqs. (3.61), by similar reasoning as above,

$$\frac{\partial}{\partial t} = 0, \quad \frac{\partial}{\partial \theta} = 0, \quad v_\theta = v_r = 0$$

From continuity, $\partial v_z/\partial z = 0$, or v_z is independent of z. From Eq. (3.61a), $\partial p/\partial r = 0$, or p is independent of r. Thus only Eq. (3.61c) remains, which simplifies to

$$\frac{dp}{dz} = \mu\left(\frac{\partial^2 v_z}{\partial r^2} + \frac{1}{r}\frac{\partial v_z}{\partial r}\right) \tag{3.63}$$

Equation (3.63) is identical to Eq. (3.12) if coordinates are appropriately redefined.

Case 3: Boundary-Layer Flow

Let us consider two-dimensional flow of an incompressible fluid over a flat plate, Fig. 3.3. Here $v_z = 0$ and $\partial/\partial z = 0$. Because the boundary layer is very thin, it is reasonable to expect changes in v_x in the x-direction to be very much smaller than in the y-direction and also that $\partial^2 v_x/\partial x^2 \ll \partial^2 v_x/\partial y^2$. Then Eq. (3.60) reduces to

$$\frac{\partial v_x}{\partial t} + v_x\frac{\partial v_x}{\partial x} + v_y\frac{\partial v_x}{\partial y} = -\frac{1}{\rho}\frac{\partial p}{\partial x} + \nu\frac{\partial^2 v_x}{\partial y^2} \tag{3.64}$$

which is the boundary-layer equation for the unsteady state with a pressure gradient. This equation applies also to curved surfaces where x and y are measured, respectively, along the surface and perpendicular to the surface. For the steady state on a flat plate where $\partial p/\partial x = 0$, Eq. (3.64) reduces to Eq. (3.18).

By similar order-of-magnitude reasoning, it can be shown (1) that the second equation of Eq. (3.60) reduces to $\partial p/\partial y = 0$; in other words, the pressure within the boundary layer is uniform at any x position.

Earlier in this chapter we derived the momentum equation for each special system. Here we see that the applicable equation can be obtained by reduction of the general, three-dimensional equations. It is advisable to become skilled in both methods of arriving at the proper form of the equations for a specific case.

REFERENCES

1. Schlichting, H., *Boundary Layer Theory*, McGraw-Hill, New York, 1955.
2. Blasius, H., *Z. Math. u. Phys.*, **56** (1908), also *Nat. Advisory Comm. Aeronaut. Tech. Mem.* 1256 (1950).

3. Howarth, L., *Proc. Roy. Soc. (London)*, **A164,** 547 (1938).

4. Schlichting, H., *Z. angew. Math. Mech.*, **14,** 368 (1934).

5. Sparrow, E., *Nat. Advisory Comm. Aeronaut. Tech. Note* 3331 (Jan. 1955).

6. Langhaar, H. L., *Jour. Appl. Mech.*, **9**:2, A55–A58 (1942).

7. Shapiro, A. H., *Dynamics and Thermodynamics of Compressible Fluid Flow*, Vol. I, Ronald Press, New York, 1953.

PROBLEMS

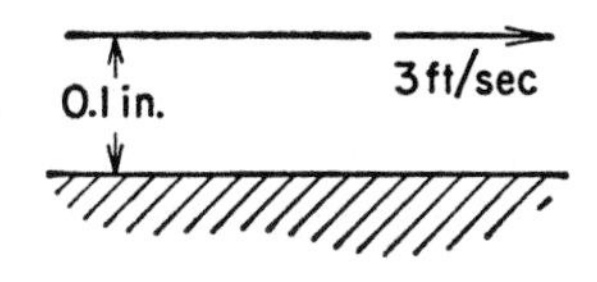

3.1 (a) In the Couette flow shown in the accompanying sketch, find the pressure gradient required for zero discharge. Assume fluid is light oil and properties may be evaluated at 100 F.
(b) Find the corresponding shear stress at each plate.
(c) Find the location and value of the minimum velocity.

3.2 Consider the same Couette flow system as in Prob. 3.1. If the shear stress at the fixed plate is zero, what is the rate of discharge?

3.3 Water at 60 F flows through a $\frac{1}{4}$-in. diameter tube, with a pressure drop of 0.01 psi/ft.
(a) Find the shear stress at the wall and plot the shear stress distribution over the cross section. Assume flow is fully developed.
(b) Find the maximum flow rate through the tube for which flow will still be laminar.

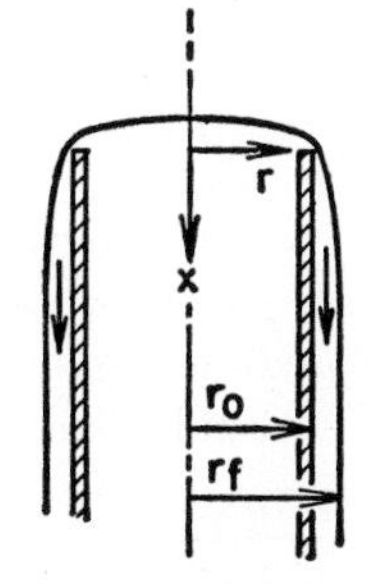

3.4 Find the pressure drop per foot of length for flow of liquid Freon-12 at 120 F through a $\frac{1}{64}$-in. diameter tube when Re = 600.

3.5 Fluid overflows down the outside of a vertical pipe of radius r_0 as shown. Derive by momentum balance the differential equation for the velocity distribution in the film. Make any assumptions that you can justify. How does your equation compare with Eq. (3.12)?

3.6 Two immiscible fluids flow between parallel plates under a pressure gradient $\Delta p/L$ at such rates that their interface coincides with the centerline. Assume flow is steady and both fluids are incompressible.
(a) Write the momentum balance for each fluid.
(b) Write the four boundary conditions necessary to determine the constants of integration.
(c) Find the velocity at the interface of the two fluids.
(d) Check your results by assuming that the two fluids are identical. What shape velocity profile would you expect?

3.7 A cable is cooled after a coating operation by being pulled through an open-ended, oil-filled tube immersed in a cooling bath (see diagram).

(a) In a region in the tube far from the open ends so that end effects are negligible, obtain an expression for the velocity profile.

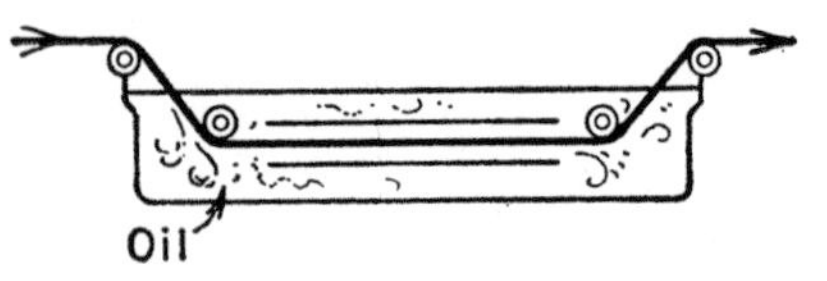

(b) List the assumptions necessary to obtain this expression.

Cable diam = 0.5 in., tube diam = 0.625 in., cable velocity = 5 fps. Oil properties: $\mu = 0.0005\ \text{lb}_f\ \text{sec/ft}^2$, $\rho = 55\ \text{lb/ft}^3$.

3.8 Air at 60 F flows over a flat plate with a velocity of 32 ft/sec.

(a) Find the Reynolds number and the boundary-layer thickness 4 in. from the leading edge.

(b) Find the rate of growth of the boundary layer at that point.

3.9 Solve Eq. (3.37) for δ and $C_{f,x}$ assuming a linear velocity profile: $v_x/V_\infty = y/\delta$.

(b) Solve Eq. (3.37) for δ and $C_{f,x}$ assuming

$$\frac{v_x}{V_\infty} = \sin\frac{\pi}{2}\frac{y}{\delta}$$

3.10 Consider steady flow of air over a flat plate. Measurement of the velocity profile at a particular location on the plate gave $v_x = 3 \times 10^4 y - 3 \times 10^6 y^2$. Find the rate of growth of the displacement thickness of the boundary layer at that point.

3.11 Calculate the boundary-layer development near the stagnation line of a cylinder with circular cross section, assuming that the stream velocity in this range can be approximated by $V_1 = 4V_\infty x/D$. (V_∞ is the free-stream velocity, D is the diameter, and x is distance measured around the circumference from the stagnation line.) The boundary-layer thickness for $x = 0$ can be determined from the condition that at this location $d\delta/dx = 0$. (Hint: Assume that surface can be treated as flat plate.)

3.12 Air at 1 fps flows between two parallel flat plates spaced 2 in. apart. Estimate the distance from the entrance where the boundary layers meet and sketch the velocity profile at that point. (Density: $\rho = 0.0765\ \text{lb/ft}^3$; viscosity: $\mu = 0.045\ \text{lb/hr ft}$.)

3.13 Derive the momentum integral relation for the boundary layer over a flat surface when $dp/dx \neq 0$. (*Hint*: V_∞ varies in the x-direction.)

3.14 By assuming a series solution of Eq. (3.26) of the form

$$f(\eta) = A_0 + A_1\eta + \frac{A_2\eta^2}{2!} + \frac{A_3\eta^3}{3!} + \cdots$$

and applying the two boundary conditions at the wall, show that Eq. (3.28) results, with all nonvanishing coefficients expressed in terms of a single unknown α. Comment briefly on how α may be determined.

3.15 The differential equation of an infinite flat plate oscillating parallel to itself is given by Eq. (3.53). For the boundary condition at the plate

$$y = 0, \qquad v_x = V_0 \cos nt$$

The solution is known from analogous heat conduction problems to be of the form

$$v_x = V_0 e^{-ky} \cos (nt - ky)$$

Evaluate the parameter k that would satisfy Eq. (3.53). Letting $v^* = v_x/V_0$ and $\eta = ky$, plot the velocity distribution v^* vs. η for $nt = 0, \pi/2, \pi$. Interpret your result.

3.16 An incompressible fluid is passing through a two-dimensional converging nozzle in which, owing to the increasing velocity, the pressure decreases linearly along the nozzle surface according to the equation:

$$p = p_0 - Cx$$

Set up a solution to the boundary-layer equation using the momentum integral equation technique, accounting for this decreasing pressure.

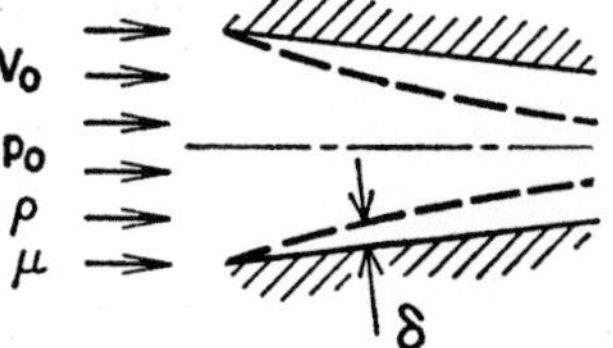

CHAPTER 4

Momentum Transfer in Turbulent Flow and Experimental Results

In the previous chapter, we solved a number of rather simple laminar flow problems. In each instance, a solution was found by defining the problem in terms of a simple model which satisfied the basic conditions and which yielded to analytical attack. There is no assurance, however, that a solution so obtained is valid, except through experimental verification. When the choice of a simple model is not apparent, as when the flow configuration is complex or when the flow field is turbulent, our only recourse then is to experimental information. A solution may be completely empirical if the information we seek is directly provided by experiment, or semi-empirical if provided by a combination of experimental data and analytic techniques.

In all cases, we note that ultimate answers are found in the laboratory—a forceful reminder that experiment is not an adjunct to, but the very foundation of macroscopic analysis. It is therefore essential for an engineer to be able to interpret and evaluate old and current experimental data.

In this chapter we shall selectively examine some of the empirical and semiempirical information on laminar and turbulent flow systems that has accumulated over the years. It is customary to classify the information broadly into two groups: flow through closed conduits and flow over submerged bodies. In the former group, we generally seek pressure drop data; correlations are given in the literature in terms of a friction factor, defined by Eq. (4.3). In the latter group, forces acting on submerged bodies may be reported in terms of a drag coefficient, defined by Eq. (4.46). The effects of fluid compressibility and rarefaction will be discussed in Chap. 11 and are ignored here. Familiarity with the methods of dimensional analysis is assumed; those wishing to review the subject will find a brief outline in Chap. 17. The analysis of a turbulent boundary layer on a flat plate (Sec. 4.8) is particularly noteworthy in that it illustrates how empirical data may be utilized to complete a theoretical analysis.

4.1 Flow in Closed Conduits—Friction Factor

In fully developed flow in a pipe—either laminar or turbulent, we assume Δp is proportional to the length L and that the following functional relationship is valid:

$$\frac{\Delta p}{L} = \phi(V, D, \rho, \mu, e) \tag{4.1}$$

The quantity e is a statistical measure of surface roughness of the pipe and has the dimensions of length. With force F, mass M, length L, and time θ as fundamental dimensions, and V, D, ρ as the set of maximum number of quantities which in themselves cannot form a dimensionless group, the pi theorem (Chap. 17) leads to

$$\frac{\Delta p}{4(L/D)(\rho V^2/2)} = \psi\left(\frac{VD\rho}{\mu}, \frac{e}{D}\right) \tag{4.2}$$

where the dimensionless numerical constants 4 and 2 are added here for convenience.

The above dimensionless group involving Δp has been defined as a *friction factor, f.**

$$f \equiv \frac{\Delta p}{4(L/D)(\rho V^2/2)} \tag{4.3}$$

Then, Eq. (4.2) becomes

$$f = \psi(\text{Re}, e/D) \tag{4.4}$$

Figure 4.1 shows this relationship as deduced by Moody (5) from experimental data for fully developed flow. In the laminar region, existing

* Other definitions of f appear in the literature. Some multiply the right side of Eq. (4.3) by 2 or 4, changing the numerical value by a factor.

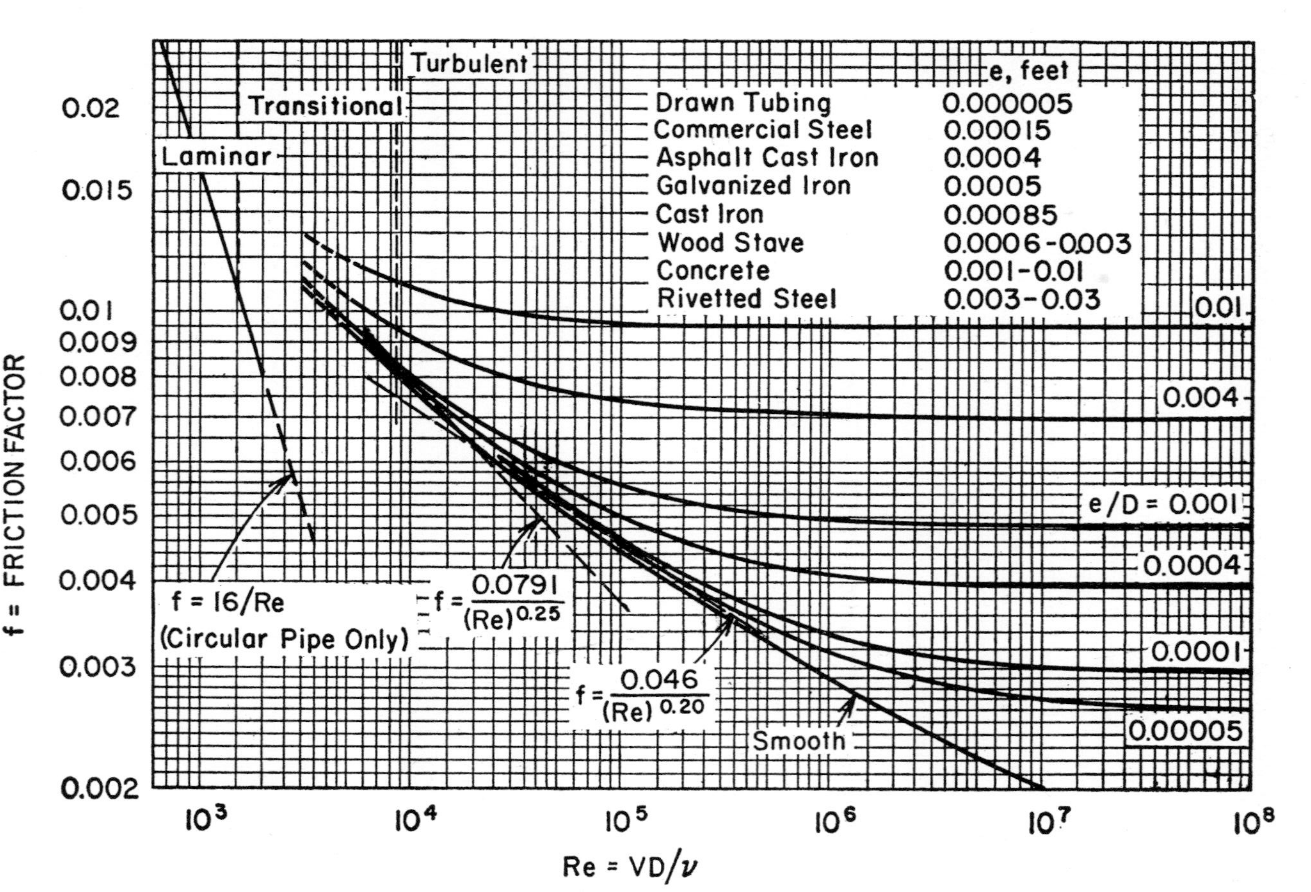

Fig. 4.1. Friction factor for flow in circular pipes. [Moody (5)].

empirical data on pressure drop within round pipes can all be correlated by a simple relation between f and Re, independent of the surface roughness. This statement is, of course, verified by the Hagen-Poiseuille solution. Combining Eq. (3.16), where $\Delta x = L$, with Eq. (4.3) results in

Laminar:
$$f = \frac{16}{\text{Re}} \tag{4.5}$$

The transition from laminar to turbulent flow is somewhere in the neighborhood of 2300–4000 for Re. Values of f for commercial pipes are difficult to determine precisely; even for so-called smooth pipes, f may vary as much as ± 5 per cent. In addition, the change of roughness with age is difficult to predict.

The f-vs.-Re relation for smooth pipes in turbulent flow has a slight curvature on a log-log plot, Fig. 4.1, and is given by Nikuradse as

Turbulent:
$$\frac{1}{\sqrt{f}} = 4.0 \log_{10} (\text{Re}\sqrt{f}) - 0.40 \tag{4.6}$$

Two linear approximations, shown dotted in Fig. 4.1, are:

Turbulent:
$$f \cong \frac{0.046}{\text{Re}^{0.2}} \tag{4.7}$$

$$f \cong \frac{0.0791}{\text{Re}^{0.25}} \tag{4.8}$$

For calculation purposes f should be read from the graph, but these approximate equations are often useful in showing functional relationships of various quantities.

For fully developed flow in a tube, a simple force balance yields

$$\Delta p \frac{\pi}{4} D^2 = \tau_0(\pi DL)$$

which may be combined with Eq. (4.3) to get an equivalent form for the friction factor:

$$f = \frac{\tau_0}{\rho V^2/2} \tag{4.9}$$

Curved pipes. In curved pipes the friction factor may rise considerably above the values shown in Fig. 4.1 and the transition Reynolds number may

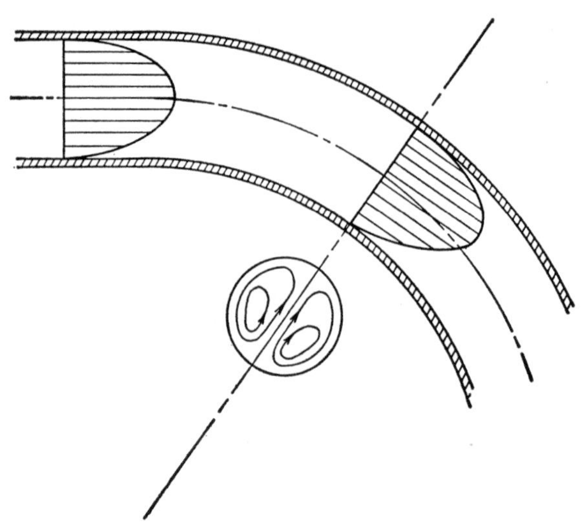

Fig. 4.2. Distortion of velocity profile in a curved pipe.

be considerably higher than 2300. This results from the distortion of the velocity profile by centrifugal effects producing secondary flow components transverse to the flow direction. Figure 4.2 shows the effect of curvature on velocity distribution in laminar flow, and Fig. 4.3 shows the effect of curvature on f in helical pipe coils, D_H being the mean diameter of the helical coil.

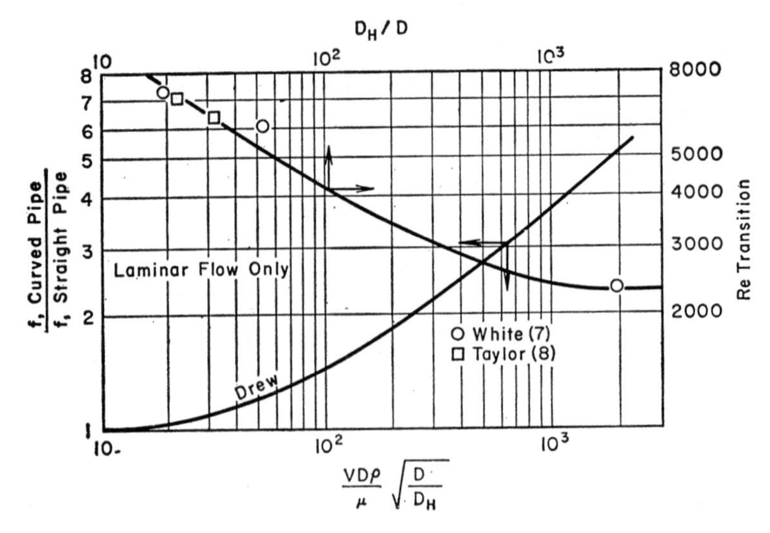

Fig. 4.3. Effect of curvature on f in laminar flow in circular pipes. [McAdams (6)].

In turbulent flow Ito (26) found experimentally that the ratio of curved-to-straight pipe friction factors, f_c/f_s, could be expressed as follows:

$$\frac{f_c}{f_s} = \left[\text{Re}\left(\frac{D}{D_H}\right)^2\right]^{1/20} \tag{4.10}$$

for Re $(D/D_H)^2 > 6$.

Nonisothermal flow. In nonisothermal flow, the effect of temperature on viscosity produces a distortion of the velocity distribution pattern. In liquids μ decreases with increasing T, and in gases μ increases with T. In Fig. 4.4, curve (a) shows distortion of the parabolic profile of laminar flow when a liquid is heated or a gas is cooled at the wall, since μ near the wall would be decreased. Curve (b) represents cooling a liquid or heating a gas.

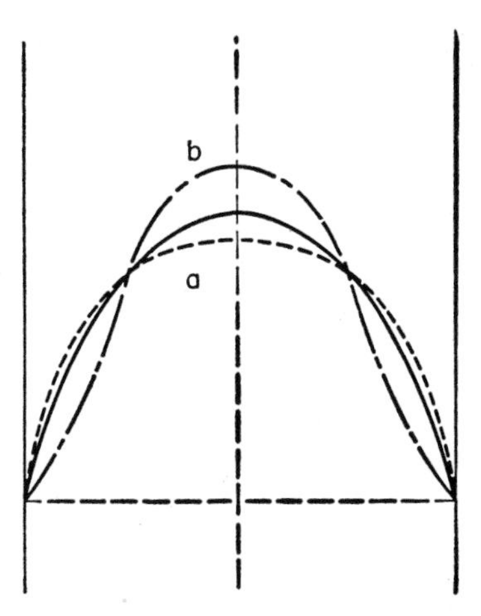

Fig. 4.4. Velocity distribution in non-isothermal flow.

Data for heating or cooling petroleum oils in laminar flow has been correlated (9a) by Eq. (4.5), $f = 16/\text{Re}$, if μ is evaluated at the quarter point between bulk (b) and surface (w) temperature.

Petroleum oils:

$$T_\mu = T_b + \tfrac{1}{4}(T_w - T_b) \tag{4.11}$$

For gases and liquid metals, Deissler (9b) analytically determined the following results:

Gases: $$T_\mu = T_b + 0.58(T_w - T_b) \tag{4.12a}$$

Liquid metals: $$T_\mu = T_b + 0.54(T_w - T_b) \tag{4.12b}$$

In turbulent flow in smooth pipes, data for liquids have been correlated (9) by Fig. 4.1, where μ is evaluated at the mean temperature, $T_m = \frac{1}{2}(T_b + T_w)$, and f read directly; or alternatively (10), μ in Re is evaluated at T_b and the ordinate is considered to be $f(\mu_b/\mu_w)^{0.14}$. For nonisothermal flow of air in smooth pipes with $(T_w/T_b)_{\text{abs}}$ as high as 2.5, data were correlated (11) by Fig. 4.1 where the ordinate and abscissa are considered to be

Ordinate:

$$f(T_m/T_b)_{\text{abs}}$$

Abscissa: (4.13)

$$\frac{VD\rho_b}{\mu_b}\frac{\nu_b}{\nu_m}$$

Example 4.1: Calculate the axial pressure gradient in a 1-in. diameter tube with air at $T_b = 200$ F, $p = 14.7$ psia flowing at $V = 20$ fps with $T_w = 800$ F.

Air properties at:	$T_b = 200$ F	$T_m = 500$ F	$T_w = 800$ F
μ (lb/ft hr)	0.052	0.068	0.081
ρ (lb/ft^3)	0.0601	0.0413	0.0315
ν (ft^2/hr)	0.864	1.63	2.56

$$\frac{D\rho_b V}{\mu_b}\frac{\nu_b}{\nu_m} = \frac{\frac{1}{12}(0.061)(20)(3600)}{(0.052)}\frac{(0.864)}{(1.63)} = 3660, \quad \text{turbulent}$$

From Fig. 4.1, $f(T_m/T_b)_{\text{abs}} = 0.01$. Then,

$$f = 0.01\left(\frac{660}{960}\right) = 0.0069$$

From Eq. (4.3)

$$\frac{\Delta p}{L} = 4\left(\frac{12}{1}\right)0.0069\frac{0.060(20)^2}{2(32.2)} = 0.124\,\frac{\text{lb/ft}^2}{\text{ft}}$$

4.2 Noncircular Cross Sections

A duct of noncircular cross section is not geometrically similar to a circular pipe; hence, dimensional analysis does not relate the performance of these two geometries. However, in *turbulent flow*, f for noncircular cross sections (annular spaces, rectangular and triangular ducts, etc.) may be evaluated from the data for circular pipes if D is replaced by an "equivalent diameter," D_e, defined by

$$D_e = \frac{4A}{P} = \frac{4\ (\text{flow area})}{\text{wetted perimeter}} \tag{4.14}$$

The equivalent diameter of an annulus of inner and outer diameter D_i, D_0 is

$$D_e = \frac{4(\pi/4)(D_o^2 - D_i^2)}{\pi(D_o + D_i)} = (D_o - D_i)$$

For a circular pipe, Eq. (4.14) reduces to $D_e = D$.

The transition Reynolds number $VD_e\rho/\mu$ is also found to be approximately 2300, as for circular ducts.

For *laminar flow*, however, the results for noncircular cross sections are not universally correlated. In a thin annulus in which spacing Z is very much less than the mean diameter of the annulus, the flow has a parabolic velocity distribution perpendicular to the wall and has this same distribution at every circumferential position. If we treat this as flow between parallel flat plates, Eq. (3.6) applies in the following form:

$$\frac{\Delta p}{\Delta x} = \frac{12\mu V}{Z^2} \tag{4.15}$$

Here $D_e = 2Z$ and Eq. (4.15) can be written in the form

$$f = \frac{24}{\text{Re}} \tag{4.16}$$

with D_e replacing D in the definitions of f and Re. This equation is obviously different from Eq. (4.5) which applied to laminar flow in circular pipes.

Flow in a rectangular duct (dimensions $Z_1 \times Z_2$) in which $Z_2 \ll Z_1$ is similar to this annular flow. For rectangular ducts of other aspect ratios (Z_1/Z_2),

$$f = \frac{16}{\phi \text{ Re}} \tag{4.17}$$

where

$$D_e = \frac{2(Z_1 Z_2)}{(Z_1 + Z_2)} \tag{4.18}$$

and ϕ is given by Fig. 4.5.

For laminar flow in ducts of triangular and trapezoidal cross section, Nikuradse (12b) showed that f is approximated by 16/Re with D_e given by Eq. (4.14), and transition occurs at approximately Re = 2300.

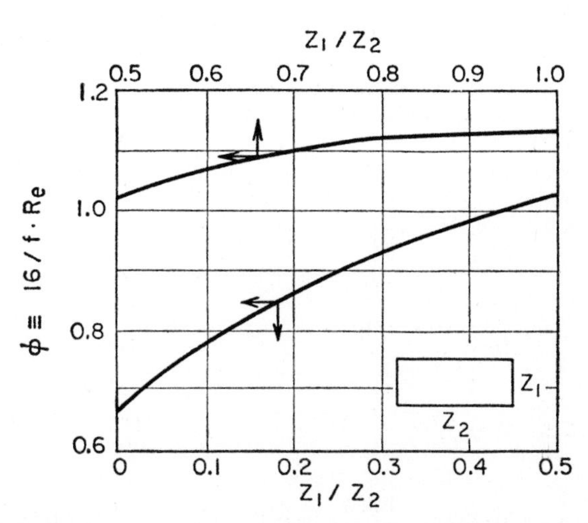

Fig. 4.5. Values of ϕ for rectangular ducts [computed from (12a)].

4.3 Valves and Fittings

Flow through valves and the various pipe fittings is very difficult to describe and analyze. A generally acceptable procedure involves evaluating, based on experimental data, the length of the same diameter pipe, L_e, which would have the same pressure drop as the fitting at a particular Reynolds number.

$$\Delta p = 4f \frac{L_e}{D} \rho \frac{V^2}{2} \tag{4.19}$$

where f is evaluated at Re for the pipe diameter equal to the fitting diameter, and L_e/D for various fittings is given in Table 4.1.

TABLE 4.1 L_e/D FOR SCREWED FITTINGS, TURBULENT FLOW ONLY*

Fitting	L_e/D	Fitting	L_e/D
45-deg elbow	15	Tee (as el, entering branch)	90
90-deg elbow, standard radius	31	Couplings, unions	Negligible
90-deg elbow, medium radius	26	Gate valve, open	7
90-deg elbow, long sweep	20	Gate valve, 1/4 closed	40
90-deg square elbow	65	Gate valve, 1/2 closed	190
180-deg close return bend	75	Gate valve, 3/4 closed	840
Swing check valve, open	77	Globe valve, open	340
Tee (as el, entering run)	65	Angle valve, open	170

* Calculated from Crane Co. Tech Paper 409, May 1942.

Example 4.2: What is the pressure drop in 100 ft length of a 3/4-in diameter standard wrought iron pipe (I.D. = 0.824 in.) which contains two 90 deg medium radius elbows, a swing check valve, and a wide open gate valve when water at 70 F flows at 2.5 gpm?

The total equivalent length of pipe is

$$100 + (2 \times 26 + 77 + 7)(0.824/12) = 109.3 \text{ ft.}$$

$$V = \frac{(2.5)(231/1728)}{(60)(\pi/4)(0.824/12)^2} = 1.5 \text{ fps}$$

$$\text{Re} = \frac{(0.824/12)(1.5)(62.4)(3600)}{2.36} = 9800$$

From Fig. 4.1 with $e/D = (0.00015)(12)/(0.824) = 0.00218$, $f = 0.0086$. From Eq. (4.3)

$$\Delta p = 4(0.0086)\frac{109.3}{(0.824/12)}(62.4)\frac{(1.5)^2}{(2)(32.2)}\frac{1}{144} = 0.83 \text{ psi}$$

4.4 One-Dimensional Flow—Correction for Nonuniform Velocity Distribution

Flow in closed conduits such as a heat-exchanger tube can often be analyzed as a steady, one-dimensional flow. The approximate continuity equation and momentum equation can, of course, be obtained by reduction of Eq. (1.11) and Eq. (3.59); however, for this simple system, it is more informative to derive the equations directly by mass and momentum balances. Consider a control volume shown in Fig. 4.6.

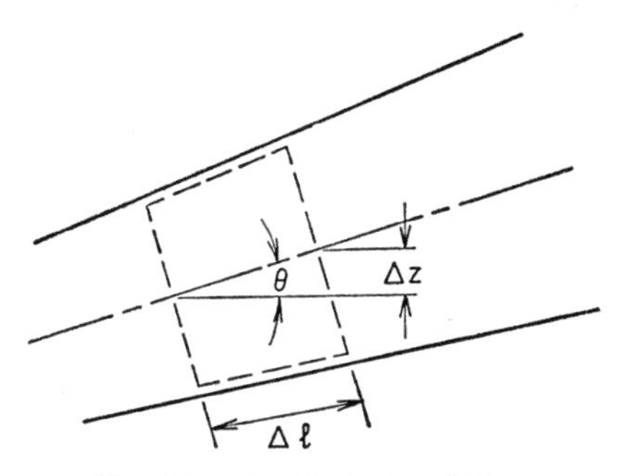

Fig. 4.6. One-dimensional flow.

Continuity of matter requires simply

$$d(\rho A V) = 0 \quad \text{or} \quad \rho A V = \text{const.} \tag{4.20}$$

where ρ is the density, A is the control surface area normal to flow, V is the average velocity.

A momentum balance in the flow or streamline direction involves (a) net average momentum flux associated with mass flow, (b) pressure and viscous forces acting on the control surface, and (c) gravity body force acting on the control volume. At steady state,

$$\rho A V \frac{dV}{dl}\,\Delta l + A\,\frac{dp}{dl}\,\Delta l + \tau_0\,(\text{per.})\,\Delta l + g\rho A\,\Delta l \sin\theta = 0 \tag{4.21}$$

or

$$-dp = \rho V\,dV + \frac{4\tau_0\,dl}{D_e} + \rho g\,dZ \tag{4.22}$$

Equation (4.22) is exactly true only if the velocity is uniform over the flow cross section. If the velocity varies in some arbitrary manner as shown in Fig. 4.7, the average momentum flux entering section 1 is not equal to $\rho A V^2$, but may be expressed as $\rho A V^2/\alpha$ where α is defined by Eq. (4.23) to account for the effect of varying velocity across the section. Referring to Fig. 4.7, we can write

$$\frac{1}{\alpha}(\rho A V^2) = \int_A v(\rho v\,dA)$$

or, with ρ uniform,

$$\frac{1}{\alpha} = \frac{1}{A}\int_A \left(\frac{v}{V}\right)^2 dA \tag{4.23}$$

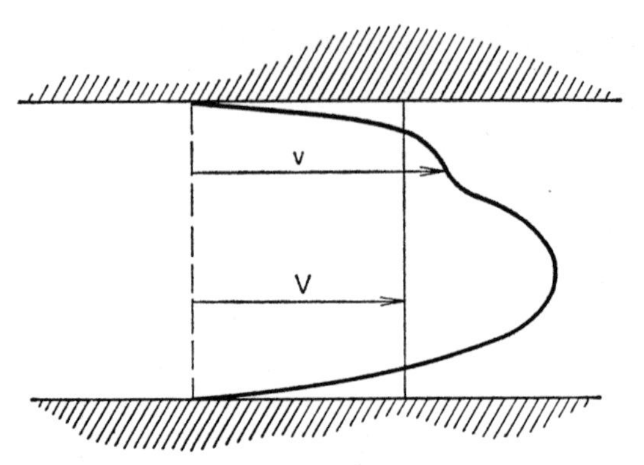

Figure 4.7

For uniform velocity, e.g., highly turbulent sluglike flow, $\alpha = 1$; for parabolic velocity distribution in a circular pipe (Hagen-Poiseuille flow), $\alpha = 0.75$.

For flow of fluid in a tube of uniform cross section, Eq. (4.22), modified by factor α if necessary, can be readily integrated to find the pressure drop between any two sections 1 and 2 along the tube. We can replace τ_0 in Eq. (4.22) by the friction factor f using Eq. (4.9), noting further that ρ in Eq. (4.9) may be evaluated at

$$\frac{1}{\rho_m} = \frac{1}{2}\left(\frac{1}{\rho_1} + \frac{1}{\rho_2}\right)$$

Letting $G = \rho V$, Eq. (4.22) when integrated for $\alpha \neq 1$ becomes

$$p_1 - p_2 = \frac{G^2}{\alpha}\left(\frac{1}{\rho_2} - \frac{1}{\rho_1}\right) + 4f\frac{L}{D_e}\frac{G^2}{2\rho_m} + g\rho_m(Z_2 - Z_1) \qquad (4.24)$$

For liquids $\rho_1 \cong \rho_2 \cong \rho_m$ making the first term on the right-hand side of Eq. (4.24) essentially zero.

For a fluid with nearly uniform velocity ($\alpha = 1$) at any cross section, Eq. (4.22) becomes

$$d(H) = d\left(\frac{p}{\rho} + \frac{V^2}{2} + gZ\right) = dF \qquad (4.25)$$

where dF is the frictional effect. The quantity $[(p/\rho + (V^2/2) + (gZ)]$ is known as the *Bernoulli head*, H, and Eq. (4.25) for idealized frictionless flow ($dF = 0$) is known as the *Bernoulli equation*.

Similar to α, a factor β which we shall find useful in the next section may be defined to account for the fact that if velocity is nonuniform at a section, the average kinetic energy of fluid at that section is not simply

$\rho AV \cdot V^2/2$ but may be expressed as $(1/\beta)(\rho AV \cdot V^2/2)$. Referring again to Fig. 4.7,

$$\frac{1}{\beta}\left(\rho AV\frac{V^2}{2}\right) = \int_A (\rho v\, dA)\frac{v^2}{2}$$

uniform ρ:

$$\frac{1}{\beta} = \frac{1}{A}\int_A \left(\frac{v}{V}\right)^3 dA \tag{4.26}$$

For uniform velocity, $\beta = 1$, and for a parabolic distribution, $\beta = 0.5$.

4.5 Sudden Enlargement and Contraction

The pressure drop associated with sudden enlargements and sudden contractions, Fig. 4.8, are reported in terms of a decrease in Bernoulli

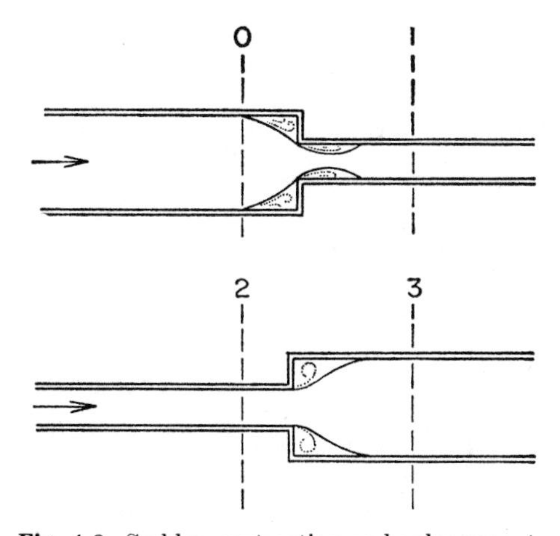

Fig. 4.8. Sudden contraction and enlargement.

head H, Eq. (4.25), and a loss coefficient K referred to the kinetic energy of the flow in the smaller cross section. Since $\Delta Z = 0$ for each of the cases,

$$\frac{p_0 - p_1}{\rho_{01}} + \frac{V_0^2 - V_1^2}{2} = K_c\frac{V_1^2}{2} \tag{4.27}$$

$$\frac{p_2 - p_3}{\rho_{23}} + \frac{V_2^2 - V_3^2}{2} = K_e\frac{V_2^2}{2} \tag{4.28}$$

From continuity since $\rho AV = \text{const.}$, and $\rho_0 \cong \rho_1$ and $\rho_2 \cong \rho_3$,

$$A_0V_0 = A_1V_1 \quad \text{and} \quad A_2V_2 = A_3V_3 \tag{4.29}$$

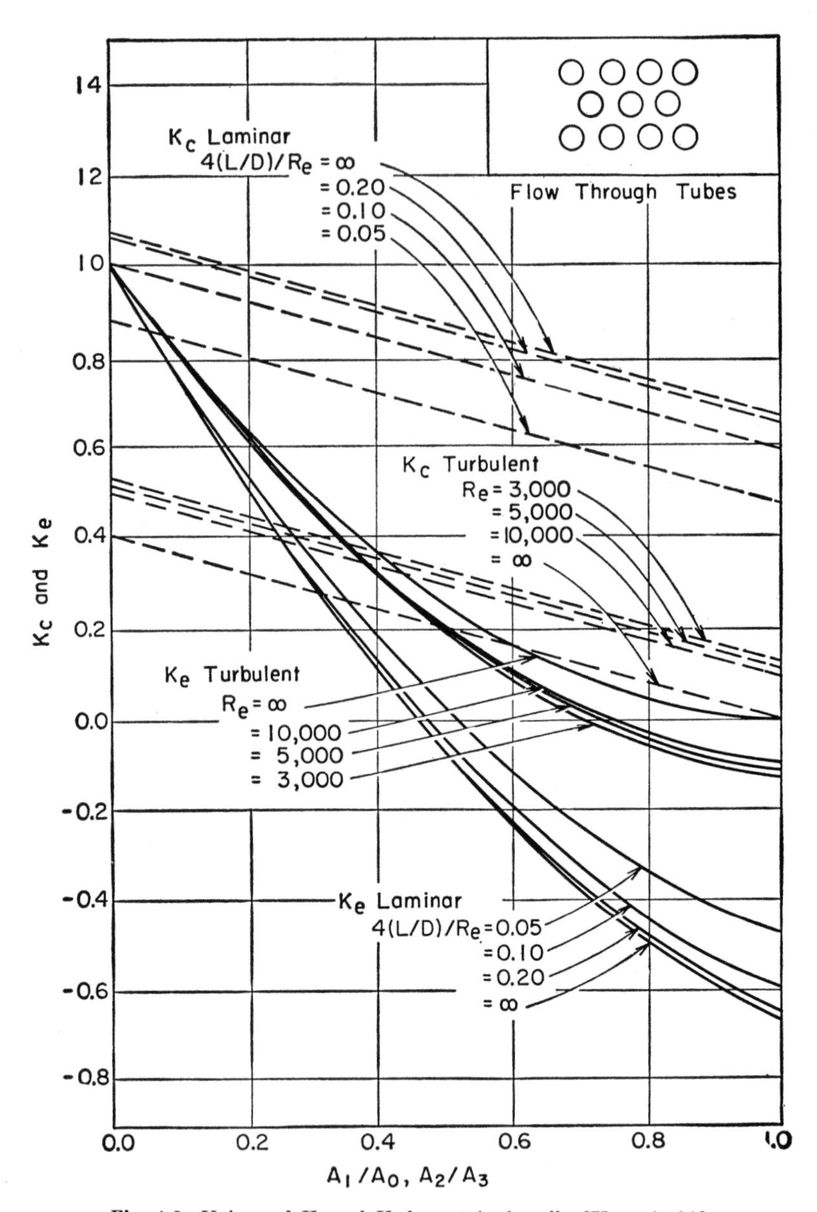

Fig. 4.9. Values of K_c and K_e for a tube bundle. [Kays (14b)].

Values of K_c and K_e have been determined by Kays (14) for a number of different geometries and checked by his test data. Figure 4.9 presents K_c and K_e for a bundle of tubes with headers at either end where the velocity distribution is essentially uniform.

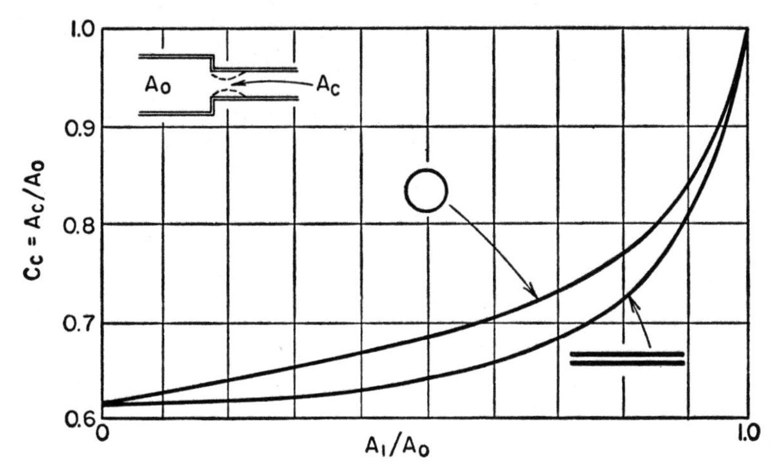

Fig. 4.10. Magnitude of C_c.

For the more general case of any shape of cross section, Kays derived the following equations:

$$K_e = 1 - \frac{2}{\alpha_2}\frac{A_2}{A_3} + \left(\frac{A_2}{A_3}\right)^2\left(\frac{2}{\alpha_3} - 1\right) \tag{4.30a}$$

$$K_c = \frac{1 - (A_1/A_0)^2(C_c^2/\beta_0) - 2C_c + 2(C_c^2/\beta_1)}{C_c^2} - 1 + \left(\frac{A_1}{A_0}\right)^2 \tag{4.30b}$$

where α and β are defined by Eqs. (4.23) and (4.26), and C_c is the contraction coefficient at the tube entrance. Magnitudes of C_c are shown in Fig. 4.10 for circular tubes and for parallel plates.

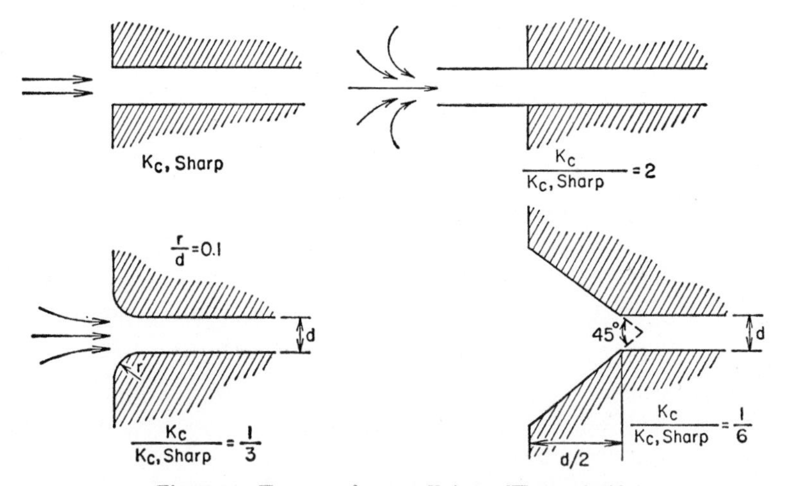

Fig. 4.11. Entrance loss coefficients [Henry (16)].

Equations (4.30) apply to a sharp-edged abrupt change in cross section for single tubes or for tube bundles. For tube bundles with headers, the velocity in the headers is usually nearly uniform; then $\alpha_3 = \beta_0 = 1.0$. Further, for very high Reynolds number flow, the velocity distribution approaches uniformity and $\alpha_2 = \beta_1 = 1.0$. Then Eq. (4.30a) reduces to $K_e = (1 - A_2/A_3)^2$, a result obtained earlier by Schutt (15).

The preceding results for K_e and K_c are for sharp-edged inlets and exits. The results for K_e apply equally well to practically all the shapes at the tube end; however, K_c may be considerably reduced by rounding or tapering the inlet. Figure 4.11 shows the ratio of K_c for a particular inlet shape to that for a sharp-edged entry.

Example 4.3: Air flowing at $w =$ 1000 lb/hr in smooth tubes of a shell-and-tube heat exchanger. There are 40 tubes, 1.0-in. I.D., 10 ft long, and the diameter of the entrance and exit headers is 0.75 ft. Air enters at 200 F, 14.7 psia, and leaves at 600 F. Calculate the pressure drop between the two headers.

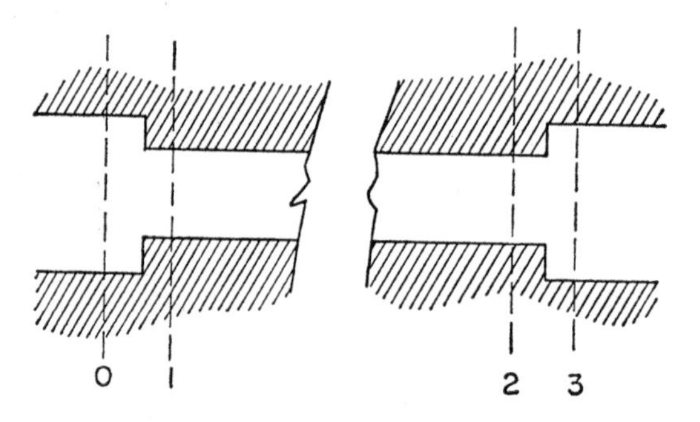

Solution:

$$A_0 = A_3 = \frac{\pi}{4}(0.75)^2 = 0.441 \text{ ft}^2$$

$$A_1 = A_2 = 40\frac{\pi}{4}\left(\frac{1}{12}\right)^2 = 0.218 \text{ ft}^2$$

$$\frac{A_1}{A_0} = \frac{A_2}{A_3} = \frac{0.218}{0.441} = 0.495$$

$$G = \frac{1000}{0.218} = 4590 \text{ lb/hr ft}^2$$

$$\rho_0 = \rho_1 = \frac{p}{RT} = \frac{(14.7)(144)}{(53.34)(660)} = 0.060 \text{ lb/ft}^3$$

$$\rho_3 = \rho_2 = 0.0374 \text{ lb/ft}^3$$

$$\rho_m = 0.0460 \text{ lb/ft}^3$$

$$\mu_{400} = 0.063 \text{ lb/ft hr}; \qquad \text{Re}_m = \frac{GD}{\mu} = \frac{4590(\frac{1}{12})}{0.063} = 6060, \quad \text{turbulent}$$

From Fig. 4.1, $f = 0.0088$. From Fig. 4.9, $K_e = 0.21$, $K_c = 0.31$.

$V_0 = w/A_0\rho_0 = 10.5$ ft/sec; $V_1 = 21.2$, $V_2 = 34.0$, $V_3 = 16.8$

From Eq. (4.24) with $\alpha \cong 1$ for turbulent flow and $Z_1 = Z_2$,

$$p_1 - p_2 = \left(\frac{4590}{3600}\right)^2 \frac{1}{32.2}\left[\left(\frac{1}{0.0374} - \frac{1}{0.06}\right) + 4(0.0088)\frac{10}{(\frac{1}{12})}\frac{1}{(2)(0.046)}\right]$$

$$p_1 - p_2 = 0.0503[10.0 + 46.0] = 0.503 + 2.32 = 2.82 \text{ lb/ft}^2$$

From Eqs. (4.27) and (4.28)

$$p_0 - p_1 = \frac{(0.060)}{2(32.2)}[0.31(21.2)^2 + (21.2)^2 - (10.5)^2]$$

$$= 0.000932[139 + 340] = 0.130 + 0.318 = 0.448 \text{ lb/ft}^2$$

$$p_2 - p_3 = \frac{(0.0374)}{2(32.2)}[0.21(34.0)^2 + (16.8)^2 - (34.0)^2]$$

$$= 0.000581[242 - 875] = 0.141 - 0.509 = -0.368 \text{ lb/ft}^2$$

Then

$$p_0 - p_3 = 0.448 + 2.82 - 0.368 = 2.02 \text{ lb/ft}^2$$

4.6 Entrance Effects

The developing velocity profile in the entrance region of a tube, Fig. 3.8, and between parallel plates, Fig. 3.7, is accompanied by a varying wall shear stress or friction factor. The analytical results of Langhaar (33) for laminar flow in the entrance region of round pipes are shown in Fig. 4.12. f is the average friction factor given by Eq. (4.3) where Δp is the static pressure drop between $x = 0$ and $x = L$. The results were verified experimentally in the range $10^{-5} < (L/D)/\text{Re} < 10^{-3}$ by Kline and Shapiro (25). Turbulent friction factors in the entrance region of round pipes were deter-

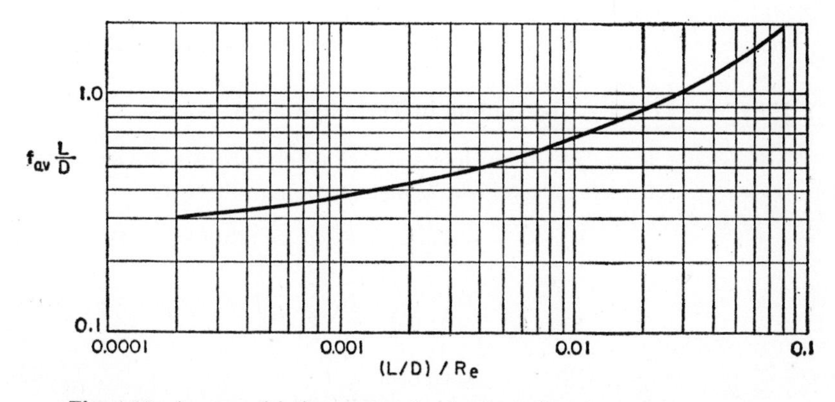

Fig. 4.12. Average friction factor for laminar flow in entrance region of circular tubes [Langhaar (33)].

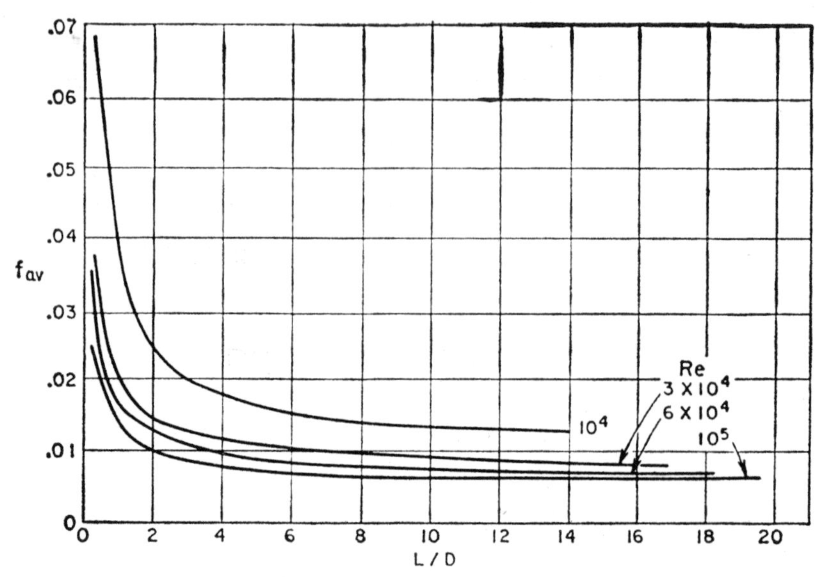

Fig. 4.13. Average friction factor based on static pressure drop for turbulent flow in entrance region of circular tubes [Deissler (27)].

mined analytically by Deissler (27) using the integral method. His results indicate that the local friction factors attain their fully developed values in much shorter lengths (roughly ten diameters from the inlet) than those required for the full development of the velocity profiles (known from experiment to exceed 50 diameters). Figure 4.13 shows his results, where f is again the average friction factor defined by Eq. (4.3). Hartnett (28) showed close agreement with these experimental results. It is worth noting that, in the entrance region, the friction factor defined by Eq. (4.3) is not identical to that defined by Eq. (4.9); the former includes the effects of momentum change associated with velocity profile development whereas the latter does not.

Similar results are available for flow between parallel plates (27).

4.7 Universal Velocity Distribution in Turbulent Flow

A dimensional analysis for velocity distribution for fully developed flow in circular pipes leads to the following formulation and result:

$$v = \phi_1(y, r_0, \tau_0, \rho, \mu, \text{roughness}) \tag{4.31}$$

$$\frac{v}{\sqrt{\tau_0/\rho}} = \phi_2\left(\frac{y}{\nu}\sqrt{\frac{\tau_0}{\rho}}, \frac{r_0}{\nu}\sqrt{\frac{\tau_0}{\rho}}, \frac{e}{D}\right) \tag{4.32a}$$

or

$$v^+ = \phi_2\left(y^+, r_0^+, \frac{e}{D}\right) \tag{4.32b}$$

where y in Eq. (4.32a) is the radial distance from the tube wall ($y = r_0 - r$).

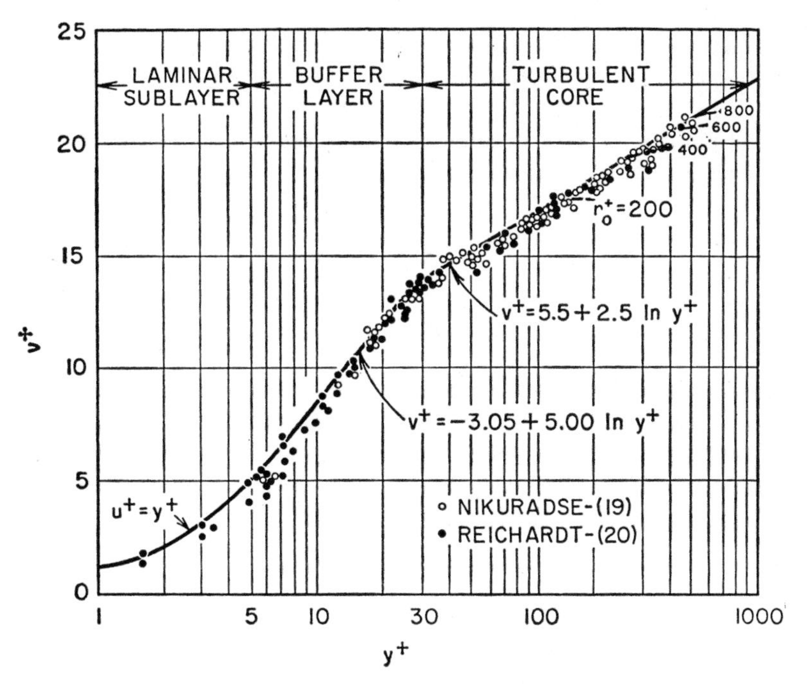

Fig. 4.14. Universal velocity distribution in smooth pipes.

For smooth pipes ($e/D \cong 0$), the velocity distribution is found to be independent of r_0^+ in the region near the wall. Figure 4.14 shows the agreement with test data. The solid line represents Kármán's (21) linear approximations to the test data:

$$\begin{aligned} 0 \le y^+ \le 5, &\quad v^+ = y^+ \\ 5 \le y^+ \le 30, &\quad v^+ = -3.05 + 5.0 \ln y^+ \\ 30 \le y^+, &\quad v^+ = 5.5 + 2.5 \ln y^+ \end{aligned} \tag{4.33}$$

At the centerline of the pipe ($y = r_0$) the velocity gradient is zero. The dotted lines in Fig. 4.14 represent estimates of this effect for various values of r_0^+.

Actually the velocity distribution changes smoothly from the center of the pipe out to the wall, and the turbulence level very near the wall approaches zero because of the rigidity of the wall surface. Very near the wall, as close as 0.000025 in., Fage and Townend (2) showed the colloidal particles in tap water moved in turbulent-like sinuous motion transverse to the flow direction parallel to the wall; however, no fluctuations were observed perpendicular to the wall. The three-zone mathematical approximation, Eq. (4.33), has led to the somewhat erroneous but commonly accepted naming of these arbitrary zones as *laminar sublayer*, *transition* or *buffer region*, and *turbulent core*.

It has been found also (22) that the curve of Fig. 4.14 and Eq. (4.33) describes the velocity distribution in the turbulent boundary layers on flat plates.

The quantity $\sqrt{\tau_0/\rho}$ is called a *friction velocity*, v_τ, since substituting for τ_0 from Eq. (4.9) there results

$$v_\tau \equiv \sqrt{\frac{\tau_0}{\rho}} = V\sqrt{\frac{f}{2}} \tag{4.34}$$

and

$$v^+ = \frac{v}{v_\tau}$$

In the central turbulent portion of the flow in a pipe, the velocity distribution depends more on the turbulent fluctuations than on viscosity. Also, wall roughness has little effect in this region. Therefore, the difference between centerline velocity, $v_{℄}$, and velocity at any y away from the wall may be represented by the functional relation

$$v_{℄} - v = \psi_1(\tau_0, \rho, r_0, y) \tag{4.35a}$$

or

$$\frac{v_{℄} - v}{v_\tau} = v_{℄}^+ - v^+ = \psi_2\left(\frac{y}{r_0}\right) \tag{4.35b}$$

and, hence, upon integration the average velocity is expressible as

$$\frac{v_{℄} - V}{v_\tau} = v_{℄}^+ - V^+ = \text{constant} \tag{4.36a}$$

or, with Eq. (4.34)

$$\frac{v_{℄}}{V} = 1 + \psi_3\left(\sqrt{\frac{f}{2}}\right) = \psi_4\,(\text{Re}) \tag{4.36b}$$

which is shown for the experimental data of Stanton and Pannel (23) in Fig. 4.15. Note that in the laminar region, $V/v_{℄} = 0.5$ as predicted by

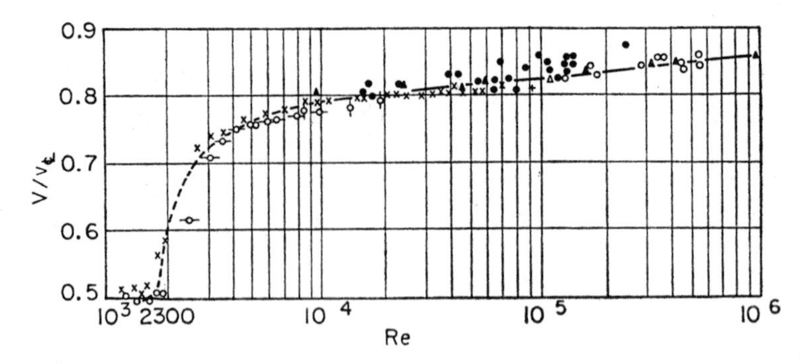

Fig. 4.15. $V/v_{℄}$ for pipe flow.

Eq. (3.17), whereas in the turbulent region $V/v_{\text{¢}}$ attains a value in the neighborhood of 0.8, indicating a flattened profile characteristic of turbulent flow.

Nikuradse suggested the following approximation for expressing the radial velocity distribution:

$$\frac{v_x}{v_{\text{¢}}} = \left(\frac{y}{r_0}\right)^{1/n} \tag{4.37}$$

His experiments showed excellent agreement when n varied with Re as follows, Table 4.2:

TABLE 4.2 EXPONENT IN EQ. (4.37)

Re	n	$V/v_{\text{¢}}$
4,000	6	0.791
110,000	7	0.817
3,240,000	10	0.865

Equation (4.37) is, of course, not valid right at the tube wall since it gives an infinite velocity gradient there.

We may express τ_0 in the following way in the neighborhood of Re $\cong 10^5$ by combining Eqs. (4.8) and (4.9) with $V/v_{\text{¢}} = 0.8$ and $D = 2r_0$

$$\tau_0 = 0.0225\rho v_{\text{¢}}^2 (\nu/v_{\text{¢}} r_0)^{1/4} \tag{4.38}$$

This result will be used in the next section.

Historically the universal velocity distribution, Fig. 4.14, Eq. (4.33), and the approximate relation, Eq. (4.37), were developed for flow in circular tubes. Subsequently it has been found that it represents quite satisfactorily the velocity distribution for flow between parallel plates and in boundary layers on flat plates. Also, Eq. (4.33) applies to flow in annuli if the respective shear stresses on the inner and outer walls are used.

4.8 Turbulent Boundary Layer on Flat Plate

The simplest case of a turbulent boundary layer occurs on a flat plate at zero incidence. Skin friction calculations for flat plates apply almost equally to many other cases such as drag on ships, airplane bodies, and many types of airfoils for airplane wings, turbines, compressors, and propellers, provided separation does not occur.

The only method available, at present, for the mathematical analysis of turbulent boundary layers is the approximate integral technique discussed in Sec. 3.4. In the following discussion we assume the laminar boundary-layer region is small and that the turbulent region can be assumed to begin at $x = 0$. Here the sketch in Fig. 3.6 applies equally well and Eq. (3.37) is valid.

$$\frac{d}{dx}\int_0^\delta \rho(V_\infty - v_x)v_x\,dy = \tau_0 \tag{4.39}$$

To illustrate the type of results obtainable, we adopt Prandtl's suggestion that the 1/7 power velocity distribution, $n = 7$ in Eq. (4.37), is applicable to the boundary layer on a flat plate. Equation (4.37) written for the flat plate is

$$\frac{v_x}{V_\infty} = \left(\frac{y}{\delta}\right)^{1/7} \tag{4.40}$$

Prandtl further assumed for the flat plate that the wall shear stress, τ_0, is given by an expression identical with Eq. (4.38):

$$\tau_0 = 0.0225\rho V_\infty^2\left(\frac{\nu}{V_\infty\delta}\right)^{1/4} \tag{4.41}$$

The assumptions made above surely are not exactly correct, but may be adequate for the present problem. We substitute Eqs. (4.40) and (4.41) into (4.39) and perform the necessary operations to get

$$\frac{7}{72}\frac{d\delta}{dx} = 0.0225\left(\frac{\nu}{V_\infty\delta}\right)^{1/4}$$

Separating variables and integrating from $\delta = 0$ at $x = 0$ gives

$$\frac{\delta}{x} = 0.37\left(\frac{V_\infty x}{\nu}\right)^{-1/5} \tag{4.42}$$

In the turbulent boundary layer, $\delta \sim x^{4/5}$, whereas in the laminar boundary layer, $\delta \sim x^{1/2}$.

The local skin friction coefficient $C_{f,x}$ is obtained from Eqs. (4.41) and (4.42):

$$C_{f,x} = \frac{\tau_0}{\frac{1}{2}\rho V_\infty^2} = 0.0576\left(\frac{\nu}{V_\infty x}\right)^{1/5} \tag{4.43}$$

or the average value over length L is

$$C_f = 0.072\left(\frac{\nu}{V_\infty L}\right)^{1/5} \tag{4.44}$$

The last equation agrees well with experimental data, Fig. 4.16, being low by only approximately 2 to 3 per cent in the range $5 \times 10^5 < \text{Re}_L < 10^7$. The graph in Fig. 4.16 also includes the line for laminar flow, Eq. (3.33). When Re_L exceeds 10^7, the discrepancy between Eq. (4.44) and experimental data gets large. A more elaborate analysis was performed by

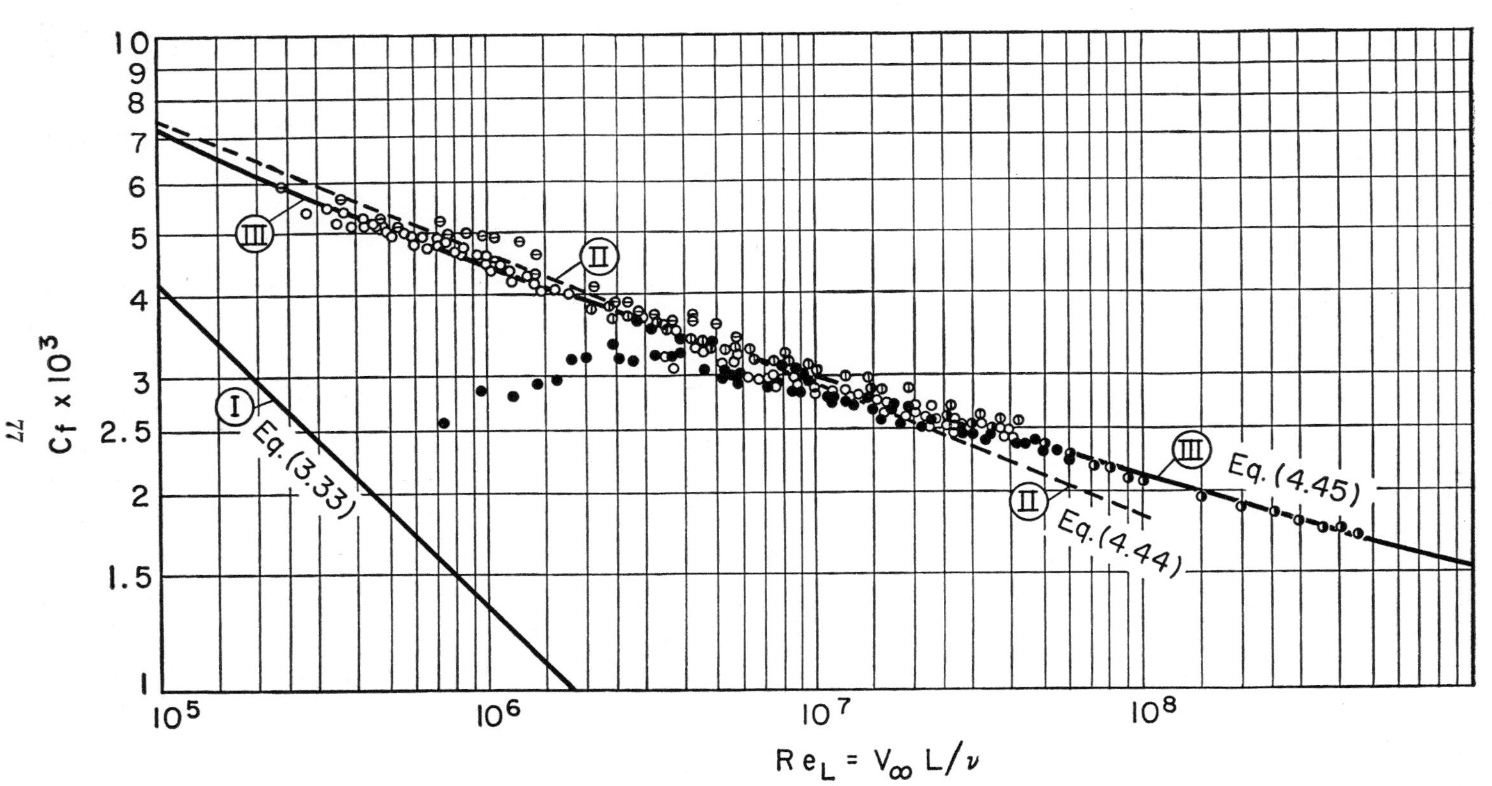

Fig. 4.16. Friction coefficient for smooth flat plate.

Prandtl using the logarithmic velocity distribution of Eq. (4.33) instead of the 1/7 power distribution. His results are plotted as Curve III in Fig. 4.16; for numerical work, the curve may be approximated by the following empirical expression proposed by Schlichting:

$$C_f = \frac{0.455}{(\log \mathrm{Re}_L)^{2.58}} \tag{4.45}$$

Example 4.4: A pair of water skis 8 in. wide and 6 ft long is towed through still water at 65 F with a velocity of 30 ft/sec. Find the drag on the bottom of the skis.

Solution:

$$\nu_{65} = 1.08 \times 10^{-5}\ \mathrm{ft^2/sec}$$

$$\mathrm{Re}_L = \frac{(6)(30)(10^5)}{1.08} = 1.67 \times 10^7$$

If we assume $(\mathrm{Re}_x)_{\mathrm{tr}} \cong 5 \times 10^5$, $x_{\mathrm{tr}} = 0.18$ ft or the boundary layer is turbulent over approximately the whole length of the skis. From Eq. (4.44),

$$C_f = 0.072(\mathrm{Re}_L)^{-1/5} = 0.00259$$

$$\mathrm{Drag} = (C_f)(A)\left(\frac{\rho V_\infty^2}{2}\right) = (0.00259)\left(\frac{2}{3} \times 6\right)\left(\frac{62.3 \times 30 \times 30}{32.2 \times 2}\right)$$

$$= 9\ \mathrm{lb}_f/\mathrm{ski}$$

4.9 Flow Past Submerged Bodies—Drag Coefficient

The forces acting on a body moving steadily through a fluid of infinite extent are the same as those acting on a stationary body in an infinite fluid moving at uniform velocity, V_∞. The pressure distribution over a cylinder with separated flow is shown in Fig. 2.9. The integrated pressure over the front projected area of the body is greater than that on the back resulting in a net force called the *form drag*. This force plus the drag force due to frictional effects over the surface comprise the total drag force, F_D.

For a convenient presentation of information, a drag coefficient is defined as

$$C_D = \frac{F_D/A}{\rho V_\infty^2/2} \tag{4.46}$$

For shapes such as cylinders and spheres, the area A is usually chosen, arbitrarily, to be the projected area normal to the flow; for airfoil sections, A is the plan area. Thus, though F_D/A has dimensions of a stress, it is not a true stress.

Dimensional analysis for geometrically similar shapes leads to

$$C_D = \phi\left(\frac{V_\infty L\rho}{\mu}\right)$$

where L is some characteristic dimension of the body. For high-speed compressible flow C_D also depends upon the Mach number, M, defined as the ratio of fluid velocity to velocity of sound in the fluid medium.

Figure 4.17 shows C_D for a few common shapes with subsonic flow of $M < 0.5$.

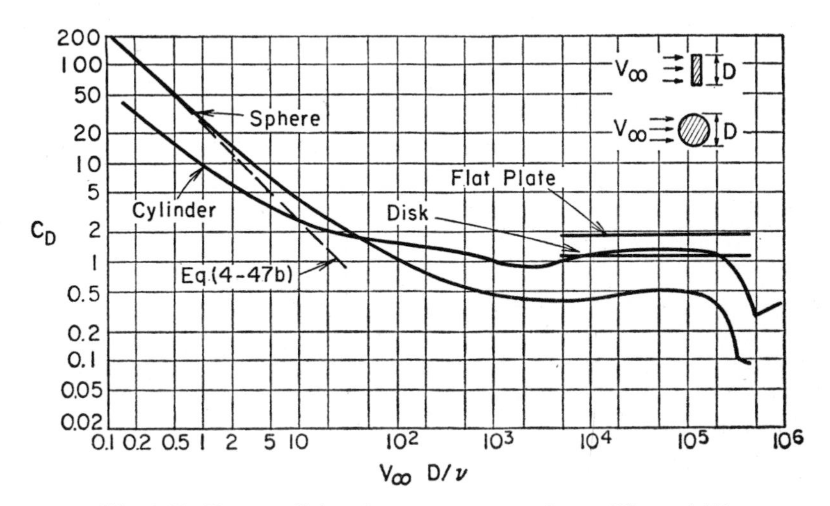

Fig. 4.17. Drag coefficient for some common shapes [Eisner (17)].

The phenomenon of boundary-layer separation (Fig. 2.7) has a major effect on the form drag, since pressure recovery downstream of the separation point is prevented. The effect is particularly noticeable at large Re where the form drag comprises most of the total drag. It explains the characteristic sudden decrease in C_D at $\mathrm{Re}_D \approx 10^5$ in Fig. 4.17. At this high Reynolds number, flow transition to turbulent flow occurs before separation, moving the separation point farther downstream and thereby reducing the form drag.

At very low Reynolds number, flow separation becomes unimportant and the inertial terms on the left of Navier-Stokes equations (3.62) become negligibly small compared to the pressure and the viscous terms on the right. Stokes (29) solved the resulting equation for a sphere of diameter D, and found the drag force to be

$$F_D = 3\pi D\mu V_\infty \tag{4.47a}$$

or

$$C_D = \frac{24}{\mathrm{Re}_D} \tag{4.47b}$$

Equation (4.47b) appears as a straight line in Fig. 4.17 and shows good agreement with experimental data when $\mathrm{Re}_D < 1$.

Example 4.5: Find the terminal velocity of a small metal sphere of 0.005-in. diameter, *s.g.* = 8.0, settling in still water at 68 F.

Solution:

$$\text{Weight} = \text{bouyancy force} + \text{drag force}$$

$$\tfrac{1}{6}\pi D^3\gamma_s = \tfrac{1}{6}\pi D^3\gamma_f + 3\pi D\mu V_\infty$$

or

$$V_\infty = \frac{D^2}{18\mu}(\gamma_s - \gamma_f)$$

$$= \frac{(0.005)^2(8-1)(62.4)}{(144)(18)(2.1\times 10^{-5})}$$

$$= 0.2 \text{ ft/sec}$$

Turbulent flow across banks of tubes is treated empirically because the flow field is too complicated to analyze. Frictional pressure drop is expressed as follows:

$$\Delta p_f = 4f'N\frac{G_{\max}^2}{2\rho}\left(\frac{\mu_w}{\mu_b}\right)^{0.14} \tag{4.48}$$

where for turbulent flow ($D_0G_{\max}/\mu$ from 2000 to 40,000), Jakob (30) correlated the data as follows:

For tubes in line with $1.5 < x_T < 4.0$,

$$f' = \left[0.044 + \frac{0.08x_L}{(x_T - 1)^n}\right]\left(\frac{D_0G_{\max}}{\mu}\right)^{-0.15} \tag{4.49}$$

where $n = 0.43 + 1.13/x_L$.

For staggered tubes with $1.5 < x_T < 4.0$,

$$f' = \left[0.23 + \frac{0.11}{(x_T - 1)^{1.08}}\right]\left(\frac{D_0G_{\max}}{\mu}\right)^{-0.15} \tag{4.50}$$

In these equations $G_{\max}$ is mass velocity at the *minimum* cross-sectional area, either transverse or diagonally. x_L is ratio of longitudinal pitch to tube diameter, and x_T is ratio of transverse pitch to tube diameter, where pitch is center-to-center distance.

For results outside the range of $D_0G_{\max}/\mu$ of 2000 to 40,000, suggested correlations are given in graph form by Grimison (31) and Bergelin *et al.* (32).

4.10 Flow in Packed Beds

A packed bed consists of a bundle of spheres, cylinders, wires, steel wool, or other odd-shaped bodies. It is assumed that the packing is uniform, thus avoiding any "channeling" of the flow. Semitheoretical equations can be developed based on a friction factor for flow through a representative devious passage through the bed but ultimately the final answer must be modified empirically. Ergun (24) suggests the following dimensionless correlation of experimental data which agrees well with the data when D_p is small compared with the container diameter.

$$\left(\frac{\Delta p\rho}{G_0^2}\right)\left(\frac{D_p}{L}\right)\left(\frac{\epsilon^3}{1-\epsilon}\right) = 150\,\frac{(1-\epsilon)}{D_pG_0/\mu} + 1.75 \tag{4.51}$$

where G_0 = mass flow rate per unit cross-sectional area of empty bed; ϵ = void fraction; D_p = mean particle diameter $= 6/S_0$ (for spherical particles, this definition gives D_p = diameter of sphere); S_0 = total particle surface/volume of particles.

Equation (4.51) is valid in the range $D_pG/\mu(1-\epsilon)$ from 1 to 2500, and ϵ from 0.40 to 0.65. Other correlations for various other operating conditions may be found in the various handbooks.

REFERENCES

1. Reynolds, O., *Trans. Royal Soc. (London)*, **174**:3, 935 (1883).
2. Fage, A., and H. C. H. Townend, *Proc. Royal Soc. (London)*, **A135,** 656 (1932).
3. Couch, W. H., and C. E. Herrstrom, Chem. Engineering M.S. Thesis, M.I.T., 1924.
4. Schlichting, H., *Boundary Layer Theory*, p. 33, McGraw-Hill, New York, 1955.
5. Moody, L. F., *Trans. ASME*, **66,** 671 (1944) and *Mech. Eng.*, **69,** 1005 (1947).
6. McAdams, W. H., *Heat Transmission*, 3rd Ed., p. 151, McGraw-Hill, New York, 1954.
7. White, C. M., *Proc. Roy. Soc. (London)*, **A123,** 645 (1929).
8. Taylor, G. I., *Proc. Roy. Soc. (London)*, **A124,** 243 (1929).

9a. Keevil, C. S., and W. H. McAdams, *Chem. & Met. Eng.*, **36,** 464 (1929).

9b. Deissler, R. G., *Nat. Advisory Comm. Aeronaut. Tech. Note* 2410 (July 1951).

10. Seider, E. N., and G. E. Tate, *Ind. Eng. Chem.*, **28,** 1429 (1936).
11. Humble, L. V., W. H. Lowdermilk, and L. G. Desmon, *Nat. Advisory Comm. Aeronaut. Rept.* 1020 (1951).

12a. McAdams, *ibid.*, Fig. 6–5, p. 150.

12b. Nikuradse, J., *Ingenieur-Archiv*, **1,** 306 (1930).

13. Keenan, J. H., *Thermodynamics*, p. 35, Wiley, New York, 1941.

14a. Kays, W. M., *Trans. ASME*, **72**:8, 1067 (Nov. 1950).

14b. Kays, W. M., and A. L. London, *Trans. ASME*, **74:7**, 1179 (Oct. 1952).

15. Schutt, H. C., *Trans. ASME*, HYD, **51,** 83 (1929).

16. Henry, John R., *Nat. Advisory Comm. Aeronaut.*, ARR No. L4F26 (June 1944).

17. Eisner, F., *3rd Int. Cong. App. Mech.*, Stockholm, 1930.

18. Hunsaker, J. C., and B. G. Rightmire, *Fluid Mechanics*, p. 192, McGraw-Hill, New York, 1947.

19. Nikuradse, J., *Forschungsheft*, **361,** 1–22 (1933), *Petroleum Eng.*, **11** (6), 164 (1940); **11** (8), 75 (1940); **11** (9), 124 (1940); **11** (11), 38 (1940); **11** (12), 83 (1940).

20. Reichardt, H., "Vollständige Darstellung der turbulenten Geschwindigkeitsverteilung in glatten Lecitungen," *Zeitschrift für angewandte Math. und Mech.*, **31**:2, 208–19 (1951).

21. Von Kármán, T., *Trans. ASME*, **61,** 705 (1939).

22. Dryden, H. L., von Kármán, T., (eds.), *Advances in Applied Mechanics*, Vol. IV, p. 7, Academic, New York, 1956.

23. Stanton, T. E., and J. R. Pannell, *Trans. Roy. Soc.* (*London*), **A214,** 199 (1914).

24. Ergun, S., *Chem. Eng. Prog.*, **48,** 93 (1952).

25. Kline, S., and A. H. Shapiro, *Nat. Advisory Comm. Aeronaut. Tech. Note* 3048, 1953.

26. Ito, H., *Trans. ASME, Jour. Engrg.*, **81,** D, 2 (June 1959), pp. 123–34.

27. Deissler, R. G., *Trans. ASME*, **77:7** (Nov. 1955); also *Nat. Advisory Comm. Aeronaut. Tech. Note* 3016.

28. Hartnett, J. P., *Trans. ASME*, **77:7**, 1211 (Nov. 1955).

29. Stokes, G., *Trans. Cambridge Phil. Soc.*, **8** (1845) and **9** (1851).

30. Jakob, M., *Trans. ASME*, **60,** 384–86 (1938).

31. Grimison, E. D., *Trans. ASME*, **59,** 583–94 (1937).

32. Bergelin, O. P., G. A. Brown, and S. C. Doberstein, *Trans. ASME*, **74** (1952).

33. Langhaar, H. L., *Jour. Appl. Mech.*, **9:**2, A55 (1942).

34. Schlichting, H., *Boundary Layer Theory*, *ibid.* p. 439.

PROBLEMS

4.1 (a) Derive the dimensionless groups of Eq. (4.2) from Eq. (4.1) by dimensional analysis (pi theorem).

(b) Obtain Eq. (4.4) and (4.9) from dimensional analysis by replacing $\Delta P/L$ by τ_0 in Eq. (4.1).

4.2 Calculate the pressure drop in a 100-ft length of 1-in. diameter standard galvanized pipe with water flowing at the rate of 20 gpm at 60 F.

4.3 Calculate the pressure drop in the pipe of Prob. 4.2 if the flow rate is 0.20 gpm.

4.4 If the 100 ft of 1-in. pipe of Prob. 4.2 and 4.3 is wound in a helical coil with a mean diameter of 8 in., what is the pressure drop in the 100-ft pipe length for 0.20 and 20 gpm flow?

4.5 Calculate the axial pressure gradient in a 0.407-in. I.D. smooth tube with oil (sp. gr = 0.85) flowing at 2000 lb/hr at T_b = 200 F where T_w = 400 F. (μ_{200F} = 14.5 lb/hr ft; μ_{400F} = 1.97 lb/hr ft.)

4.6 (a) Determine the pressure drop in a 100-ft length of smooth rectangular duct of dimensions 1 in. $\times$ 3 in. for an average air flow velocity of 1.5 ft/sec at 100 F and 15 psia.
(b) Determine Δp for part (a) if V = 15 ft/sec.

4.7 Calculate the axial pressure gradient in a $\frac{3}{4}$-in. diameter tube with water at T_b = 200 F, p = 15 psia flowing at V = 10 ft/sec with T_w = 60 F.

4.8 Calculate the pressure drop for air flowing in the 1 ft long heated section of a $\frac{1}{4}$-in. diameter tube. There is an unheated starting length; so the flow is fully developed. The air conditions at the beginning of the heated section are 60 F, 14.7 psia, and the air is heated to 800 F at the exit. The flow rate is 1.0 lb/hr.

4.9 How much horsepower is required to pump 100 gpm of water at 100 F from one large tank to another 20 ft higher connected by 100 ft of 2-in. pipe (2.067 in. I.D.) containing two standard radius 90-deg elbows, two 45-deg elbows, and an open globe valve. Pump efficiency = 40% and e = 0.0008 ft.

4.10 A shell-and-tube heat exchanger has a bundle of 500 smooth tubes each 0.47 in. I.D. and 8 ft long. Assuming the headers have twice the cross-sectional area of the tube flow area and that 10,000 lb/hr of air is cooled from 800 F to 300 F at an entering pressure of 20 psia, what is the pressure drop between the headers?

4.11 For fully developed laminar flow in a pipe, show that the momentum correction factor α = 0.75 and the kinetic energy correction factor β = 0.5.

4.12 Water at 60 F flows through a short tube $\frac{1}{2}$ in. in diameter, 5 in. long. Calculate the friction pressure drop in the tube with water velocity of (a) 0.4 ft/sec, (b) 9 ft/sec.

4.13 Perform the dimensional analysis which leads from Eq. (4.31) to (4.32) for velocity distribution over the cross section of a pipe with fluid flowing in turbulent flow.

4.14 Draw the velocity distribution from Eq. (4.33) for the fully developed turbulent flow of air flowing at Re = 8000 in a 1.0-in. I.D. tube for air at 60 F, 14.7 psia. Compare with the result from Eq. (4.37). Calculate the thickness of the laminar sublayer at y^+ = 5.

4.15 A flat plate 1 ft wide extends 2 ft from the body of an airplane and is parallel to the air stream which is at 60 F, 14.7 psia. When the plane velocity is 300 mph, estimate (a) the drag on the plate; (b) the horsepower required to move it through the air.

4.16 A circular billboard 10 ft in diameter is subjected to a 100-mph wind at 60 F. What force is exerted on the billboard when the wind is flowing perpendicular to the board?

4.17 Wind at 60 mph strikes a 1-in. transmission cable at right angles. Find the force exerted on the cable per foot of length, assuming air temperature of 60 F.

4.18 An automobile hood ornament may be considered to be a 1.5-in. diameter sphere in a free stream. How much horsepower is required to push this ornament through 60 F air at 50 mph? What hp is required to move the radio antenna ($\frac{3}{8}$ in. diam, 3 ft long) at this speed?

4.19 A lead shot $\frac{1}{8}$ in. in diameter is fired upward over open water. Find
(a) the drag on the shot as it leaves the gun barrel at 500 ft/sec
(b) its terminal velocity as it falls in air at 60 F
(c) its terminal velocity as it settles in still water at 60 F.

4.20 Compare the rates at which a 0.2-in. diameter drop of water would fall in air and the same size bubble of air would rise through water.

PART TWO Heat Transfer

CHAPTER 5

Introduction to Heat Transfer

We shall begin our study with brief comments on the microscopic aspects of transfer mechanism and show how they provide a rational basis for the classification of modes of heat transfer. Otherwise, our viewpoint in this and following chapters will be entirely macroscopic and phenomenological, and the validity of results that we obtain will be independent of any uncertainties in the microscopic theories.

5.1 Modes of Heat Transfer

The mechanism by which heat is transferred in physical equipment is quite complex; however, there appear to be two rather basic and distinct types of heat transfer processes: conduction and radiation.

(a) *Conduction* is the transfer by molecular motion of heat between one part of a body to another part of the same body, or between one body and another in physical contact with it. In fluids, heat is conducted by nearly elastic collisions of the molecules, or essentially by an energy diffusion process. Theory of heat conduction in solids distinguishes between conductors and nonconductors (dielectrics) of electricity. In a dielectric, heat is conducted by lattice waves caused by atomic motion (these waves transfer thermal energy but no electric charge); in electrical conductors, free electrons behaving almost like gas molecules contribute additionally to heat conduction (these free electrons transport both thermal energy and electric charge). Macroscopic theory of conduction, in contrast, is based on the definition of thermal conductivity, which simply expresses a proportionality between a heat flux and temperature gradient, independent of the various mechanisms of conduction in different media.

$$\left(\frac{q}{A}\right)_y = -k \frac{\partial T}{\partial y} \tag{5.1}$$

Equation (5.1) was suggested by Biot (1) and by Fourier (2) but is generally attributed to the latter.

(b) *Radiation*, or more precisely thermal radiation, is a phenomenon identical to the emission of light and is significant over the entire range of wave length from zero to infinity. Modern theories explain radiant energy on the basis of quantum theory and electromagnetic wave theory. Macroscopic theory of radiation is based on the Stefan-Boltzmann law (Chap. 13), which expresses a proportionality between energy flux emitted by an ideal radiator (black body) and the fourth power of its absolute temperature.

$$e_b = \sigma T^4 \tag{5.2}$$

where σ is a constant of proportionality, called the *Stefan-Boltzmann constant*. In the case of nonblack surfaces, the law is modified by several empirical factors.

Quite frequently, the processes of conduction and radiation occur simultaneously, even within solid bodies. In many engineering problems, however, the heat transferred by one of the modes is negligible compared with the other, and can be assumed with good approximation to involve only one of the processes.

(c) *Convection* is a process involving mass movement of fluids. When a temperature difference produces a density difference which results in the mass movement, the process is called *free or natural convection*. When a pump or other similar device causes the mass motion to take place, the process is called *forced convection*. Both natural and forced convection may be bounded or unbounded. Figure 5.1 represents schematically these classifications. The first two, (a) and (b), represent a hot plate in an infinite atmosphere (unbounded). In the first, (a), the fluid is in natural convection, and in (b), in forced convection provided by a fan. In (c) heat is added in one leg of a closed loop of piping and removed at the top of the other. Because of the density differences in the two legs, a natural circulation is

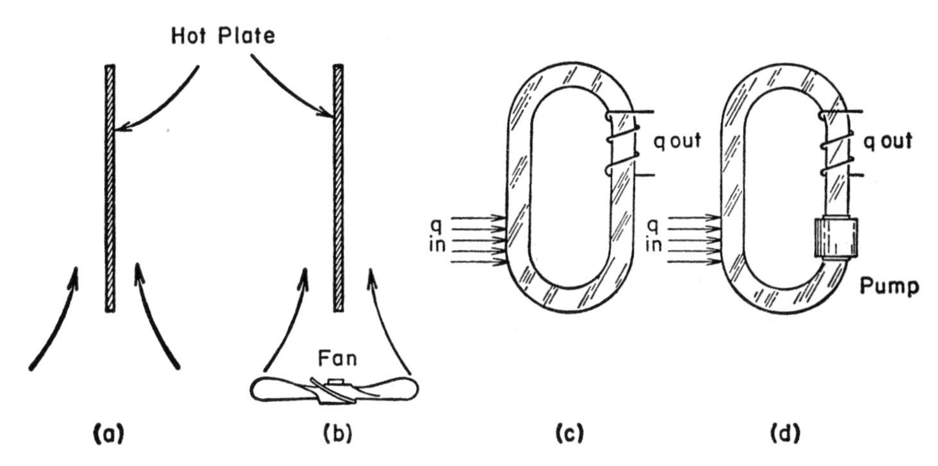

Fig. 5.1. Types of convection flow. (a) Unbounded natural convection. (b) Unbounded forced convection. (c) Bounded natural convection. (d) Bounded forced convection.

established. In (d), a pump provides the circulation in the same closed loop. Both (c) and (d) are cases of bounded convection.

We repeat, the word *convection* applies to fluid motion. The mechanisms of heat transfer anywhere in the fluid are only conduction and radiation.

A major concern of the heat transfer engineer is the prediction of the rate of heat transferred between a fluid and a solid bounding surface. Since the motion of fluid adjacent to a wall is either stationary or laminar in character, the mechanism of heat transfer at the wall-fluid interface is conduction. At a finite distance from the wall in the main fluid stream, the mechanism for heat transferred across a plane representing a locus of flow lines is also conduction. This is true whether the fluid motion is laminar or turbulent. When the fluid motion is turbulent there is superimposed on this conduction process a motion of packets of molecules which alternately pass to-and-fro across the plane of flow lines. In a thermodynamic analysis applied at any instant of time to a control surface fixed in space in the fluid, the energy quantity associated with this random fluid motion is quite properly called an *enthalpy*.

An ambiguity associated with the word *convection* has arisen in common usage of the term. The combined process of conduction and the "fluctuation" enthalpy at the flow-line plane in turbulent flow is often referred to as "heat transfer by convection." Furthermore, the heat transferred between a fluid and its bounding surface when the fluid flows either by free or forced convection is also called "heat transfer by free or forced convection." In addition, the enthalpy of the fluid entering or leaving a control volume such as a heat exchanger is sometimes quite erroneously referred to as "heat being convected into or out of the control volume." This promiscuous use of the word *convection* sometimes leads to confusion.

Imagine heat being transferred in a steady state through a hollow building wall, one face being hot and the other cold. The quantity of heat entering the hot face equals that leaving the cold face, and there results a natural convection fluid motion within the hollow wall. From this gross view it may be said that here heat is transferred by natural convection; however, on close inspection and analysis of the details of the flow and temperature distribution within the fluid, it is clear that the only mechanisms of heat transfer (energy transferred by virtue of temperature difference or temperature gradient) are those of conduction and radiation. The effects of fluid motion are included as enthalpy fluxes.

5.2 Thermal Conductivity

The magnitude of k for different materials varies from near zero for highly evacuated spaces to around 7000 Btu/hr ft F for copper crystals at very low temperatures.

On the basis of molecular physics, the absolute temperature is directly proportional to the mean kinetic energy of the molecules. Heat conduction is visualized as a transmission of energy by the more active molecules at the higher temperature to the neighboring less active molecules at the lower temperatures by successive collisions. Because molecular spacings are so very much larger in gases than in liquids, it is not surprising that gases have much lower values of conductivity than liquids. Kinetic theory leads to the following approximate relation for gases: $k = a\mu c_v$, where μ is the viscosity, c_v is the specific heat at constant volume, and approximate values of a are 2.45 for monatomic gases, 1.90 for diatomic, 1.70 for triatomic, and 1.30 for the more complex molecules. For most gases values of k are in the range of 0.005 to 0.05, and for most liquids, 0.05 to 0.50, except for the liquid metals, which have much higher values because of free electron flow ($k_{Hg} = 4.8$, $k_{Na} = 49$). In general, the variation of k with temperature is far more significant than its variation with pressure.

The conductivity of crystalline solids varies approximately as the reciprocal of the absolute temperature whereas the conductivity of amorphous (glasslike) substances increases with temperature. Solid materials may be solely crystalline in structure, solely amorphous, mixtures of the two, or may be somewhat porous with air or other gas in the pores. It is then quite understandable that some solids may have high values of k and others have low values; some have k increasing with T and others have k decreasing with T. Still other solids have k-versus-T curves which have maximum and minimum values.

It is to be expected that solid bodies having pores filled with gas will have rather low values of k and the more dense nonporous materials will have larger values of k. The free electrons contained in pure crystalline metals result in a high electrical conductivity and contribute greatly to producing a very high thermal conductivity. For pure crystalline metals the ratio of thermal to electrical conductivities is found to be nearly proportional to the absolute temperature. A modified Lorenz (4) equation expressing this relation is $(k/k_e) = 783 \times 10^{-9}T$. This equation does not hold for amorphous materials or alloys of metals. Conductivities of alloys may be less than k for any constituent; e.g., constantan is an alloy of 60% Cu and 40% Ni and has $k = 13$ whereas $k = 220$ for copper and 36 for nickel.

A composite graph showing k versus T for many different types of materials is shown in Fig. E-1 of Appendix E.

For any particular solid material k increases with density and with moisture content. The moisture presents a special problem, since in the presence of a temperature gradient it usually migrates toward the colder regions, thus changing the apparent conductivity and perhaps damaging the material or the surrounding structure.

For twenty different kinds of wood k is found to increase nearly linearly with apparent density. At 75 F and 12 per cent moisture $k = 0.0095 + 0.00193\rho$ with a spread of value of about ± 0.005. The value of k parallel to the grain is around 1.7 times k across the grain.

The apparent conductivity of loosely packed fibrous materials (rock and glass wools) and some powders usually reaches minimum values as the apparent density is decreased. At the very low densities the gas spaces may be so large that internal convection may result in increasing heat transfer rates and hence values of k.

In these cellular or porous materials, internal radiation may also become important. Because of the effects of internal convection and of radiation, the apparent conductivity of these materials usually increases with temperature and may actually increase somewhat with temperature difference at a particular mean temperature. If internal radiation is very significant the curve of k versus T will be concave upward.

Tabulated values of k for various materials are found in Appendix E.

5.3 Surface Heat Transfer Coefficient

When a fluid at one temperature is in contact with a solid surface at a different temperature, heat is transferred by conduction and its rate is given by Eq. (5.1), where k is the conductivity of the fluid and $\partial T/\partial y$ the temperature gradient in the fluid normal to the wall at the wall-fluid interface. If the details of the flow and heat transfer processes were known for any convection flow situation, the temperature distribution within the fluid could be determined and the heat transfer rate at the wall calculated from Eq. (5.1). Much of the effort in heat transfer research is devoted to this problem. For many of the convection problems with surfaces having a variety of geometries, there is available very meager information regarding the velocity and temperature patterns. In these cases the heat transfer information is largely empirical. It has been found convenient to define a *surface coefficient of heat transfer*, h, by the equation

$$h = \frac{q/A}{T_s - T_f} = \frac{-k(\partial T/\partial y)_s}{T_s - T_f} \tag{5.3}$$

where T_s is the surface temperature and T_f is some temperature of the fluid. If the fluid is infinite in extent, T_f is usually the fluid temperature at a distance far removed from the surface; if the fluid is flowing in a confined space such as inside a round pipe, it is usually the mixed mean temperature which would exist if the fluid at a particular cross section were removed and allowed to mix adiabatically. The surface heat transfer coefficient, h, was first suggested by Newton (3) and Eq. (5.3) is called Newton's equation. The units of h are Btu/hr ft^2 F. Defined in this way, h is a function of properties of the fluid, including velocity, and may also depend slightly on the roughness of the surface.

5.4 Thermal Boundary Layer

A thermal boundary layer of thickness δ_T is defined in an exactly analogous manner to the velocity boundary layer δ. Within the layer, tempera-

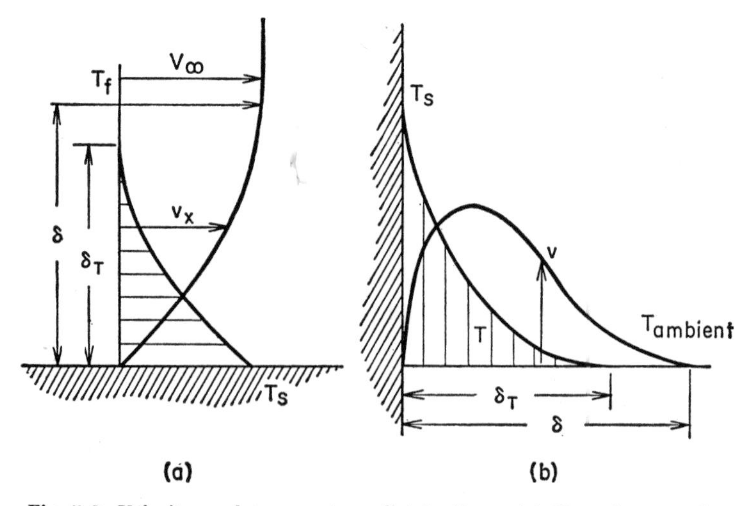

Fig. 5.2. Velocity and temperature distributions. (a) Forced convection over a heated flat plate. (b) Natural convection near a heated vertical plate.

ture varies from T_s at the wall to T_f in the undisturbed flow. Figure 5.2(a) shows the thermal boundary layer on a heated horizontal plate.

Practically, since temperature within the boundary layer approaches T_f, asymptotically, δ_T is defined as the thickness at which $T_s - T = 0.99$ $(T_s - T_f)$. In general, δ_T is not equal to δ.

Figure 5.2(b) shows the temperature—and velocity—profile on a short vertical plate under conditions of natural convection. At some distance, δ_T from the plate the temperature approaches the free-stream temperature, and at some other distance δ the velocity is nearly equal to zero. In general $\delta \neq \delta_T$.

REFERENCES

1. Biot, J. B., *Bibliothèque Britannique,* **27,** 310 (1804); and *Traité de Physique,* **4,** 669 (Paris, 1816).

2. Fourier, J. B. J., *Théorie Analytique de la Chaleur,* Paris, 1822. English translation by Freeman, Cambridge, 1878.

3. Newton, I., *Trans. Roy. Soc. (London),* **22,** 824 (1701).

4. Lorenz, L., *Ann. Physik,* **13,** 422 (1882).

CHAPTER 6

Heat Transfer in Stationary Systems

In this chapter we shall show how typical solutions of the heat conduction equations are obtained for relatively simple steady-state and transient systems. The solution gives the temperature distribution within a system in steady state and the temperature history (temperature as a function of time and space coordinates) within a transient system. Once the temperature distribution, $T(x, y, z)$, or the temperature history, $T(x, y, z, t)$, is known, the heat transfer rate in any direction at any instant of time is calculated with the Fourier rate equation

$$q_n = -kA \frac{\partial T}{\partial n} \tag{6.1}$$

where n is any direction in space and A is the area normal to n.

When a mathematical solution is too complex, numerical or analog methods can be used to obtain a solution which closely approximates the true behavior of the physical situation being investigated. These tools of analysis are presented in Part IV (Chap. 18) to avoid interrupting the continuity of presentation of the main subject matter.

6.1 Equations of Conduction

We derive the differential equation governing the temperature distribution in a solid body or a stationary fluid from energy conservation.

Figure 6.1 represents an element of space $\Delta x \, \Delta y \, \Delta z$ in an isotropic (k has the same magnitude in any direction, but may vary in space and time) stationary system.

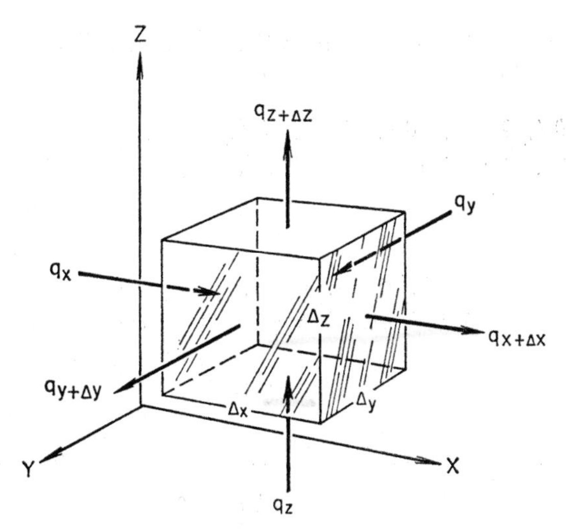

Fig. 6.1. The volume element for deriving the heat conduction equation.

The net rate of heat conduction into the element in the x-direction is

$$(q_x - q_{x+\Delta x}) = \frac{-\partial q_x}{\partial x} \Delta x = -\frac{\partial}{\partial x}\left(-k \frac{\partial T}{\partial x} \Delta y \, \Delta z\right) \Delta x$$

Similarly, in the y- and z-directions, the net rate of heat conduction into the element is

$$\frac{\partial}{\partial y}\left(k \frac{\partial T}{\partial y} \Delta z \, \Delta x\right) \Delta y \quad \text{and} \quad \frac{\partial}{\partial z}\left(k \frac{\partial T}{\partial z} \Delta x \, \Delta y\right) \Delta z$$

Further, we may consider the case in which there is an energy of the external work type introduced into the system as a result of electric current flow (I^2R type). Since this energy must be transferred away as heat, it is often called *heat generation* or distributed *heat source*. Nuclear reactions and chemical reactions when the reactants are dilute in the system can be treated as "heat sources." Define a quantity W_i as the heat generation per unit volume and time, Btu/ft^3 hr. Then the rate of heat generation within the element is

$$W_i \, \Delta x \, \Delta y \, \Delta z$$

The sum of all of these energy quantities should, by conservation, equal the rate of increase of internal energy of the system,

$$\frac{\partial e}{\partial t}\rho\ \Delta x\ \Delta y\ \Delta z = c\frac{\partial T}{\partial t}\rho\ \Delta x\ \Delta y\ \Delta z$$

where e is the internal energy per unit mass, and the internal energy change may be related to the temperature change by a specific heat, $de = c\,dT$.

Equating all of the energy quantities and dividing by the volume $\Delta x\ \Delta y\ \Delta z$ results in

$$\frac{\partial}{\partial x}\left(k\frac{\partial T}{\partial x}\right) + \frac{\partial}{\partial y}\left(k\frac{\partial T}{\partial y}\right) + \frac{\partial}{\partial z}\left(k\frac{\partial T}{\partial z}\right) + W_i = \rho c\frac{\partial T}{\partial t} \tag{6.2}$$

If the thermal conductivity k is uniform and constant, Eq. (6.2) reduces to

$$\frac{\partial^2 T}{\partial x^2} + \frac{\partial^2 T}{\partial y^2} + \frac{\partial^2 T}{\partial z^2} + \frac{W_i}{k} = \frac{1}{\alpha}\frac{\partial T}{\partial t} \tag{6.3}$$

or

$$\nabla^2 T + \frac{W_i}{k} = \frac{1}{\alpha}\frac{\partial T}{\partial t}$$

where $\alpha \equiv k/\rho c$, the thermal diffusivity. In the absence of heat sources

$$\nabla^2 T = \frac{1}{\alpha}\frac{\partial T}{\partial t} \tag{6.4}$$

and in the steady state

$$\nabla^2 T = 0 \tag{6.5}$$

In cylindrical coordinates (r, θ, z) and spherical coordinates (r, θ, ϕ) the equations comparable to Eq. (6.3) may be readily derived with the following results:

Cylinder:

$$\left[\frac{1}{r}\frac{\partial}{\partial r}\left(r\frac{\partial T}{\partial r}\right) + \frac{1}{r^2}\frac{\partial^2 T}{\partial \theta^2} + \frac{\partial^2 T}{\partial z^2}\right] + \frac{W_i}{k} = \frac{1}{\alpha}\frac{\partial T}{\partial t} \tag{6.6}$$

Sphere, where θ is the meridianal angle and ϕ is the azimuthal angle:

$$\frac{1}{r^2}\left[\frac{\partial}{\partial r}\left(r^2\frac{\partial T}{\partial r}\right) + \frac{1}{\sin\theta}\frac{\partial}{\partial \theta}\left(\sin\theta\frac{\partial T}{\partial \theta}\right) + \frac{1}{\sin^2\theta}\frac{\partial^2 T}{\partial \phi^2}\right] + \frac{W_i}{k} = \frac{1}{\alpha}\frac{\partial T}{\partial t} \tag{6.7}$$

Problems in heat conduction essentially consist of finding solutions to the appropriate partial differential equations, (6.2) through (6.7), which will satisfy certain initial and boundary conditions.

An initial condition specifies temperature distribution in a system $T = f(x, y, z)$ at time $t = 0$.

Some of the typical conditions that may be specified at the boundaries of a system are given below:

(a) Specified boundary temperature. The temperature at the boundary may be held constant; then

$$\text{at} \quad x = L, \quad T = T_L \tag{6.8a}$$

(b) Insulated surface at boundary. Then, since $q = 0$,

$$\text{at} \quad x = L, \quad \left(\frac{\partial T}{\partial x}\right)_L = 0 \tag{6.8b}$$

(c) Specified magnitude of q at boundary. In cases such as those in which heat is supplied at the boundary or solar radiation is absorbed at the boundary, the magnitude of q is specified by this external process. Then

$$\text{at} \quad x = L, \quad \left(\frac{\partial T}{\partial x}\right)_L = -\frac{1}{k}\left(\frac{q}{A}\right)_L \tag{6.8c}$$

(d) Specified fluid temperature, T_f, and h at the surface. Then

$$\text{at} \quad x = L, \quad \left(\frac{\partial T}{\partial x}\right)_L = -\frac{h}{k}(T_L - T_f) \tag{6.8d}$$

Alternatively, the magnitudes of T_L, q, or T_f may be specified as some functions of time.

6.2 Steady One-Dimensional Systems

(*a*) *Infinite Flat Plate*

Equation (6.5) for an infinite flat plate reduces to

$$\frac{\partial^2 T}{\partial x^2} = 0 \tag{6.9}$$

Boundary conditions are

$$x = 0, \quad T = T_1$$

$$x = L, \quad T = T_2$$

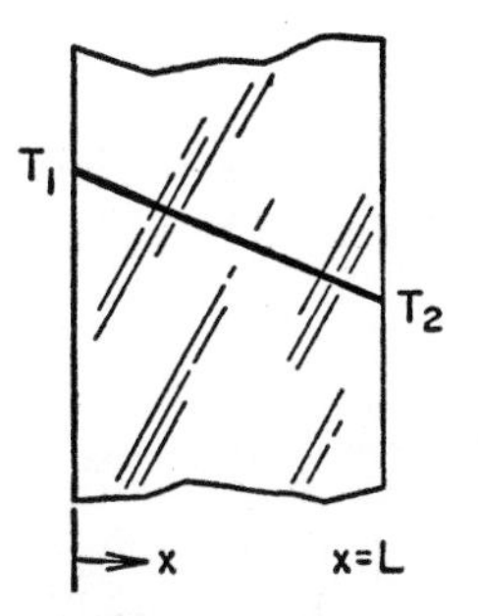

Fig. 6.2. Steady temperature distribution in a slab.

Equation (6.9) readily integrates to $T = C_1x + C_2$. Determination of constants by the specified boundary conditions gives the following solution for temperature distribution in the plate:

$$T = (T_2 - T_1)\frac{x}{L} + T_1 \tag{6.10}$$

The rate of heat transfer is given by the Fourier equation,

$$q = -kA \frac{\partial T}{\partial x} = \frac{kA}{L}(T_1 - T_2) \tag{6.11}$$

(*b*) *Infinite Cylinder*

For steady radial flow of heat through the wall of a hollow cylinder, Eq. (6.6) reduces to

$$\frac{1}{r}\frac{\partial}{\partial r}\left(r\frac{\partial T}{\partial r}\right) = 0 \tag{6.12}$$

Boundary conditions are

$$r = r_1, \quad T = T_1$$

$$r = r_2, \quad T = T_2$$

Equation (6.12) integrates to $T = C_1 \ln r + C_2$. Determination of constants by the specified boundary conditions yields

$$T = \frac{T_1 - T_2}{\ln (r_1/r_2)} \ln \left(\frac{r}{r_2}\right) + T_2 \tag{6.13}$$

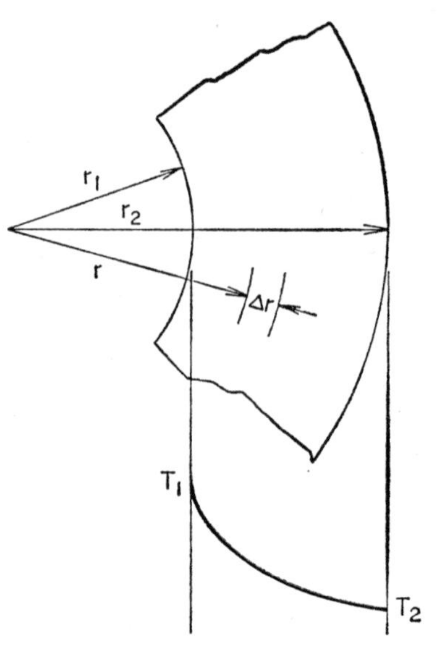

Fig. 6.3. Steady temperature distribution in a cylinder wall.

In the determination of heat flux, we note that area normal to flow is given by $2\pi rL$ where L = axial length of tube.

$$q = -k(2\pi rL) \frac{\partial T}{\partial r} = \frac{2\pi kL}{\ln (r_2/r_1)}(T_1 - T_2) \tag{6.14}$$

For these simplest systems, Eqs. (6.11) and (6.14) could have been obtained by direct integration of the Fourier equation $q/A = -k\,dT/dx$:

$$q \int_{x_1}^{x_2} \frac{dx}{A} = -\int_{T_1}^{T_2} k\,dT \equiv k_m(T_1 - T_2) \tag{6.15}$$

Here in cylindrical coordinates dx/A becomes $dr/2\pi rL$. Equation (6.15) may be integrated for any variation of k with T, but it clearly shows that k may be replaced by k_m, a mean k, defined by the right-hand parts of Eq. (6.15).

6.3 Resistance Concept

The heat transfer processes may be compared by analogy with the flow of electricity in an electrical resistance. The flow of heat, q, as a result of a temperature difference $T_1 - T_2$ is analogous to the flow of electric current, I, as a result of an electrical potential difference $E_1 - E_2$. From Ohm's law for electricity,

$$R_{\text{elect}} = \frac{E_1 - E_2}{I}$$

and by analogy a heat transfer resistance may be defined as

$$R_{\text{thermal}} = \frac{T_1 - T_2}{q} \tag{6.16}$$

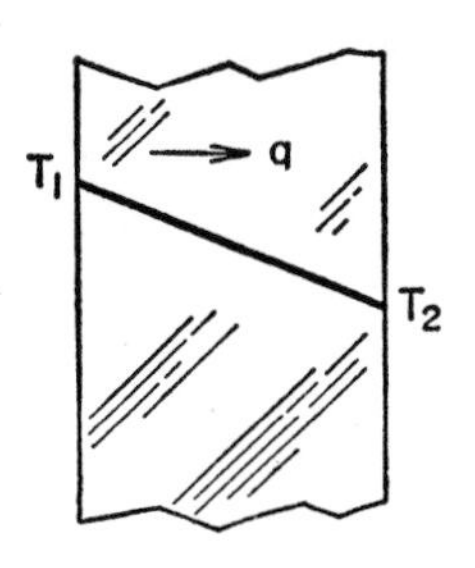

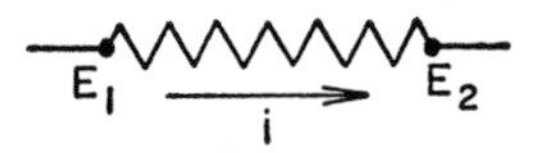

Fig. 6.4. Analogy between conduction of heat and electricity.

Comparing this definition with Eqs. (6.11), (6.14), and (5.3) results in expressions for heat transfer resistance as follows:

Infinite flat plate: $$R = \frac{L}{kA}$$

Infinite cylinder: $$R = \frac{\ln (r_2/r_1)}{2\pi Lk} \tag{6.17}$$

Surface to fluid: $$R = \frac{1}{Ah}$$

Rules for combining electrical resistances in series and parallel apply equally well to thermal resistances. An example is the composite wall. Some care must be exercised in representing thermal systems with parallel resistances since multi-dimensional effects are usually present.

6.4 Composite Walls

A wall may be constructed of more than one building material; it may be composed of successive layers of plaster and brick, shown in Fig. 6.5 as materials a and b. Equation (6.16) written for each section is

$$(T_i - T_1) = qR_i \qquad (T_2 - T_3) = qR_b$$
$$(T_1 - T_2) = qR_a \qquad (T_3 - T_o) = qR_c \tag{6.18a}$$

Adding these four equations results in

$$T_i - T_o = q(R_i + R_a + R_b + R_o) \tag{6.18b}$$

and since the over-all resistance is $(T_i - T_o)/q$, this equation shows that the over-all resistance equals the sum of the individual resistances; this is also true for electrical resistances in series.

For the flat composite wall of Fig. 6.5, the resistances in Eq. (6.18) are given by Eq. (6.17); then Eq. (6.18) becomes

$$q = \frac{T_i - T_o}{\dfrac{1}{Ah_i} + \dfrac{x_a}{k_a A} + \dfrac{x_b}{k_b A} + \dfrac{1}{Ah_o}} \tag{6.19}$$

For the cylindrical composite wall of Fig. 6.6, the reasoning leading to Eq. (6.18) is also applicable, and the resistances are given by Eq. (6.17); then Eq. (6.18) becomes

$$q = \frac{T_i - T_o}{\dfrac{1}{2\pi L r_1 h_i} + \dfrac{\ln (r_2/r_1)}{2\pi L k_a} + \dfrac{\ln (r_3/r_2)}{2\pi L k_b} + \dfrac{1}{2\pi L r_3 h_o}} \tag{6.20}$$

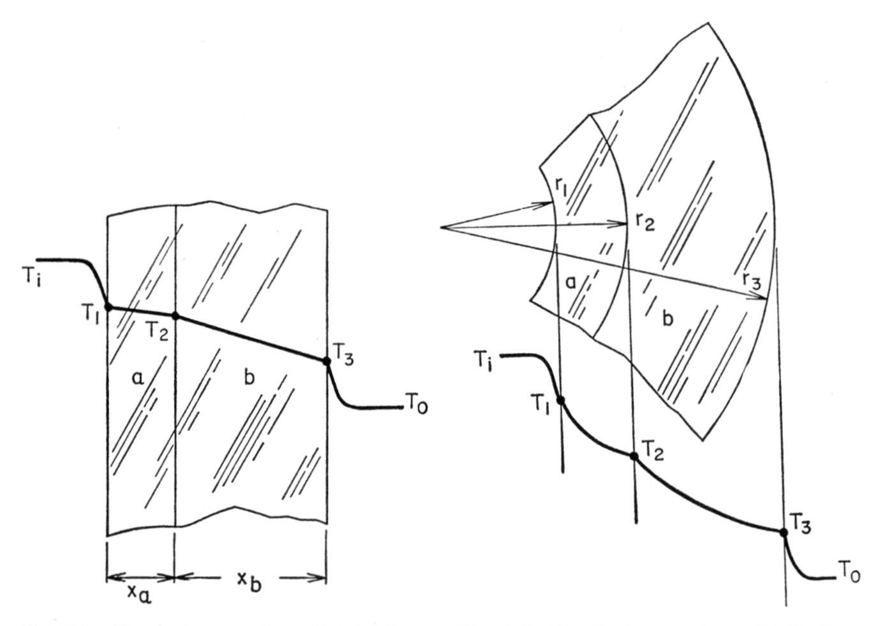

Fig. 6.5. Steady temperature distribution in a composite slab.

Fig. 6.6. Steady temperature distribution in a cylindrical composite wall.

6.5 Over-all Coefficient of Heat Transfer

For the composite wall such as those of Figs. 6.5 and 6.6, an over-all coefficient of heat transfer, U, has come into common use; it is defined by the equation

$$q = AU(T_i - T_o) \tag{6.21}$$

Comparing this equation with Eq. (6.19) for the flat composite wall,

$$U = \frac{1}{\dfrac{1}{h_i} + \dfrac{x_a}{k_a} + \dfrac{x_b}{k_b} + \dfrac{1}{h_o}} \tag{6.22}$$

For the composite cylinder, comparison of Eqs. (6.21) and (6.20) results in

$$\frac{1}{AU} = \frac{1}{2\pi L r_1 h_i} + \frac{\ln (r_2/r_1)}{2\pi L k_a} + \frac{\ln (r_3/r_2)}{2\pi L k_b} + \frac{1}{2\pi L r_3 h_o} \tag{6.23}$$

Since Eq. (6.21) defines U in terms of a heat transfer area, A, and since A varies from $2\pi L r_1$ to $2\pi L r_3$ for the cylinder, it is necessary to select arbitrarily some area for defining U.

There are no accepted practices in this matter; hence, if U is defined by Eq. (6.21) with $A = A_1 = 2\pi L r_1$, then from Eq. (6.23)

$$U_1 = \frac{1}{\dfrac{1}{h_i} + \dfrac{r_1}{k_a} \ln \dfrac{r_2}{r_1} + \dfrac{r_1}{k_b} \ln \dfrac{r_3}{r_2} + \dfrac{r_1}{r_3 h_o}} \tag{6.24}$$

This definition of U was arbitrarily based on A_1; it could also have been based on either A_2 or A_3 or any area between these. Then,

$$A_1 U_1 = A_2 U_2 = A_3 U_3$$

In addition to these resistances, heat transfer systems often have two other types of resistances. When two elements of a composite wall are in contact, there exists a thermal resistance which depends on the nature of the roughness of the two surfaces, the gas or liquid between the surfaces, and the contact pressure. A *contact coefficient of heat transfer* h_c is defined as (q/A) divided by the ΔT across the contact.

Heat exchangers often are plagued with another more elusive thermal resistance resulting from surface contamination with deposit of foreign matter or chemical reaction with the contacting fluid. These so-called "scale" or "dirt" deposits vary with time and are difficult to predict. A "*scale" coefficient* h_s is defined as before, but is based on the original clean surface area. These contact and scale coefficients are discussed more fully in Chap. 12 on heat exchangers.

6.6 Orders of Magnitude of *h*

Table 6.1 shows the usual range of values of surface coefficients under various conditions.

TABLE 6.1 ORDER OF MAGNITUDE OF h, BTU/HR FT^2 F

Gases (natural convection)	0.9–5
Flowing gases	2–50
Flowing liquids (nonmetallic)	30–1,000
Flowing liquid metals	1,000–50,000
Boiling liquids	200–50,000
Condensing vapors	500–5,000

For a wall, the equivalent h is k/x. For example, for a steel wall 0.12 in. thick and $k = 26$ Btu/hr ft F, the equivalent h is $26 \times 12/0.12$ or 2600, but for an asbestos wall 1 ft thick with $k = 0.13$ Btu/hr ft F, the equivalent h is 0.13/1 or 0.13.

For certain combinations of these various resistances in series, some may be negligible compared with others.

Example 6.1: What resistances are negligible when heat is transferred through 1 ft^2 of a 0.12-in. thick steel plate ($k_w = 26$) with a flowing liquid ($h_l = 1000$) on one side and a flowing gas on the other ($h_{g_o} = 10$)?

Assume a scale coefficient of $h_s = 1000$ on the liquid side. From Eq. (6.22)

$$\frac{1}{U} = \frac{1}{h_l} + \frac{1}{h_s} + \frac{x_w}{k_w} + \frac{1}{h_{g_o}} = \frac{1}{1000} + \frac{1}{1000} + \frac{0.12}{(26)(12)} + \frac{1}{10}$$

$$\cong \frac{1}{10} = \frac{1}{h_{g_o}}$$

In this case, the only significant resistance is at the gas side surface.

Example 6.2: Same as Example (6.1) with the flowing gas replaced by condensing steam ($h = 1000$).

$$\frac{1}{U} = \frac{1}{1000} + \frac{1}{1000} + \frac{1}{2600} + \frac{1}{1000}$$

In this case, none of the resistances is negligible.

Example 6.3: Same as Example (6.1) with the flowing liquid replaced by another flowing gas ($h_{g_i} = 5$).

$$\frac{1}{U} = \frac{1}{5} + \frac{1}{1000} + \frac{1}{2600} + \frac{1}{10} \cong \frac{1}{5} + \frac{1}{10} = \frac{1}{h_{g_1}} + \frac{1}{h_{g_o}}$$

In this case the wall and scale resistances are negligible.

6.7 Critical Radius of Insulation

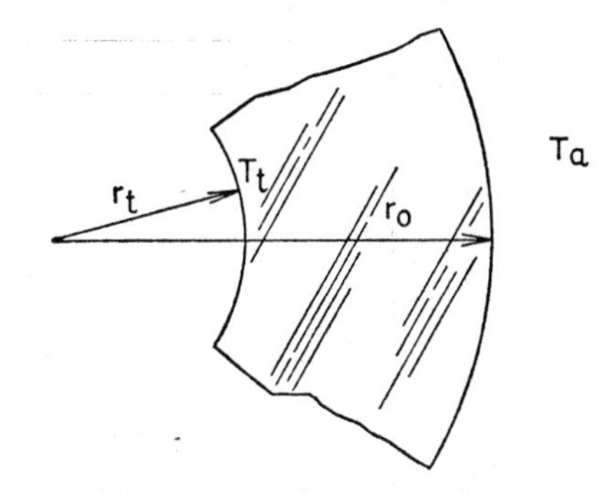

Fig. 6.7. Insulation on a tube.

Consider a tube of insulating material with inside radius r_t at constant temperature T_t, Fig. 6.7. At the outside radius of the insulating tube, r_o, a surface coefficient h_a may be assumed for heat transfer from the outside surface of the insulation to the atmosphere at temperature T_a.

From Eq. (6.20) for this case:

$$q = 2\pi L(T_t - T_a)\frac{1}{\dfrac{1}{h_a r_o} + \dfrac{1}{k}\ln\dfrac{r_o}{r_t}} \tag{6.25}$$

If L, T_t, T_a, h_a, k and r_t are all assumed to remain constant while r_o varies, then q is a function of r_o alone. As r_o increases, the term $1/(h_a r_o)$ decreases but the term $(1/k) \ln r_o/r_t$ increases; hence, it is possible that q might have a maximum value.

Differentiate Eq. (6.25) with respect to r_o; then, setting $dq/dr_o = 0$ and solving for $(r_o)_{\text{critical}}$, the critical radius for which q is a maximum,

$$(r_o)_{\text{critical}} = \frac{k}{h_a} \tag{6.26}$$

Then, if r_t is less than $(r_o)_{\text{critical}}$, the rate of heat loss from a cylinder is increased as insulation is added until

$$r_o = (r_o)_{\text{critical}}$$

Further increases in r_o cause the rate of heat loss to decrease. If, however, r_t is greater than $(r_o)_{\text{critical}}$, any addition of insulation will decrease the rate of heat loss, Fig. 6.8.

Example 6.4: Will the heat transfer rate increase if asbestos insulation (k = 0.13 Btu/hr ft F) is added to a 1-in. O.D. pipe, assuming the surface coefficient of heat transfer to the atmosphere is h_a = 1.0 Btu/hr ft^2 F?

From Eq. (6.26),

$$(r_o)_{\text{critical}} = 0.13/1.0 = 0.13 \text{ ft} = 1.56 \text{ in.}$$

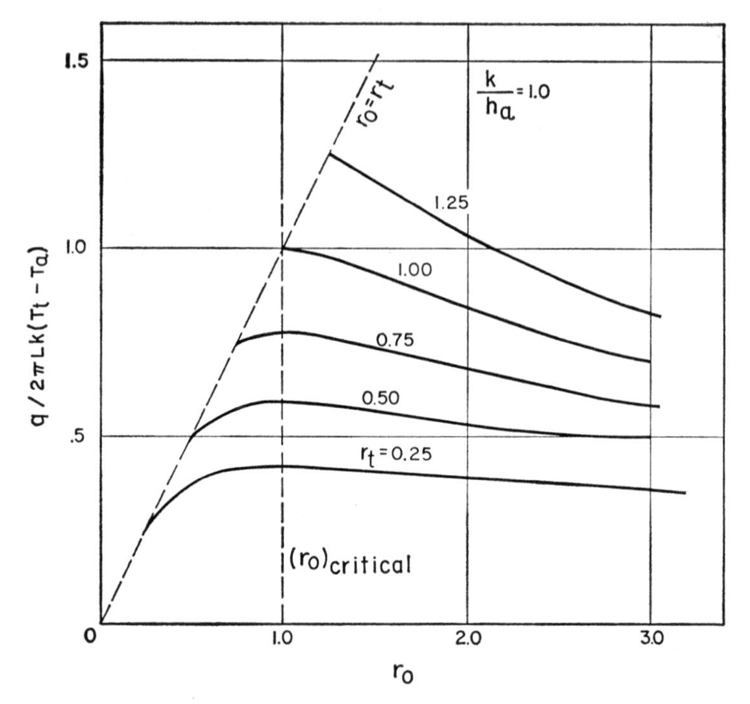

Fig. 6.8. Effect of insulation thickness on heat loss from tubes showing critical radius of insulation.

The radius of the pipe is 0.5 in.; therefore, as insulation is added, the heat transfer will increase until the insulation radius is made greater than 1.56 in.

6.8 Heat Sources, Steady State

Consider the case of a flat plate of thickness x_0 with a distributed heat source W_i resulting from such effects as an electric current or a nuclear reaction (nuclear fuel element). For this case, Eq. (6.2) reduces in the steady state to the following:

$$\frac{d^2T}{dx^2} = -\frac{W_i}{k} \tag{6.27}$$

for uniform k.

For some cases involving electric current flow or nuclear reactions in a thin fuel element, the assumption that W_i is constant is reasonable. When gamma rays are attenuated in reactor shielding material, W_i falls off exponentially. Equation (6.27) applies for any variation of W_i.

Successive integration of this equation for cases where W_i is uniform results in the following equations:

$$\frac{dT}{dx} = \frac{-W_i}{k}x + C_1 \tag{6.28}$$

$$T = -\frac{W_i}{2k}x^2 + C_1x + C_2 \tag{6.29}$$

where C_1 and C_2 are constants of integration.

If there is a symmetry around the centerline, the boundary conditions are $(\partial T/\partial x) = 0$ at $x = 0$ and $T = T_0$ at $x = x_0$. The two boundary conditions result in $C_1 = 0$ and $C_2 = T_0 + W_i x_0^2/2k$; then Eq. (6.29) becomes

$$T - T_0 = \frac{W_i}{2k}(x_0^2 - x^2) \tag{6.30}$$

which results in a parabolic temperature distribution with a centerline temperature of $T_{℄} = T_0 + W_i x_0^2/2k$. Equation (6.29) may be evaluated for any other of the boundary conditions, Eq. (6.8).

For a similar problem in cylindrical coordinates with circular symmetry $\partial T/\partial \theta = 0$ and also $\partial T/\partial z = 0$ everywhere; Eq. (6.6) reduces to the following:

$$\frac{d}{dr}\left(r\frac{dT}{dr}\right) + \frac{W_i r}{k} = 0 \tag{6.31}$$

which, when integrated twice with uniform W_i, results in

$$\frac{dT}{dr} = -\frac{W_i}{2k}r + \frac{C_1}{r} \tag{6.32}$$

$$T = -\frac{W_i}{2k}r^2 + C_1 \ln r + C_2 \tag{6.33}$$

For a solid cylinder, if $T \neq \infty$ at $r = 0$, then $C_1 = 0$. Further, if $T = T_0$ at $r = r_0$, $C_2 = T_0 + W_i r_0^2/4k$. Then Eq. (6.33) becomes

$$T - T_0 = \frac{W_i}{4k}(r_0^2 - r^2) \tag{6.34}$$

At $r = 0$, $T = T_{℄}$; so $T_{℄} - T_0 = W_i r_0^2/4k$

The quantity W_i may be expressed in terms of q/A for any given case. In the case of the flat plate, $W_i A x_0 = q$ when the plane at $x = 0$ is insulated; so

$$W_i = \frac{1}{x_0}\frac{q}{A} \tag{6.35}$$

and for the solid cylinder

$$W_i \pi r_0^2 L = \left(\frac{q}{A}\right) 2\pi r_0 L$$

$$W_i = \frac{2}{r_0}\frac{q}{A} \tag{6.36}$$

Equation (6.33) may be solved for other boundary conditions for either solid or hollow cylinders.

Example 6.5: Determine the centerline temperature of a bare steel wire (k = 13 Btu/hr ft F) of 0.75 in. diameter carrying a current of 1000 amp. The surface heat transfer coefficient is 5.0 Btu/hr ft² F and the air temperature is 70 F. The electrical resistance is 0.0001 ohm per foot of length.

The heat transfer rate is

$$I^2R = (1000)^2(0.0001) = 100/\text{watt ft} = 341.5 \text{ Btu/hr ft}$$

or

$$\frac{q}{A} = \frac{341.5}{\pi}\frac{12}{0.75} = 1740 \text{ Btu/hr ft}^2$$

From Eq. (5.3),

$$T_0 = 70 + \frac{1740}{5.0} = 418 \text{ F}$$

Combining Eqs. (6.34) and (6.36),

$$T = T_0 + \frac{q}{A}\frac{r_0}{2k} = 418 + 1740\frac{0.375/12}{2(13)} = 420 \text{ F}$$

6.9 Fins

Figure 6.9 represents a fin of uniform area transferring heat from a wall at T_b to air at T_a. We consider a steady state with k and h uniform. The assumption is that the rod is so thin and of high enough thermal conductivity that the temperature distribution in the thin direction perpendicular to x is uniform—surface temperature essentially equals centerline temperature at any x.

An energy balance applied to elemental section, Δx, of a fin of uniform cross section (Fig. 6.9) results in a differential equation for temperature distribution. In the steady state

$$q_x = q_{x+\Delta x} + q_{\Delta A} \tag{6.37}$$

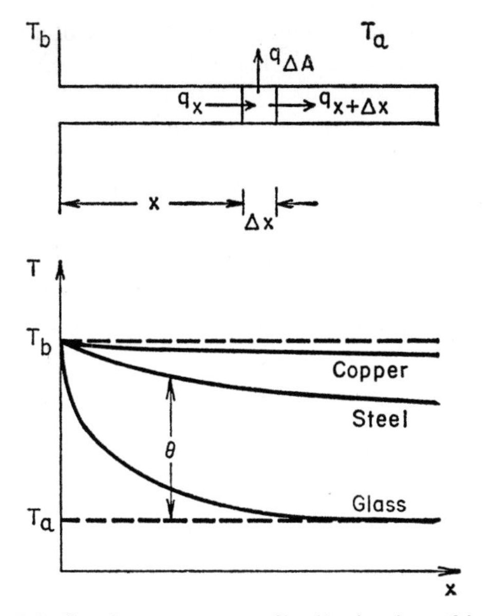

Fig. 6.9. Steady temperature distribution in a thin rod.

Then, with all of the above assumptions, this becomes

$$\frac{d}{dx}\left(kS\frac{dT}{dx}\right)\Delta x - hP\,\Delta x(T - T_a) = 0 \tag{6.38}$$

where S is the cross-sectional area and P is the perimeter of the fin. For constant S and k, and defining $\theta \equiv T - T_a$ with T_a = constant, this equation becomes

$$\frac{d^2\theta}{dx^2} - \frac{hP}{kS}\theta = 0 \tag{6.39}$$

A comparison of Eqs. (6.39) and (6.27) shows that the quantity $hP\theta/S$ is similar to a negative heat source $(-W_i)$. Equation (6.38) shows how variable area S is included and is readily modified to include terms for the effect of unsteady state and heat sources of the electric-current type.

The solution of Eq. (6.39) is of the form $\theta = e^{mx}$, which when substituted into Eq. (6.39) results in $m = \pm\sqrt{hP/kS}$ or

$$\theta = C_1e^{Bx} + C_2e^{-Bx} \tag{6.40}$$

where

$$B = \sqrt{hP/kS} \tag{6.41}$$

since the differential equation is linear.

Consider the evaluation of C_1 and C_2 for the case of a finite fin of length L with the following boundary conditions:

$$x = 0, \qquad \theta = \theta_b$$

$$x = L, \qquad (\partial\theta/\partial x) = 0 \tag{6.42}$$

Actually, there may be some heat transfer out the end of the fin at $x = L$, but usually can be neglected as a good approximation. Substituting Eq. (6.40) into Eq. (6.42) yields the following two relations:

$$\theta_b = C_1 + C_2$$

$$0 = C_1 e^{BL} - C_2 e^{-BL}$$

or solving for C_1 and C_2

$$C_1 = \frac{\theta_b e^{-BL}}{e^{BL} + e^{-BL}}$$

$$C_2 = \theta_b - C_1$$

Then, Eq. (6.40) becomes

$$\frac{\theta}{\theta_b} = \frac{e^{-B(L-x)} + e^{B(L-x)}}{e^{-BL} + e^{BL}} = \frac{\cosh B(L - x)}{\cosh BL} \tag{6.43}$$

The temperature at the tip ($x = L$) is then $\theta_L/\theta_b = 1/\cosh BL$.

The heat transfer rate through the base of the fin is calculated from either

$$q_b = -Sk\,(dT/dx)_{x=0} \quad \text{or} \quad q_b = \int_{x=0}^{L} hP\theta\,dx$$

Both equations give the same result as follows:

$$q = \sqrt{hPkS}\,\theta_b \tanh BL \tag{6.44}$$

Harper and Brown (1) suggest increasing L by the amount S/P to account for heat loss from the tip end provided $h/kB < \frac{1}{3}$.

The previous equations apply to fins of many different shapes—pins, squares, etc., and flat fins such as the one shown in Fig. 6.10(a) where $P = 2Y$ and $S = bY$. If the fin is thick and short, isotherms are curved as shown in Fig. 6.10(b). A thin fin is one for which $2k/hb > 6$, in which case the isotherms are nearly straight, Fig. 6.15.

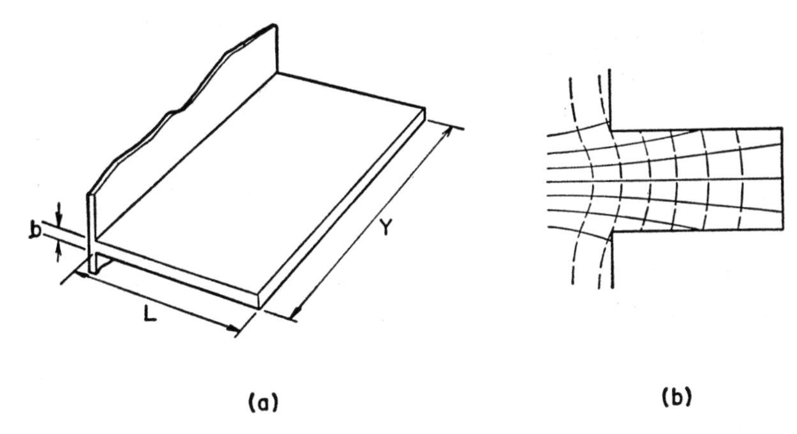

Fig. 6.10. Rectangular fin. (a) Thin plate fin. (b) Short fin showing isotherms and flow paths.

A performance factor η called *fin efficiency* is defined as the heat transfer rate from the fin divided by the heat transfer rate if the fin were uniformly at T_b:

$$\eta = \frac{q_{\text{fin}}}{hPL\theta_b} = \frac{\sqrt{hPkS}\,\theta_b \tanh BL}{hPL\theta_b} = \frac{\tanh BL}{BL} \tag{6.45}$$

From this definition, the heat transfer rate can be expressed as

$$q = h\theta_b\eta A_F \tag{6.46}$$

where A_F is the exposed surface area of the fin.

Similar analyses have been performed for fins of various shapes—circular fins on tubes, tapered fins, etc.—and the results presented graphically as η versus BL (2). For the use of fins in heat exchangers, see Chap. 12.

Example 6.6: Calculate the tip temperature and the rate of heat transfer from a pin fin, $\frac{1}{4}$ in. in diameter and 2 in. long, with $T_b = 200$ F, $T_a = 60$ F, $h_e = h = 2$, and $k = 24$.

$$B = \sqrt{\frac{hP}{kS}} = \sqrt{\frac{2\pi(0.25)(12)}{24\pi(0.25)^2/4}} = 4 \text{ ft}^{-1}$$

Take $L_e = L + S/P = 2 + (0.25)/4 = 2.0625$ in. From Eq. (6.43), with $x = L$,

$$\frac{T_L - 60}{200 - 60} = \frac{1}{\cosh[4(2.0625/12)]} = \frac{1}{1.248} = 0.801; \quad T_L = 172 \text{ F}$$

From Eq. (6.44),

$$q = \sqrt{\frac{2(24)\pi^2(0.25)^3}{1728(4)}}\,(200 - 60)\tanh(0.689) = 2.74 \text{ Btu/hr}$$

6.10 Semi-Infinite Rectangular Plate

Figure 6.11 represents a rectangular plate of thickness L and extending from $y = 0$ to ∞ and $z = -\infty$ to $+\infty$. If $T \neq f(z)$ and $W_i = 0$ and $\partial/\partial t = 0$, then Eq. (6.2) reduces to

$$\frac{\partial^2 T}{\partial x^2} + \frac{\partial^2 T}{\partial y^2} = 0 \qquad (6.47)$$

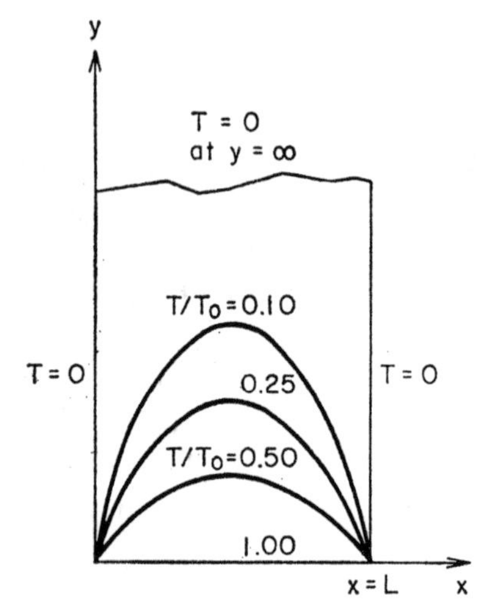

Fig. 6.11. Steady temperature distribution in a rectangular plate.

For the case shown, the boundary conditions are $T(0, y) = T(L, y) = T(x, \infty) = 0$ and $T(x, 0) = T_0$, uniform.

The solution for this case is given in detail in Appendix A where $P \equiv T$, as follows:

$$\frac{T}{T_0} = \frac{4}{\pi}\left[e^{-\pi y/L}\sin\frac{\pi x}{L} + \frac{1}{3}e^{-3\pi y/L}\sin\frac{3\pi x}{L} + \cdots\right] \qquad (6.48)$$

Lines of constant temperature are shown in Fig. 6.11.

6.11 Unsteady-State Processes

Unsteady-state heat transfer processes are those in which either or both the temperature or heat transfer rate vary with time at a particular place. Typical examples of unsteady-state heat transfer in solids include the batch heating of materials such as steel ingots, the heat transfer in a cold furnace wall when starting up or shutting down a furnace, and the hardening of steel by quenching in an oil bath. The daily periodic variations of the heat transfer between the sun and the earth's surface result in an unsteady-state temperature distribution within the earth. Some industrial apparatus such as the regenerative furnaces of a steel mill or the Ljungstrom rotary air preheaters of a steam power plant rely on the characteristics of unsteady-state heat transfer for their operation.

In subsequent sections, it will be considered that the solid body is homogeneous; its physical properties (ρ, c, k) are constant during the process; it is initially at a uniform temperature T_i, and is suddenly plunged into a fluid whose temperature is uniform at T_f; a heat transfer coefficient at the surface of the body remains constant throughout the process.

Although these conditions may seem restrictive, the solutions to such problems are quite useful because most practical heating and cooling problems may be idealized to this type. For example, the processes of removing an object from a furnace and allowing it to cool in air or of placing an object initially at room temperature into a furnace are of this type—so-called "dunking" problems.

6.12 Bodies with Negligible Temperature Gradients

When a solid such as a large flat plate with an initially uniform temperature is heated on both sides by a hot fluid at T_f the temperature distribution varies with time as sketched in Fig. 6.12(a). If the plate is thin, Fig. 6.12(b), or if the conductivity is high, the temperature gradients within the body may be negligible and a single value of temperature T may be used to describe the thermal state of the plate at any instant of time. A similar description is valid for bodies of any shape.

In time interval dt, the energy balance for a body of arbitrary shape with negligible temperature gradients may be written,

$$dQ = Ah(T_f - T)\, dt = V\rho c\, dT \tag{6.49}$$

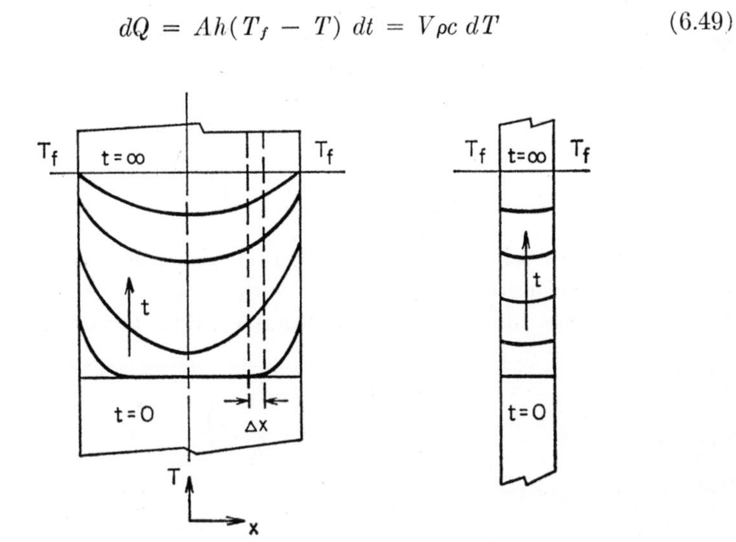

Fig. 6.12. Transient temperature distribution. (a) Large rectangular plate. (b) Thin plate.

For constant values of h, T_f, and physical properties, this equation integrates to

$$t = \frac{V\rho c}{Ah} \ln \frac{T_f - T_1}{T_f - T_2} \tag{6.50a}$$

which is the time required to heat the body from temperature T_1 to T_2. This equation is quite useful in calculating response rates of thermocouples. The time required for a thermocouple to rise 63.2 per cent of the way between an initial and final temperature after a step-change in fluid temperature is called the *time constant* of the couple. This occurs when $t = V\rho c/Ah$.

The following are values of V/A for simple shapes:

Sphere: $$\frac{V}{A} = \frac{\frac{4}{3}\pi r_0^3}{4\pi r_0^2} = \frac{r_0}{3}$$

Cylinder: $$\frac{V}{A} = \frac{r_0}{2} \quad \text{for} \quad L \gg r_0$$

Cube: $$\frac{V}{A} = \frac{L}{6}$$

The analysis in this section applies only to the cases in which internal temperature gradients are negligible. Inspection of the curves of Fig. 6.15 shows that this is true when $(k/hr_0) > 6$, approximately.

Equation (6.50) may be written in the form

$$\frac{T_f - T_2}{T_f - T_1} = \exp\left(-\frac{Ah}{V\rho c}t\right) \tag{6.50b}$$

The electric equivalent (analog) of this system is a condenser (capacitance C) discharging into a circuit with a pure resistance (R) only. The equation describing this electric behavior, ratio of voltage E at time t to initial voltage E_0 is

$$\frac{E}{E_1} = \exp\left(-\frac{1}{RC}t\right) \tag{6.51}$$

The analogy is obvious.

Example 6.7: What length of time is required to heat a 1 in. diameter, 2-in. long copper ($k = 220$ Btu/hr ft F) cylinder from 60 F to 200 F in a furnace whose temperature is 600 F? The value of h at the cylinder surface is 10 Btu/hr ft^2 F and $\alpha = 2$ ft^2/hr.

Equation (6.50) may be used because $k/hr_0 = (220)/(10)(0.5/12) = 528$ which is greater than 6.

$$\rho c = \frac{k}{\alpha} = \frac{220}{2} = 110 \text{ Btu/ft}^3 \text{ F}$$

$$\frac{V}{A} = [(\pi/4)(1)^2(2)]/[\pi(1)(2) + 2(\pi/4)(1)^2] = 0.2 \text{ in.}$$

Then from Eq. (6.50)

$$t = \left(\frac{0.2}{12}\right)\left(\frac{110}{10}\right)\left(\ln \frac{600 - 60}{600 - 200}\right) = 0.055 \text{ hr} \quad \text{or} \quad 3.3 \text{ min}$$

Another common problem is the calculation of rate of temperature change of an electrical conductor after an electric circuit is closed. Consider a thin electric wire initially at the temperature of the surrounding fluid T_f when suddenly a constant electric current begins to flow. Consider the I^2R to be equivalent to a uniform and constant heat source W_i, Btu/hr ft³. Then, at some time during the transient an energy balance is as follows:

$$W_iV - hA(T - T_f) = V\rho c \frac{dT}{dt}$$

or, integrating between T_1 and T_2,

$$t = \frac{V\rho c}{Ah} \ln \frac{(W_iV/hA) - (T_1 - T_f)}{(W_iV/hA) - (T_2 - T_f)} \tag{6.52}$$

At the final steady state $dT/dt = 0$; so

$$T_{\text{final}} = T_f + \frac{W_iV}{hA}$$

6.13 Bodies with Internal Temperature Gradients

For the flat plate of Fig. 6.12(a) the applicable differential equation from Eq. (6.2) is

$$\frac{\partial T}{\partial t} = \alpha \frac{\partial^2 T}{\partial x^2} \tag{6.53}$$

The solution $T(x, t)$ of this equation for the following boundary conditions is given in Appendix B in general symbols.

$$T(x, 0) = T_i, \quad \text{uniform;}$$

$$\frac{\partial T(0, t)}{\partial x} = 0 \quad \text{and} \quad \frac{\partial T(r_0, t)}{\partial x} = -\left(\frac{h}{k}\right)[T(r_0, t) - T_f]$$

In Appendix B,

$$P \equiv (T - T_f), \quad P_i \equiv (T_i - T_f), \quad \mathfrak{D} \equiv \alpha \quad \text{and} \quad N \equiv \left(\frac{h}{k}\right)$$

Then for this heat transfer problem

$$\frac{T - T_f}{T_i - T_f} = 2 \sum_{n=1}^{\infty} \frac{\sin (\lambda_n r_0)}{\lambda_n r_0 + \sin (\lambda_n r_0) \cos (\lambda_n r_0)} \exp (-\lambda_n^2 \alpha t) \cos (\lambda_n x) \tag{6.54a}$$

where λ_n are roots of the following equation:

$$\cot (\lambda_n r_0) = \frac{k}{h r_0} (\lambda_n r_0) \tag{6.54b}$$

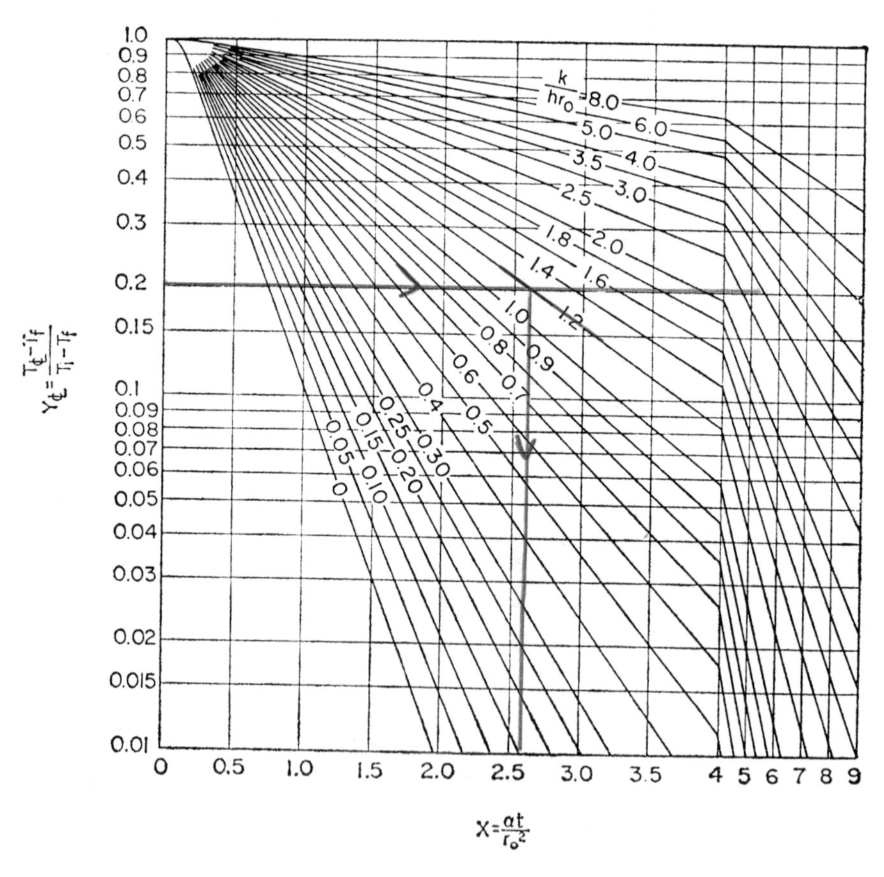

Fig. 6.13. Temperature history at the center of an infinite slab [Hottel (10)].

This result is of the following dimensionless form (see also Sec. 17.2):

$$\frac{T - T_f}{T_i - T_f} = \phi\left(\frac{\alpha t}{r_0^2}, \frac{hr_0}{k}, \frac{x}{r_0}\right)$$

The dimensionless group $\alpha t/r_0^2$ is known as the *Fourier number*, Fo, and the group hr_0/k as the *Biot number*, Bi.

Gurnie-Lurie (3), Adams-Williamson (4), Shack (5), Groeber (6), and Heisler (7) present solutions of the above equation in the form of graphs. Figures 6.13 and 6.14 are solutions of this equation for $x/r_0 = 0$ for the centerline temperature and $x/r_0 = 1.0$ for the surface temperature. Similar charts for intermediate values of x/r_0 are found in Ref. 8. Figure 6.15 shows curves for some intermediate magnitudes of x/r_0 for various values of k/hr_0. Temperature distribution curves may be integrated to determine the

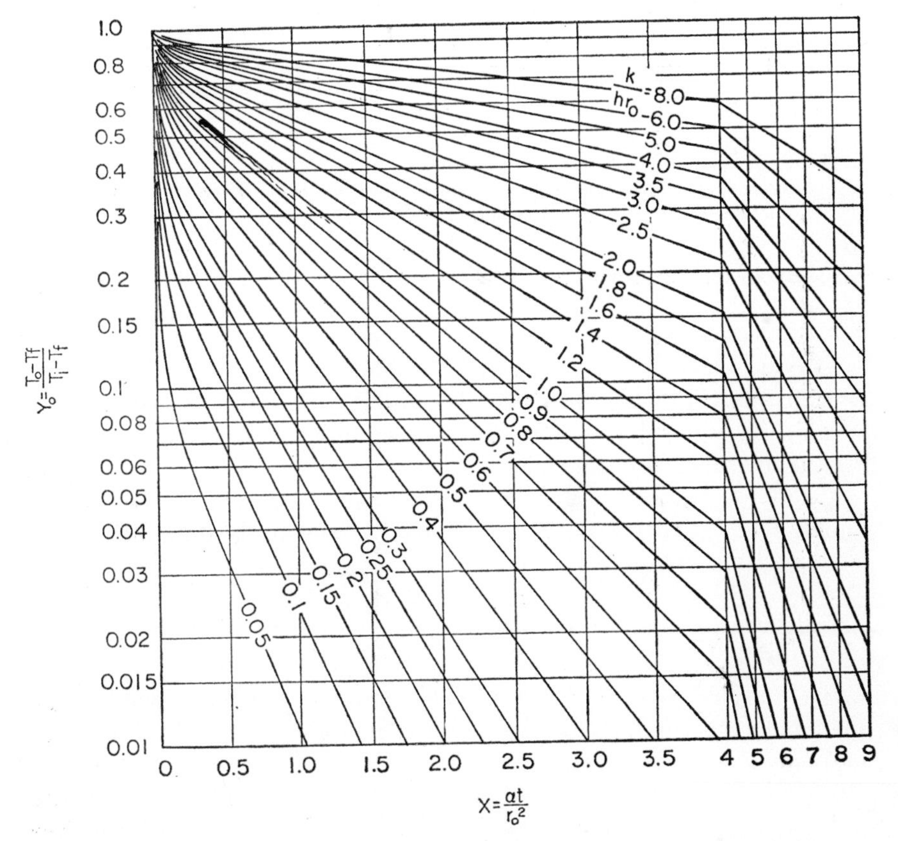

Fig. 6.14. Temperature history at the surface of an infinite slab [Hottel (10)].

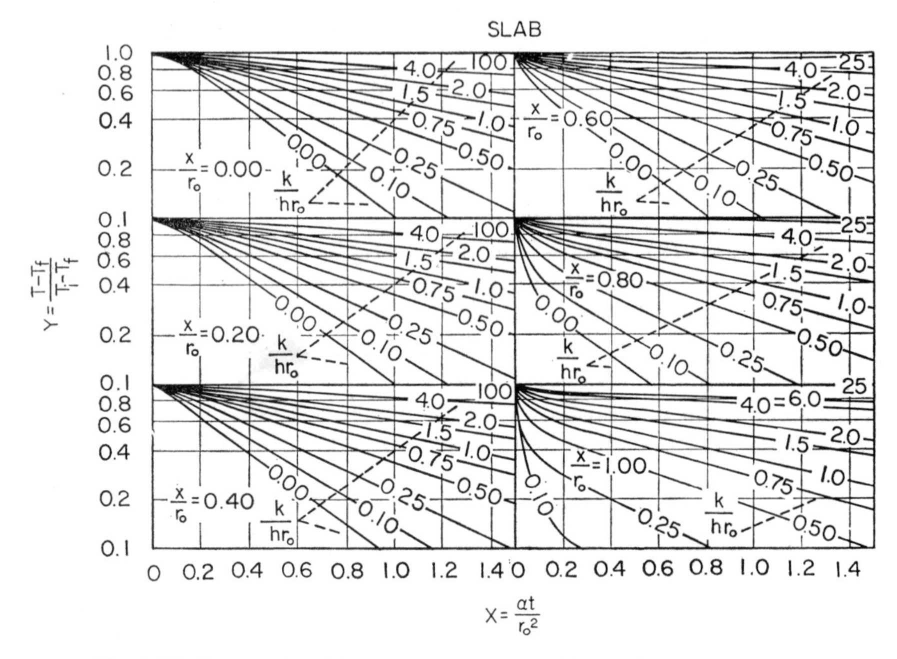

Fig. 6.15. Temperature history at some intermediate points in an infinite slab [Boelter, Cherry and Johnson (11)].

space mean temperature

$$T_m = \frac{1}{r_0}\int_0^{r_0} T\,dx$$

which is plotted in Fig. 6.16. The curves for $k/hr_0 = 0$ represent the case in which $h \rightarrow \infty$. Here the surface temperature immediately changes to T_f, the temperature of the fluid.

Similar solutions exist for infinite cylinders and spheres. For an infinitely long cylinder without heat sources, the unsteady-state differential equation from Eq. (6.6) is as follows:

$$\frac{1}{r}\frac{\partial}{\partial r}\left(r\frac{\partial T}{\partial r}\right) = \frac{1}{\alpha}\frac{\partial T}{\partial t} \tag{6.55a}$$

which considers the existence of radial temperature gradients only. The solution of this equation involves the use of Bessel functions and is left to more advanced treatment.

For the case in which a solid cylinder is at a uniform initial temperature T_i and is suddenly exposed to a fluid at temperature T_f with a finite surface heat transfer coefficient, the solution involves the dimensionless groups as above with x/r_0 replaced by r/r_0 and is shown plotted in Fig. 6.17.

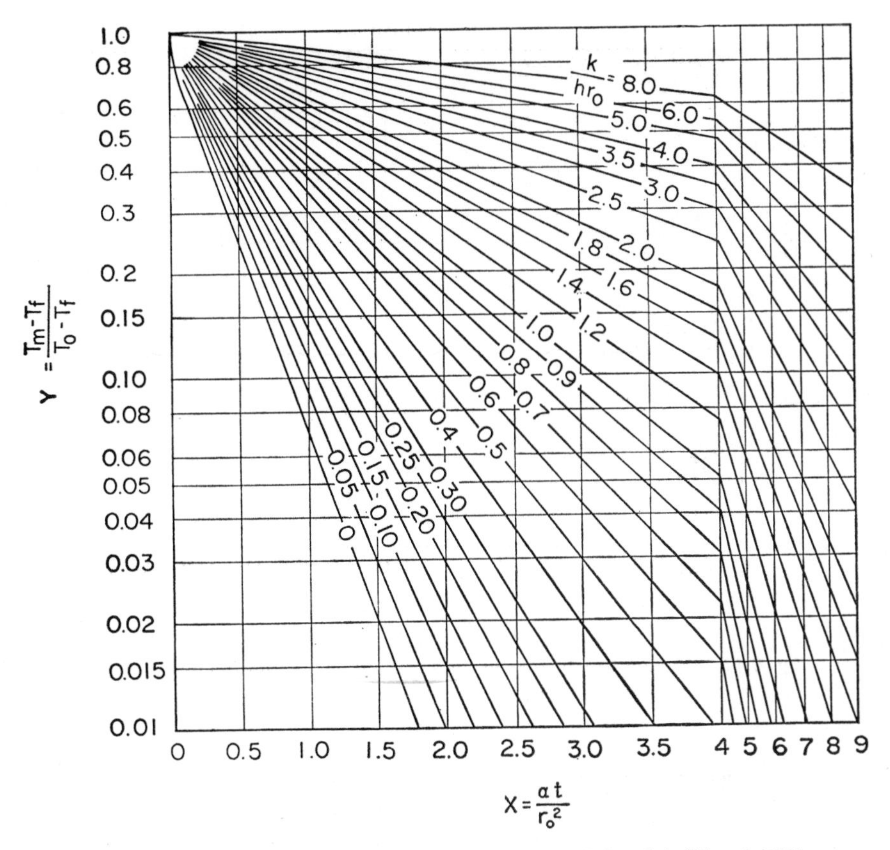

Fig. 6.16. Space-mean-temperature history in an infinite slab [Hottel (10)].

Similar analyses for spheres result in the differential equation,

$$\frac{1}{r^2}\frac{\partial}{\partial r}\left(r^2\frac{\partial T}{\partial r}\right) = \frac{1}{\alpha}\frac{\partial T}{\partial t} \tag{6.55b}$$

whose solution for the preceding type problem is shown in chart form in Fig. 6.18.

Example 6.8: The wall of a rocket-motor combustion chamber is $\frac{1}{4}$-in. thick alloy steel (k = 20 Btu/hr ft F, ρ = 480 lb/ft³, c = 0.25 Btu/lb F). With a flame side surface coefficient of h = 1000 Btu/hr ft² F, how long may this chamber operate with a wall temperature below 2000 F when with an initial uniform wall temperature of 80 F, the combustion gases at 4500 F begin suddenly to flow immediately following light-off?

Solution: The wall is thin compared with the diameter and may be assumed to be an infinite flat plate. The outside of the wall if exposed to atmospheric air would have a very low value of h. An answer on the safe side (e.g., shorter

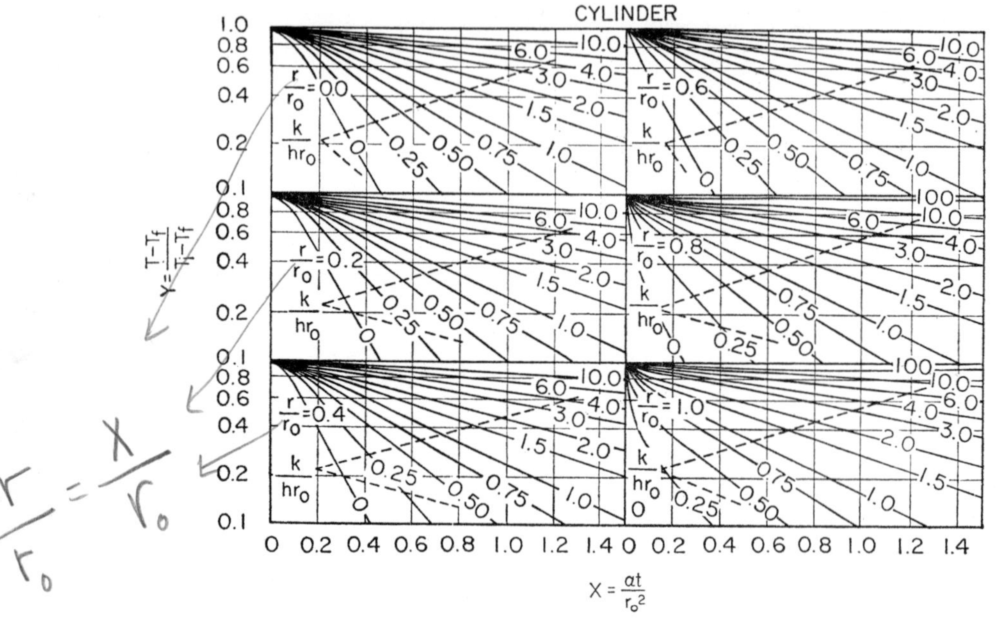

Fig. 6.17. Temperature history in a circular cylinder subjected to a sudden change in environmental temperature [Boelter, Cherry and Johnson (11)].

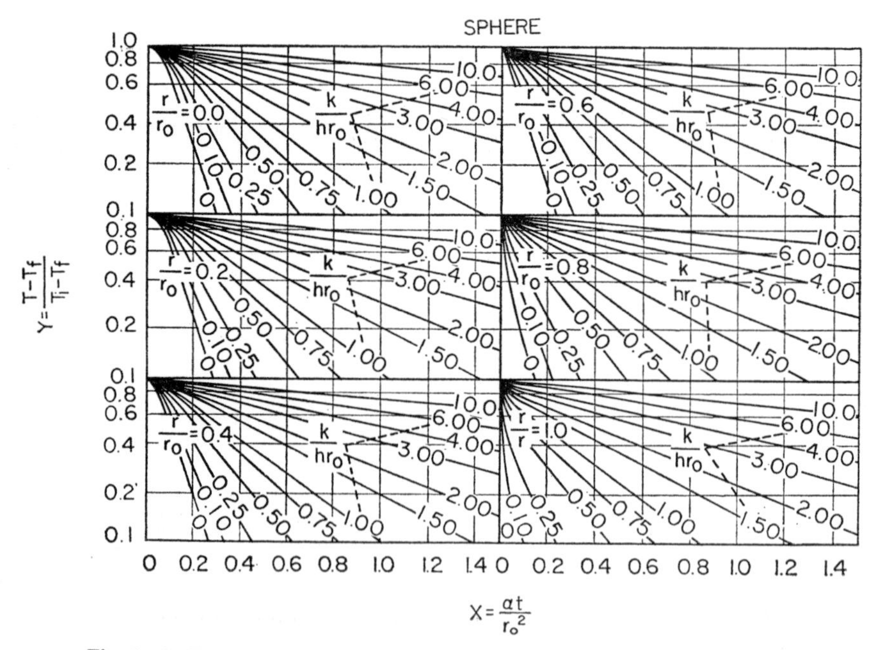

Fig. 6.18. Temperature history in a sphere subjected to a sudden change in environmental temperature [Boelter, Cherry and Johnson (11)].

allowable time) would result from assuming the outer surface to be insulated. During the transient, the inner surface temperature is the hottest place in the wall; so Fig. 6.14 for $x/r_0 = 1.0$ represents the solution. Here

$$r_0 = \frac{0.25}{12} = 0.0208 \text{ ft} \quad \text{and} \quad \alpha = \frac{k}{\rho c} = \frac{20}{(480)(0.25)} = 0.167 \text{ ft}^2/\text{hr}$$

$$\frac{k}{hr_0} = \frac{20}{(1000)(0.0208)} = 0.961 \quad \text{and} \quad Y = \frac{4500 - 2000}{4500 - 80} = 0.565$$

From Fig. 6.14,

$$\alpha t/r_0 = 0.31 \quad \text{or} \quad t = \frac{(0.31)(0.0208)^2}{0.167} = 0.0008 \text{ hr} = 2.9 \text{ sec}$$

6.14 Induction Heating

A solid being heated by low-frequency electric induction heating experiences a heat-generation effect distributed throughout the body. At very high frequencies this effect becomes concentrated in a thin skin at the surface. Define W_E'' as the heat generation rate per unit surface, Btu/hr ft². Consider the case of the infinite flat plate, Fig. 6.19, with an initially uniform temperature T_i in a fluid at temperature $T_f - T_i$. When suddenly W_E'' Btu/hr ft² begins to be generated at the skin, the temperature distribution changes as sketched on Fig. 6.19. The solution for the temperature distribution is given exactly in Figs. 6.13 through 6.16 with $(T - T_f)/(T_i - T_f)$ replaced by $1 - (h/W_E'')(T - T_i)$. The final temperature attained by the body is calculated from $W_E'' = h(T - T_i)$ since at steady state all of the heat generated must be transferred to the fluid.

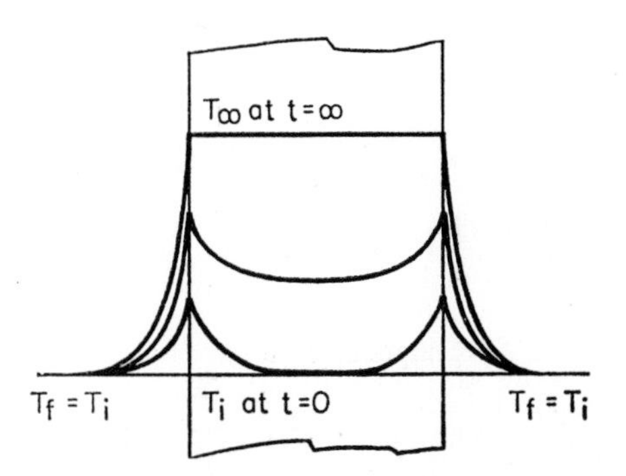

Fig. 6.19. Transient temperature distribution in an infinite slab with heat generation in a thin skin at the surface (induction heating).

6.15 Infinite and Semi-Infinite Bodies

The transient temperature distributions for a number of cases approximate those existing in a solid of infinite extent in all directions having an initial temperature distribution $f(x)$, with planes parallel to the yz plane remaining isothermal surfaces. The following physical examples approximate this idealized case.

In thermit welding, hot molten metal is poured between and around the broken junction of a rod such as a railroad rail. Sand, which is a thermal insulator compared with the metal rail, is packed around the rail for some distance in either direction from the break. When the hot molten metal is poured in the joint heat conduction is essentially unidirectional along the rails and approximates the heat flow unidirectionally in the infinite solid.

Similarly, the cooling of a concrete mix at 60 F poured into a deep narrow trench in the ground at 30 F also approximates unidirectional transient heat flow in the infinite solid.

We consider here only the special case in which the initial temperature distribution in an infinite solid is everywhere T_∞ except in the range $a \leq x \leq b$ where it is uniform at T_i. For this case, the solution of Eq. (6.53) for T as a function of x and t becomes, from Appendix C

$$\frac{P}{P_i} \equiv \frac{T - T_\infty}{T_i - T_\infty} \quad \text{and} \quad \mathfrak{D} \equiv \alpha$$

$$\frac{T - T_\infty}{T_i - T_\infty} = \frac{1}{2}\left[\operatorname{erf}\left(\frac{b - x}{2\sqrt{\alpha t}}\right) - \operatorname{erf}\left(\frac{a - x}{2\sqrt{\alpha t}}\right)\right] \tag{6.56}$$

which is valid over the range of x between $-\infty$ and $+\infty$. A more general solution is given in Eqs. (C.2) and (C.4) of Appendix C.

Note:

$$\operatorname{erf} w = \frac{2}{\sqrt{\pi}} \int_0^w e^{-\mu^2}\, du$$

and is known as the *error function* or *Gauss' error integral*. Note erf $w = -$ erf $(-w)$. Values of erf are given in Appendix C.

Example 6.9: Hot molten lava at 2000 F flows into a 10-ft wide crevice in a large volume of rock initially at 80 F uniformly. What will be the temperature at the center of the lava after one year? To solve this problem by the above method, it must be assumed that the lava and the rock have the same value of α, say 0.032 ft²/hr.

In Eq. (6.56) take $a = 0$, $b = 10$ ft, and $x = 5$ ft. Then

$$\frac{b - x}{2\sqrt{\alpha t}} = \frac{(10 - 5)}{2\sqrt{(0.032)(365)(24)}} = 0.15 \quad \text{and} \quad \frac{a - x}{2\sqrt{\alpha t}} = -0.15$$

and

$$\frac{T - T_\infty}{T_i - T_\infty} = \frac{1}{2}[\operatorname{erf} 0.15 - \operatorname{erf}(-0.15)] = \frac{1}{2}[0.1676 + 0.1676] = 0.1676$$

or

$$T = 80 + (2000 - 80)(0.1676) = 402 \text{ F}$$

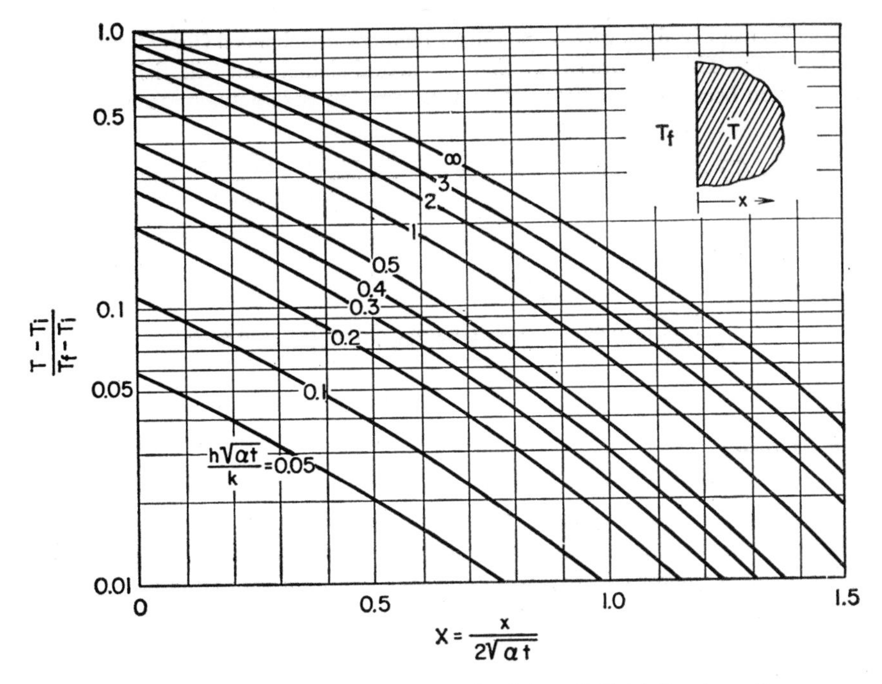

Fig. 6.20. Temperature history in a semi-infinite solid initially at a uniform temperature and suddenly exposed at its surface to a fluid at constant temperature [Schneider (12)].

Another useful idealized case is that of the semi-infinite solid initially at a uniform temperature T_i and suddenly exposed at its surface to a fluid at constant temperature T_0 and constant surface heat transfer coefficient h. The heating of a thick body by a hot fluid at the surface approximates this case during the early stages of the transient. The transient temperature distribution for the idealized semi-infinite body is

$$\frac{T - T_i}{T_f - T_i} = \left[\operatorname{erfc}\left(\frac{x}{2\sqrt{\alpha t}}\right) - \exp\left(\frac{xh}{k} + \frac{\alpha t}{(k/h)^2}\right)\operatorname{erfc}\left(\frac{x}{2\sqrt{\alpha t}} + \frac{\sqrt{\alpha t}}{(k/h)}\right)\right] \tag{6.57}$$

where erfc $w \equiv 1 - \operatorname{erf} w$, the complementary error function. For the case in which the surface temperature changes suddenly to T_f (e.g., $h = \infty$), the second term above reduces to zero. Equation (6.57) is shown graphically in Fig. 6.20.

Example 6.10: If the ground near the surface of the earth is at a nearly uniform temperature of 40 F and during the night the air temperature drops rapidly to 10 F, how deeply will the temperature of 32 F penetrate into the ground in six hours if $h = 1$, $k = 0.2$, and $\alpha = 0.02$?

To use Fig. 6.20, we assume the air temperature changed from 40 F to 10 F suddenly. This is more severe than the actual change, but will yield a "safe" solution because it predicts a deeper penetration than actually occurs.

$$\frac{h}{k}\sqrt{\alpha t} = \frac{1}{0.2}\sqrt{(0.02)(6)} = 1.73; \qquad \frac{T_i - T}{T_i - T_f} = \frac{40\text{–}32}{40\text{–}10} = 0.267$$

From Fig. 6.20:

$$\frac{x}{2\sqrt{\alpha t}} = 0.52; \qquad x = 0.52(2)\sqrt{(0.02)(6)} = 0.36 \text{ ft}$$

6.16 Ablation

Ablating heat shields have proved successful in satellite and missile re-entry to the earth's atmosphere and represent a major advance over the heat sink as a means of protecting surfaces from aerodynamic heating. In this application the high rate of heat transfer experienced at the surface first causes an initial transient temperature rise until the surface reaches the melting temperature, T_m. Ablation (melting of the surface) starts after this heat-up period, and there follows a second short transient period after which a steady-state ablation velocity is reached. The melted material is assumed to run off immediately—for example, around the sides of a nose cone.

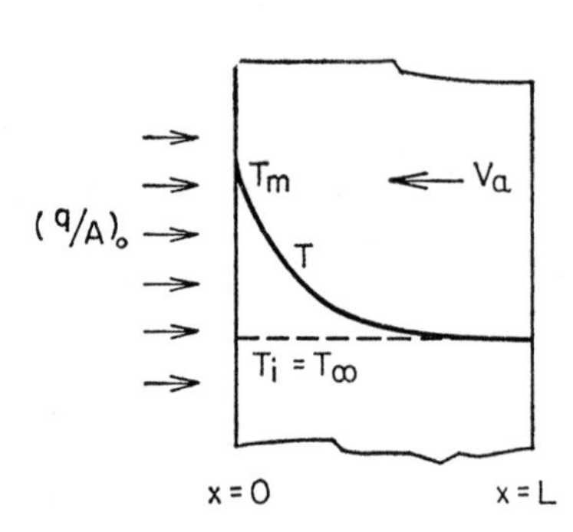

Fig. 6.21. Ablation at surface of flat plate.

We idealize the problem at the stagnation point of a missile nose as a one-dimensional conduction problem with the origin $x = 0$ at liquid-solid interface. Then the wall is imagined to move to the left, Fig. 6.21, at the ablation velocity, V_a. The differential equation for this reference frame is Eq. (6.53) with an added enthalpy flux term associated with the negative V_a:

$$\frac{\partial}{\partial x}\left(k\,\frac{\partial T}{\partial x}\right) = -V_a \rho c\,\frac{\partial T}{\partial x} + \rho c\,\frac{\partial T}{\partial t} \tag{6.58}$$

The initial transient before the surfaces reach T_m is solved as in Sec. 6.15. In the second transient period, Eq. (6.58) applies and may be solved formally or by finite differences.

After the steady-state ablation velocity has been reached, the temperature distribution in this reference frame is steady; so Eq. (6.58) becomes

$$\frac{d}{dx}\left(k\frac{dT}{dx}\right) = -V_a\rho c\frac{dT}{dx} \tag{6.59}$$

Materials such as glasses and plastics have proved the most successful. With the accompanying low thermal conductivity the temperature gradient at the surface is very steep so that $x = L$ may be considered as $x = \infty$. Satisfactory boundary conditions are as follows:

$$\begin{aligned} x &= 0, \qquad T = T_m, \\ x &= \infty, \qquad T = T_\infty = T_i \quad \text{and} \quad dT/dx = 0 \end{aligned} \tag{6.60}$$

For constant properties k, ρ, c, and for the case in which a steady $(q/A)_0$ net heat flux is transferred to the surface, Eq. (6.59) is solved by integrating twice and evaluating the integration constants with Eq. (6.60). The result is

$$\frac{T - T_\infty}{T_m - T_\infty} = \exp\left(-\frac{V_a x}{\alpha}\right) \tag{6.61}$$

where $\alpha = k/\rho c$.

If F is the heat of ablation (Btu/lb of material), an energy balance requires the following relations to hold:

$$\left(\frac{q}{A}\right)_0 - \rho V_a F = -k\left(\frac{\partial T}{\partial x}\right)_{x=0} = \rho V_a c(T_m - T_\infty) \tag{6.62}$$

which states that the net heat flux into the solid material equals the net rate of change in enthalpy of the solid "flowing" through the coordinate system. From this the ablation velocity follows:

$$V_a = \frac{(q/A)_0}{\rho F\left[1 + \dfrac{c(T_m - T_\infty)}{F}\right]} \tag{6.63}$$

The total heat conducted into the solid evaluated with the temperature distribution, Eq. (6.61), is

$$\left(\frac{Q}{A}\right)_c = \rho c\int_{x=0}^{\infty}(T - T_\infty)\,dx = \frac{k(T_m - T_\infty)}{V_a} \tag{6.64}$$

The total heat transferred to the surface in time t is $(q/A)_0 \cdot t$. Then for this period of time, the fraction of the total heat transferred which was conducted into the solid body is obtained by combining Eqs. (6.63) and (6.64) with the following result:

$$\frac{(Q/A)_c}{(Q/A)_0} = \frac{\rho k(T_m - T_\infty)[F + c(T_m - T_\infty)]}{(q/A)_0^2 t} \tag{6.65}$$

Comparing Eq. (6.63) with (6.65), one sees an apparent conflict. A large magnitude of $[F + c(T_m - T_\infty)]$ is desirable to reduce the amount of material ablated, but a small magnitude is desirable to reduce the fraction of Q_0 which is conducted into the solid body. A compromise obviously is necessary.

Example 6.11: (a) For a heat flux $(q/A)_0 = 10^6$ Btu/ft^2 hr and the following material properties, what is the steady-state ablation velocity and the fraction Q_c/Q_0 in 20 seconds?

$$c = 0.3 \text{ Btu/lb F}, \quad k = 0.5 \text{ Btu/hr ft F}, \quad \rho = 100 \text{ lb/ft}^3$$

$$F = 4000 \text{ Btu/lb}, \quad T_m = 3000 \text{ F}, \quad T_\infty = 60 \text{ F}$$

From Eq. (6.63)

$$V_a = \frac{10^6}{(100)[(4000) + (0.3)(2940)]} = 2.05 \text{ ft/hr}$$

From Eq. (6.65)

$$\frac{Q_c}{Q_0} = \frac{(100)(0.5)(2940)[4000 + (0.3)(2940)]}{10^{12}(20)/(3600)} = 0.129$$

(b) At what distance x is the temperature equal to 100 F?

$$\frac{100 - 60}{3000 - 60} = \exp\left(-\frac{2.05x}{0.0167}\right) \quad \text{or} \quad x = 0.035 \text{ ft} = 0.42 \text{ in.}$$

(c) In 20 seconds how much material has ablated?

$$x_{\text{abl}} = 2.05(20/3600) = 0.0114 \text{ ft} = 0.137 \text{ in.}$$

REFERENCES

1. Harper, D. K., and W. B. Brown, *Nat. Advisory Comm. Aeronaut. Rept.* 158 (1922).
2. Gardner, K. A., *Trans. ASME,* **68**:8, 621–31 (Nov. 1945).
3. Gurnie, H. P., and J. Lurie, *Ind. Eng. Chem.* **15,** 1170–72 (1923).
4. Williamson, E. D., and L. H. Adams, *Phys. Rev.,* **14,** 99–114 (1919).
5. Shack, A., *Industrial Heat Transfer,* Wiley, New York, 1933.
6. Groeber, H., *Z. ver. deut. Ing.,* **69,** 705 (1925).

7. Heisler, M. P., *Trans. ASME*, **69,** 227–36 (1947).

8. Rohsenow, W. M., M. J. Aronstein, and A. C. Frank, *Trans. ASME*, **68:2,** 135 (1946).

9. Newman, A. B., *Ind. Eng. Chem.*, **28,** 545–48 (1936).

10. Hottel, H. C., Personal Communication.

11. Boelter, L. M. K., V. H. Cherry, and H. A. Johnson, "Heat Transfer Notes," U. of Cal. Press, 1942.

12. Schneider, P. J., "Conduction Heat Transfer," Addison Wesley, 1955, p. 266.

PROBLEMS

6.1 Derive Eq. (6.6), the energy conservation equation, in cylindrical coordinates for a solid body with distributed heat sources.

6.2 Derive the energy conservation equation in spherical coordinates for the case in which $\partial T/\partial \theta = \partial T/\partial \phi = 0$. Integrate the equation for the wall of a hollow sphere (r_i, r_o) in the steady state without heat sources to obtain expressions for $T(r)$, q, and R_T.

6.3 If $k = 2$ Btu/hr ft F, what is the value of k in the following units:

(a) Btu/hr ft^2 (F/in.),

(b) kg cal/hr m C,

(c) watts/cm C?

6.4 Determine the conductivity of a test panel 6 in. by 6 in., and 0.50 in. thick, if during a two-hour test period 80 Btu are conducted through the panel when the temperatures of the two surfaces are 67 F and 79 F.

6.5 A furnace wall is constructed of 9 in. of fire brick ($k = 0.60$), 6 in. of red brick ($k = 0.40$), 2 in. of insulation ($k = 0.04$), and $\frac{1}{8}$ in. steel plate ($k = 26$) on the outside. The surface coefficients on the inside and outside surfaces are $h_i = 5$ and $h_o = 1$ Btu/hr ft^2 F, and the contact resistances between the various materials are negligible. The gas temperature inside the furnace is 2000 F and the outside air temperature is 90 F.

(a) Calculate the heat transfer rate per square foot of wall area.

(b) Determine the temperatures at all interfaces in the wall.

6.6 If $k = 2 + 0.001T$, calculate the rate of heat transferred per square foot of wall area for a thickness of 0.75 in. if h_i and h_o are each 1 Btu/hr ft^2 F and the fluid temperatures on either side of the wall are 200 F and 100 F.

6.7 Write an expression for the over-all coefficient of heat transfer based on the inside area, U_1, for the composite cylinder shown in Fig. 6.6. Assume the presence of a contact resistance at the interface between materials a and b and a scale or dirt resistance on the inner surface at r_1.

6.8 A tube of a heat exchanger has a 1.00 in. I.D. and 1.30 in. O.D. with liquid at 200 F ($h_i = 200$) on the inside and air at 40 F ($h_o = 2$) on the outside. What is the percentage increase in heat transfer rate if a copper ($k = 220$) tube is used instead of a steel ($k = 26$) tube? What are the values of U_i referred to the inside tube area for each case?

6.9 A power plant steam condenser has steel tubes ($k = 26$), 1.25 in. I.D., 1.45 in. O.D., with treated cooling tower water flowing inside ($h_i = 900$) and steam condensing on the outside ($h_o = 700$). Suppose the designer of the condenser had neglected to include allowance for scale deposit ($h_s = 500$) in his design. How much under size would the condenser be?

6.10 (a) Steam is supplied to the inside of a hollow sphere ($r_i = 4$ in., $r_o = 5$ in.). The wall material has $k = 2.0$. Because of heat loss through the walls, steam condenses and drains out of the sphere. At what rate is saturated steam (supplied at 14.7 psia) condensed in the sphere when $h_i = 800$ and $h_o = 1$ Btu/hr ft^2 F and $T_{\text{air}} = 80$ F on the outside?

(b) Equations (6.14), (6.20), and (6.24) apply to a cylinder; obtain similar expressions for a sphere.

6.11 Determine the expression for critical radius of insulation, corresponding to Eq. (6.26), for a sphere.

6.12 Explain why an insulated small-diameter electric wire has a higher current-carrying capacity than an uninsulated one.

6.13 At what radius of asbestos insulation ($k = 0.13$ Btu/hr ft F) will the heat transfer rate be the same as for an uninsulated pipe whose radius is $\frac{1}{2}$ inch? Assume the surface heat transfer coefficient $h_a = 1.0$ Btu/hr ft^2 F.

6.14 An 8-in. steam line (7.891 in. I.D., 8.625 in. O.D.) of steel ($k = 26$) supplies steam at 250 psia, 500 F, to a chemical process unit. Economics (minimum cost) governs the selection of the thickness of 85% magnesia ($k = 0.045$) insulation to be used. Assume the room temperature is 80 F, at the outside surface $h = 1.5$, and the inner surface of the pipe is $T_{\text{sat}} = 401$ F. The cost of generating the steam from 80 F water is 40 cents per million Btu. The cost of lagging per foot installed is: 1 in. = $2.00, 2 in. = 4.50, 3 in. = 7.50, 4 in. = 11.10, 5 in. = 15.50. Annual fixed charges on lagging (interest, repairs, insurance, depreciation) are 25% of first cost. The line operates 8000 hours per year.

(a) Recommend thickness of insulation.

(b) Calculate outer surface temperature for this thickness.

(c) Calculate the critical radius of the insulation.

6.15 A flat slab of radioactive material may be considered to have a uniform heat source function W_i when it is active. Determine the equations for temperature distribution in the material of thickness L when one face is insulated and heat is transferred to a fluid medium at the other side. Also determine the equation for the rate of heat transfer to the fluid. Consider the fluid temperature T_f and h to be known.

6.16 A rocket motor is to be constructed of an annulus of radioactive material having an equivalent heat source of W_i Btu per hr per ft³. The surface at the outer radius of the annulus is insulated while the inner surface transfers heat to a flowing gas through a film coefficient, h_i. Derive the equations for determining temperature distribution and rate of heat transfer to the gas.

6.17 Gamma radiation arrives at the inner surface ($x = 0$) of the shell of a nuclear reactor and is partially absorbed. This absorption has an effect similar to that caused by distributed internal sources, W_i, and may be expressed by

$$W_i = Ke^{-ax}$$

where K is a constant and a is the absorption coefficient for gamma radiation and is a property of the wall material for a particular radiation energy level. For the case in which the wall surfaces are cooled to $T = 0$ at $x = 0$ and x_w, derive expressions for the:

(a) temperature distribution.

(b) location and magnitude of the maximum temperature in the wall

(c) heat transfer rate, q/A, at $x = 0$ and $x = x_w$.

(d) Obtain numerical answers for parts (a), (b), and (c) if $K = 10^5$ Btu/hr-ft³, $a = 8$/ft, $k = 23$ Btu/hr ft F, and $x_w = 5$ in.

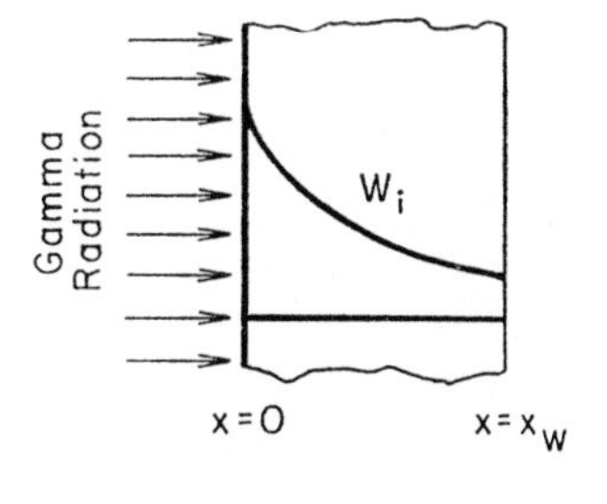

6.18 A cylindrical rod ($\frac{3}{4}$ in. diameter, 6 in. long) extends into an air stream at 60 F from a wall whose temperature is 200 F. Neglect the heat loss effect at the tip (insulated end) and calculate the tip temperature if the rod is (a) copper ($k = 220$), (b) steel ($k = 26$), and (c) glass ($k = 0.6$). In each case assume $h = 1.5$.

6.19 An air heater is composed of $\frac{3}{4}$-in. diameter standard steel pipes on which are wound spirally crimped steel fins ($k = 26$) of rectangular cross section, 0.025 in. thick by 0.50 in. wide. The crimping operation retains the fin width such that the fin extends 0.50 in. from the tube surface and results in five fins per inch of pipe length in a continuous spiral around the tube. For values of $h = 10.0$, air temperature $= 70$ F, and pipe metal temperature $= 330$ F:

(a) Calculate the fin efficiency based on fin area.

(b) Calculate the heat loss from a 1-ft length of finned pipe.

(c) Calculate the heat loss from a 1-ft length of unfinned pipe under the same conditions.

In each case neglect the heat loss from the tip of the fin.

6.20 Two 1-in. diameter, very long rods, one of steel ($k = 26$) and the other of a material whose conductivity is to be determined, are heated at one end to the same temperature, 100 F above the atmosphere. The measured temperatures along the two rods indicate that the temperature in the unknown rod at $x = 2$ in. is equal to the temperature in the steel rod at $x = 3.5$ in. If $h = 1.0$ in each case, what is the magnitude of k for the unknown material? This is a secondary method of measuring conductivity.

6.21 Consider a thin rod of cross-sectional area S, perimeter P, and thermal conductivity k, supported at its ends by two plates. The temperatures of the ends of the rod are held constant at T_1 and T_2 as shown. The rod is exposed to a fluid at T_a, and the film coefficient of heat transfer at the surface is h. Arrive at an expression for T as a function of x and the other dimensions and properties by evaluating C_1 and C_2 of Eq. (6.40) from the boundary conditions.

6.22 A thermometer well mounted through the wall of a steam pipe is a steel tube with a wall thickness of 0.1 in., outside diameter = 0.5 in., and length $L = 2$ in. ($k = 26$). The steam flow produces $h = 60$ on the outside of the tube. If $T_T = 300$ F and $T_w = 150$ F, what is the temperature of the steam?

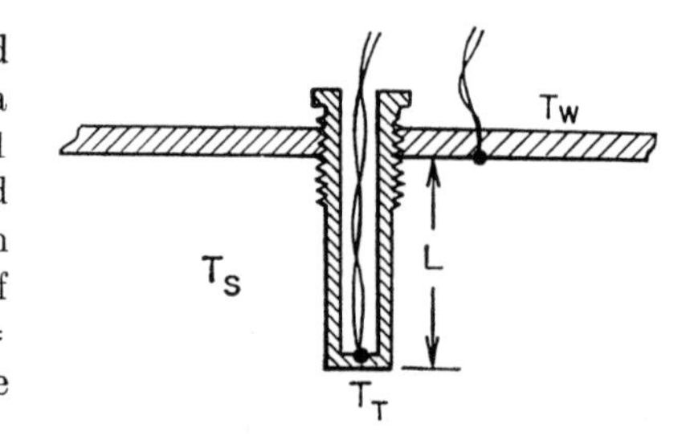

6.23 In Prob. 6.21, if an electric current is passing through the wire, it may be assumed that the effect is similar to having a uniformly distributed heat source of W_i Btu/hr ft^3.

(a) Derive the differential equation (similar to Eq. 6.39) for the temperature distribution in the rod.

(b) If T_a, T_1, and T_2 each equal zero, solve the differential equation obtaining T as a function of x and other properties. (*Note*: solution of $d^2T/dx^2 - a^2T + b = 0$ is $T = C_1e^{-ax} + C_2e^{ax} + b/a^2$.)

6.24 Thin wire is extruded at a fixed velocity through dies and the wire temperature at the die is a fixed value T_0, high enough to make the metal extrudable. The wire then passes through air for some distance before it is rolled onto large spools where the temperature has been reduced to a value T_L. It is desired to investigate the relationship between wire velocity v and the spacing distance L to obtain specific values of T_0 and T_L. Derive the differential equation for

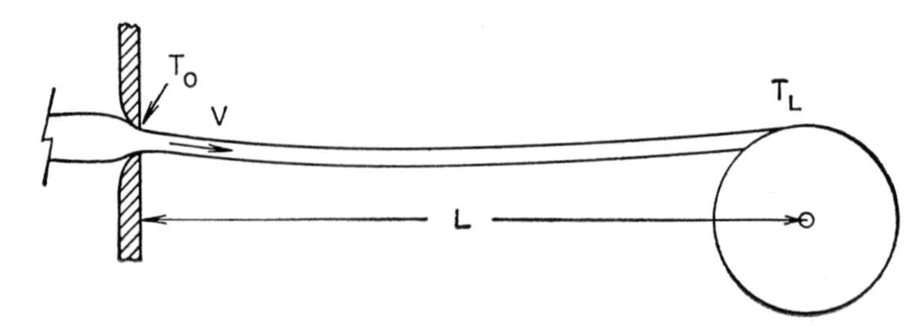

determining the wire temperature as a function of distance from the extrusion nozzle; state the boundary conditions. [*Suggestion*: Analyze problem by considering the wire to pass through a control surface fixed in space. (Similar to flow problem in thermodynamics.)]

6.25 Derive an expression for temperature distribution in a long rectangular bar for the following conditions:

$$\frac{\partial^2 T}{\partial x^2} + \frac{\partial^2 T}{\partial y^2} = 0$$

$$T(0, y) = T(a, y) = T(x, b) = 0$$

$$T(x, 0) = T_0, \quad \text{uniform}$$

Follow the method outlined in Appendix A.

6.26 Calculate the temperature at the following points in the semi-infinite plate of Fig. 6.11:

	(a)	(b)	(c)	(d)	(e)
x/L:	0.25	0.25	0.50	0.25	0.75
y/L:	0.25	0.50	0.25	1	1

Why is the temperature distribution independent of k, ρ, and c?

6.27 One theory suggests that the hardness of steel is correlated with the "half-temperature quenching time" which is the time from the start of cooling when $Y = 1$ to $Y = 0.5$. Determine the "half-temperature quenching time" for a bar of steel 1 in. in diameter and 6 in. long if it is initially at 1200 F and is quenched in oil at 100 F with $h = 20$ Btu/hr ft^2 F.

6.28 Air at a velocity of 10 ft/sec and a temperature of 100 F is flowing across a copper wire (diam. = 0.1 in., $k = 220$ Btu/hr ft F, $\rho = 500$ lb/ft^3, $c = 0.1$ Btu/lb F). An electric current is passed through the wire to raise its temperature to 130 F. Then the current is stopped and the wire is observed to cool to 110 F in 10 seconds. Calculate the heat transfer coefficient h on the wire surface.

6.29 A thermocouple probe initially at 60 F is suddenly immersed in an air stream at 200 F. How long will it take for the probe to rise to 190 F? Consider the probe to be a hollow steel tube with I.D. = 0.125 in., O.D. = 0.165 in., $k = 12$, $\rho = 450$ lb/ft^3, $c = 0.12$, and $h = 10$ Btu/hr ft^2 F. Neglect conduction in the probe and neglect radiation effects.

6.30 An electric copper wire (0.01 in. diam., $k = 220$, $\rho = 500$, $c = 0.1$) is initially at the air temperature 60 F. Suddenly an electric current of 4 amps starts flowing through it. The wire resistance is constant at 1.5 ohms per ft. Calculate the final steady-state temperature and the time required to reach a temperature 63.2% of the way between 60 F and the final steady-state temperature. Assume $h = 4$ Btu/hr ft^2 F.

6.31 A large flat 4-in. thick slab of steel ($k = 30$, $\rho = 460$, $c = 0.15$) initially at 80 F and immersed in a well-stirred tank of water at 200 F reaches a temperature of 150 F at its mid-plane before being removed from the tank ($h = 1000$).

(a) Calculate the length of the heating period.

(b) Calculate the temperature of the plate at a plane 1 in. from the surface at the end of the run.

(c) Calculate the time required for the steel at the 1-in. depth plane to reach 150 F.

6.32 Steel ball bearings ($\frac{1}{2}$ in. diam.) are hardened by being heated to 1600 F uniformly and then quenched in a large tank of oil at 100 F. Here $h = 600$, $k = 25$, $\rho = 450$ lb/ft³, and $c = 0.15$ Btu/lb F. Calculate:

(a) Time to cool center of sphere to 400 F.

(b) Temperature at surface when center is 400 F.

(c) Space-mean temperature when center is at 400 F.

(d) If 10,000 balls are quenched per hour, what rate of heat removal must be accomplished from the oil bath to maintain its temperature at 100 F?

6.33 Obtain a solution of $\partial^2T/\partial x^2 = (1/\alpha)\,\partial T/\partial t$, for the following boundary conditions: $T(x, 0) = T_i$, constant; $\partial T(0, t)/\partial x = 0$; $T(r_0, t) = T_f$, constant. Follow the procedure of Appendix B. Show how this identical solution applies to a momentum transfer problem.

6.34 A brick of ceramic (2 by 4 by 6 in.) is annealed by heating to 800 F uniformly and allowed to cool in air at 80 F. Here $h = 2$, $k = 0.8$, $c = 0.20$ Btu/lb F, $\rho = 100$ lb/ft³.

(a) Calculate the time required for the center of the brick to reach 150 F.

(b) At this time what are the location and temperature of the coldest place in the brick?

(*Note*: Newman (9) shows that this problem can be treated as superposition of three flat slabs calculated separately, one 2 in. thick, one 4 in. and one 6 in. Calculate Y for each slab; then Y for the brick is $Y = Y_x \cdot Y_y \cdot Y_z$.)

6.35 A large roast of beef has a shape approximated by a short cylinder of dimensions 8 in. diam., 12 in. long. The meat is considered to be properly cooked "rare" when its temperature is 140 F and "well done" at 170 F. For this meat, $k = 0.4$, $c = 1.0$ Btu/lb F, $\rho = 80$ lb/ft³, and $h = 4$.

(a) In an oven at 300 F, how long must this roast (initially at a uniform temperature of 40 F) be cooked to have the center "rare"? Neglect effect of any evaporation.

(b) How long to have the center "well done"?

(c) When the center is at 140 F, what is the temperature at the center of either flat face of the roast?

(d) Most cook books suggest that rolled roast be cooked 30 minutes per lb in an oven at 300 F. Discuss the applicability of such a general cooking rule.

(*Note*: As in Prob. 6.34, Newman (9) shows $Y = Y_r \cdot Y_z$ where Y_r is calculated for an infinite cylinder 8 in. in diameter and Y_z for a slab 12 in. thick.

6.36 A water pipe is buried 3 ft below the surface of the earth in dry soil ($k = 0.5$, $\alpha = 0.02$ ft²/hr) which is at a uniform temperature of 25 F. It is proposed to cover the ground with a 2-ft layer of hot ashes at 200 F and to cover these ashes with a deep layer of leaves which may be assumed to act as a perfect insulator to prevent heat transfer to the air. Calculate the time required for the soil at the level of the water pipe to reach 32 F. Assume the physical properties of the soil and the ashes to be identical. (*Hint*: treat top surface of ashes as a plane of symmetry.)

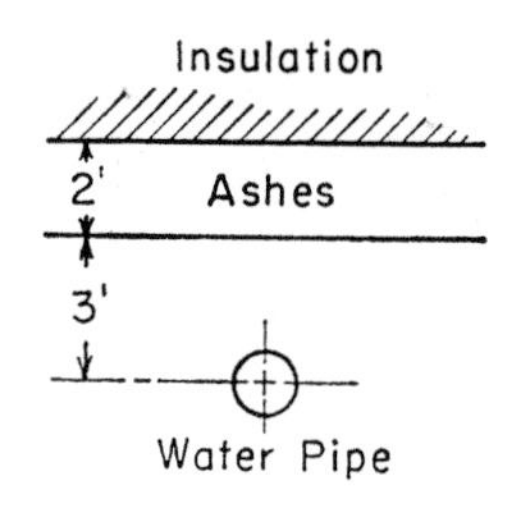

6.37 Snow was removed from parts of a city from which it previously had never been removed. The removal of this protective coating of snow caused water lines located 40 inches below the ground surface to freeze. Assume the ground near the earth's surface to be uniformly at 40 F when the snow was removed. What would be the maximum average air temperature which would just cause freezing of these pipes at the end of a cold period at constant air temperature for five weeks' duration? Properties of the ground material are $k = 0.03$, $\rho = 70$ lb/ft³, $c = 0.12$ Btu/lb F, and at the ground surface, $h = 2.0$.

CHAPTER 7

Heat Transfer in Laminar Flow

Convection is not a separate mode of heat transfer. It describes a fluid system in motion and heat transfer occurs by the mechanism of conduction alone. Obviously, we must allow for the motion of the fluid system in writing the energy balance, but there is no new basic mechanism of heat transfer involved.

Problems of heat transfer in laminar flow systems with simple boundary conditions can be treated by exact analysis. This is true whether the laminar motion is the result of free convection or of forced convection. As in Chap. 3 on momentum transfer in laminar flow, we begin our study with a few simple examples of nonisothermal, fully developed flow (Secs. 7.1–7.4) and boundary-layer flow (Secs. 7.5 and 7.6), and show how an energy balance is set up for each physical situation. Next, we briefly examine a number of entrance-region problems (Sec. 7.7). In Sec. 7.8, the general energy equation is presented and shown to reduce to the simple forms independently derived in each of the foregoing cases. Also, as is to be expected, the general energy equation reduces to Eq. (6.2) in the case of stationary or solid systems.

7.1 Forced Convection between Parallel Plates

First, consider the case of two-dimensional, steady flow between parallel plates. If the system is nonisothermal, say a uniform wall temperature with the fluid initially at a uniform but lower temperature, the temperature distribution changes along the flow path as shown in Fig. 7.1(b). In addition to the velocity boundary layer δ, a thermal boundary layer δ_T develops in the entrance region.

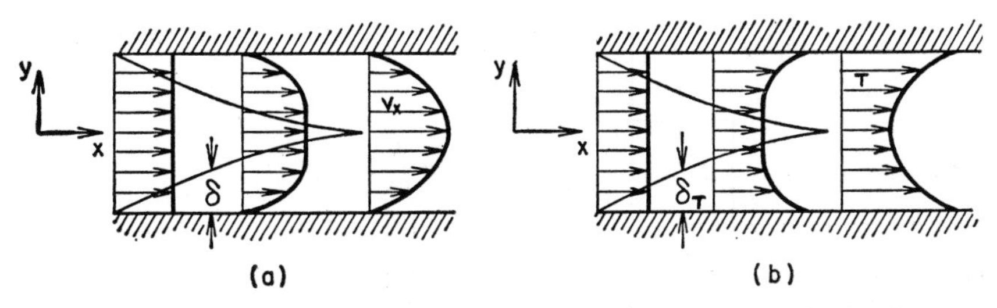

Fig. 7.1. The development of velocity and temperature profiles in the entrance region between parallel plates.

Generally, $\delta_T \neq \delta$. The relative rate of development of the two layers depends on the Pr of the fluid. This might have been expected since Pr measures relative rates at which momentum and energy are transferred by molecular diffusion through the fluid. At some position along the flow path, the boundary layers in either side essentially meet. Beyond this point the velocity or temperature profiles are said to be *fully developed.* Let us consider a differential control volume of dimensions Δx, Δy, and unity in the z-direction within the developing boundary layers. In Fig. 7.2, we have shown for simplicity that $\delta = \delta_T$ everywhere.

A momentum balance applied to the control surface gave the following equation for steady motion and constant ρ and μ:

$$v_x \frac{\partial v_x}{\partial x} + v_y \frac{\partial v_x}{\partial y} = -\frac{1}{\rho}\frac{\partial p}{\partial x} + \nu \frac{\partial^2 v_x}{\partial y^2} \tag{3.3}$$

We shall next derive the energy equation by a similar bookkeeping procedure of balancing all significant energy quantities involved. We shall assume steady state and constant properties ρ, k, and c_p in the derivation. The element of space $\Delta x\ \Delta y$ shown in Fig. 7.2 is a control volume for which an energy balance is now written for the steady state.

Net heat transfer by conduction into the control volume in the x-direction:

$$-\frac{\partial}{\partial x}\left(\frac{q}{A}\right)_x \Delta x\ \Delta y = \frac{\partial}{\partial x}\left(k\ \frac{\partial T}{\partial x}\right) \Delta x\ \Delta y$$

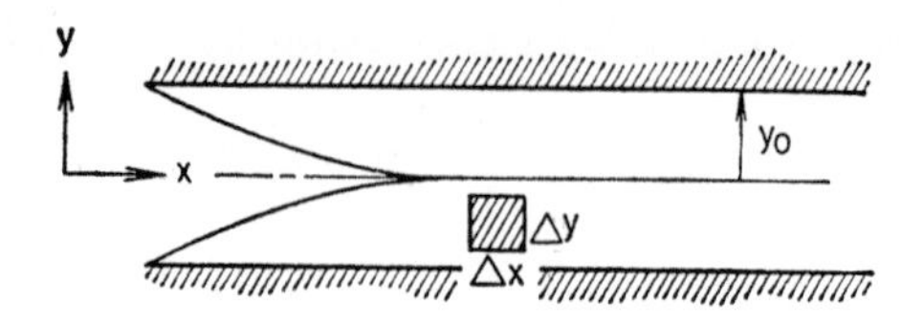

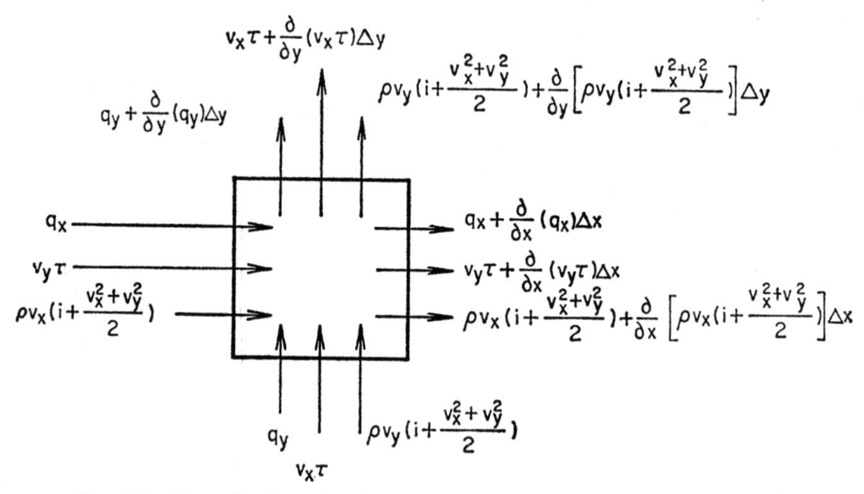

Fig. 7.2. Control volume in the boundary layer showing energy quantities crossing the control surface.

Net heat transfer by conduction in y-direction:

$$-\frac{\partial}{\partial y}\left(\frac{q}{A}\right)_y \Delta y\ \Delta x = \frac{\partial}{\partial y}\left(k\ \frac{\partial T}{\partial y}\right)\Delta y\ \Delta x$$

Net energy associated with fluid motion transferred into the control volume in the x-direction is the net flux of enthalpy, i, and kinetic energy:

$$-\frac{\partial}{\partial x}\left[\rho v_x\left(i + \frac{v_x^2 + v_y^2}{2}\right)\right]\Delta x\ \Delta y$$

Similarly in the y-direction:

$$-\frac{\partial}{\partial y}\left[\rho v_y\left(i + \frac{v_x^2 + v_y^2}{2}\right)\right]\Delta y\ \Delta x$$

Finally, we account for the energy associated with the frictional effects, or equivalently, the net rate of work added* to the control volume:

$$-\left[\frac{\partial(v_x\tau)}{\partial y} + \frac{\partial(v_y\tau)}{\partial x}\right] \Delta x \, \Delta y$$

where $\tau_{xy} = \tau_{yx} = \tau$.

Adding all these quantities and equating to zero we get the following equation for the case of uniform properties, k, ρ, μ, and c_p:

$$k\left(\frac{\partial^2 T}{\partial x^2} + \frac{\partial^2 T}{\partial y^2}\right) - \rho v_x \frac{\partial}{\partial x}\left(i + \frac{v_x^2 + v_y^2}{2}\right) - \rho v_y \frac{\partial}{\partial y}\left(i + \frac{v_x^2 + v_y^2}{2}\right)$$

$$-v_x \frac{\partial \tau}{\partial y} - v_y \frac{\partial \tau}{\partial x} - \tau \frac{\partial v_x}{\partial y} - \tau \frac{\partial v_y}{\partial x} = 0 \qquad (7.1\text{a})$$

since from continuity $\partial v_x/\partial x + \partial v_y/\partial y = 0$.

From Eq. (3.3) and $\tau = -\mu \, \partial v_x/\partial y$,

$$-v_x \frac{\partial \tau}{\partial y} = \mu v_x \frac{\partial^2 v_x}{\partial y^2} = \rho v_x^2 \frac{\partial v_x}{\partial x} + \rho v_y v_x \frac{\partial v_x}{\partial y} + v_x \frac{\partial p}{\partial x}$$

Equation (3.3) with x and y interchanged applies to the y-direction momentum balance. This with $\tau = -\mu \, \partial v_y/\partial x$ results in the following:

$$-v_y \frac{\partial \tau}{\partial x} = \mu v_y \frac{\partial^2 v_y}{\partial x^2} = \rho v_y^2 \frac{\partial v_y}{\partial y} + \rho v_x v_y \frac{\partial v_y}{\partial x} + v_y \frac{\partial p}{\partial y}$$

Substituting these quantities into Eq. (7.1a), and simplifying, we get the following equation:

$$k\left(\frac{\partial^2 T}{\partial x^2} + \frac{\partial^2 T}{\partial y^2}\right) - \rho\left(v_x \frac{\partial i}{\partial x} + v_y \frac{\partial i}{\partial y}\right) + v_x \frac{\partial p}{\partial x}$$

$$+ v_y \frac{\partial p}{\partial y} + \mu\left(\frac{\partial v_x}{\partial y}\right)^2 + \mu\left(\frac{\partial v_y}{\partial x}\right)^2 = 0 \qquad (7.1\text{b})$$

For most two-dimensional problems the terms involving p and $\mu \, (\partial v_y/\partial x)^2$ are negligible. Then Eq. (7.1b) reduces to the following form for fluids which permit replacing di by $c_p \, dT$:

$$\rho c_p\left(v_x \frac{\partial T}{\partial x} + v_y \frac{\partial T}{\partial y}\right) = k\left(\frac{\partial^2 T}{\partial x^2} + \frac{\partial^2 T}{\partial y^2}\right) + \mu\left(\frac{\partial v_x}{\partial y}\right)^2 \qquad (7.2)$$

which is the energy equation for steady-state, two-dimensional flow.

* When τ at the control volume surface is in the direction of motion (velocity component) work is added, and when τ is opposite to the direction of motion, work is removed from the control volume.

If both velocity and temperature profiles are fully developed, quite simple forms of Eqs. (3.3) and (7.2) are obtained. We showed in Sec. 3.1 that when the velocity profile is fully developed, both terms on the left of Eq. (3.3) are zero, so that we can write

$$\frac{dp}{dx} = \mu \frac{d^2v_x}{dy^2} \tag{3.4}$$

Except for the region very near the entrance, a few diameters' distance downstream, the axial conduction term $k\,(\partial^2 T/\partial x^2)$ may be neglected. For fully developed flow, $v_y = 0$. Then Eq. (7.2) reduces to

$$\rho c_p v_x \frac{\partial T}{\partial x} = k \frac{\partial^2 T}{\partial y^2} + \mu \left(\frac{\partial v_x}{\partial y}\right)^2 \tag{7.3}$$

Note that this equation is linear in T for constant property flow (ρ and μ) since here the velocity distribution is independent of the temperature distribution; so $\mu\,(\partial v_x/\partial y)^2$ is not a function of temperature. Equation (7.3) can, therefore, be solved by superposition, first obtaining the solution neglecting the viscous term and then including it. For low and moderate subsonic velocities, the frictional term is negligible.

For the case of flow of viscous fluids, such as oils, between closely spaced plates, the term on the left side of Eq. (7.3) often is negligible compared with the other terms. This is particularly true in journal bearings where oil flowing in the small clearances serves the purposes of both lubricant and coolant.

Example 7.1: Consider the case of steady, laminar flow between parallel plates with the upper plate in motion (as in Example 3.2) and where only the right side of Eq. (7.3) is significant. Then

$$k \frac{d^2T}{dy^2} = -\mu \left(\frac{dv_x}{dy}\right)^2 \tag{7.4}$$

The boundary conditions are

$$\begin{aligned} &\text{at } y = 0, \qquad v_x = 0, \qquad T = T_0 \\ &\text{at } y = l, \qquad v_x = V_1, \qquad T = T_1 > T_0 \end{aligned} \tag{7.5}$$

Solution of Eq. (3.4) for this Couette problem was obtained previously:

$$\frac{v_x}{V_1} = \frac{y}{l} - \frac{l^2}{2\mu V_1}\frac{dp}{dx}\frac{y}{l}\left(1 - \frac{y}{l}\right) \tag{3.8}$$

If dp/dx can be neglected, the solution gives a simple linear velocity distribution:

$$v_x = V_1 \frac{y}{l} \tag{7.6}$$

Substitution of dv_x/dy from Eq. (7.6) into Eq. (7.4) gives

$$k\frac{d^2T}{dy^2} = -\mu\frac{V_1^2}{l^2}$$

Integration of this equation for the specified thermal boundary conditions gives:

$$T = T_0 + (T_1 - T_0)\frac{y}{l} + \frac{\mu V_1^2}{2k}\frac{y}{l}\left(1 - \frac{y}{l}\right) \tag{7.7a}$$

or in dimensionless form:

$$\frac{T - T_0}{T_1 - T_0} = \frac{y}{l} + \frac{\mu V_1^2}{2k(T_1 - T_0)}\frac{y}{l}\left(1 - \frac{y}{l}\right) \tag{7.7b}$$

The temperature profile given by Eq. (7.7b) is a superposition of a linear profile and a parabolic profile. The group $\mu V_1^2/2k(T_1 - T_0)$ in Eq. (7.7b) is dimensionless and with ordinary units for the various quantities $J = 778$ ft lb$_f$/Btu must be included. Figure 7.3 shows $T - T_0/T_1 - T_0$ plotted against y/l for various values of the parameter $\mu V_1^2/2k(T_1 - T_0)$ represented by the symbol N.

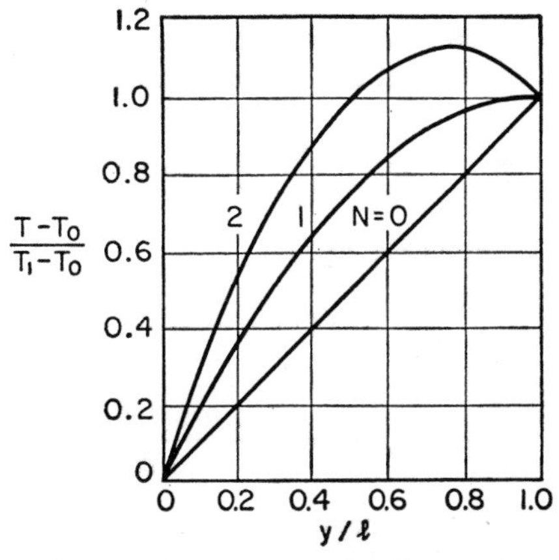

Fig. 7.3. Temperature distribution in Couette flow with viscous effect.

From Eq. (7.7b)

$$\frac{\partial T}{\partial y} = \left(\frac{T_1 - T_0}{l}\right)\left(1 + N - 2N\frac{y}{l}\right)$$

Then, since $\dfrac{q}{A} = -k\left(\dfrac{\partial T}{\partial y}\right)$,

$$\left(\frac{q}{A}\right)_{y=0} = -\frac{k}{l}(T_1 - T_0)(1 + N)$$

$$\left(\frac{q}{A}\right)_{y=l} = -\frac{k}{l}(T_1 - T_0)(1 - N) \tag{7.8}$$

The case of $N = 0$ represents the linear temperature distribution of pure conduction with stationary plate and fluid. Most lubricated bearings operate with $0 < N < 1$. When $N > 1$ the frictional effects are so large that heat is removed at the moving plate as well as at the stationary one. When $N = 1$, no heat is removed at the upper plate where $\partial T/\partial y = 0$. The parameter $\mu V_1^2/2k$ which measures the significance of viscous effects will be encountered again in Chap. 11 where heat transfer to high-velocity compressible fluids is discussed.

For the case of less viscous fluids, such as water or gases at low Mach number, the last term of Eq. (7.3) often is negligible compared with the other two. We shall illustrate this solution in the next section for the case of flow in round tubes. Since Eq. (7.3) is linear, the effect of $\mu\,(\partial v_x/\partial y)^2$ can be determined separately and superimposed on the following solutions. This superimposed solution is found in Sec. 11.2.

Example 7.2: The clearance space in a lubricated bearing may be considered to be bounded by two flat plates. The surface velocity of the shaft is $V_1 = 100$ ft/sec and for the oil, $\mu = 30$ lb/hr ft, $k = 0.076$ Btu/hr ft F and the radial clearance is 0.002″. If the temperature of the shaft surface is maintained at 150 F and the bearing surface at 100 F, calculate the rate of heat transfer to the shaft and to the bearing.

$$N \equiv \frac{\mu V_1^2}{2k(T_1 - T_0)} = \frac{30\,(100)^2}{2(0.076)(150 - 100)(778)(32.2)} = 1.58$$

From Eq. (7.8)

$$(q/A)_{\text{shaft}} = -\frac{0.076(150 - 100)}{0.002/12}(1 - 1.58) = 13{,}200 \text{ Btu/hr ft}^2$$

$$(q/A)_{\text{bearing}} = -\frac{0.076(150 - 100)}{0.002/12}(1 + 1.58) = -58{,}900 \text{ Btu/hr ft}^2$$

The temperature distribution in the oil is given by Eq. (7.7) or interpolated at $N = 1.58$ from Fig. 7.3.

7.2 Forced Convection in Round Tubes

Proceeding as in the case of parallel plates (Sec. 7.1), an energy balance on an annular control volume bounded by r, $r + \Delta r$, and Δx (Fig. 3.3), there results

$$2\pi\left[r\frac{\partial}{\partial x}(\rho v_x i) + \frac{\partial}{\partial r}(r\rho v_r i)\right]\Delta r\,\Delta x$$

$$= 2\pi\left[r\frac{\partial}{\partial x}\left(k\frac{\partial T}{\partial x}\right) + \frac{\partial}{\partial r}\left(rk\frac{\partial T}{\partial r}\right)\right]\Delta r\,\Delta x + 2\pi\mu\left(\frac{\partial v_x}{\partial r}\right)^2 r\,\Delta x\,\Delta r \qquad (7.9)$$

For the cases in which $di = c_p\,dT$, and also when properties (ρ, c_p, k, μ) are constant, this reduces with Eq. (1.14) to

$$\rho c_p\left(v_x\frac{\partial T}{\partial x} + v_r\frac{\partial T}{\partial r}\right) = k\left[\frac{1}{r}\frac{\partial}{\partial r}\left(r\frac{\partial T}{\partial r}\right) + \frac{\partial^2 T}{\partial x^2}\right] + \mu\left(\frac{\partial v_x}{\partial r}\right)^2 \qquad (7.10)$$

The terms on the left represent the enthalpy fluxes; on the right are the heat conduction and viscous work effects.

The differential equation for determining velocity distribution in this case is Eq. (3.11).

7.3 Fully Developed Flow in Round Tubes

For any set of boundary conditions, we define a fully developed temperature profile to exist when $(T_0 - T)/(T_0 - T_m)$ is a unique function of r/r_0, independent of x; then

$$\frac{T_0 - T}{T_0 - T_m} = f\left(\frac{r}{r_0}\right) \tag{7.11}$$

or

$$\frac{\partial}{\partial x}\left(\frac{T_0 - T}{T_0 - T_m}\right) = 0 \tag{7.12}$$

Alternatively fully developed flow could be defined by a similar equation with $(T_0 - T_{℄})$ replacing $(T_0 - T_m)$. These definitions are identical since one conclusion follows from the other.

The heat transfer coefficient for flow in tubes is defined in terms of a mean temperature of the fluid

$$T_m \equiv \frac{\int_0^{r_0} v_x T 2\pi r \, dr}{\int_0^{r_0} v_x 2\pi r \, dr} \tag{7.13}$$

and

$$h = \frac{q/A}{T_0 - T_m} = -\frac{k}{r_0}\frac{\partial}{\partial (r/r_0)}\left(\frac{T_0 - T}{T_0 - T_m}\right) \tag{7.14}$$

From Eqs. (7.11) and (7.12), the derivative* in Eq. (7.14) has a unique value at the wall $(r = r_0)$, independent of x. Then since k/r_0 is independent of x, h must be a uniform along the pipe in fully developed flow.

Now expand Eq. (7.12) where, in general, each quantity may vary:

$$\left(\frac{\partial T_0}{\partial x} - \frac{\partial T}{\partial x}\right) - \frac{T_0 - T}{T_0 - T_m}\left(\frac{\partial T_0}{\partial x} - \frac{\partial T_m}{\partial x}\right) = 0 \tag{7.15}$$

* Note that if y is measured from the surface into the fluid and $y \equiv r_0 - r$, then $q/A = -k\,(\partial T/\partial y)_{r_0} = +k\,(\partial T/\partial r)_{r_0}$ in Eq. (7.14); for example, if $T_0 > T_m$, the direction of (q/A) is from the surface into the fluid and $(\partial T/\partial r) > 0$ while $(\partial T/\partial y) < 0$.

The following two cases bracket the range of interest:

(a) *Uniform* (q/A). Here, since $q/A = h(T_0 - T_m)$ and since q/A and h are uniform, $T_0 - T_m$ is also uniform along the pipe. Then

$$\frac{\partial T_0}{\partial x} = \frac{\partial T_m}{\partial x}$$

and from Eq. (7.15)

$$\frac{\partial T}{\partial x} = \frac{\partial T_0}{\partial x} = \frac{\partial T_m}{\partial x}, \quad \text{independent of } r \tag{7.16}$$

(b) *Uniform* T_0. From Eq. (7.15) with $\partial T_0/\partial x = 0$,

$$\frac{\partial T}{\partial x} = \frac{T_0 - T}{T_0 - T_m}\frac{\partial T_m}{\partial x} = f\left(\frac{r}{r_0}\right)\frac{\partial T_m}{\partial x} \tag{7.17}$$

which depends upon radial position in the pipe.

We now solve the case in which the viscous term in Eq. (7.10) is negligible. Generally, although $v_x\, \partial T/\partial x$ may not be negligible, the term $(k/\rho c_p)\, \partial^2 T/\partial x^2$ is very small except very near the entrance of a tube. In fully developed flow $v_r = 0$. With these assumptions, Eq. (7.10) reduces to

$$v_x \frac{\partial T}{\partial x} = \frac{k}{\rho c_p}\left[\frac{1}{r}\frac{\partial}{\partial r}\left(r\frac{\partial T}{\partial r}\right)\right] \tag{7.18}$$

For this case of fully developed flow, the momentum balance equation (3.11) reduces to Eq. (3.12) whose solution is a parabolic velocity distribution:

$$v_x = 2V\left[1 - \left(\frac{r}{r_0}\right)^2\right] \tag{3.17}$$

where V is the average velocity given by Eq. (3.14).

For the case of uniform heat flux,

$$\begin{aligned} r &= 0, & \frac{\partial T}{\partial r} &= 0 \\ r &= r_0, & T &= T_0 \\ 0 \le r &\le r_0, & \frac{\partial T}{\partial x} &= \frac{\partial T_m}{\partial x} = \text{uniform} \end{aligned} \tag{7.19}$$

Note that T_0 at the wall is not a constant but varies with x. Substituting

Eq. (3.17) into Eq. (7.18) and integrating with conditions of Eq. (7.19), the result is

$$T_0 - T = \frac{V\rho c_p}{8r_0^2 k}\left(\frac{\partial T_m}{\partial x}\right)(3r_0^4 - 4r_0^2 r^2 + r^4) \tag{7.20}$$

The heat transfer coefficient [see footnote on page 139] at any x is

$$h = \frac{+k\left(\dfrac{\partial T}{\partial r}\right)_{r=r_0}}{T_0 - T_m}$$

where

$$T_0 - T_m = \frac{\displaystyle\int_0^{r_0} (T_0 - T)v_x 2\pi r\, dr}{\displaystyle\int_0^{r_0} v_x 2\pi r\, dr}$$

Introducing Eqs. (3.17) and (7.20) into these equations leads to the asymptotic value of the Nusselt number for uniform heat flux which is

$$\mathrm{Nu}_\infty = \frac{hD}{k} = 4.36 \tag{7.21}$$

For the case of uniform wall temperature, Eq. (7.18) is solved with Eq. (7.17). Then, since the temperature derivatives are expressed in terms of the temperature distribution, a trial-and-error solution is used; e.g., a temperature distribution is assumed [perhaps Eq. (7.20) initially] and then Eq. (7.18) integrated to find a new temperature distribution. When they agree, the solution has been accomplished. Usually one or two trials

TABLE 7.1 NUSSELT NUMBER FOR FULLY DEVELOPED LAMINAR FLOW

Geometry	Velocity distribution	Condition at wall	$\frac{hD_e}{k}$
Circular tube	Parabolic	$(q/A)_0$ Uniform	4.36
Circular tube	Parabolic	T_0 Uniform	3.66
Circular tube	Slug	$(q/A)_0$ Uniform	8.00
Circular tube	Slug	T_0 Uniform	5.75
Parallel plates	Parabolic	$(q/A)_0$ Uniform	8.23
Parallel plates	Parabolic	T_0 Uniform	7.60
Triangular duct	Parabolic (ref 2)	$(q/A)_0$ Uniform	3.00
Triangular duct	Parabolic (ref 2)	T_0 Uniform	2.35

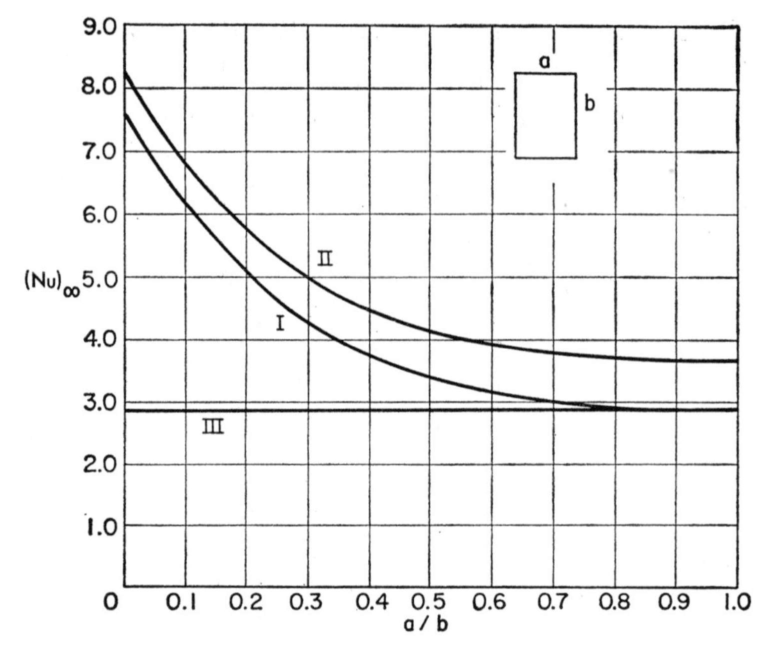

Fig. 7.4. Nusselt numbers for fully developed laminar flow in rectangular ducts. Curve I, Uniform wall temperature [Clark and Kays (2)]; curve II, Uniform heat flux per unit length but uniform peripheral wall temperature [Clark and Kays (2)]; curve III, Uniform heat flux, axially and peripherally [Cheng (26)].

are sufficient, the result for uniform wall temperature is

$$\text{Nu}_\infty = \frac{hD}{k} = 3.66$$

For liquid metals (low Pr), the thermal boundary layers, Fig. 7.1, fill the tube very near the entrance while the velocity distribution is still very nearly uniform (slug flow). An approximation for this case is obtained by solving Eq. (7.18) with v_x as a constant. The results are shown in Table 7.1 where D_e is the hydraulic diameter for each case. In the table the conditions of uniform $(q/A)_0$ are for the cases with uniform heat flux per unit length with uniform peripheral wall temperature at any position along the duct.

Similar results for rectangular ducts have been calculated by Clark and Kays (2) and Cheng (26), Fig. 7.4.

7.4 Free Convection between Parallel Plates

In the examples of forced convection considered so far, velocity distributions were independent of the temperature fields. Consequently, we solved first for velocity distributions, and used the information in the energy equation to obtain the temperature distributions. In free convection, motion results from temperature variation, and, therefore, velocity distributions cannot be determined independently of the temperature distributions.

Let us first consider air contained between essentially infinite parallel horizontal plates. If the upper plate is heated, less dense layers of air are above the denser layers and the system is in stable equilibrium. There is no air motion by free convection. At steady state, therefore, the temperature profile between the plates is linear, as in the case of conduction through a slab. However, if the lower plate is heated sufficiently, convection occurs in a distinctive cellular flow pattern (Benard cells). Heated air rises in the interior of these cells, and denser air moves down at the rims. This problem is difficult to analyze quantitatively.

Of greater interest is the problem of free convection between parallel vertical plates, one of which is heated and the other cooled. Let the heated plate be maintained at temperature T_1 and the cold plate at temperature T_0. The velocity and thermal boundary layers will grow upward at the hot plate and downward at the cold plate, in accordance with the direction of free convective air motion. Note that the direction of circulation would be reversed if the plates contained a fluid, the density of which increased rather than decreased with temperature. At a short distance from the ends where the boundary layers meet, the velocity and temperature profiles are fully developed.

Proceeding as in Sec. 7.1, an energy balance on a control volume $\Delta x\ \Delta y$ within the boundary layer at steady state leads to

$$\rho c_p\left(v_x\frac{\partial T}{\partial x} + v_y\frac{\partial T}{\partial y}\right) = k\left(\frac{\partial^2 T}{\partial x^2} + \frac{\partial^2 T}{\partial y^2}\right) + \mu\left(\frac{\partial v_x}{\partial y}\right)^2 \tag{7.2}$$

In the fully developed region, if we neglect axial conduction and frictional effects in this energy balance, and assume T is independent of x, the equation above reduces to

$$\frac{d^2T}{dy^2} = 0 \tag{7.22}$$

Thermal boundary conditions are

$$y = y_0, \qquad T = T_0$$

$$y = -y_0, \qquad T = T_1$$

Integration of Eq. (7.2) gives

$$T = \frac{T_0 - T_1}{2y_0}y + \frac{T_0 + T_1}{2} \tag{7.23}$$

or in dimensionless form, where $T_m \equiv (T_0 + T_1)/2$,

$$\frac{T - T_m}{T_1 - T_m} = -\frac{y}{y_0} \tag{7.24}$$

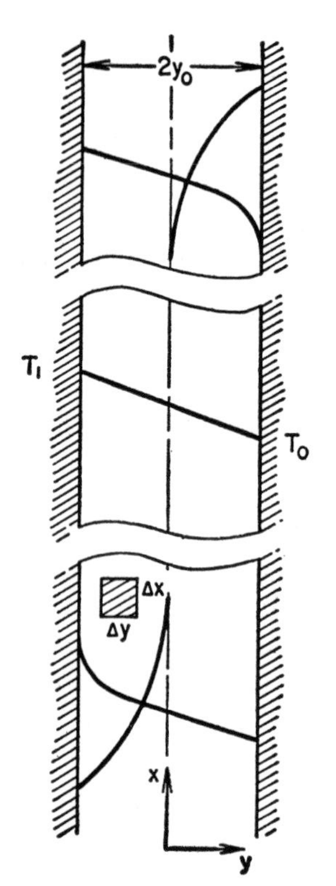

Fig. 7.5. Boundary-layer development and temperature distributions for free convection between parallel vertical plates.

According to Eq. (7.24), the temperature distribution is linear in the fully developed region between the parallel plates. Clearly, this is an approximation valid only if all the stated conditions of the solution are satisfied. Variation of temperature profile in x-direction is shown qualitatively in Fig. 7.5.

In the derivation of momentum equation, Sec. 3.1, for steady forced convection between parallel horizontal plates, we ignored external force fields such as gravity. We cannot ignore gravity effects in free convection, and, therefore, the momentum balance for the present case will contain an additional gravity body force term as follows:

$$v_x \frac{\partial v_x}{\partial x} + v_y \frac{\partial v_x}{\partial y} = -\frac{1}{\rho}\frac{\partial p}{\partial x} + \nu \frac{\partial^2 v_x}{\partial y^2} - g_x \tag{7.25}$$

where g_x is the gravity force per unit mass of fluid, in the negative x-direction.

For relatively low rates of flow, a good approximation is $p \neq f(y)$, particularly in the fully developed flow region. Then the vertical pressure gradient is expressed by the following approximation:

$$\frac{dp}{dx} = -\rho_m g_x \tag{7.26}$$

where ρ_m is the fluid density at the mean fluid temperature, T_m. Then in Eq. (7.25)

$$-\frac{1}{\rho}\frac{dp}{dx} - g_x = \frac{1}{\rho} g_x(\rho_m - \rho) = \beta g_x(T - T_m) \tag{7.27}$$

where the volume expansion coefficient β is defined as

$$\beta = -\frac{1}{\rho}\left(\frac{\partial \rho}{\partial T}\right)_p \tag{7.28}$$

The quantity $g_x(\rho_m - \rho)$ is often called *buoyancy force per unit volume.*

For the fully developed flow region, $v_y = 0$ and $\partial v_x/\partial x = 0$; so, Eq. (7.25) with (7.27) becomes

$$\mu \frac{\partial^2 v_x}{\partial y^2} + \rho \beta g_x (T - T_m) = 0 \tag{7.29}$$

With $T - T_m$ given by Eq. (7.24), the velocity distribution is obtained by integrating Eq. (7.29) with $v_x = 0$ at $y = \pm y_0$. The result is

$$v_x = \frac{g_x \beta y_0^2 (T_1 - T_m)}{6\nu} \left[\left(\frac{y}{y_0}\right)^3 - \left(\frac{y}{y_0}\right) \right] \tag{7.30}$$

Equation (7.30) can be put in a dimensionless form by introducing the following definitions:

$$v^* = \frac{v_x \rho c_p y_0}{k}, \qquad \text{dimensionless velocity}$$

$$\text{Gr} = \frac{g_x \beta y_0^3 (T_1 - T_m)}{\nu^2}, \qquad \text{Grashof number}$$

$$y^* = \frac{y}{y_0}, \qquad \text{dimensionless length}$$

Then Eq. (7.30) becomes

$$v_x^* = \tfrac{1}{6}(\text{Gr} \cdot \text{Pr})(y^{*3} - y^*) \tag{7.31}$$

Equation (7.31) is shown plotted in Fig. 7.6 for various magnitudes of the parameters. Note that because of the assumption of linear temperature

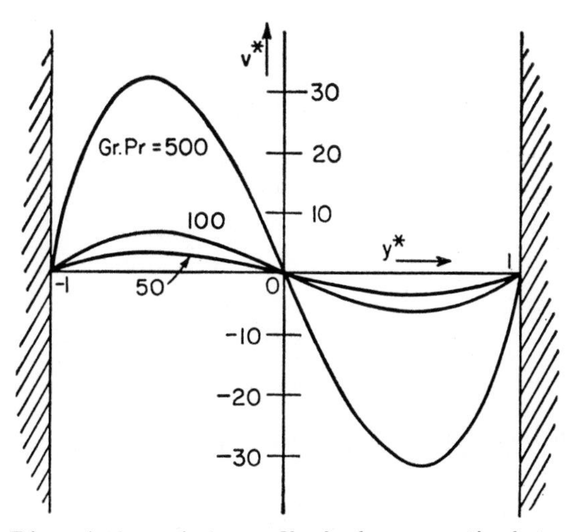

Fig. 7.6. Dimensionless velocity profiles for free convection between parallel vertical plates for very low values of Gr·Pr.

gradient, Eq. (7.22), the solutions are valid only for very low values of Gr·Pr. Eckert (22) suggests $\mathrm{Gr}\cdot\mathrm{Pr} < 62(1 + \mathrm{Pr})(L/y_0)/\mathrm{Pr}$.

7.5 Forced-Convection Laminar Boundary-Layer on Flat Plate

The simplest boundary-layer solution is that of forced convection over a flat plate. In Chap. 3, we determined the velocity distribution within the boundary layer. We now calculate the temperature distribution in the thermal boundary layer, first by exact solution of the boundary-layer equations and next by an approximate integral method.

(a) *Exact Solution*

Consider steady, two-dimensional flow over a plate maintained at a constant temperature. The energy equation of a control volume $\Delta x\ \Delta y$

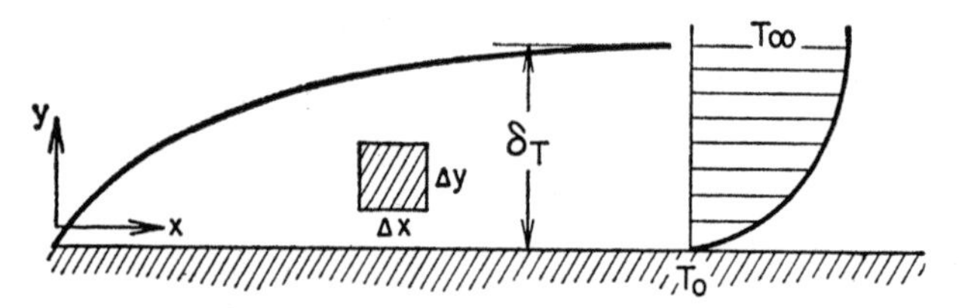

Fig. 7.7. Thermal boundary layer on a flat plate.

and unity depth, Fig. 7.7 within the boundary layer was derived in Sec. 7.1:

$$\rho c_p \left(v_x \frac{\partial T}{\partial x} + v_y \frac{\partial T}{\partial y} \right) = k \left(\frac{\partial^2 T}{\partial x^2} + \frac{\partial^2 T}{\partial y^2} \right) + \mu \left(\frac{\partial v_x}{\partial y} \right)^2 \tag{7.2}$$

Since the boundary layer is thin, the temperature gradient in the y-direction is much steeper than that in the x-direction, and the x-direction conduction term may be neglected. In addition, the frictional effect as an energy quantity may be neglected for flows at low Mach numbers. Then, Eq. (7.2) simplifies to

$$v_x \frac{\partial T}{\partial x} + v_y \frac{\partial T}{\partial y} = \alpha \frac{\partial^2 T}{\partial y^2} \tag{7.32}$$

The momentum boundary-layer equation, solved in Chap. 3, was of the form

$$v_x \frac{\partial v_x}{\partial x} + v_y \frac{\partial v_x}{\partial y} = \nu \frac{\partial^2 v_x}{\partial y^2} \tag{7.33}$$

Equations (7.32) and (7.33) are analogous, and we see immediately that if $\nu = \alpha$ (Pr = 1) and if the velocity and thermal boundary conditions are similar, the temperature and velocity profiles are identical and $\delta_T = \delta$.

The general solution, first obtained by Pohlhausen (3), parallels that of Eq. (3.18). By defining a stream function as in Eq. (3.20) and η and $f(\eta)$ as in Eqs. (3.22) and (3.23), the partial differential equation (7.32) reduces to the following ordinary differential equation.

$$\frac{d^2\theta}{d\eta^2} + \frac{1}{2} \Pr f(\eta) \frac{d\theta}{d\eta} = 0 \tag{7.34}$$

where $$\theta = \frac{T - T_0}{T_\infty - T_0}$$

For the case of a uniform-temperature plate, the following boundary conditions apply:

$$y = 0,\ T = T_0; \quad \text{or} \quad \eta = 0,\ \theta = 0$$

$$y = \infty,\ T = T_\infty; \quad \text{or} \quad \eta = \infty,\ \theta = 1.0 \tag{7.35}$$

Pohlhausen solved Eq. (7.34) by integrating twice—first setting $p(\eta) \equiv d\theta/d\eta$ and integrating between $p(0)$ and $p(\eta)$ for $\eta = 0$ and η, then integrating between $\theta = 0$ and θ for $\eta = 0$ and η. The result is

$$\theta(\eta) = \frac{d\theta(0)}{d\eta} \int_0^\eta \exp\left[-\frac{1}{2} \Pr \int_0^\eta f\, d\eta\right] d\eta \tag{7.36a}$$

where $d\theta(0)/d\eta$ is evaluated by setting $\theta = 1$ at $\eta = \infty$. Then

$$\frac{d\theta(0)}{d\eta} = \frac{1}{\int_0^\infty \exp\left[-\frac{1}{2} \Pr \int_0^\eta f\, d\eta\right] d\eta} \tag{7.36b}$$

Using results for $f(\eta)$ such as Eq. (3.28), Pohlhausen integrated Eq. (7.36b) numerically. The following empirical equation expresses these results within a few per cent agreement for Pr > 0.5.

$$\frac{d\theta(0)}{d\eta} = 0.332 \Pr^{0.343} \tag{7.37}$$

For simplicity, however, the exponent 0.343 is usually taken as $\frac{1}{3}$, (23).

Similarly, Eq. (7.36a) was evaluated with Eq. (3.28) with results as shown in Fig. 7.8. Note that the dimensionless temperature distribution curve for Pr = 1 is identical with the dimensionless v_x-velocity distribution curve of Fig. 3.5. For Pr < 1, $\delta_T > \delta$ and for Pr > 1, $\delta_T < \delta$.

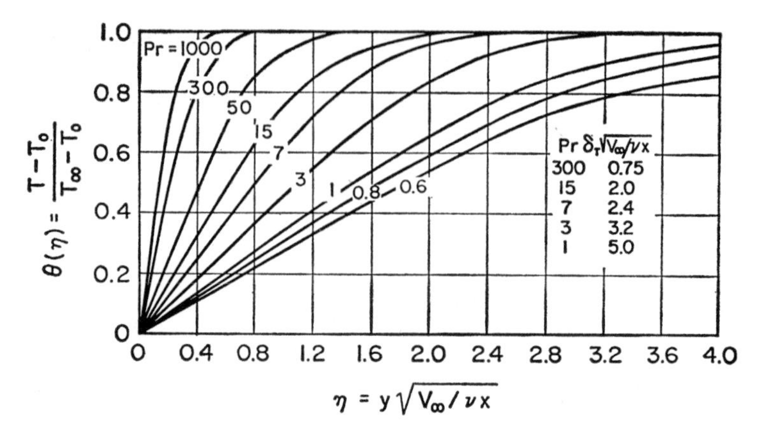

Fig. 7.8. Dimensionless temperature profiles in the laminar boundary layer of a flat plate for various Pr [Pohlhausen (3)].

The local heat transfer coefficient h_x is

$$h_x = \frac{-k\left(\frac{\partial T}{\partial y}\right)_{y=0}}{T_0 - T_\infty} = k\sqrt{\frac{V_\infty}{\nu x}}\,\frac{d\theta(0)}{d\eta} \tag{7.38}$$

From the results of Fig. 7.8 or Eq. (7.37),

$$h_x = 0.332k\ \mathrm{Pr}^{1/3}\sqrt{V_\infty/\nu x} \tag{7.39}$$

where exponent 0.343 has been simplified to $\frac{1}{3}$. Or,

$$\frac{h_x x}{k} = 0.332\ \mathrm{Pr}^{1/3}\sqrt{\frac{V_\infty x}{\nu}}$$

$$\mathrm{Nu}_x = 0.332\ \mathrm{Pr}^{1/3}\ \mathrm{Re}_x^{1/2} \tag{7.40}$$

The average heat transfer coefficient is

$$h = \frac{1}{L}\int_0^L h_x\,dx = 0.664k\ \mathrm{Pr}^{1/3}\sqrt{\frac{V_\infty}{\nu L}}$$

or

$$\mathrm{Nu} = 0.664\ \mathrm{Pr}^{1/3}\ \mathrm{Re}_L^{1/2} \tag{7.41a}$$

All these results in Eqs. (7.37) through (7.41) are valid for $\mathrm{Pr} > 0.5$. Sparrow and Gregg (5) solved Eq. (7.34) for uniform wall temperature with no unheated starting length with the aid of a computing machine for $0.006 \leq \mathrm{Pr} \leq 0.03$, the liquid metal range, and plotted the results. A good approximation to the resulting curve is as follows:

$$\frac{\mathrm{Nu}_x}{\sqrt{\mathrm{Re}_x\,\mathrm{Pr}}} = \frac{0.564}{1 + 0.90\sqrt{\mathrm{Pr}}} \tag{7.41b}$$

The right side of the equation does not vary much over this range of Pr, so in this region $\mathrm{Nu}_x \sim \sqrt{\mathrm{Pr}}$ instead of $\sqrt[3]{\mathrm{Pr}}$.

For uniform heat flux at the wall, the following results were obtained (10):

$$\mathrm{Pr} > 0.5 \quad \mathrm{Nu}_x = 0.458\,\mathrm{Pr}^{1/3}\sqrt{\mathrm{Re}_x} \tag{7.42a}$$

$$0.006 \leq \mathrm{Pr} \leq 0.03, \quad \frac{\mathrm{Nu}_x}{\sqrt{\mathrm{Re}_x\,\mathrm{Pr}}} = \frac{0.880}{1 + 1.317\sqrt{\mathrm{Pr}}} \tag{7.42b}$$

Comparison of Eq. (7.42a) with experimental data is given in Fig. 8.16.

(b) *Approximate Integral Method*

An alternate solution as in Sec. 3.4 involves writing the energy equation not for a small element, but for energy quantities integrated over the

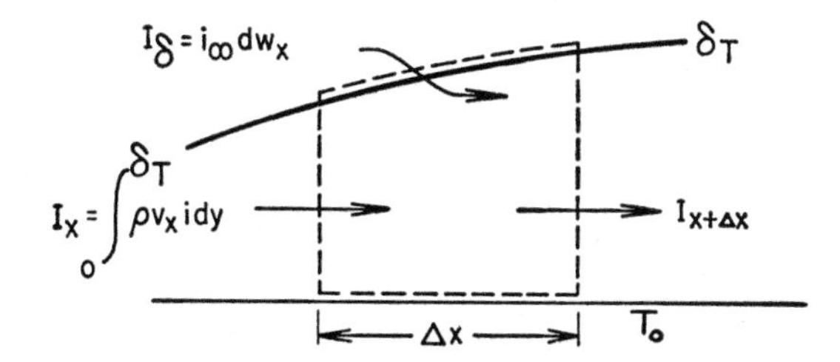

Fig. 7.9. Integrated enthalpy quantities in the boundary layer on flat plate.

entire boundary layer, Fig. 7.9. For the control surface shown dotted, the energy equation (neglecting frictional effects) is

$$I_{x+\Delta x} - I_x - I_\delta = dq$$

where I represents the enthalpy flux crossing a given plane. Then

$$\frac{d}{dx}\left[\int_0^{\delta_T} \rho i v_x\, dy\right] - i_\infty \frac{d}{dx}\left[\int_0^{\delta_T} \rho v_x\, dy\right] = -k\left(\frac{\partial T}{\partial y}\right)_{y=0} \tag{7.43}$$

or when $di = c_p\, dT$ and ρ and c_p are constant,

$$\frac{d}{dx}\left[\int_0^{\delta_T} (T_\infty - T) v_x\, dy\right] = \frac{k}{\rho c_p}\left(\frac{\partial T}{\partial y}\right)_{y=0} \tag{7.44}$$

Equation (3.37) is the equivalent relation for the momentum balance. As before, we arbitrarily select Eq. (3.38) as the velocity distribution and by similarity assume a parabolic temperature distribution

$$\frac{T - T_0}{T_\infty - T_0} = \theta = \frac{3}{2}\left(\frac{y}{\delta_T}\right) - \frac{1}{2}\left(\frac{y}{\delta_T}\right)^3 \tag{7.45}$$

which satisfies the boundary conditions $\theta = 0$ at $y = 0$ and $\theta = 1$ at $y = \delta_T$. Substituting these into Eq. (7.44) and performing the operations indicated, there results the following relation for the case in which $\zeta \equiv \delta_T/\delta \leq 1$, (Pr > 1):

$$\frac{d}{dx}\left[\theta_\infty V_\infty \delta\left(\frac{3}{20}\zeta^2 - \frac{3}{280}\zeta^4\right)\right] = \frac{3}{2}\alpha\frac{\theta_\infty}{\zeta\delta}$$

For $\zeta \leq 1$, we may neglect $\frac{3}{280}\zeta^4$ as small compared with $\frac{3}{20}\zeta^2$. Then, with Eq. (3.39), eliminate δ from the equation above to get

$$\zeta^3 + \frac{4}{3}x\frac{d(\zeta^3)}{dx} = \frac{13}{14\ \mathrm{Pr}} \tag{7.46}$$

The solution of this equation is of the form $\zeta^3 = a + cx^n$ where $a = 13/(14\ \mathrm{Pr})$ and $n = -\frac{3}{4}$. Now if δ begins growing at $x = 0$ and there exists an unheated starting length with the wall temperature maintained at T_0 for $x > x_0$, then $\delta_T = 0$ at $x = x_0$ is a boundary condition which determines the magnitude of c. The solution is

$$\frac{\delta_T}{\delta} = \zeta = \frac{1}{1.026\sqrt[3]{\mathrm{Pr}}}\sqrt[3]{1 - \left(\frac{x_0}{x}\right)^{3/4}} \tag{7.47a}$$

If the entire plate is heated $x_0 = 0$, or

$$\frac{\delta_T}{\delta} = \zeta = \frac{1}{1.026\sqrt[3]{\mathrm{Pr}}} \tag{7.47b}$$

From Eq. (7.45), the local heat transfer coefficient becomes

$$h_x = -\frac{k(\partial T/\partial y)_{y=0}}{T_0 - T_\infty} = \frac{3}{2}\frac{k}{\delta_T} \tag{7.48}$$

and with Eqs. (7.47) and (3.40), becomes

$$h_x = 0.323\frac{k}{x}\frac{\sqrt[3]{\mathrm{Pr}}\sqrt{V_\infty x/\nu}}{\sqrt[3]{1 - (x_0/x)^{3/4}}} \tag{7.49}$$

or in terms of dimensionless groups

$$\mathrm{Nu}_x = 0.323\frac{\sqrt[3]{\mathrm{Pr}}\sqrt{\mathrm{Re}_x}}{\sqrt[3]{1 - (x_0/x)^{3/4}}} \tag{7.50a}$$

and with the entire plate heated

$$\mathrm{Nu}_x = 0.323\sqrt[3]{\mathrm{Pr}}\sqrt{\mathrm{Re}_x} \tag{7.50b}$$

These solutions are valid when ζ does not greatly exceed unity; e.g., they are valid for all fluids except liquid metals. For liquid metals $\delta_T > \delta$ and

$\zeta = \frac{\delta_T}{\delta}$ if Pr = 1; $\zeta = 1$

the integration of Eq. (7.44) must be done in two parts—$y = 0$ to δ and $y = \delta$ to δ_T. Eckert (4) obtained the following result for the entire plate heated:

$$\mathrm{Nu}_x = \frac{\sqrt{\mathrm{Re}_x \cdot \mathrm{Pr}}}{1.55\sqrt{\mathrm{Pr}} + 3.09\sqrt{0.372 - 0.15\,\mathrm{Pr}}} \tag{7.51}$$

Sparrow and Gregg (5) obtained the exact solution for this case, approximated closely by Eq. (7.41b). Equation (7.51) agrees well, within ± 5 per cent for $0.005 < \mathrm{Pr} < 0.05$, with the exact solution.

For the case in which properties vary with temperature, Eckert (6) showed that Eq. (7.50b) gave correct answers for air when the property values were determined at a reference temperature halfway between wall and stream temperature. Experimental results verify this.

The average heat transfer coefficient, averaged over 0 to L, is

$$h = \frac{1}{L}\int_0^L h_x\,dx$$

and from Eq. (7.50b), the average Nusselt number is

$$\mathrm{Pr} > 0.5, \quad \mathrm{Nu} = 0.646\sqrt[3]{\mathrm{Pr}}\,\sqrt{\mathrm{Re}_L} \tag{7.52}$$

Solutions of Eq. (7.44) for velocity and temperature profiles other than those of Eq. (3.38) and (7.45) yield the results shown in Table 7.2.

TABLE 7.2 RESULTS OF INTEGRAL METHOD SOLUTION FOR LAMINAR BOUNDARY LAYER ON A FLAT PLATE ($\delta_T \leq \delta$)

Velocity and temperature distribution:

$\frac{V_x}{V_\infty}$ and $\frac{\theta}{\theta_\infty}$	$\frac{\delta}{x}\sqrt{\mathrm{Re}_x}$	$C_{fx}\sqrt{\mathrm{Re}_x}$	$\frac{\mathrm{Nu}_x}{\sqrt{\mathrm{Re}_x}\,\sqrt[3]{\mathrm{Pr}}}$
Exact solution:	5.0	0.664	0.332
$\frac{y}{\delta}$	3.46	0.578	0.289
$\frac{3}{2}\frac{y}{\delta} - \frac{1}{2}\left(\frac{y}{\delta}\right)^3$	4.64	0.646	0.323
$\sin\frac{\pi}{2}\frac{y}{\delta}$	4.80	0.654	—

(where δ = either δ or δ_T)

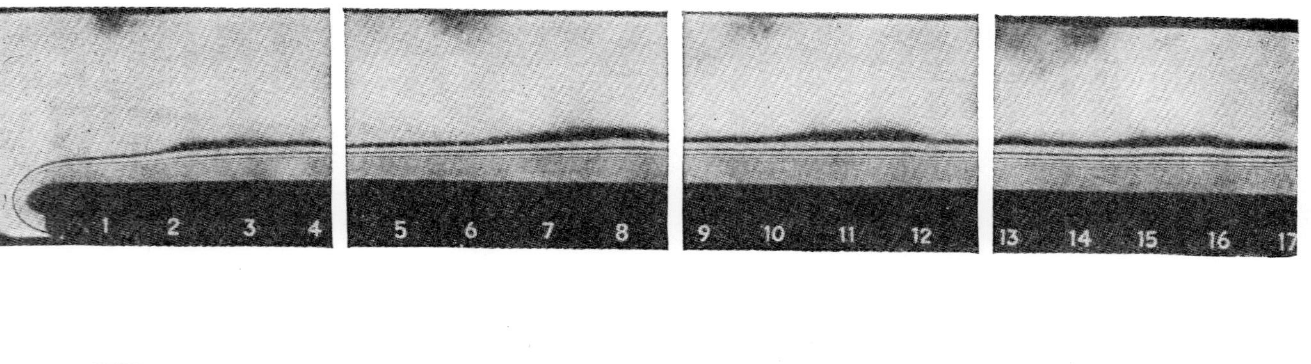

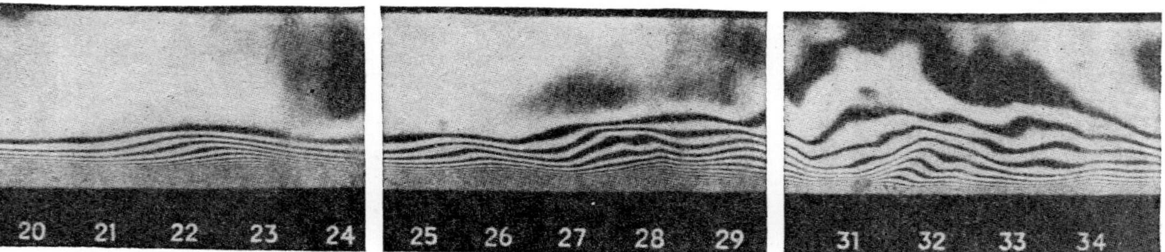

Fig. 7.10. Interferometer photographs showing laminar and turbulent free-convection flow on a vertical plate. Numbers give distance in inches from the lower edge of the plate [Eckert and Soehngen (24)].

Example 7.3: Calculate the average heat flux to a wide flat plate 3″ long in an air stream at 200 F, 14.7 psia flowing at 200 ft/sec parallel to the plate when the plate surface is uniformly at 100 F. For the air

$$\mu = 0.049 \text{ lb}_m/\text{hr ft} \qquad k = 0.017 \text{ Btu/hr ft F}$$

$$c_p = 0.24 \text{ Btu/lb F} \quad \text{and} \quad \rho = p/RT = 0.060 \text{ lb/ft}^3$$

$$\text{Re}_L = \frac{VL\rho}{\mu} = \frac{(200)(3600)(3/12)(0.06)}{0.049} = 220{,}000, \text{ Laminar}$$

$$\text{Pr} = \frac{c_p\mu}{k} = \frac{(0.24)(0.049)}{0.017} = 0.69$$

$$\text{Nu} = 0.664\sqrt{\text{Re}_L}\,\sqrt[3]{\text{Pr}} = 0.664\sqrt{220{,}000}\,\sqrt[3]{0.69} = 274$$

$$h = 274\,k/L = 274\,\frac{(0.017)}{(3/12)} = 18.6 \text{ Btu/hr ft}^2 \text{ F}$$

$$q/A = h\,(\Delta T) = 18.6\,(100) = 1860 \text{ Btu/hr ft}^2$$

7.6 Free-Convection Laminar Boundary-Layer on a Vertical Plate

Free convection currents in the neighborhood of a heated or cooled object in a fluid of essentially infinite extent are familiar phenomena.

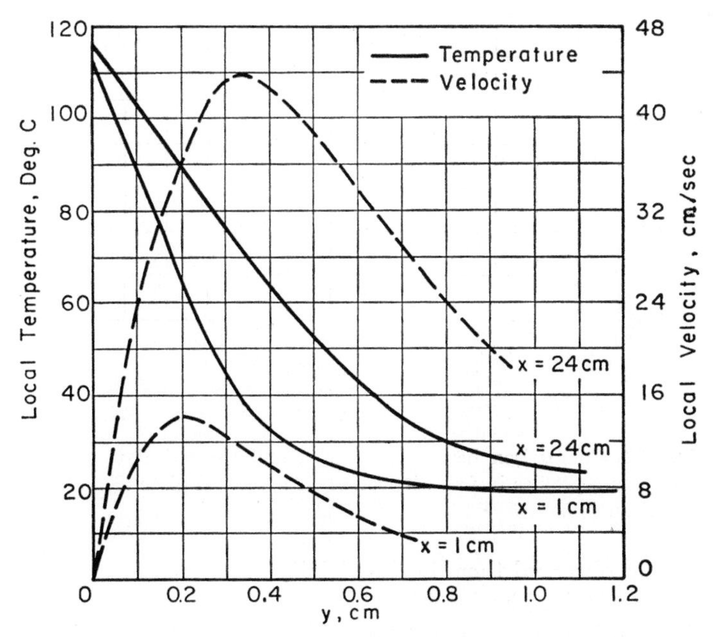

Fig. 7.11. Velocity and temperature distributions at two locations (x = 1 cm and x = 24 cm) near the surface of a heated vertical plate exposed to still air [Schmidt (25)].

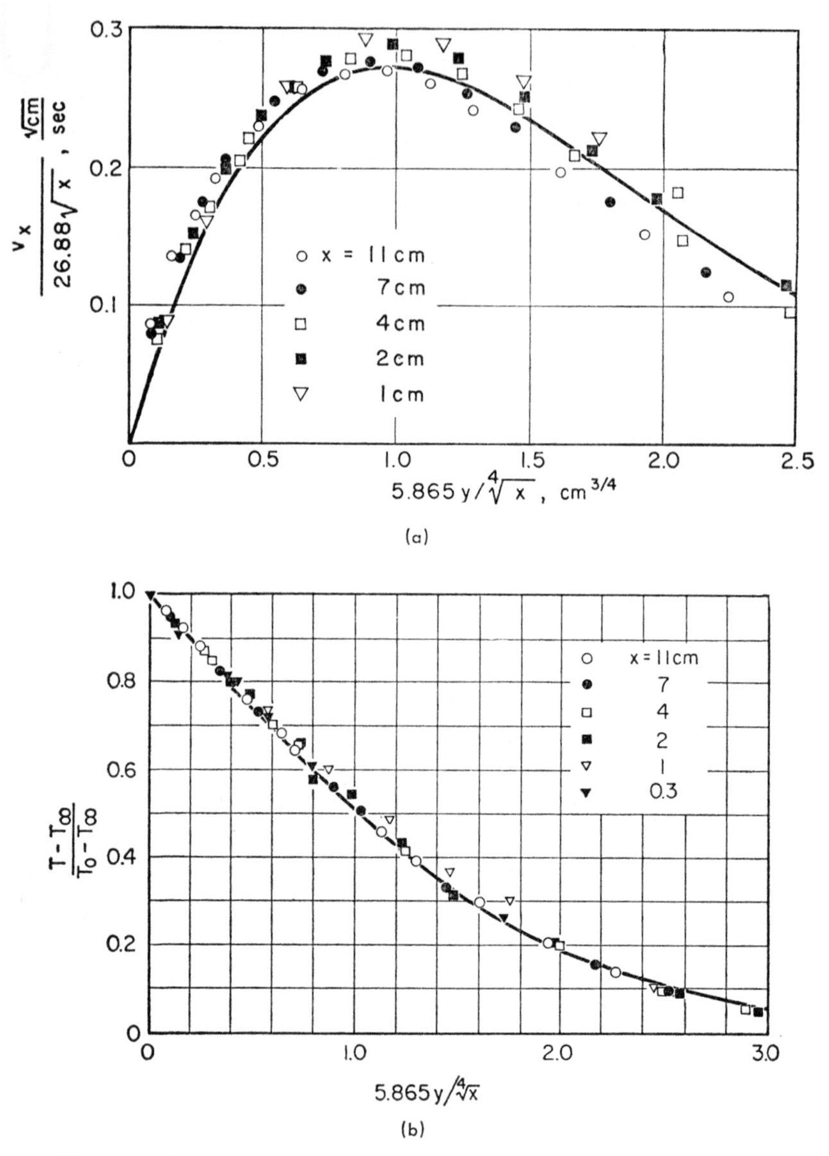

Fig. 7.12. Comparison between measured and theoretical velocity and temperature profiles in free convection on a vertical plate [Schmidt and Beckmann (8)].

Everyone has observed them in the vicinity of a "radiator" of the household heating system. The density of the fluid near the hot or cold surface is changed, resulting in a buoyancy effect causing circulation of the fluid. Here the velocity and temperature distributions are interrelated; the temperature distribution, in effect, produces the velocity distribution.

Figure 7.10 is an interferometer picture of natural or free convection over a heated vertical plate in air, showing the development of a boundary layer along the plate. Near the leading edge the flow is laminar; farther up the plate it changes to turbulent within the boundary layer. Figure 7.11 shows some measurements of the velocity and temperature distributions at two positions along the plate in air. Figure 7.12 shows how these data can be correlated at various points on single curves by the variable $y/\sqrt[4]{x}$, suggested by Pohlhausen to Schmidt and Beckmann (8).

The free convection flows thus far described result from gravitational body forces. They may also result from centrifugal or Coriolis forces in rotating systems such as hollow gas-turbine blades. Here we confine ourselves to gravitation-induced free convection.

(a) *Exact Solution*

The vertical flat plate of Fig. 7.13 shows the significant quantities. We consider the case of a steady state with uniform wall temperature.

The energy equation applied to the control volume $\Delta x \, \Delta y$ in this case is identical with Eq. (7.32).

$$v_x \frac{\partial T}{\partial x} + v_y \frac{\partial T}{\partial y} = \alpha \frac{\partial^2 T}{\partial y^2} \tag{7.32}$$

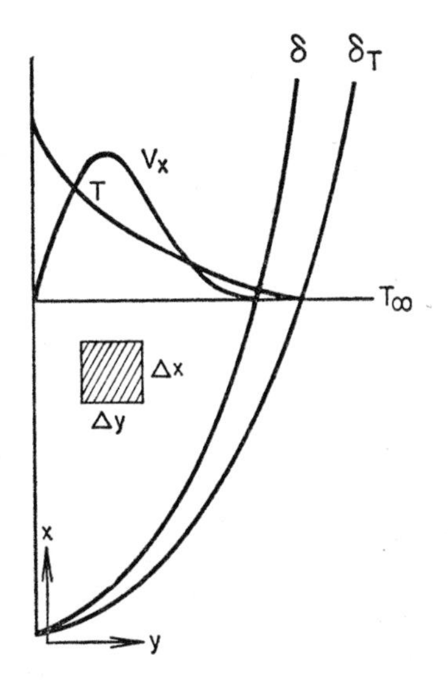

Fig. 7.13. Hydrodynamic and thermal boundary layers for natural convection on a vertical plate.

In free convection the frictional effects in the energy equation are almost always negligible.

The momentum equation in the x-direction is identical with Eq. (7.25). In the y-direction, the momentum equation reduces to $(\partial p/\partial y) = 0$, or at any x the pressure is uniform.

When $T_0 = T_\infty$, there is no convection and equilibrium requires that $dp/dx = -\rho_\infty g_x$. When $T_0 > T_\infty$, $\rho < \rho_\infty$ for most fluids, and fluid motion exists due to the presence of the bouyancy force. In Eq. (7.27), we replace ρ_m by ρ_∞ for this case and Eq. (7.25) becomes

$$v_x \frac{\partial v_x}{\partial x} + v_y \frac{\partial v_x}{\partial y} = \nu \frac{\partial^2 v_x}{\partial y^2} + g_x \beta (T - T_\infty) \tag{7.53}$$

which is identical with the forced-convection boundary-layer equation except for the buoyancy term. In Eq. (7.53), it has been assumed that the density is essentially constant, except in the buoyancy term.

The general solution of these equations proceeds as before for forced convection over a plate, Secs. 3.3 and 7.5, first introducing the stream function ψ from Eq. (3.20) into Eqs. (7.32) and (7.53). For the free-convection problem, Schmidt and Beckmann (8) showed for the uniform-wall-temperature case that η and ψ should have the following form:

$$\eta \equiv C\frac{y}{\sqrt[4]{x}}$$

$$F(\eta) \equiv \frac{\psi}{4\nu C x^{3/4}} \tag{7.54}$$

$$C \equiv \left[\frac{g_x\beta(T_0 - T_\infty)}{4\nu^2}\right]^{1/4}$$

Then, Eq. (7.32) and (7.53) become

$$\frac{d^3F}{d\eta^3} + 3F\frac{d^2F}{d\eta^2} - 2\frac{dF}{d\eta} + \theta = 0 \tag{7.55}$$

$$\frac{d^2\theta}{d\eta^2} + 3\text{ Pr }F\frac{d\theta}{d\eta} = 0 \tag{7.56}$$

where $\theta \equiv (T - T_\infty)/(T_0 - T_\infty)$.

The boundary conditions for the uniform-wall-temperature case are as follows:

$$y = 0,\quad v_x = v_y = 0,\quad T = T_0,\quad \text{or}\quad \eta = 0,\quad \frac{dF}{d\eta} = 0,\ \theta = 1$$

$$(7.57)$$

$$y = \infty,\ v_x = v_y = 0,\quad T = T_\infty,\quad \text{or}\quad \eta = \infty,\quad \frac{dF}{d\eta} = 0,\ \theta = 0$$

These equations were first solved in this form by Schmidt, Beckmann, and Pohlhausen during the period 1930–32 for only one Prandtl number, 0.733. The advent of the computing machine permitted Ostrach (7) and Sparrow (9) to extend the solution over a wide range of Pr, $0.00835 \leq \text{Pr} \leq 1000$. The results of Ostrach for velocity and temperature distribution are shown in Fig. 7.14. Inspection of these curves shows that for $\text{Pr} \leq 1$, $\delta_T \cong \delta$, but for $\text{Pr} > 1$, $\delta_T < \delta$. The heat transfer rate at the wall can be expressed as follows

$$\left(\frac{q}{A}\right)_0 = -k\left(\frac{\partial T}{\partial y}\right)_{y=0} = -k(T_0 - T_\infty)\frac{C}{\sqrt[4]{x}}\left(\frac{d\theta(0)}{d\eta}\right) \tag{7.58a}$$

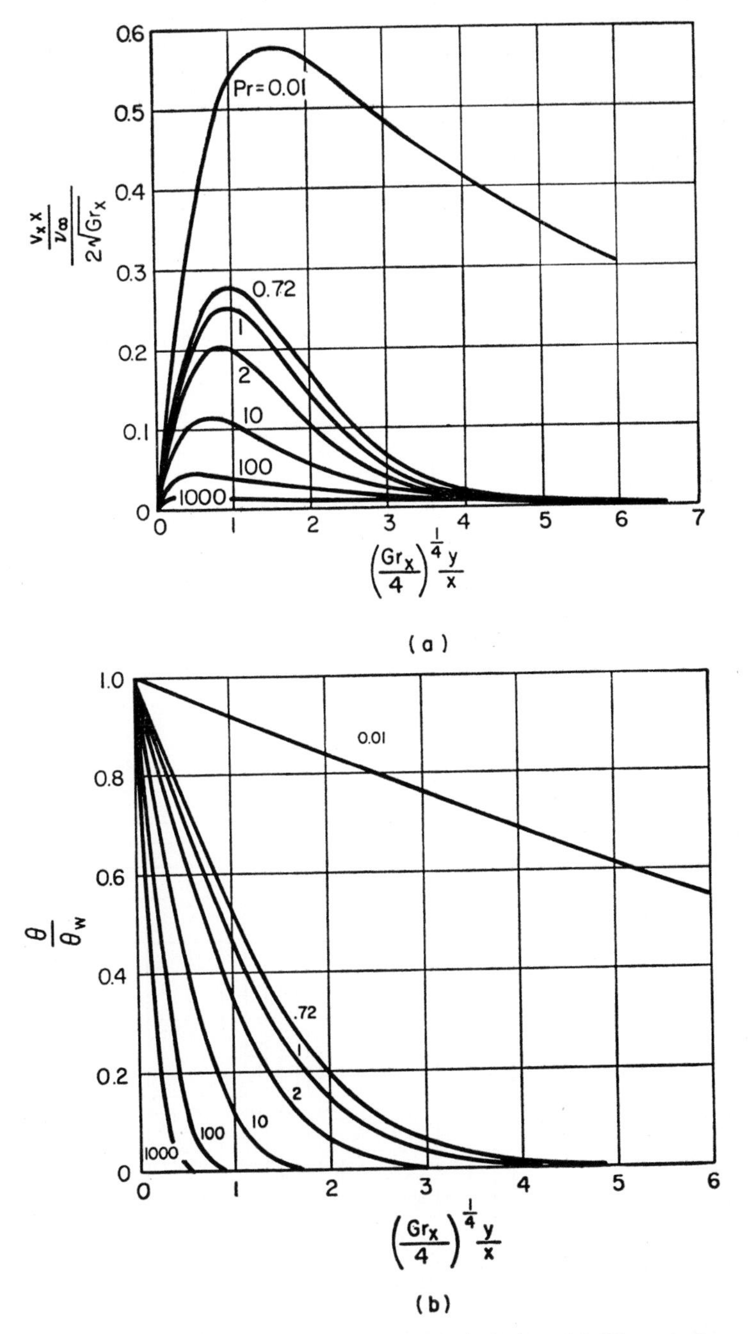

Fig. 7.14. Laminar free convection on a vertical plate (a) Dimensionless velocity profiles, and (b) Dimensionless temperature profiles [Ostrach (7)].

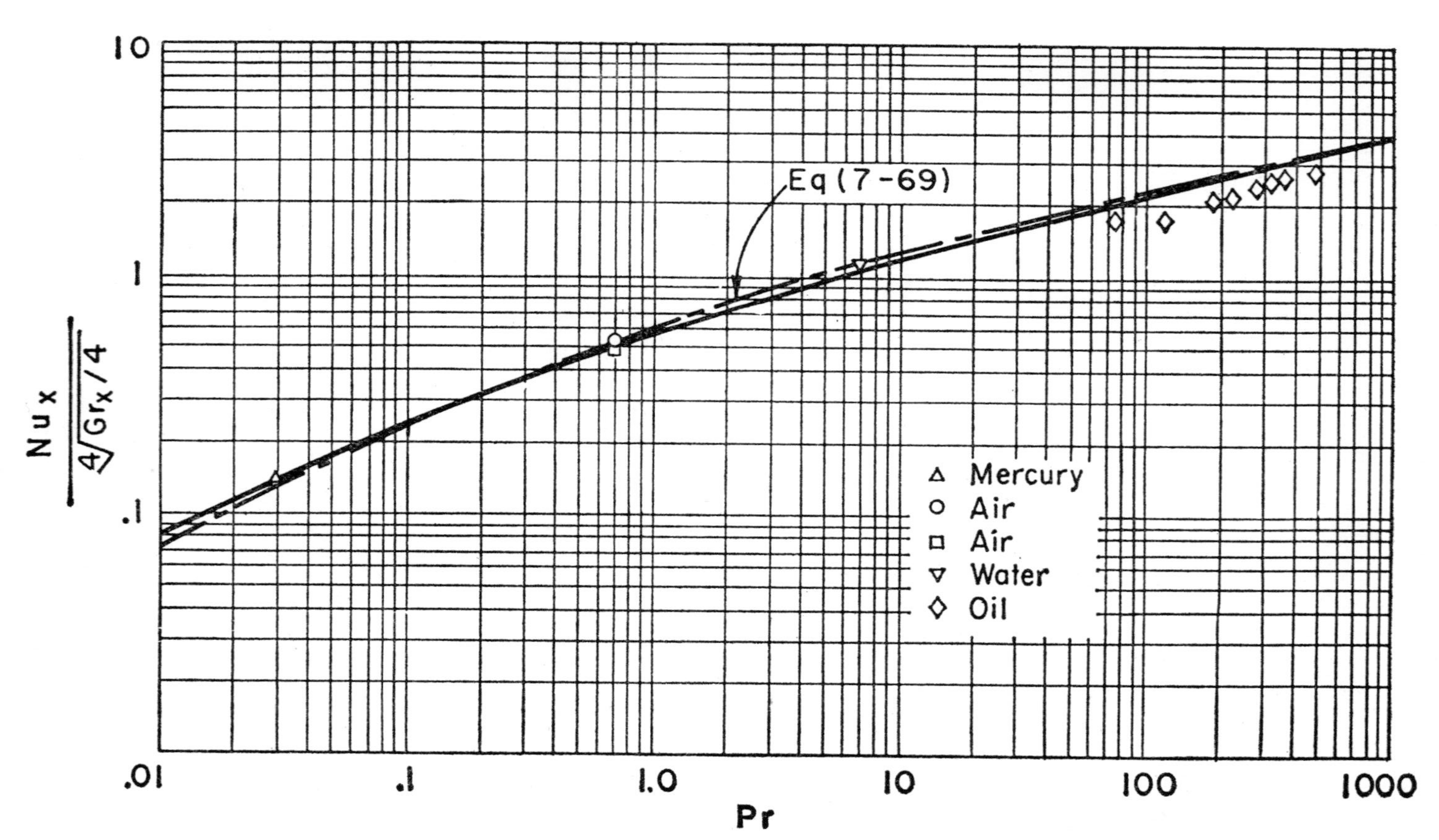

Fig. 7.15. Local Nusselt number for free convection on vertical plate with uniform wall temperature (7). Solid curve represents exact solution.

or

$$\mathrm{Nu}_x = \frac{h_x x}{k} = -\left(\frac{\mathrm{Gr}_x}{4}\right)^{1/4} \frac{d\theta(0)}{d\eta} \tag{7.58b}$$

where the Grashoff number is defined as $\mathrm{Gr}_x \equiv \dfrac{g_x \beta (T_0 - T_\infty) x^3}{\nu^2}$

The results for Nu_x are shown graphically as the solid line in Fig. 7.15. An empirical equation approximating the calculated results is as follows:

$$\frac{\mathrm{Nu}_x}{\sqrt[4]{\mathrm{Gr}_x/4}} = \frac{0.676\ \mathrm{Pr}^{1/2}}{(0.861 + \mathrm{Pr})^{1/4}} \tag{7.59a}$$

The curve representing this equation follows the exact solution very closely. Since h_x varies as $x^{-1/4}$, then the average h is $\frac{4}{3}h_x$, or the average Nusselt number over a plate of height L is

$$\frac{\mathrm{Nu}}{\sqrt[4]{\mathrm{Gr}/4}} = \frac{0.902\ \mathrm{Pr}^{1/2}}{(0.861 + \mathrm{Pr})^{1/4}} \tag{7.59b}$$

For air with $\mathrm{Pr} = 0.733$, the right side of Eq. (7.59b) is 0.685. Equation (7.59) agrees well with experimental results.

We may also calculate the shear stress on the plate. From Eqs. (3.20) and (7.54), it follows that

$$\frac{v_x x/\nu_\infty}{2\sqrt{\mathrm{Gr}_x}} = \frac{\partial F(\eta)}{\partial \eta}$$

Then, since $(\tau_0)_x = \mu_0(\partial v_x/\partial y)_{y=0}$, where here $(\tau_0)_x$ is the shear stress acting *on* the plate, we obtain

$$\frac{(\tau_0)_x}{(4\ \mathrm{Gr}_x^3)^{1/4}(\nu_\infty \mu_0/x^2)} = \frac{\partial^2 F(0)}{\partial \eta^2} \tag{7.60}$$

The results of this calculation are shown in Fig. 7.16. Since $(\tau_0)_x \sim x^{1/4}$, the average shear stress τ over $x = 0$ to L is $\frac{4}{5}$ of the local $(\tau_0)_x$ at $x = L$.

The preceding results have applied to a laminar boundary layer. Experimentally, it has been found for gases and for liquid water that transition to turbulence occurs in the neighborhood of $\mathrm{Gr}_x \cdot \mathrm{Pr} \approx 5 \times 10^8$. Actually, the data show a range from 10^8 to as high as 10^{10}.

The preceding solutions also are limited to cases with relatively small temperature difference in the few hundreds of degrees F for air, since the properties such as μ and k were assumed constant. The solution also fails at $\mathrm{Gr}_x\ \mathrm{Pr} < 10^4$, where the boundary-layer assumptions are no longer valid.

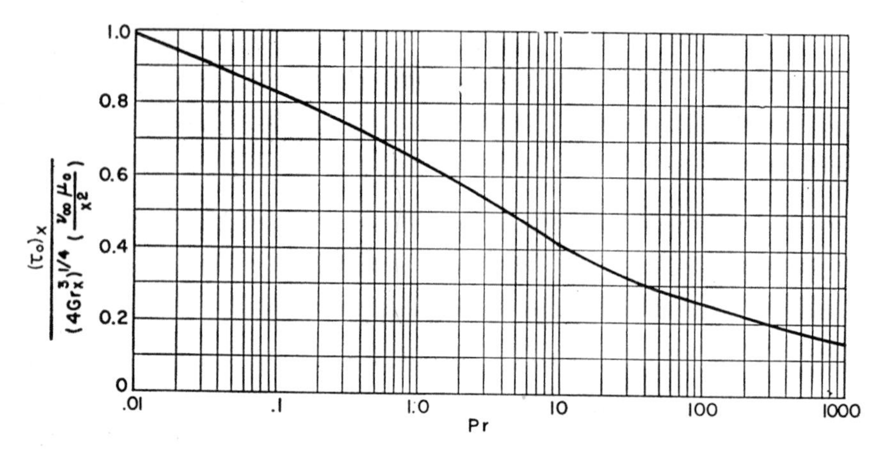

Fig. 7-16. Local shear stress for free convection on vertical plate with uniform wall temperature (7).

(b) *Approximate Integral Method*

Proceeding as in the flat-plate problem, we define a control surface 1234 of unity depth, Fig. 7.17, and write the integral momentum and energy balances for the surface of unit depth in the z-direction.

Momentum balance:

$$\frac{d}{dx}\left[\int_0^\delta \rho v_x^2\, dy\right]\Delta x = (\tau_{yx})_0\,\Delta x - \left(\frac{dp}{dx}\right)(\delta\ \Delta x) - \Delta x \int_0^\delta \rho g_x\, dy \qquad (7.61)$$

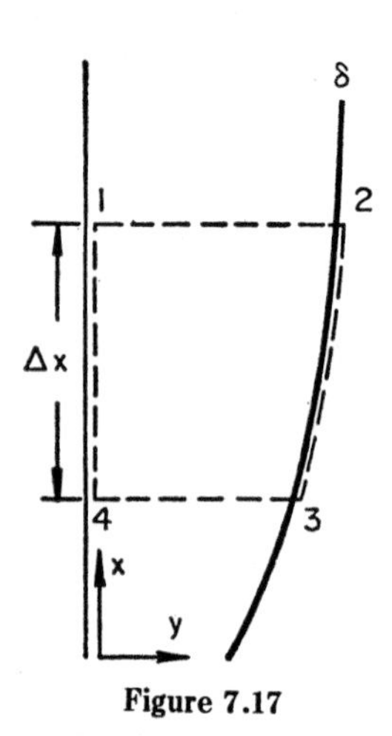

Figure 7.17

No momentum crosses boundary 23 because $V_\infty = 0$. The last term on the right is the gravity force acting on the volume Δx by δ, the mean density of which is

$$\frac{1}{\delta}\int_0^\delta \rho\, dy$$

Noting as before;

$$\frac{dp}{dx} = -\rho_\infty g_x$$

$$(\tau_{yx})_0 = -\mu\left(\frac{dv_x}{dy}\right)_0$$

$$\rho - \rho_\infty = -\rho\beta(T - T_\infty)$$

and assuming temperature differences are small so that $\rho/\rho_\infty \simeq 1$, Eq.

(7.61) reduces to:

$$\frac{d}{dx}\int_0^\delta v_x^2\,dy = -\nu\left(\frac{dv_x}{dy}\right)_0 + g_x\beta\int_0^\delta (T - T_\infty)\,dy \tag{7.62}$$

The energy balance for this case is identical with that of the forced-convection, flat-plate problem:

$$\frac{d}{dx}\left[\int_0^\delta (T_\infty - T)v_x\,dy\right] = \alpha\left(\frac{dT}{dy}\right)_0 \tag{7.44}$$

Equations (7.62) and (7.44) can also be obtained by integrating Eqs. (7.32) and (7.53) with respect to y between the limits 0 and δ.

Before we can solve Eqs. (7.62) and (7.44), we must assume some suitable functions for the velocity and temperature distributions within the boundary layer that satisfy the known boundary conditions of the problem.

Let us assume

$$\frac{T - T_\infty}{T_0 - T_\infty} = \left(1 - \frac{y}{\delta_T}\right)^2 \tag{7.63}$$

$$v_x = \Gamma\frac{y}{\delta}\left(1 - \frac{y}{\delta}\right)^2 \tag{7.64}$$

where Γ is a function of x and has the dimensions of velocity. The form of these equations closely approximates the experimental and analytical results of Figs. 7.11, 7.12, and 7.14. In the following solution we take $\delta_T = \delta$.

Substitute Eqs. (7.63) and (7.64) into Eqs. (7.62) and (7.44), with the following results:

$$\begin{aligned}\frac{1}{105}\frac{d}{dx}(\Gamma^2\delta) &= \frac{1}{3}g_x\beta\,\Delta T\delta - \frac{\nu\Gamma}{\delta}\\ \frac{1}{30}\,\Delta T\frac{d}{dx}(\Gamma\delta) &= \frac{2\alpha\,\Delta T}{\delta}\end{aligned} \tag{7.65}$$

where $\Delta T \equiv T_0 - T_\infty$.

We assume solutions of the following form

$$\begin{aligned}\Gamma &= Ax^m\\ \delta &= Bx^n\end{aligned} \tag{7.66}$$

Substitute these into the above equations and perform the indicated differentiations. The magnitudes of A, B, m, and n are then determined by equating coefficients and exponents of the resulting expressions. The results are:

$$m = \frac{1}{2} \qquad n = \frac{1}{4}$$

$$A = 5.17\nu\left(\frac{\nu}{\alpha} + \frac{20}{21}\right)^{-1/2}\left(\frac{g_x\beta\,\Delta T}{\nu^2}\right)^{1/2} \tag{7.67}$$

$$B = 3.93\left(\frac{\alpha}{\nu}\right)^{1/2}\left(\frac{\nu}{\alpha} + \frac{20}{21}\right)^{1/4}\left(\frac{g_x\beta\,\Delta T}{\nu^2}\right)^{-1/4}$$

Then, the expression for δ becomes

$$\frac{\delta}{x} = 3.93\ \mathrm{Pr}^{-1/2}(0.952 + \mathrm{Pr})^{1/4}\ \mathrm{Gr}_x^{-1/4} \tag{7.68}$$

The heat transfer rate where T is expressed by Eq. (7.63) is calculated as follows:

$$\frac{q}{A} = -k\left(\frac{\partial T}{\partial y}\right)_{y=0} = 2k\frac{\Delta T}{\delta} = h_x\,\Delta T$$

Then with Eq. (7.68), the Nusselt number becomes

$$\mathrm{Nu}_x = \frac{h_x x}{k} = \frac{2x}{\delta} = \frac{0.508\ \mathrm{Pr}^{1/2}}{(0.952 + \mathrm{Pr})^{1/4}}\mathrm{Gr}_x^{1/4} \tag{7.69}$$

which agrees well with Eq. (7.59a) as shown in Fig. 7.15. The average h over $x = 0$ to L is $\frac{4}{3}$ times the h_x at $x = L$.

The case of laminar free convection from a vertical flat plate with uniform surface heat flux was solved by Sparrow and Gregg (21).

Example 7.4: A 1 ft-high flat plate of glass at 200 F is removed from an annealing furnace and hung vertically in air at 60 F, 14.7 psia. Calculate the initial rate of heat transfer to the air. For air,

$$\mu = 0.047\ \mathrm{lb}_m/\mathrm{hr\ ft} \qquad k = 0.016\ \mathrm{Btu/hr\ ft\ F}$$

$$c_p = 0.24\ \mathrm{Btu/lb\ F} \qquad \rho = \frac{p}{RT} = 0.076\ \mathrm{lb/ft^3}$$

$$\beta = \frac{1}{T} = \frac{1}{590} = 0.00169\ (\mathrm{F\ abs})^{-1}$$

$$\mathrm{Gr}_L = \frac{g\beta(T_0 - T_\infty)L^3}{(\mu/\rho)^2} = \frac{(32.2)(3600)^2(0.00169)(140)(1)^3}{(0.047/0.076)^2} = 2.6 \times 10^8$$

$$\mathrm{Pr} = \frac{c_p\mu}{k} = \frac{(0.24)(0.047)}{0.016} = 0.705$$

From Eq. (7.59b)

$$\mathrm{Nu} = \frac{0.902(0.705)^{1/2}}{(0.861 + 0.705)^{1/4}} \sqrt[4]{\frac{2.6 \times 10^8}{4}} = 60.8$$

$$h = 60.8 \frac{k}{L} = \frac{(60.8)(0.016)}{1} = 0.97$$

$$\frac{q}{A} = h\,(T_0 - T_\infty) = 0.97\,(200 - 60) = 136 \text{ Btu/hr ft}^2$$

In addition to this heat flux there will be heat transfer by radiation as discussed in Chap. 13.

7.7 Entrance-Region Solutions for Laminar Flow in Channels

In Secs. 7.1 through 7.3, we obtained solutions for laminar flow in channels when both the velocity and the temperature profiles are fully developed. Here we consider the entrance-region problem when both profiles may be developing. In the entrance region of a heated section, the temperature profile develops as shown in Fig. 7.1. Corresponding idealized hydrodynamical conditions may be any one of the types shown in Fig. 7.18.

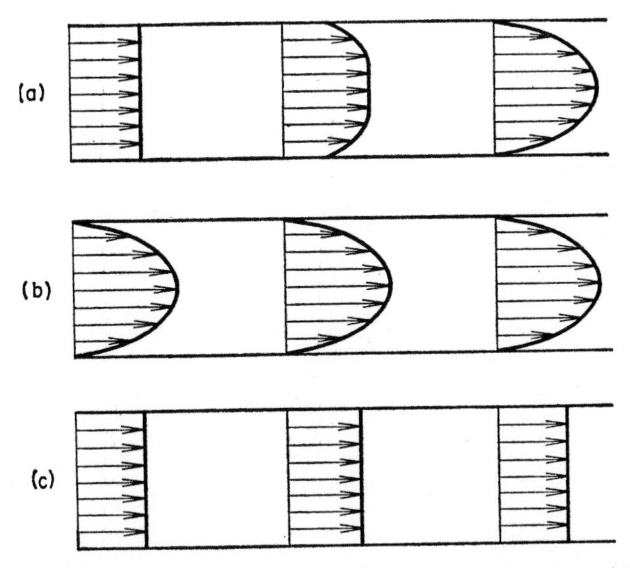

Fig. 7.18. Different hydrodynamic conditions for laminar flow in the entrance region of heated section of a channel (a) Developing velocity profile, (b) Fully developed velocity profile, and (c) Uniform velocity profile.

Case (a) shows the velocity profile developing from an initially uniform velocity; case (b) shows the velocity profile already fully developed at the entrance to the heated section, as will be the case if a long length of unheated section precedes the heated section of the channel; and case (c) shows slug flow with uniform velocity across the channel. The last case approximates the entrance region with liquid metals where the thermal boundary layer develops in a far shorter distance than the velocity boundary layer. In this region, the velocity distribution is nearly uniform.

In each of the cases above, a variety of thermal boundary conditions may be specified. Conditions most frequently studied are the following:

(a) uniform wall temperature, T_0 = constant

(b) uniform heat flux at the wall, $(q/A)_0$ = constant

The applicable differential energy equation representing energy conservation in the developing thermal region in round tubes is Eq. (7.10). The following solutions are for cases in which the viscous shear work term is negligible, low Mach number flows, and the terms involving $v_r\,\partial T/\partial r$ and $\partial^2 T/\partial x^2$ are negligible, which are valid assumptions beyond a short distance from the beginning of the heated section. Then, Eq. (7.10) reduces to

$$v_x \frac{\partial T}{\partial x} = \alpha \frac{1}{r} \frac{\partial}{\partial r}\left(r \frac{\partial T}{\partial r}\right) \tag{7.70}$$

(a) *Slug Flow*

Equation (7.70) has been solved by Graetz (11) for the case in which v_x is constant, as represented in Fig. 7.18(c). The differential equation is then identical in form with the transient heat conduction equation in a cylinder and may be solved by separation of variables, similar to the solution outlined in Appendix A. The resulting solution is shown plotted in Fig. 7.19 as the upper curve.

(b) *Parabolic Velocity Distribution*

The case of the unheated starting length, Fig. 7.18(b), is solved by substituting the parabolic velocity distribution, Eq. (3.17), for v_x in Eq. 7.70. The result for the case of uniform wall temperature T_0 is shown as the bottom curve of Fig. 7.19 and Fig. 7.20.

(c) *Developing Velocity and Temperature Distribution*

The velocity distribution in the entrance region was determined analytically by Langhaar (14) with the results shown in Fig. 3.8. With these velocity profiles, Eq. (7.70) was solved by the finite difference method by Kays (13) for the case of Pr = 0.7 and by a computer by Goldberg (12) for

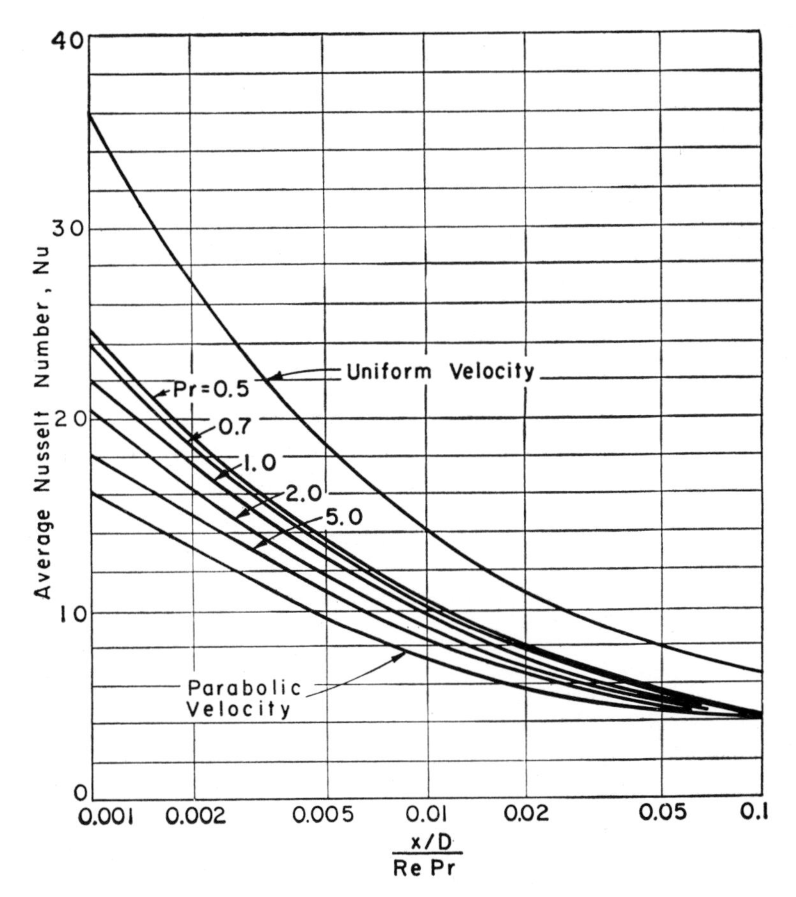

Fig. 7.19. Variation of mean Nusselt number for laminar flow in entrance region of tubes, uniform wall temperature [Goldberg (12)].

other magnitudes of Pr in the range 0.5 to 5.0. The results are shown in Fig. 7.19. Figure 7.20 shows a comparison for the cases of parabolic and developing velocity distributions and for the wall boundary conditions of uniform heat flux and uniform wall temperature.

Other solutions of the entrance effect are given by Nusselt (15), Sellars, Tribus, and Klein (16), Millsaps and Pohlausen (17), and Norris and Streid (18).

Most existing experimental data are not suitable for comparison with these analytical results for the following reasons. Much data is taken using viscous fluids such as oils whose properties change markedly with temperature; secondly, the presence of superimposed natural convection effects produces departures from these analytical results. Hausen (19) presents a comparison with some experimental data. Data of Nusselt (15) for air lie

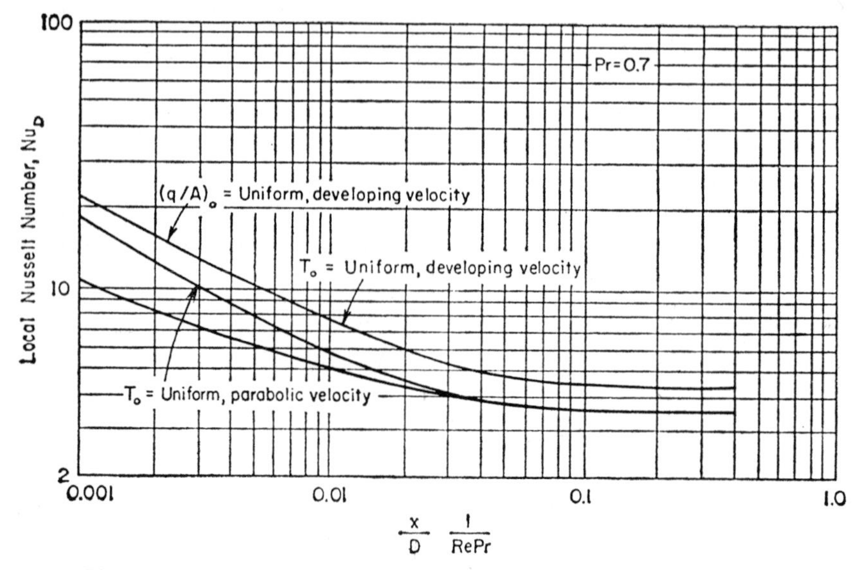

Fig. 7.20. Variation of local Nusselt number for laminar flow in entrance region of tubes [Kays (13)].

10 to 15 percent above these analytical results. To account for the effect of viscosity variation with temperature, the right side of Eq. (7.71) may be multiplied by $(\mu_m/\mu_0)^{0.14}$ where μ_m is viscosity at the bulk fluid temperature and μ_0 at the wall temperature.

For numerical work, Kays (13) presents the following empirical equation—suggested by Hausen (19)—for the cases listed in the following table.

$$\mathrm{Nu} = \mathrm{Nu}_\infty + \frac{K_1[(D/x)\ \mathrm{Re}_D\ \mathrm{Pr}]}{1 + K_2[(D/x)\ \mathrm{Re}_D\ \mathrm{Pr}]^n} \tag{7.71}$$

Wall condition	Velocity	Pr	Nu	Nu_∞	K_1	K_2	n
Unif. T_0	Parab. V	any	avg.	3.66	0.0668	0.04	$\frac{2}{3}$
Unif. T_0	Developing V	0.7	avg.	3.66	0.104	0.016	0.8
Unif. $(q/A)_0$	Parab. V	any	local	4.36	0.023	0.0012	1.0
Unif. $(q/A)_0$	Developing V	0.7	local	4.36	0.036	0.0011	1.0

A similar graph for laminar flow between parallel plates may be drawn, based on Sparrow's work (20). He solved the problem of developing velocity and temperature profiles in the entrance region between parallel plates at uniform temperature, using the Kármán-Pohlhausen integral method. The solution is summarized in Fig. 7.21.

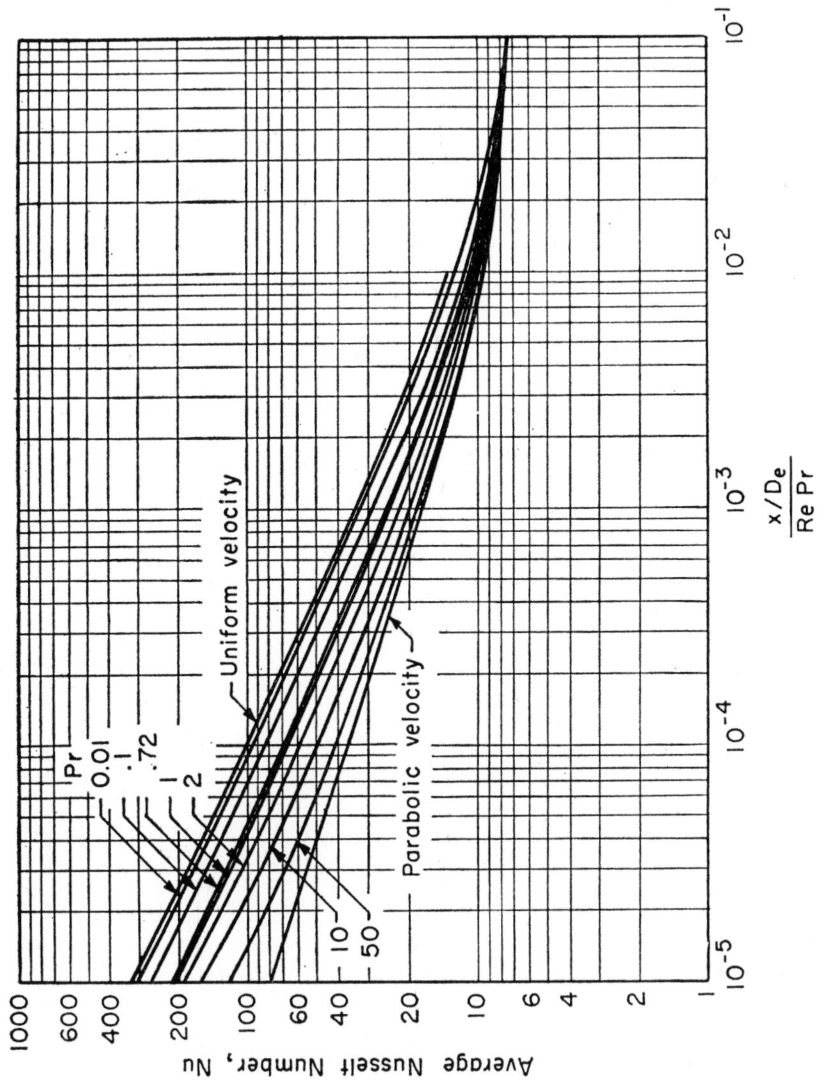

Fig. 7.21. Variation of mean Nusselt number for laminar flow in entrance region between parallel plates, uniform wall temperature [Sparrow (20)].

Example 7.5: Calculate the heat transferred to air at 14.7, 100 F flowing at 4 ft/sec inside a round tube (1 in. dia, 5 in. long) with a uniform tube wall temperature of 200 F. Use the properties of air given in Example 7.3.

$$\mathrm{Re}_D = \frac{VD\rho}{\mu} = \frac{(4)(3600)(1/12)(0.06)}{0.049} = 1470, \text{ Laminar}$$

$$\mathrm{Pr} = 0.69$$

$$\frac{x/D}{\mathrm{RePr}} = \frac{(5/1)}{(1470)(0.69)} = 0.0049$$

From Fig. 7.19, $\mathrm{Nu}_D = 13.2$

$$h = 13.2\frac{k}{D} = 13.2(0.017)/(1/12) = 2.70 \text{ Btu/hr ft}^2 \text{ F}$$

as a first estimate

$$\frac{q}{A} < h(\Delta T) = (2.70)(100) = 270 \text{ Btu/hr ft}^2$$

$$q < \pi DL\,(q/A) = \pi\frac{1}{12}\frac{5}{12}270 = 29.5 \text{ Btu/hr}$$

$$w_{\text{air}} = \frac{\pi}{4}D^2\rho V = \frac{\pi}{4}\left(\frac{1}{12}\right)^2(0.06)(4)(3600) = 4.72 \text{ lb/hr}$$

$$\text{Air temperature rise} < \frac{q}{w_{\text{air}}c_p} = \frac{29.5}{(4.72)(0.24)} = 26 \text{ F}$$

Recalculate (q/A) using new approximation for ΔT as $100 - \frac{1}{2}(26)$ or 87 F.

7.8 General Differential Energy Equation

Consider an elementary system $\Delta x\ \Delta y\ \Delta z$ of a pure substance in a moving medium. For an observer moving with the medium, the First Law of thermodynamics may be expressed as:

$$\begin{Bmatrix}\text{rate of increase of}\\ \text{internal energy of system}\end{Bmatrix} = \begin{Bmatrix}\text{rate of heat}\\ \text{transfer into system}\end{Bmatrix} - \begin{Bmatrix}\text{rate of work done by surface}\\ \text{forces (pressure and shear}\\ \text{forces) on the surroundings}\end{Bmatrix}$$

Rate of change of internal energy of system:

$$\rho\ \Delta x\ \Delta y\ \Delta z\ \frac{De}{Dt}$$

$$\frac{D}{Dt} = \frac{\partial}{\partial t} + V_x\frac{\partial}{\partial x} + V_y\frac{\partial}{\partial y} + V_z\frac{\partial}{\partial z}$$

Heat may be transferred across the system boundaries by conduction. Net rate of heat flow by conduction in x-direction:

$$-\frac{\partial}{\partial x}\left(\frac{q}{A}\right)_x \Delta x(\Delta y\ \Delta z)$$

Similar expressions may be written for the y- and the z-directions. The energy associated with a heat source W_i, defined in Sec. 6.1, is as follows:

$$W_i\ \Delta x\ \Delta y\ \Delta z$$

Net rate of work done by normal and tangential forces in the x-direction:

$$\sigma_x \frac{\partial v_x}{\partial x} \Delta x(\Delta y\ \Delta z) + \tau_{yx} \frac{\partial v_x}{\partial y} \Delta y(\Delta z\ \Delta x) + \tau_{zx} \frac{\partial v_x}{\partial z} \Delta z(\Delta x\ \Delta y)$$

Similar expressions may be written for work crossing boundaries in the y- and the z-directions.

The First Law may be written, dividing through by $\Delta x\ \Delta y\ \Delta z$.,

$$\rho \frac{De}{Dt} = -\frac{\partial}{\partial x}\left(\frac{q}{A}\right)_x - \frac{\partial}{\partial y}\left(\frac{q}{A}\right)_y - \frac{\partial}{\partial z}\left(\frac{q}{A}\right)_z + W_i - \sigma_x \frac{\partial v_x}{\partial x} - \sigma_y \frac{\partial v_y}{\partial y}$$

$$-\sigma_z \frac{\partial v_z}{\partial z} - \tau_{xy}\left(\frac{\partial v_x}{\partial y} + \frac{\partial v_y}{\partial x}\right) - \tau_{yz}\left(\frac{\partial v_y}{\partial z} + \frac{\partial v_z}{\partial y}\right) - \tau_{zx}\left(\frac{\partial v_z}{\partial x} + \frac{\partial v_x}{\partial z}\right) \qquad (7.72)$$

Substitution of appropriate rate equations for σ and τ [Eq. (3.56)] gives:

$$\rho \frac{De}{Dt} = -\frac{\partial}{\partial x}\left(\frac{q}{A}\right)_x - \frac{\partial}{\partial y}\left(\frac{q}{A}\right)_y - \frac{\partial}{\partial z}\left(\frac{q}{A}\right)_z + W_i - p\left(\frac{\partial v_x}{\partial x} + \frac{\partial v_y}{\partial y} + \frac{\partial v_z}{\partial z}\right) + \mu\Phi$$

where

$$\Phi = 2\left[\left(\frac{\partial v_x}{\partial x}\right)^2 + \left(\frac{\partial v_y}{\partial y}\right)^2 + \left(\frac{\partial v_z}{\partial z}\right)^2\right] + \left[\left(\frac{\partial v_x}{\partial y} + \frac{\partial v_y}{\partial x}\right)^2 + \left(\frac{\partial v_y}{\partial z} + \frac{\partial v_z}{\partial y}\right)^2 + \left(\frac{\partial v_z}{\partial x} + \frac{\partial v_x}{\partial z}\right)^2\right] - \frac{2}{3}\left(\frac{\partial v_x}{\partial x} + \frac{\partial v_y}{\partial y} + \frac{\partial v_z}{\partial z}\right)^2 \qquad (7.73a)$$

or in vector notation

$$\rho \frac{De}{Dt} = -\text{div}\ \frac{\mathbf{q}}{A} + W_i - p\ \text{div}\ \mathbf{V} + \mu\Phi \qquad (7.73b)$$

From the definition of enthalpy, $i = e + p/\rho$ or

$$\frac{De}{Dt} = \frac{Di}{Dt} - \frac{1}{\rho}\frac{Dp}{Dt} + \frac{p}{\rho^2}\frac{D\rho}{Dt}$$

The continuity equation (1.11) may be written in the following equivalent form:

$$\frac{1}{\rho}\frac{D\rho}{Dt} + \text{div } \mathbf{V} = 0$$

Therefore,

$$\rho \frac{De}{Dt} = \rho \frac{Di}{Dt} - \frac{Dp}{Dt} - p \text{ div } \mathbf{V} \tag{7.74}$$

Substitution of Eq. (7.74) into Eq. (7.73b) gives:

$$\rho \frac{Di}{Dt} = -\text{div}\left(\frac{\mathbf{q}}{A}\right) + W_i + \frac{Dp}{Dt} + \mu\Phi \tag{7.75}$$

Caution must be exercised in writing the energy equation (7.75) in terms of temperature. From thermodynamic relations of properties,

$$di = T\,ds + \frac{dp}{\rho} \tag{7.76}$$

For a pure substance

$$ds = \left(\frac{\partial s}{\partial T}\right)_p dT + \left(\frac{\partial s}{\partial p}\right)_T dp \tag{7.77}$$

From Maxwell's relation

$$\left(\frac{\partial s}{\partial p}\right)_T = \frac{1}{\rho^2}\left(\frac{\partial \rho}{\partial T}\right)_p = -\frac{\beta}{\rho} \tag{7.78}$$

Combining Eqs. (7.76), (7.77), and (7.78), and noting $T(\partial s/\partial T)_p = c_p$:

$$di = c_p\,dT + \frac{1}{\rho}(1 - T\beta)\,dp$$

or

$$\rho \frac{Di}{Dt} = \rho c_p \frac{DT}{Dt} + (1 - T\beta)\frac{Dp}{Dt} \tag{7.79}$$

For pure conduction, heat flux $\mathbf{q}/A$ in Eq. (7.75) is given by the Fourier rate equation:

$$\frac{\mathbf{q}}{A} = -k\,\nabla T \tag{7.80}$$

Substitution of Eqs. (7.79) and (7.80) into Eq. (7.75) finally gives:

$$\rho c_p \frac{DT}{Dt} = \text{div } (k\,\nabla T) + W_i + T\beta\frac{Dp}{Dt} + \mu\Phi \tag{7.81}$$

Equation (7.81) is the general differential energy equation and reads as follows:

> Rate of increase of temperature of a fluid element moving with the stream is governed by heat conduction across its boundaries, heat source effects, compressibility effects, and viscous effects.

If k is independent of the space coordinates:

$$\rho c_p \frac{DT}{Dt} = k \nabla^2 T + W_i + T\beta \frac{Dp}{Dt} + \mu\Phi \tag{7.82}$$

Further, if the system is an ideal gas, $\beta = 1/T$:

$$\rho c_p \frac{DT}{Dt} = k \nabla^2 T + W_i + \frac{Dp}{Dt} + \mu\Phi \tag{7.83}$$

We could have obtained Eq. (7.83) directly from Eq. (7.75) by noting that for an ideal gas, $di = c_p\, dT$ where $c_p =$ constant, so that

$$\frac{Di}{Dt} = c_p \frac{DT}{Dt}$$

These equations are valid whether motion is the result of free or of forced convection. The following examples illustrate how the energy equation may be reduced to the simple forms obtained previously by direct energy balance.

Case 1: Conduction in Stationary System

Assume that the system is incompressible and that viscous effects can be neglected. Since the system is stationary, the convective terms in the substantial derivative are zero. If k is constant, Eq. (7.82) reduces to

$$\rho c_p \frac{\partial T}{\partial t} = k\left(\frac{\partial^2 T}{\partial x^2} + \frac{\partial^2 T}{\partial y^2} + \frac{\partial^2 T}{\partial z^2}\right)$$

which is identical with Eq. (6.4) in Chap. 6.

Case 2: Two-Dimensional Boundary Layer over a Plate

Equation (7.82) in two-dimensional case for negligible W_i and constant p becomes:

$$\rho c_p \left[\frac{\partial T}{\partial t} + v_x \frac{\partial T}{\partial x} + v_y \frac{\partial T}{\partial y}\right] = k \left[\frac{\partial^2 T}{\partial x^2} + \frac{\partial^2 T}{\partial y^2}\right] + \mu\Phi \tag{7.84}$$

where $$\Phi = 2\left[\left(\frac{\partial v_x}{\partial x}\right)^2 + \left(\frac{\partial v_y}{\partial y}\right)^2\right] + \left[\frac{\partial v_y}{\partial x} + \frac{\partial v_x}{\partial y}\right]^2$$

Order-of-magnitude analysis reduces Eq. (7.84) to

$$\rho c_p \left(\frac{\partial T}{\partial t} + v_x \frac{\partial T}{\partial x} + v_y \frac{\partial T}{\partial y}\right) = k \frac{\partial^2 T}{\partial y^2} + \mu \left(\frac{\partial v_x}{\partial y}\right)^2 \tag{7.85}$$

and for steady state and negligible frictional effect

$$\rho c_p \left(v_x \frac{\partial T}{\partial x} + v_y \frac{\partial T}{\partial y}\right) = k \frac{\partial^2 T}{\partial y^2} \tag{7.86}$$

Eq. (7.86) is identical with Eq. (7.32).

REFERENCES

1. Schlichting, H., *Boundary Layer Theory*, McGraw-Hill, New York, 1955.
2. Clark, S. H., and W. M. Kays, *Trans. ASME*, **75,** 859 (1953).
3. Pohlhausen, E., *Z. angew. Math. u. Mech.*, **1,** 115 (1921).
4. Eckert, E. R. G., and R. M. Drake, *Heat and Mass Transfer*, 2nd Ed., p. 178, McGraw-Hill, New York, 1959.
5. Sparrow, E. M., and J. L. Gregg, *J. Aero. Sc.*, **24,** 852 (1957).
6. Eckert, E. R. G., *Trans. ASME*, **56,** 1273–83 (1956).
7. Ostrach, S., *Nat. Advisory Comm. Aeronaut. Tech. Note* 2635 (Feb. 1952).
8. Schmidt, E., and W. Beckmann, *Forsch. Gebiete Ingenieurw.*, **1,** 391 (1930).
9. Sparrow, E. M., Ph.D. Thesis, Mech. Engrg. Dept., Harvard Univ., 1956.
10. Nickerson, R. J., and H. P. Smith, Term Project, Course 2.521, M.I.T., Cambridge, Mass., 1958.
11. Graetz, L., *Ann. Physik* **18,** 79 (1883), and **25,** 337 (1885).
12. Goldberg, P., M.S. Thesis, Mech. Engrg. Dept., M.I.T., Jan. 1958.
13. Kays, W. M., *Trans. ASME*, **77,** 1265 (1955).
14. Langhaar, H. L., *Jour. Appl. Mech.*, **64,** A-55 (1942).
15. Nusselt, W., *Z. ver. deut. Ing.*, **54,** 1154 (1910).
16. Sellars, Tribus, and Klein, *Trans. ASME*, **78,** 441 (1956).
17. Millsaps, K., and K. Pohlhausen, "Heat Transfer to Hagen-Poiseuille Flow," Proc. Conf. on Diff. Eqns., U. of Md., University Book Store, College Park, Md., March 1955.
18. Norris, R. H., and D. P. Streid, *Trans. ASME*, **62,** 525 (1940).
19. Hausen, H., *Z. ver. deut. Ing.*, Beih, Verfarenstech **4,** 91 (1943).

20. Sparrow, E. M., *Nat. Advisory Comm. Aeronaut. Tech. Note* 3331 (Jan. 1955).

21. Sparrow, E. M., and J. L. Gregg, ASME Paper No. 55-SA-4.

22. Eckert, E. R. G., and R. M. Drake, *op. cit.*, p. 328.

23. Nickerson, R. J., Personal Communication, 1957.

24. Eckert, E. R. G., and E. Soehnghen, USAF Tech. Rept. 5747, Dec. 1948.

25. Schmidt, E., Z. ges. Kälte-Ind, **35,** 213, (1928).

26. Cheng, H., M.S. Thesis, Dept. Nav. Arch., Sept. 1957.

PROBLEMS

7.1 The temperature distribution between parallel, horizontal plates is given by

$$k \frac{d^2T}{dy^2} = -\mu \left(\frac{dv_x}{dy}\right)^2$$

State clearly all assumptions underlying this simplified ordinary differential equation.

7.2 (a) Determine expressions for the velocity distribution, temperature distribution, and heat transfer rates at the wall for very viscous fluids for fully developed flow between stationary parallel plates; e.g., solve Eqs. (3.4) and (7.4) with the following boundary conditions: at $y = \pm b$, $v_x = 0$, $T = T_0$, the mean fluid temperature T_m is greater than T_o, and the mean fluid velocity is V_m.

(b) Obtain an expression for heat transfer coefficient h as defined by Eqs. (7.14) and (7.13).

(c) Oil flows at a mean velocity of 10 ft/sec between parallel plates. For oil, μ = 210 lb_m/hr ft, k = 0.077 Btu/hr ft F, and ρ = 57 lb/ft^3. For fully developed flow calculate the difference between the centerline temperature and the wall temperature if the plate spacing is 0.05 in.; if plate spacing is 0.10 in.

(d) In part (c), calculate the heat flux to the bounding walls.

7.3 Solve Prob. 7.2 (a)(b) for the following boundary conditions

$$\text{at } y = 0 \qquad v_x = 0, \quad T = T_0$$

$$\text{at } y = b \qquad v_x = 0, \quad T = T_1 > T_0 > T_m$$

7.4 In a system composed of two parallel plates, the upper plate is maintained at the temperature T_1 and moves at the velocity V_1, and the lower, stationary plate is insulated. Find the temperature increase of the lower plate in terms of T_1, V_1, and the properties of the fluid between the plates. Neglect any pressure gradient in the flow field.

7.5 Derive Eq. (7.10) by applying energy balance to an annular element control volume.

7.6 Solve Prob. 7.2 (a)(b) for flow in a small-diameter tube where the term involving $\partial T/\partial x$ is negligible but the viscous term is large; e.g., the following form of the energy equation is applicable:

$$k\left[\frac{1}{r}\frac{d}{dr}\left(r\frac{dT}{dr}\right)\right] = -\mu\left(\frac{\partial v_x}{\partial r}\right)^2$$

The fully developed velocity distribution is given by Eq. (3.17) and at $r = r_0$, $T = T_0 < T_m$.

7.7 Consider the Couette flow of light oil in the clearance space of a bearing. Assume the walls have equal temperatures $T_0 = T_1 = 100$ F. If the shaft surface speed is $V_1 = 30$ ft/sec, find the location and value of the maximum temperature in the bearing. Assume the radial clearance is 0.003 in.

7.8 For laminar flow, calculate the results given in Table 7.1 for Nu_∞ for the following cases:

(a) circular tube with parabolic velocity distribution and uniform wall temperature ($\text{Nu}_\infty = 3.66$)

(b) circular tube with slug flow (v_x = uniform) and uniform heat flux axially ($\text{Nu}_\infty = 8.00$)

(c) parallel plates with parabolic velocity distribution and uniform heat flux axially ($\text{Nu}_\infty = 8.23$).

7.9 Air at 14.7 psia and 70 F flows at $V_\infty = 50$ ft/sec over a 1-ft-long flat plate at 200 F.

(a) Calculate the velocity and the thermal boundary-layer thicknesses halfway down the plate.

(b) Calculate the heat transfer rate from the plate per foot of plate width.

7.10 A heated thin wing section (approximately a flat surface) at $T_0 = 255$ F is placed in a wind tunnel of air velocity $V_\infty = 33$ ft/sec and air temperature $T_\infty = 125$ F.

(a) Plot the temperature distribution in the boundary layer 4 in. from the leading edge.

(b) Calculate the local heat transfer coefficient and the Nusselt number at that point.

7.11 A plate of glass (4 ft by 4 ft) heated uniformly to 180 F is cooled in a horizontal position by air at 64 F and free-stream velocity of 3 fps. Calculate the rate of cooling of the plate in Btu/hr.

7.12 Liquid sodium at 700 F flows at a velocity of $V_\infty = 2$ ft/sec over a flat plate, 3 in. long in the flow direction. The plate temperature is uniform at 400 F. Calculate the average heat flux to the plate.

7.13 A vertical wall at 150 F is exposed to air at 60 F. Find the Grashof number 2 ft from the lower edge (for an ideal gas, $\beta = 1/T$).

7.14 Compare the laminar boundary-layer equations for a heated plate in horizontal and vertical positions. Comment briefly on how they differ.

7.15 (a) A film of liquid of thickness b is in steady laminar motion down an inclined surface as shown. Derive the differential equation of velocity distribution in the film and solve for appropriate boundary conditions.

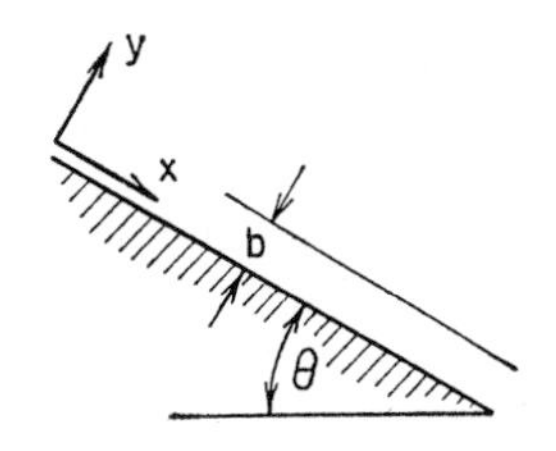

(b) If the wall is at T_0 and the liquid surface temperature is at T_s, and if the convective terms in the energy equation can be neglected, find the heat flux at the wall in terms of known quantities.

7.16 A sheet of paper 1 ft square is suspended by a calibrated spring in still air at 70 F. Initial weight of the paper is 0.1 oz. When sun shines on the paper so that it is uniformly heated to 140 F, do you expect the weight of the paper as indicated by the spring to increase or decrease? Estimate the per cent change in weight. For simplicity, evaluate properties such as β at the arithmetic mean temperature, unless otherwise specified by the solution.

7.17 A vertical, flat, stainless steel plate 2 in. square and 0.05 in. thick is suspended in liquid bismuth. The plate temperature initially is 600 F and the bismuth temperature remains at 800 F. How long does it require to heat the plate from 600 F to 750 F?

7.18 A nuclear reactor is to be built with ten fuel elements in the shape of vertical plates 1 ft square.

(a) For a maximum heat transfer, the plates should be spaced closer than 2δ where δ is the boundary-layer thickness. True or false?

(b) If the plate assembly were to be immersed in water at 50 F and the maximum allowable plate surface temperature were 100 F, at what power level could the reactor operate safely? Assume constant fluid properties as given below.

$$\mathrm{Pr} = 5.0$$

$$(\mathrm{Gr})_L = 4 \times 10^8$$

$$k = 0.35 \text{ Btu/hr ft}^2 \text{ F/ft}$$

$$\mu = 2.7 \text{ lb}_m\text{/hr ft}$$

$$\rho = 62.4 \text{ lb}_m/\text{ft}^3$$

7.19 Air at atmospheric pressure and 70 F enters a 1-in. pipe at 10 ft/sec. If the pipe wall temperature is 180 F how long must the pipe be to increase the bulk temperature of the air to (a) 80 F, (b) 160 F?

7.20 A flow of 100 lb/hr of molasses at 100 F is pumped through a 2-in. I.D. pipe. After a very long unheated length, the fluid passes through a 3-ft-long heated section where the tube wall is maintained at 180 F by condensing steam. Calculate the mean temperature of the molasses leaving the heated section. Molasses properties: $\rho = 70$ lb/ft^3, $\mu = 150$ lb/ft hr, $c = 0.4$ Btu/lb, and $k = 0.5$ Btu/hr ft F.

7.21 For liquid metals flowing between parallel flat plates, the velocity boundary layer builds up much less rapidly than the thermal boundary layer; consequently, the temperature profile becomes fully developed, while the velocity profile is still uniform across the cross section. Consider the laminar flow of a liquid metal between flat plates, each of which has a different uniform heat transfer per unit area to the fluid. Assume the velocity to be uniform throughout, and the temperature profile to be fully developed, and the heat transfer per unit area on one plate to be four times that of the other plate. Find the temperature difference between the two plates as a function of the heat transfer per unit area, the geometry, and properties of the fluid.

7.22 Consider the problem of natural convection over a flat plate in vertical position with constant heat flux at the wall. Solve the momentum and the energy integral equations, Eqs. (7.62) and (7.44), and obtain an expression for the average Nusselt number over a height L in terms of the fluid properties and $(q/A)_0$. Assume that the velocity and the temperature profiles can be approximated by Eqs. (7.63) and (7.64) and proceed with the solution of the boundary-layer equations by specifying Γ and δ as simple power functions of x, Eq. (7.66).

CHAPTER 8

Heat Transfer in Turbulent Flow and Experimental Results

The complex nature of turbulent motion and its inaccessibility to mathematical treatment was mentioned in Chap. 2. The materials in Chap. 4 were limited to gaining familiarity with some of the more important correlations on the momentum transfer in turbulent flow systems. However, if attention were centered on the temporal mean motion in turbulent flow, it is possible to establish certain theoretical bases for the analysis of turbulent transfer processes and thereby to provide some logic in the interpretation of the experimental results. This chapter begins with a study of the effect of turbulent fluctuations on the mean motion. Starting in Section 8.4, heat transfer in turbulent flow systems, including liquid metals, is analyzed by the postulation of an analogy between the transfer of momentum and heat. The remainder of the chapter deals systematically with empirical and semiempirical correlations of heat transfer data on forced and free convection in channels and over submerged bodies.

8.1 Heat and Momentum Transfer Equations in Turbulent Flow

It was pointed out in Chap. 2 that the instantaneous velocity at a point in a turbulent flow field can be expressed as

$$v_x = \bar{v}_x + v_x' \tag{8.1}$$

where the barred quantities represent mean components independent of time, and the primed quantities, the high-frequency fluctuating components. Likewise, instantaneous temperature and pressure at a point in turbulent flow can be expressed as a superposition of a temporal mean and a fluctuating component:

$$\begin{aligned} T &= \bar{T} + T' \\ p &= \bar{p} + p' \end{aligned} \tag{8.2}$$

Since momentum and energy must be conserved in turbulent as well as in laminar flow, the conservation equations derived in Chaps. 3 and 7 hold equally for turbulent flow, provided the velocities and temperatures in these equations are interpreted as instantaneous velocities and temperatures of the turbulent field. From a practical standpoint, such equations are of little value because we are not usually interested in knowing the complete and complex history of instantaneous velocities and temperatures, but only their time-averaged mean values that can be measured or observed. Of course, if we could solve the instantaneous differential equations, the mean distributions could be found by integrating the solutions over an appropriate time interval; since this is not the case, we shall attempt to obtain useful information from time-averaging the differential equations themselves. Before discussing the merits of this procedure, we shall illustrate it by a simple example of fully developed turbulent flow between parallel plates with energy transfer in the direction normal to flow. We first write the governing equations for flow over a flat plate or in the entrance region between parallel plates and simplify them to the case of fully developed flow.

The continuity equation and the momentum and energy equations for a two-dimensional, incompressible flow in a boundary layer in the entrance region between parallel plates were previously derived:

$$\frac{\partial v_x}{\partial x} + \frac{\partial v_y}{\partial y} = 0 \tag{8.3}$$

$$\frac{\partial v_x}{\partial t} + v_x \frac{\partial v_x}{\partial x} + v_y \frac{\partial v_x}{\partial y} = \nu \frac{\partial^2 v_x}{\partial y^2} - \frac{1}{\rho}\frac{\partial p}{\partial x} \tag{8.4a}$$

$$\frac{\partial T}{\partial t} + v_x \frac{\partial T}{\partial x} + v_y \frac{\partial T}{\partial y} = \alpha\left(\frac{\partial^2 T}{\partial y^2} + \frac{\partial^2 T}{\partial x^2}\right) \tag{8.5a}$$

For our purpose, we shall rewrite Eq. (8.4a) and Eq. (8.5a) in the following equivalent forms by using Eq. (8.3):

$$\frac{\partial v_x}{\partial t} + \frac{\partial}{\partial x}(v_x^2) + \frac{\partial}{\partial y}(v_x v_y) = \frac{1}{\rho}\frac{\partial}{\partial y}\left(\mu\frac{\partial v_x}{\partial y}\right) - \frac{1}{\rho}\frac{\partial p}{\partial x} \tag{8.4b}$$

$$\frac{\partial T}{\partial t} + \frac{\partial}{\partial x}(v_x T) + \frac{\partial}{\partial y}(v_y T) = \frac{1}{\rho c_p}\left[\frac{\partial}{\partial y}\left(k\frac{\partial T}{\partial y}\right) + \frac{\partial}{\partial x}\left(k\frac{\partial T}{\partial x}\right)\right] \tag{8.5b}$$

When flow is turbulent, the velocities and temperatures in the equations above are instantaneous quantities. In the substitution of Eqs. (8.1) and (8.2) into Eqs. (8.4b) and (8.5b), it should be noted that for this special case of two-dimensional boundary-layer flow, the mean components $\bar{v}$, $\bar{T}$, and $\bar{p}$ vary with (x, y, t), but the fluctuating components v', T', and p' vary with (x, y, z, t). In other words, the mean flow may be two-dimensional, but the turbulent fluctuations at a point are still three-dimensional.

The substitution yields

$$\frac{\partial}{\partial t}(\bar{v}_x + v_x') + \frac{\partial}{\partial x}[(\bar{v}_x + v_x')^2] + \frac{\partial}{\partial y}[(\bar{v}_x + v_x')(\bar{v}_y + v_y')]$$
$$= \frac{1}{\rho}\frac{\partial}{\partial y}\left[\mu\frac{\partial}{\partial y}(\bar{v}_x + v_x')\right] - \frac{1}{\rho}\frac{\partial}{\partial x}(\bar{p} + p') \tag{8.4c}$$

$$\frac{\partial}{\partial t}(\bar{T} + T') + \frac{\partial}{\partial x}[(\bar{v}_x + v_x')(\bar{T} + T')] + \frac{\partial}{\partial y}[(\bar{v}_y + v_y')(\bar{T} + T')]$$
$$= \frac{1}{\rho c_p}\frac{\partial}{\partial y}\left[k\frac{\partial}{\partial y}(\bar{T} + T')\right] + \frac{1}{\rho c_p}\frac{\partial}{\partial x}\left[k\frac{\partial}{\partial x}(\bar{T} + T')\right] \tag{8.5c}$$

Taking the time average of these equations is the same as taking the time average of each term. In performing these operations, we note from definition that the time average of the fluctuation quantities must be zero, namely $\overline{v'} = 0, \overline{T'} = 0$, and $\overline{p'} = 0$. Also, $\overline{\bar{v}v'} = \bar{v}\overline{v'} = 0$ and $\overline{\bar{v}T'} = \bar{v}\overline{T'} = 0$. However, $\overline{v_x'v_y'} \neq 0$, as can be seen as follows. If at any instant of time a packet of fluid particles moves with positive v_y', Fig. 8.1, from a region of

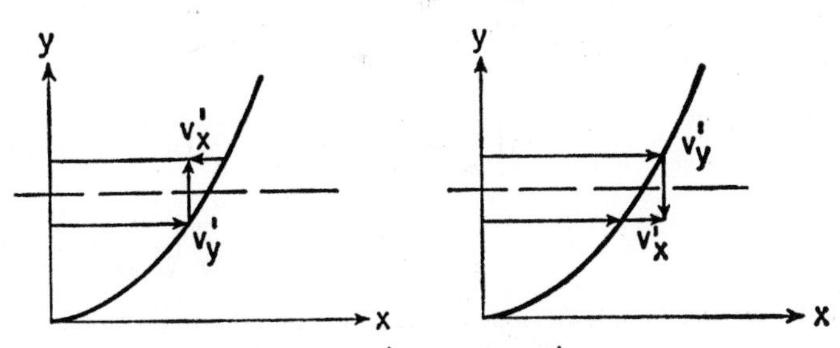

Fig. 8.1. v_x' induced by v_y'.

lower velocity to one of higher velocity, it induces a negative v_x'. Similarly a negative v_y' induces a positive v_x'; so the product $v_x'v_y'$ always has the same sign and $\overline{v_x'v_y'}$ is not zero. Likewise, $\overline{v'T'}$ is not zero. We assume the fluctuations are far more rapid than the transient effects. Averaging over a sufficient number of fluctuations but a short time interval for the transient effect reduces the average of $\partial(\bar{v}_x + v_x')/\partial t$ to $\partial\bar{v}_x/\partial t$ and $\partial(\bar{T} + T')/\partial t$ to $\partial\bar{T}/\partial t$.

The instantaneous continuity equation is given by

$$\frac{\partial}{\partial x}(\bar{v}_x + v_x') + \frac{\partial}{\partial y}(\bar{v}_y + v_y') = 0$$

and, when averaged, becomes

$$\frac{\partial \bar{v}_x}{\partial x} + \frac{\partial \bar{v}_y}{\partial y} = 0 \tag{8.6}$$

When all terms of Eqs. (8.4c) and (8.5c) have been averaged, some rearrangement and use of Eq. (8.6) yield the following equations:*

$$\frac{\partial \bar{v}_x}{\partial t} + \bar{v}_x \frac{\partial \bar{v}_x}{\partial x} + \bar{v}_y \frac{\partial \bar{v}_x}{\partial y} = \frac{1}{\rho}\left[\frac{\partial}{\partial y}\left(\mu \frac{\partial \bar{v}_x}{\partial y} - \rho\overline{v_x'v_y'}\right) - \frac{\partial}{\partial x}(\rho\overline{v_x'^2})\right] - \frac{1}{\rho}\frac{\partial \bar{p}}{\partial x} \tag{8.7}$$

$$\frac{\partial \bar{T}}{\partial t} + \bar{v}_x \frac{\partial \bar{T}}{\partial x} + \bar{v}_y \frac{\partial \bar{T}}{\partial y} = \frac{1}{\rho c_p}\left[\frac{\partial}{\partial y}\left(k \frac{\partial \bar{T}}{\partial y} - \rho c_p \overline{v_y'T'}\right) + \frac{\partial}{\partial x}\left(k \frac{\partial \bar{T}}{\partial x} - \rho c_p \overline{v_x'T'}\right)\right] \tag{8.8}$$

There are two noteworthy features. First, we note that the terms associated with mean motion in turbulent flow are strikingly similar to the corresponding terms in laminar flow. We next note that both Eq. (8.7) and Eq. (8.8) contain two additional terms on the right, expressing re-

* Generally, the turbulent stress terms are derived from the three-dimensional Navier-Stokes equations, (3.60). The x-direction momentum equation for turbulent flow then becomes

$$\frac{\partial \bar{v}_x}{\partial t} + \bar{v}_x \frac{\partial \bar{v}_x}{\partial x} + \bar{v}_y \frac{\partial \bar{v}_x}{\partial y} + \bar{v}_z \frac{\partial \bar{v}_x}{\partial z} = \frac{1}{\rho}\left[\frac{\partial}{\partial x}\left(\mu \frac{\partial \bar{v}_x}{\partial x} - \rho\overline{v_x'^2}\right) + \frac{\partial}{\partial y}\left(\mu \frac{\partial \bar{v}_x}{\partial y} - \rho\overline{v_x'v_y'}\right) + \frac{\partial}{\partial z}\left(\mu \frac{\partial \bar{v}_x}{\partial z} - \rho\overline{v_x'v_z'}\right)\right] - \frac{1}{\rho}\frac{\partial \bar{p}}{\partial x}$$

However, the simpler form is given here in Eq. (8.7) in order to relate the derivation to the momentum balance treated in Sec. 3.1.

spectively momentum and enthalpy fluxes associated with fluctuating motion in the x- and the y-directions. These terms embody a key concept of the turbulent transfer phenomena, namely that the property fluctuations in turbulent systems produce an apparent increase in the transfer fluxes. Since we are interested in the y-direction transfer, we shall confine our attention to the total apparent fluxes in this direction, namely

$$(\tau_{yx})_{\text{app}} = -\left(\mu \frac{\partial \bar{v}_x}{\partial y} - \rho \overline{v_x' v_y'}\right) \tag{8.9}$$

$$\left(\frac{q}{A}\right)_{y,\text{app}} = -\left(k \frac{\partial \bar{T}}{\partial y} - \rho c_p \overline{v_y' T'}\right). \tag{8.10}$$

Equations (8.9) and (8.10) may be regarded as the rate equations for turbulent flow systems; the negative sign is introduced to conform to the sign convention adopted for rate equations in Chap. 1. Each apparent flux consists of a molecular diffusion flux (as in laminar flow) and an eddy flux. Various theories, both mathematical and empirical, have been postulated in an attempt to express the eddy fluxes in terms of measurable quantities, as for instance the mean potential gradients. None of these theories has proven adequate, and the problem remains a major challenge to engineers and scientists working in this field. The time-averaged conservation equations (8.7) and (8.8) must still be solved in order to know, for instance, the rate of heat transfer at the wall. Since exact solution of these equations is not possible, we shall develop in Secs. 8.4 through 8.7 semiempirical solution techniques based upon a momentum-heat transfer analogy. Extensive literature on friction factor (Chap. 4) can thereby be utilized to solve heat transfer problems. The technique suffers from a common defect of all empirical methods in that the true nature of the problem gets somewhat submerged by oversimplification, but it has proved the most useful and powerful tool in turbulent heat transfer work.

8.2 Eddy Diffusivities of Momentum and Heat

The eddy momentum flux $\rho \overline{v_x' v_y'}$ in Eq. (8.9) may be interpreted as a turbulent shear stress in the same manner that the molecular shear stress was interpreted. From Fig. 8.1, it is clear also that the greater the velocity gradient $\partial \bar{v}_x / \partial y$, the larger will be the v_x' induced by a v_y', and hence the larger will be $\overline{v_x' v_y'}$. The above suggests defining an eddy diffusivity for momentum, ϵ_m, such that

$$\overline{v_x' v_y'} \equiv -\epsilon_m \frac{\partial \bar{v}_x}{\partial y} \tag{8.11}$$

Then Eq. (8.9) may be written

$$\tau_{\text{app}} = \tau_{\text{molec}} + \tau_{\text{eddy}} = -(\mu + \rho\epsilon_m)\frac{\partial \bar{v}_x}{\partial y} \tag{8.12a}$$

$$\frac{\tau_{\text{app}}}{\rho} = -(\nu + \epsilon_m)\frac{\partial \bar{v}_x}{\partial y} \tag{8.12b}$$

The quantity $(\rho\epsilon_m)$ is usually interpreted as an eddy viscosity analogous to μ, the molecular viscosity, but whereas the latter is a fluid property, the former is a parameter of fluid motion. In a similar way, the enthalpy flux $\rho c_p \overline{v_y' T'}$ in Eq. (8.10) suggests the concept of an eddy diffusivity for energy transfer, ϵ_h. Then

$$\left(\frac{q}{A}\right)_{\text{app}} = \left(\frac{q}{A}\right)_{\text{molec}} + \left(\frac{q}{A}\right)_{\text{eddy}} = -(k + \rho c_p \epsilon_h)\frac{\partial \bar{T}}{\partial y} \tag{8.13a}$$

$$\frac{(q/A)_{\text{app}}}{\rho c_p} = -(\alpha + \epsilon_h)\frac{\partial \bar{T}}{\partial y}. \tag{8.13b}$$

We may interpret $\rho c_p \epsilon_h$ as an eddy conductivity, again noting that it is not only a fluid property but a flow parameter as well. For a given turbulent system, ϵ_m and ϵ_h might be expected to be closely related.

Although the equations introducing the concepts of τ_{app}, $(q/A)_{\text{app}}$, ϵ_m, and ϵ_h—Eqs. (8.9) through (8.13)—were obtained from the boundary-layer equations—(8.3), (8.4), and (8.5)—they are equally applicable to turbulent flows in general. They apply in external flows over bodies and in internals flows in tubes, ducts, and between parallel plates.

The quantity $\overline{v_x' v_y'}$ has been interpreted alternatively in terms of a "mixing length" concept introduced by Prandtl and later modified by Kármán, among others. The mixing length concept and the eddy diffusivity concept are empirical to the same degree. We proceed here with the eddy diffusivity concept.

8.3 Distribution of τ_{app} and $(q/A)_{\text{app}}$

Let us consider the simple case in which flow between parallel plates, Fig. 3.1, is in steady state and fully developed ($\bar{v}_y = 0$). From Eq. (8.6), $\partial \bar{v}_x / \partial x = 0$. Also,

$$\frac{\partial \overline{(\rho v_x'^2)}}{\partial x}$$

is zero for steady fully developed flow. We have so far not written the momentum equation in the y-direction, but it is not difficult to verify that for the case under consideration $\partial \bar{p}/\partial x$ is independent of y; therefore

$$\frac{\partial \bar{p}}{\partial x} = \frac{d\bar{p}}{dx}$$

Equation (8.7) combined with Eq. (8.9) then reduces to the following simple form:

$$\frac{dp}{dx} = -\frac{d\tau_{app}}{dy} \tag{8.14}$$

Since in fully developed flow dp/dx is uniform across the stream, clearly τ_{app} varies linearly with y, so with $y = 0$ at the wall, the shear stress distribution between the centerline and the wall is as follows:

$$\frac{\tau_{app}}{(\tau_{yx})_0} = \left(1 - \frac{y}{y_0}\right) \tag{8.15}$$

where $(\tau_{yx})_0$ is the shear stress at the wall, $y = 0$.

The result in Eq. (8.15) also applies to round tubes, with r_0 replacing y_0. Thus τ_{app} varies linearly with the radius.

Similarly, Eqs. (8.8) and (8.10) for steady, fully developed flow between parallel plates lead to the following energy equation:

$$\rho c_p \bar{v}_x \frac{\partial \bar{T}}{\partial x} = -\frac{\partial (q/A)_{app}}{\partial y} \tag{8.16}$$

Integration of Eq. (8.16) between y and y_0 and between 0 and y_0 gives the following expression:

$$\frac{(q/A)_{app}}{(q/A)_0} = M\left(1 - \frac{y}{y_0}\right) \tag{8.17}$$

where

$$M = \frac{\dfrac{1}{y_0 - y}\displaystyle\int_y^{y_0} \bar{v}_x \frac{\partial \bar{T}}{\partial x}\, dy}{\dfrac{1}{y_0}\displaystyle\int_0^{y_0} \bar{v}_x \frac{\partial \bar{T}}{\partial x}\, dy} \tag{8.18}$$

since ρc_p is taken as constant and $\bar{v}_x$ is not a function of x.

For a circular tube the following results are obtained: in Eq. (8.17) r_0 replaces y_0 and Eq. (8.18) has the form

$$M = \frac{\dfrac{1}{\pi r^2}\displaystyle\int_0^{r} \bar{v}_x \frac{\partial \bar{T}}{\partial x}\, 2\pi r\, dr}{\dfrac{1}{\pi r_0^2}\displaystyle\int_0^{r_0} \bar{v}_x \frac{\partial \bar{T}}{\partial x}\, 2\pi r\, dr} \tag{8.19}$$

The analysis leading to Eqs. (7.16) and (7.17) holds for fully developed turbulent flow as well as laminar flow. Then for the case of uniform $(q/A)_0$ along the tube $\partial T/\partial x = \partial T_m/\partial x$, independent of r; so M becomes

$$M = \frac{\bar{v}_{x\,avg\,(0-r)}}{\bar{v}_{x\,avg\,(0-r_0)}} \tag{8.20}$$

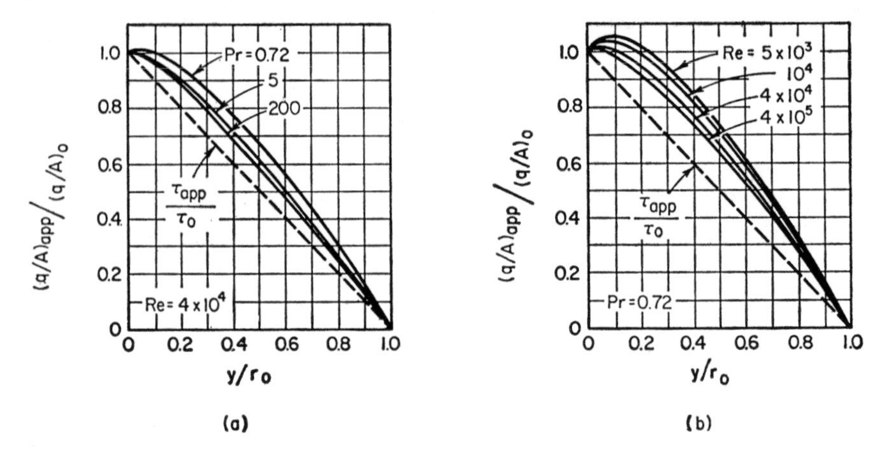

Fig. 8.2. Distribution of $(q/A)_{app}$ in a round tube: (a) for Re = 4×10^4 with Pr at the parameter; and (b) for Pr = 0.72 with Re as the parameter [Reichardt (1)].

For various magnitudes of Pr and Re, Reichardt (1) shows the distribution of $(q/A)_{app}$ as plotted in Fig. 8.2.

The rise of the curves above $(q/A)_{app}/(q/A)_0 = 1.0$ results from the decreasing area with radius in a tube. For flow between flat plates, similar curves do not rise above 1.0.

8.4 Analysis of Heat Transfer in Turbulent Flow Using the Momentum-Heat Transfer Analogy

Our knowledge of turbulence is so limited that we are unable to determine analytically the velocity and temperature distributions. By means of an analogy between heat and momentum transfer, we have been able to predict heat transfer coefficients from friction factor data. This analogy has led to successful prediction of heat transfer coefficients for liquid metals and for fluids at supercritical pressures.

Most analyses of heat transfer in turbulent flow start with Eqs. (8.12b) and (8.13b) with Eq. (8.15) and Eq. (8.17) in which M is taken as unity. The errors resulting from this assumed linear variation of $(q/A)_{app}$ are quite small.

$$\frac{\tau_0}{\rho}\left(1 - \frac{y}{r_0}\right) = (\nu + \epsilon_m)\,\frac{dv}{dy} \tag{8.21}$$

$$\frac{(q/A)_0}{\rho c_p}\left(1 - \frac{y}{r_0}\right) = -(\alpha + \epsilon_h)\,\frac{dT}{dy} \tag{8.22}$$

where the notation is simplified by writing v for $\bar{v}_x$ and T for $\bar{T}$, and τ_0

here is the shear stress of the fluid acting *on* the wall,* and $(q/A)_0$ is the heat flux from the wall *into* the fluid.

Inspection of Eqs. (8.21) and (8.22) shows them to be quite similar. The similarity between the two processes—heat transfer and momentum transfer—was noted as early as 1874 by Reynolds (3) when he suggested that the heat transferred to a surface divided by the maximum amount which could be transferred in bringing the fluid to the temperature of the surface equals the momentum transferred in passing over the surface divided by the maximum amount of momentum which could be transferred in stopping the fluid relative to the surface.

A modern version of the analogy between heat and momentum transfer states that for any particular fluid the ratio, $E \equiv \epsilon_h/\epsilon_m$, is a constant and, in general, is at most a function of Pr.

8.5 The Reynolds Result

The Reynolds result can be obtained from Eqs. (8.21) and (8.22) by applying them to a fluid in highly turbulent flow, in which case $\nu \ll \epsilon_m$ and $\alpha \ll \epsilon_h$. Since, from Eqs. (8.21) and (8.22), v and T have the exact same shape radial distribution with the assumptions above, the location at which $v = V$, the average velocity, is the same as the location at which $T = T_{\text{mean}}$. If we divide Eq. (8.21) by Eq. (8.22) and integrate between the wall and this location, we get the following result:

$$\frac{\tau_0 c_p}{(q/A)_0} = \frac{1}{E} \frac{V}{(T_0 - T_m)} \tag{8.23}$$

We now introduce the definitions $h \equiv (q/A)_0/(T_0 - T_m)$ and $\tau_0 \equiv f(\rho V^2/2)$ into this equation to get

$$\frac{h}{\rho c_p V} = E\frac{f}{2} \tag{8.24}$$

This result, Eq. (8.24), with $E = 1$ is known as the *Reynolds analogy equation* and is reasonably valid for practically all gases.

The magnitudes of ϵ_h and ϵ_m and hence E have been determined from Eqs. (8.21) and (8.22) and from velocity and temperature distribution measurements. Such experimental results were determined mostly for air and for water and have provided magnitudes for E in the range 0.9 to 1.7. All of the earlier analogy calculations have been made with $E = 1.0$.

The Reynolds result was obtained assuming the fluid to be a single zone of highly turbulent flow. Prandtl (1910) (4) and Taylor (1916) (5)

* If y is measured from the surface into the fluid and $y \equiv r_0 - r$, then the shear stress of the fluid acting *on* the wall is $\tau_0 = (\tau_{rx})_0 = -(\tau_{yx})_0$; so, a minus sign does not appear in Eq. (8.21). Further, since $(q/A)_0$ is the heat flux from the wall *into* the fluid, it is proportional to $-(\partial T/\partial y)$, but to $+(\partial T/\partial r)$; so, a minus sign appears in Eq. (8.22).

solved Eqs. (8.21) and (8.22) for two zones, a laminar zone next to the wall and a highly turbulent core. Later Kármán (1939) (6) solved the problem for three zones using the universal velocity distribution of Fig. 4.14. In the turbulent core Kármán neglected ν and α. Martinelli (1941) (7) directed his analysis to liquid metals and hence neglected only ν but not α in the turbulent core. Gazely (1947) (8) included both ν and α in the turbulent core.

8.6 Martinelli Result

Equations (8.21) and (8.22) may be integrated for a particular fluid (c_p, ρ, μ, k are known), with a particular value of τ_0 and $(q/A)_0$ and a particular tube size r_0. Under these conditions ϵ_m vs. y or r is determined from Eq. (8.21) with the velocity distribution given by Fig. 4.14; then since $\epsilon_h = E\epsilon_m$, Eq. (8.22) is integrated to determine the following expressions for the temperature distribution, where the subscripts 1 and 2 refer to conditions at $y^+ = 5$ and 30, respectively.

$y^+ < 5$

$$T_0 - T = \frac{(q/A)_0}{c_p\rho\sqrt{\tau_0/\rho}} \Pr y^+ \tag{8.25a}$$

$5 < y^+ < 30$

$$T_1 - T = \frac{5(q/A)_0}{Ec_p\rho\sqrt{\tau_0/\rho}} \ln\left[1 + E \Pr\left(\frac{y^+}{5} - 1\right)\right] \tag{8.25b}$$

$y^+ > 30$

$$T_2 - T = \frac{1.25(q/A)_0}{Ec_p\rho\sqrt{\tau_0/\rho}}\left\{\ln\left[\frac{5\lambda + \frac{y}{r_0}\left(1 - \frac{y}{r_0}\right)}{5\lambda + \frac{y_2}{r_0}\left(1 - \frac{y_2}{r_0}\right)}\right] + \frac{1}{\sqrt{1+20\lambda}}\ln\left[\left(\frac{\left(\frac{2y}{r_0} - 1\right) + \sqrt{1+20\lambda}}{\left(\frac{2y}{r_0} - 1\right) - \sqrt{1+20\lambda}}\right)\left(\frac{\left(\frac{2y_2}{r_0} - 1\right) - \sqrt{1+20\lambda}}{\left(\frac{2y_2}{r_0} - 1\right) + \sqrt{1+20\lambda}}\right)\right]\right\} \tag{8.25c}$$

where

$$1/\lambda \equiv E \Pr \mathrm{Re} \sqrt{f/2} \tag{8.26}$$

Equation (8.25c) provides an expression for $(T_2 - T_c)$, where T_c is the centerline temperature, by setting $y = r_0$. Then adding this equation and Eqs. (8.25a) and (8.25b), there results the following expression for surface minus centerline temperature:

$$(T_0 - T_c) = \frac{5(q/A)_0}{Ec_p\rho\sqrt{\tau_0/\rho}}\left[E \text{ Pr} + \ln(1 + 5E \text{ Pr}) + 0.5F \ln \frac{\text{Re}}{60}\sqrt{\frac{f}{2}}\right] \tag{8.27}$$

where F is the ratio of the thermal resistance for molecular and eddy conduction to that for eddy conduction alone. This quantity F equals the term in the braces { } of Eq. (8.25c) divided by $0.5 \ln(y/y_2)$. See Table 8.1.

TABLE 8.1 MAGNITUDE OF F IN EQ. (8.30) FOR $E = 1$

Re·Pr \ Re	10^4	10^5	10^6
10^2	0.18	0.098	0.052
10^3	0.55	0.45	0.29
10^4	0.92	0.83	0.65
10^5	0.99	0.985	0.980
10^6	1.00	1.00	1.00

Equation (8.25) can be put in a dimensionless form by dividing by Eq. (8.27). For the region $y^+ < 5$ this becomes

$$\frac{T_0 - T}{T_0 - T_c} = \frac{E \text{ Pr } (y/y_1)}{E \text{ Pr} + \ln(1 + 5E \text{ Pr}) + 0.5F \ln[(\text{Re}/60)\sqrt{f/2}]} \tag{8.28}$$

Equation (8.28) is shown plotted in Fig. 8.3 for the special case of Re = 10,000 and $E = 1.0$. Note the change to flatter temperature profiles for the higher Prandtl-number fluids. The velocity profile coincides with the temperature profile for Pr = 1. The Nusselt number from Eq. (5.1) may be rewritten as follows:

$$\text{Nu} = \frac{hD}{k} = \left[\frac{\partial\left(\dfrac{T_0 - T}{T_0 - T_m}\right)}{\partial\left(\dfrac{y}{D}\right)}\right]_{y=0} \tag{8.29}$$

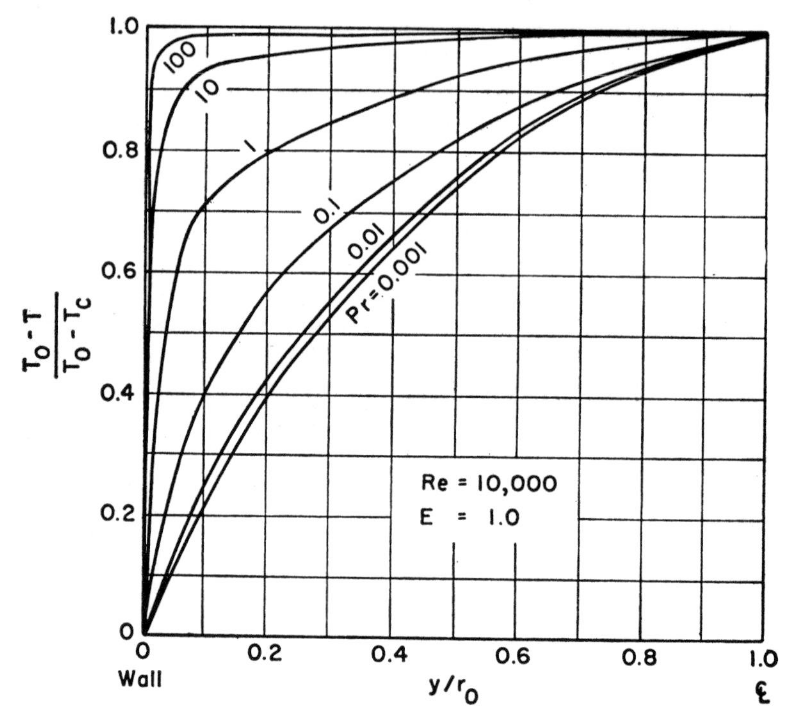

Fig. 8.3. Radial temperature distribution in a tube [Martinelli (7)].

Then by differentiating Eq. (8.28), the expression for the Stanton number (St) becomes:

$$\text{St} = \frac{\text{Nu}}{\text{Re Pr}} = \frac{E\sqrt{\dfrac{f}{2}}\left(\dfrac{T_0 - T_c}{T_0 - T_m}\right)}{5\left[E \text{ Pr} + \ln\,(1 + 5E \text{ Pr}) + \dfrac{F}{2} \ln \dfrac{\text{Re}}{60}\sqrt{\dfrac{f}{2}}\right]} \tag{8.30}$$

Here the mean temperature T_m is defined by

$$\frac{T_0 - T_m}{T_0 - T_c} = \frac{1}{\pi r_0^2} \int_0^{r_0} \frac{v}{V}\left(\frac{T_0 - T}{T_0 - T_c}\right) 2\pi r \, dr \tag{8.31}$$

which may be evaluated from Eq. (8.28) and the universal velocity distribution, Fig. 4.14. In these equations $(T_0 - T_m)/(T_0 - T_c)$ is a function of Re, Pr, and E; further, f is a function of Re. Then St or Nu is a function of Re, Pr, and E.

Martinelli integrated these equations with $E = 1$. Experimental results for $\text{Pr} > 0.5$ agree very well with Eq. (8.30), but most data for liquid metals fall well below the results calculated in this way.

8.7 Modified Martinelli Analogy for Liquid Metals

It was suggested by Jenkins (9) and Deissler (10) that an eddy might not break up soon enough in a liquid metal to discount the effect of internal conduction, causing ϵ_h to be less than ϵ_m or $E < 1.0$. Lykoudis and Touloukian (11) evaluated liquid metal heat transfer data postulating that E is a function of Prandtl number alone.

Rohsenow and Cohen (12) modified these results. Equation (8.30) was solved plotting Nu vs. (Re·Pr) for various magnitudes of E for a given Prandtl number, as shown for Pr = 0.02 in Fig. 8.4. Experimental data superimposed on similar plots for the appropriate Prandtl number show that E is essentially independent of Reynolds number except at fairly low magnitudes of Reynolds number where axial conduction into the upstream fluid becomes significant. The range of magnitudes of E for the experimental data for a given fluid, and hence, given Prandtl number, is shown by the short vertical lines in Fig. 8.5.

Imagine a spherical eddy travelling at velocity v_y' (Fig. 8.1) in a fluid whose temperature varies linearly with time; the conduction equation for a sphere is solved to determine the mean temperature T_m of the eddy after a short time interval t_0, as shown in Fig. 8.6. Deissler (10) showed* that the magnitude of E is

$$E = \frac{T_f - T_m}{T_f - T_i} \tag{8.32}$$

where the temperature subscripts are explained in Fig. 8.6. With this idealized model, with the surface heat transfer coefficient on the spherical

* Imagine a spherical eddy traveling at velocity v_y' (Fig. 8.1) from a region where the temperature is T_i to a region a distance λ away where the temperature is T_f. In traveling the distance λ in time t_0 the mean temperature of the eddy is raised to T_m (Fig. 8.6) by conduction. Then the eddy breaks up and mixes with the fluid at T_f. We define the $(q/A)_\epsilon$ associated with the eddy break-up by

$$\left(\frac{q}{A}\right)_\epsilon = \rho v_y c_p (T_f - T_m) = \rho c_p \epsilon_h \frac{dT}{dy} \tag{A}$$

Also, the shear stress associated with the eddy motion is the momentum change $\rho v_y (\Delta v_x)$ or

$$\tau_\epsilon = \rho v_y \frac{dv_x}{dy} \lambda = \rho \epsilon_m \frac{dv_x}{dy} \tag{B}$$

From this equation $\epsilon_m = v_y \lambda$.

Since $dT/dy = (T_f - T_i)/\lambda$, write Eq. (A) in the form

$$\left(\frac{q}{A}\right)_\epsilon = \rho v_y c_p \frac{T_f - T_m}{T_f - T_i} \lambda \frac{dT}{dy} = \rho c_p \epsilon_h \frac{dT}{dy} \tag{C}$$

Now, since $v_y \lambda = \epsilon_m$, this equation reduces to Eq. (8.32).

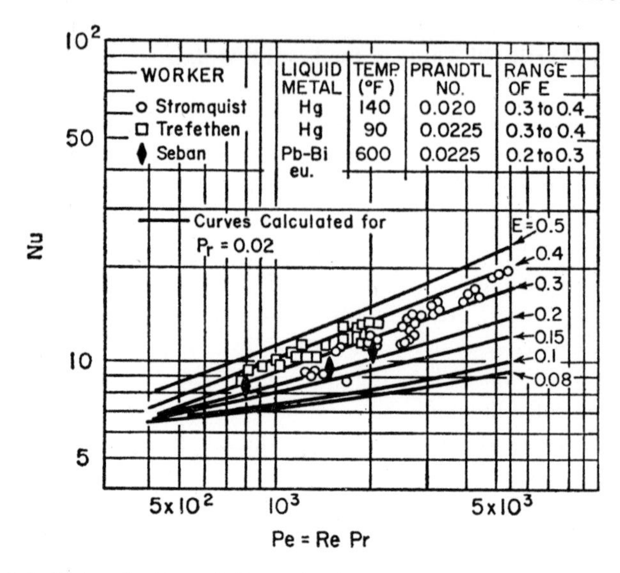

Fig. 8.4. Determination of $E \equiv \epsilon_h/\epsilon_m$ from experimental data and Eq. (8.30).

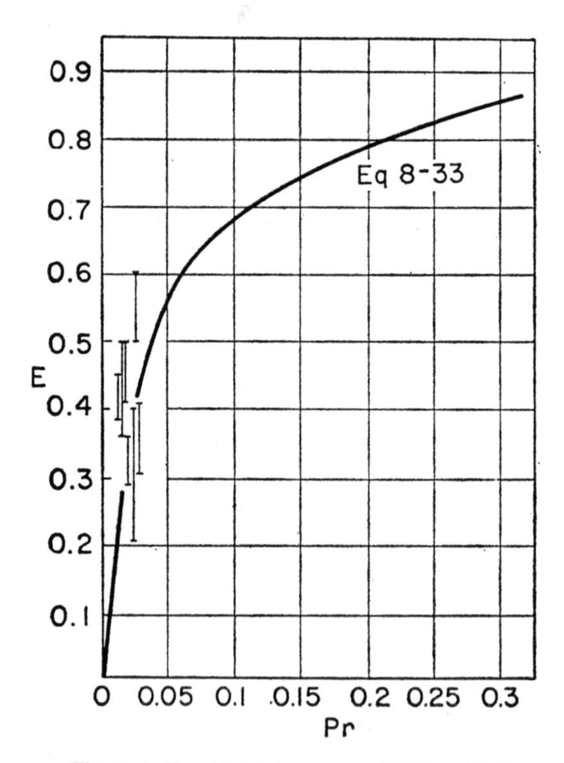

Fig. 8.5. E vs Pr [Rohsenow and Cohen (12)].

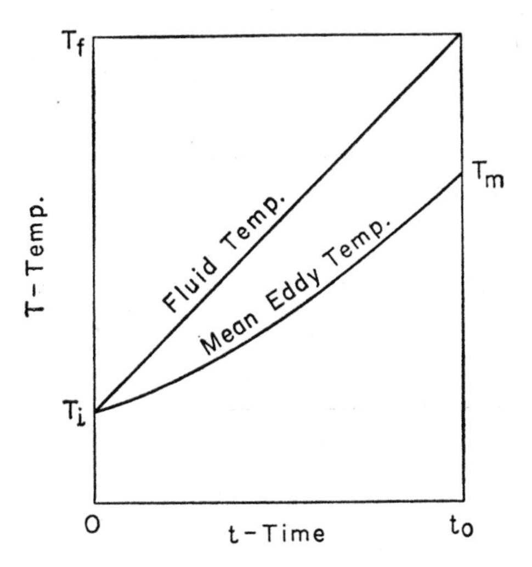

Fig. 8.6. Mean eddy temperature.

eddy taken as ∞, and t_0 taken as being proportional to D^2/ν, Rohsenow and Cohen (12) arrived at the following expression for E:

$$E = 416 \text{ Pr} \left[\frac{1}{15} - \frac{6}{\pi^4} \sum_{n=1}^{\infty} \frac{1}{n^4} \exp \left(-\frac{0.0024\pi^2 n^2}{\text{Pr}} \right) \right] \tag{8.33}$$

which passes through the center of the data points, Fig. 8.5.

With Eq. (8.33), E may be eliminated from Eq. (8.30). This result is shown graphically in Fig. 8.7 where for Pr > 0.1, E was taken as 1.0 and for $0.005 \leq \text{Pr} \leq 0.05$, E was taken from Eq. (8.33). Most liquid metals have Prandtl numbers in the range of 0.003 to 0.03, with no known fluids between 0.05 and 0.5. The calculated or predicted heat transfer results shown in Fig. 8.7 compared reasonably well with available liquid metal

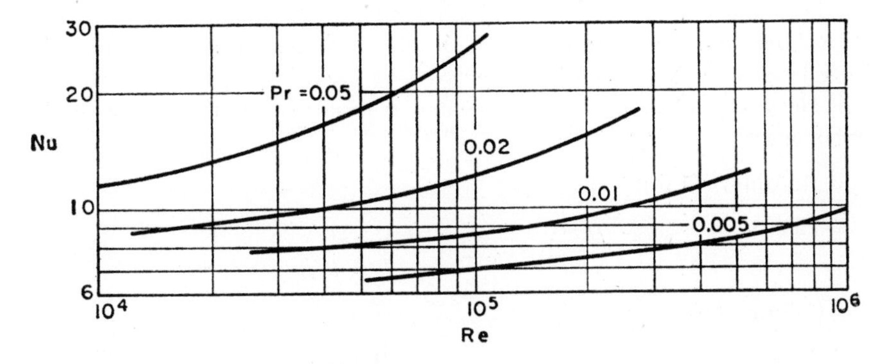

Fig. 8.7. Predicted heat transfer results for turbulent flow in round tubes [Rohsenow and Cohen (12)].

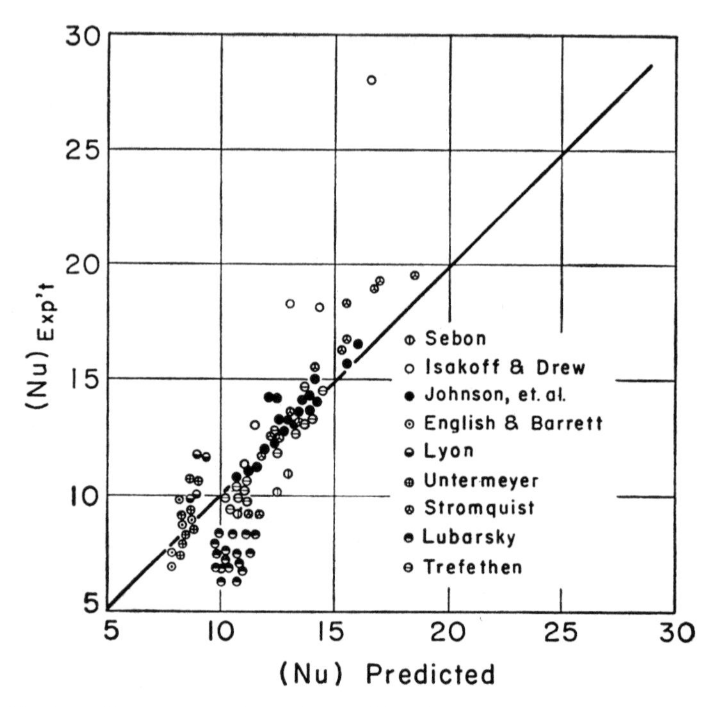

Fig. 8.8. Comparison of experimental liquid metal data with predicted Nu from Fig. 8.7.

data in the range $0.01 < \mathrm{Pr} < 0.024$. Figure 8.8 shows the agreement. Empirical equations representing these results are given in Sec. 8.8.

8.8 Experimental Results for Forced-Convection, Turbulent Flow in Tubes

For fully developed flow in round tubes, dimensional analysis (Chap. 17) shows that heat transfer data in the turbulent-flow region may be correlated with $\mathrm{Nu} = \phi_1(\mathrm{Re}, \mathrm{Pr})$ or $\mathrm{St} = \phi_2(\mathrm{Re}, \mathrm{Pr})$.

Three of the more widely accepted correlations of the data are the following:

McAdams (15):

$$\left(\frac{hD}{k_b}\right) = 0.023\left(\frac{GD}{\mu_b}\right)^{0.8}\left(\frac{\mu c_p}{k}\right)_b^{0.4} \tag{8.34}$$

Colburn (16):

$$\frac{h}{c_{p_b}G}\left(\frac{\mu_f c_{p_b}}{k_b}\right)^{2/3} = \frac{0.023}{(GD/\mu_f)^{0.2}} \tag{8.35}$$

Sieder and Tate (17):

$$\frac{h}{c_{p_b}G}\left(\frac{\mu c_p}{k}\right)_b^{2/3}\left(\frac{\mu_w}{\mu_b}\right)^{0.14} = \frac{0.027}{(GD/\mu_b)^{0.2}} \tag{8.36}$$

Here the subscripts refer to the temperature at which the particular fluid property is evaluated; w refers to wall temperature, b to bulk or mixed mean fluid temperature, and f to a "film" temperature defined by $\frac{1}{2}(T_w+T_b)$. These equations include the effect of temperature-dependent properties and agree well with both experimental data and the results of the analyses in Secs. 8.6 and 8.7 when $0.5 < \mathrm{Pr} < 120$, $2300 < \mathrm{Re} < 10^7$ and $L/D > 50$.

For very large temperature difference $(T_w - T_b)$ with air, Desmon and Sams (13) found the following equation to correlate experimental data:

$$\frac{hD}{k_b} = 0.020\left(\frac{V_b D \rho_f}{\mu_f}\right)^{0.8}\left(\frac{c_p \mu}{k}\right)_b^{0.4} \tag{8.37}$$

This was found to be valid for $\mathrm{Re} > 10^4$ and (T_w/T_b) up to 3.55.

In the liquid metal range,* $0.005 < \mathrm{Pr} < 0.050$, the following equation (12) represents the calculated results of Fig. 8.7, which agree well with experimental data as shown in Fig. 8.8:

$$\mathrm{Nu} = 6.7 + 0.0041\ (\mathrm{Re}\cdot\mathrm{Pr})^{0.793}\ e^{41.8\ \mathrm{Pr}} \tag{8.38}$$

for $\mathrm{Re} > 10{,}000$.

Because of the very high magnitudes of h, the temperature differences $(T_w - T_b)$ are not very large and properties will not vary significantly. Touloukian and Viscanta (14) showed analytically that the effect of property variation could be included as follows:

$$\frac{hD}{k_b} = \mathrm{Nu}_w\left(\frac{\alpha_w}{\alpha_b}\right) \tag{8.39}$$

where α is thermal diffusivity and Nu_w is the Nusselt number evaluated with properties at T_w. Here we recommend Nu_w be calculated from Eq. (8.38).

The equation above for liquid metals applies to the case of uniform heat flux along the tube and is based on actual fluid and wall temperatures. At low Reynolds numbers axial conduction along both the tube wall and the liquid metal itself can introduce departures from this calculated result.

* Earlier Lyons (18) proposed the following equation which correlated the calculated results of Eq. (8.30) only when E was assumed equal to unity: $\mathrm{Nu} = 7 + 0.025\ (\mathrm{Re}\cdot\mathrm{Pr})^{0.8}$. Most of the experimental data, however, fall between 40 and 80 per cent below the values calculated from this equation.

The crosses in Fig. 8.9 represent locations of thermocouples. From these temperature readings one would deduce that the fluid temperature might follow the dotted line; however, the effect of axial conduction upstream would cause the fluid temperature actually to be represented by the solid line. This is perhaps the major cause of disagreement between experimental data and the analytical equations.

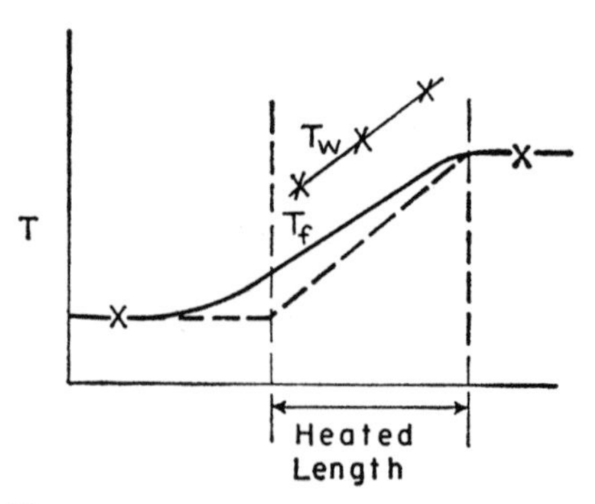

Fig. 8.9. Axial conduction effect on temperature distribution for fluid flowing in a tube.

It has been suggested that wetting and nonwetting of the wall is another cause of disagreement between experimental and analytical results for liquid metals; however, the evidence in the literature is not convincing. Seban and Shimazaki (19) solved the equations of Secs. 8.2 and 8.3 for the case of uniform wall temperature with $E = 1.0$ and obtained the results shown in Fig. 8.10. For Pr > 0.5 the difference between Nu_T (uniform wall temperature) and Nu_q (uniform heat flux) is negligible, but in the liquid-metal region the difference is significant. The results for the case in which E is given by Eq. (8.33) are not currently available.

In the entrance region of round tubes where the flow is developing, Fig. 7.1, the measured experimental magnitudes of h are much greater than in the fully developed flow region. Figure 8.11 shows the nature of experi-

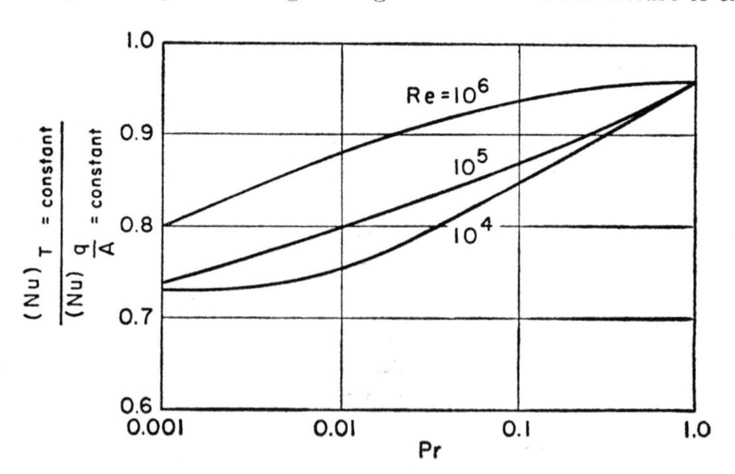

Fig. 8.10. Comparison of effect of uniform wall temperature and uniform heat flux on turbulent flow heat transfer in tubes with $E = 1$ [Seban and Shimazaki (19)].

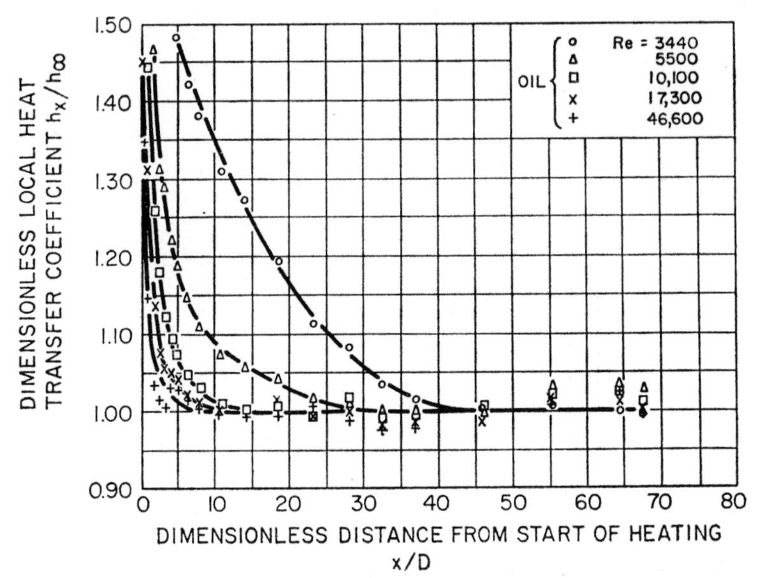

Fig. 8.11. Entrance effect on heat transfer coefficient [Hartnett (20)].

mental results for an oil (Hartnett (20)) with fully developed velocity distribution at the entrance. Results obtained with water and oil agree well with the analytical expressions of Latzko (21) and Deissler (22). McAdams (15) suggests correlating this information with equations of the form

$$\frac{h_m}{h_\infty} = 1 + \frac{C}{(L/D)^n} \tag{8.40}$$

where C is a function of Re_D and, possibly, of Pr.

In addition, the entrance effect varies with the type of entrance condition—fully developed flow at entrance, bell-mouthed entrance, sharp-edged entrance, etc. For any specific use, the papers of Hartnett (20), Latzko (21), Deissler (22), and Boelter *et al.* (23) should be consulted. In the range Re_D of 26,000 to 56,000 experimental data for air (23) with a bell-mouthed entrance was correlated by Eq. (8.40) with $C = 1.4$ and $n = 1$. With Re > 10,000 the entrance effects are limited to the region of $L/D < 20$.

For fully developed turbulent flow of air in a helical coil, Jeschke (24) obtained the following relation for the ratio of h_c (in the coil) to h (in a straight pipe):

$$\frac{h_c}{h} = 1 + 3.5\frac{D}{D_{He}} \tag{8.41}$$

where D is the pipe diameter, and D_{He} the helix diameter.

The effect of roughness (rings of various shapes attached to the inside surface of the pipe) for air in fully developed turbulent flow was investigated by Nunner (25). The results shown in Fig. 8.12 relate the heat transfer coefficient with Re_D and the ratio of friction factor in the rough pipe, f, to friction factor in a smooth pipe, f_s, of the same diameter.

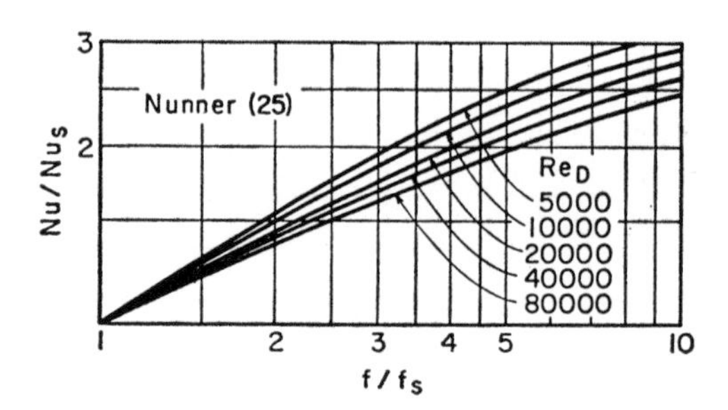

Fig. 8.12. Effect of roughness on heat transfer in turbulent flow [Nunner (25)].

For flow in noncircular channels—rectangular, triangular, etc.—in turbulent flow a good approximation is to multiply Eqs. (8.34) through (8.38) by 0.75 and replace D by the hydraulic diameter D_e of Eq. (4.14).

It should be noted that the dimensionless equations of this section can be simplified for specific fluids. For example, Eq. (8.34) for water in the neighborhood of 40 to 200 F may be simplified as follows (15):

$$h = \frac{150(1 + 0.011T_b)(V')^{0.8}}{(D')^{0.2}} \tag{8.42}$$

where T_b is water temperature in °F, V' in ft/sec, and D' in inches.

An interesting observation results from comparing Eq. (4.7) for f with Eq. (8.35) for Nu:

$$\frac{h}{c_pG}\,\mathrm{Pr}^{2/3} = \frac{f}{2} \tag{8.43}$$

this equation is known as the *Colburn analogy between heat transfer and fluid friction*.

8.9 Combined Free and Forced Convection in Tubes

A fluid flowing in either a horizontal or vertical tube with a radial temperature distribution will have natural convection superimposed on the forced flow. In a horizontal tube with hot walls, natural convection occurs as shown in Fig. 8.13. This motion superimposed on the axial flow distorts the axial velocity distributions of Fig. 7.18. At low flow rates, laminar flow range, these effects are more pronounced than at high flow rates, turbulent flow. In addition, the effect of temperature on viscosity distorts the familar velocity distribution as discussed in connection with Fig. 4.4. These variable-property effects are significant with liquids in laminar flow.

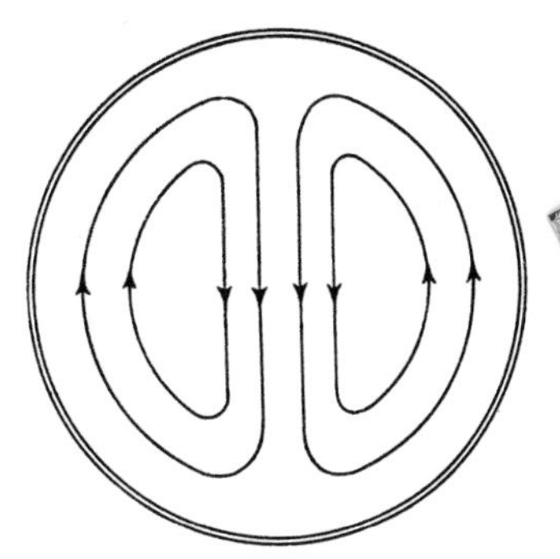

Fig. 8.13. Natural convection in horizontal hot tubes.

The laminar-flow, heat-transfer results in Secs. 7.2 and 7.7 are correct only for the cases in which D and ΔT are very small, where natural convection effects are negligible.

Martinelli and Boelter (26) treated the cases of fluids flowing in vertical tubes. The results were correlated by the following equation:

$$\frac{h_a D}{k_b} = 1.75 F_1 \sqrt[3]{\frac{\pi}{4} (\text{Re}_D \cdot \text{Pr})_b \frac{D}{L} \pm 0.0722 \left[\frac{D}{L} \text{Gr}_D \text{Pr}\right]_w^{0.84} F_2} \quad (8.44)$$

where h_a is based on ΔT_a, the arithmetic mean temperature difference between entrance and exit; Gr_D is evaluated at inlet ΔT; and functions F_1 and F_2 are plotted in Fig. 8.14. The $+$ sign applies to heating upward flow and cooling downward flow; the $-$ sign applies to cooling upward flow and heating downward flow. Figure 8.15 shows agreement between data and Eq. (8.44). Additional results are presented by Pigford (27) for a wider temperature range.

The data for horizontal tubes was correlated empirically by Eubank and Proctor (28) by the following equation:

$$\frac{h_a D}{k}\left(\frac{\mu_w}{\mu_b}\right)^{0.14} = 1.75 \sqrt[3]{\frac{\pi}{4} (\text{Re}_D \text{Pr})_b \frac{D}{L} + 0.04 \left[\frac{D}{L} \text{Gr}_D \text{Pr}\right]_b^{0.75}} \quad (8.45)$$

where Gr_D is based on D and ΔT_a. The data correlated covered the range $3 \times 10^5 < \text{Gr}_D \text{Pr} < 9 \times 10^8$ and $140 < \text{Pr} < 15{,}200$.

Other analyses and data for combined free and forced convection in a variety of flow arrangements are presented by Eckert and Diagula (29).

8.10 Turbulent Flow over Flat Plates

For turbulent flow over a flat plate, an analysis similar to the one outlined in Sec. 8.6 can be performed. This was done for $E = 1$ by Kármán (30). Rather than repeat this type of analysis, we now show that the simple analogy of Colburn can be extended to the flat-plate case.

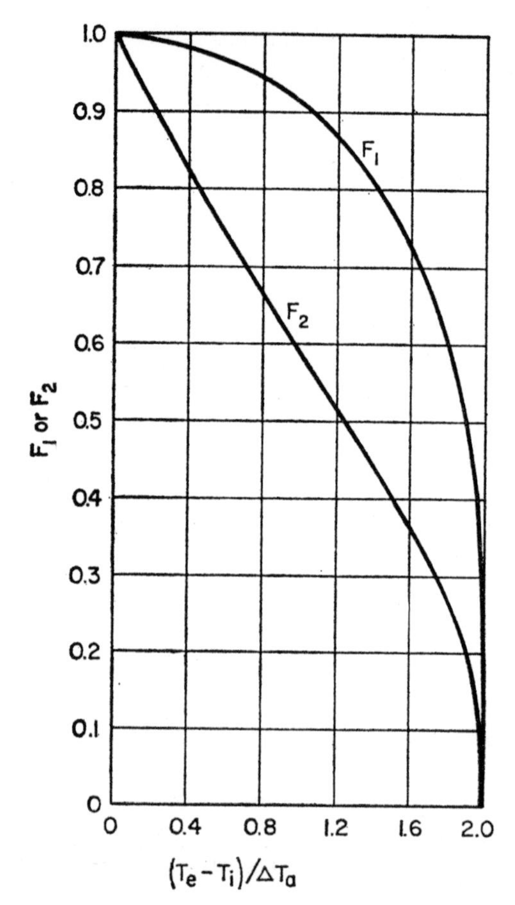

Fig. 8.14. F_1 and F_2 for Eq. (8.44) [Martinelli and Boelter (26)].

We assume Eq. (8.43) can represent the relation between local heat transfer coefficient h_x and local friction factor $C_{f,x}$ for a flat plate:

$$\frac{h_x}{c_p \rho V_\infty} \mathrm{Pr}^{2/3} = \frac{C_{f,x}}{2} \tag{8.46a}$$

From* Eq. (4.43) $C_{f,x} = 0.0592(V_\infty x/\nu)^{-1/5}$; then Eq. (8.46a) becomes:

$$\frac{h_x}{c_p \rho V_\infty} \mathrm{Pr}^{2/3} = 0.0296 \sqrt[5]{\frac{\nu}{V_\infty x}} \tag{8.46b}$$

or

$$\frac{h_x x}{k} = 0.0296 \left(\frac{V_\infty x}{\nu}\right)^{0.8} \mathrm{Pr}^{1/3} \tag{8.46c}$$

* Here the coefficient 0.0592 replaces 0.0576 since 0.0592 is in better agreement with the experimental data.

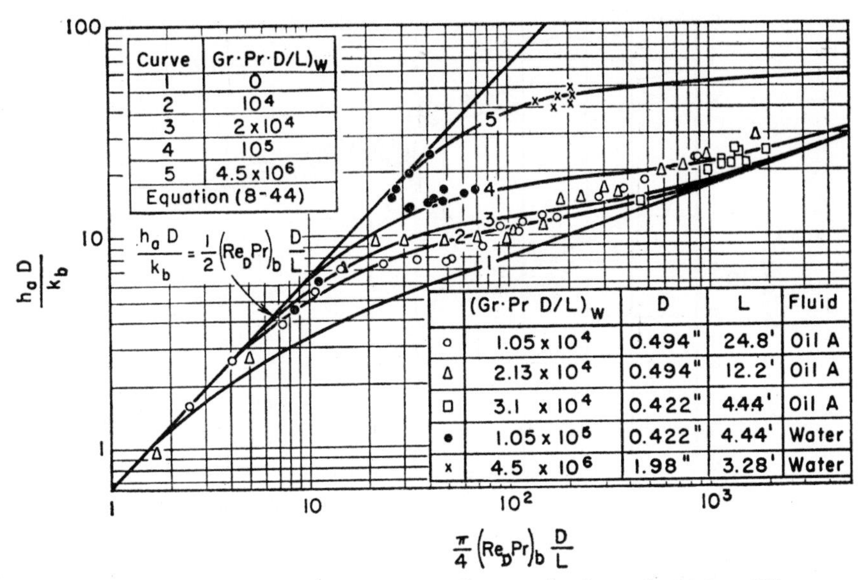

Fig. 8.15. Superimposed free and forced convection in vertical tubes (26).

Since $h_x \sim x^{-0.2}$, the average coefficient h between $x = 0$ and $x = L$ becomes

$$\frac{hL}{k} = 0.037\left(\frac{V_\infty L}{\nu}\right)^{0.8} \mathrm{Pr}^{1/3} \tag{8.47}$$

Experimental data (31) agree well with Eqs. (8.46) and (8.47) as shown in Fig. 8.16. Also shown is the agreement with the laminar-boundary-layer result, Eq. (7.42a). Johnson and Rubesin (32) suggest accounting for properties variation by evaluating the properties in Eq. (8.47) at $T_f = \frac{1}{2}(T_w + T_\infty)$.

Over a flat plate a laminar boundary layer exists over the forward portion between 0 and L_{tr}, and the turbulent boundary layer exists beyond L_{tr}. With this simple model the average h over a plate of length $L > L_{tr}$ is calculated as follows (32) with Eq. (7.39) and Eq. (8.46), assuming the turbulent boundary layer extrapolates to $\delta = 0$ at $x = 0$:

$$h = \frac{1}{L}\left[\int_0^{L_{tr}} h_{x,\ \mathrm{lam}}\, dx + \int_{L_{tr}}^{L} h_{x,\ \mathrm{turb}}\, dx\right] \tag{8.48}$$

Since $L_{tr}/L = \mathrm{Re}_{tr}/\mathrm{Re}_L$, for $\mathrm{Re}_{tr} = 3.2 \times 10^5$ (Sec. 2.2a), Eq. (8.48) becomes

$$\frac{hL}{k} = 0.037\ \mathrm{Pr}^{1/3}\left[(\mathrm{Re}_L)^{0.8} - 15{,}500\right] \tag{8.49}$$

A more precise formulation is given by Reynolds, *et al.*, (39). Note that Eqs. (8.46) through (8.49) are valid for $\mathrm{Pr} > 0.5$ but not for liquid metals.

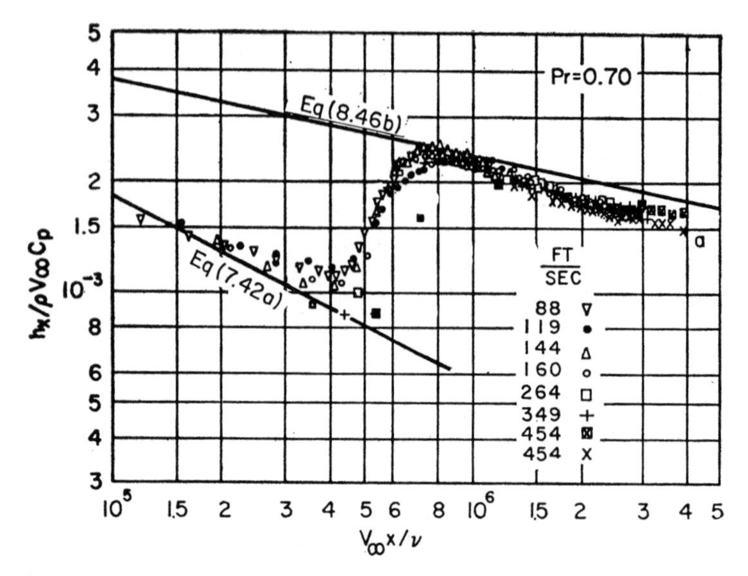

Fig. 8.16. Local heat transfer coefficient for air-flow over plate with turbulent boundary layer and uniform heat flux [Seban and Doughty (31)].

An analysis similar to the one in Sec. 8.7 could be performed, but to date is not available. The equations are valid for either uniform wall temperature or uniform heat flux.

8.11 Experimental Results for Flow over Submerged Bodies

The flow and drag characteristics for flow across cylinders, spheres, and banks of tubes are discussed in Secs. 2.2a and 4.9. Figure 8.17 shows data of Giedt (33) for local values of Nu_θ at any position around the cylinder. At Re < 100,000, separation occurs near the 80-deg position. Beyond this point h_θ increases because of the increasing turbulence in the wake. At higher Re where the laminar boundary layer changes to turbulent before separation occurs, the Nu_θ curve has two minimum points—the first at transition to turbulence and the second at the separation point.

The more generally useful result is the average value of h or Nu over the cylinder surface. McAdams (15) correlated the data for water and four hydrocarbon oils with the following equation:

$$1 < \frac{D_0 G}{\mu_f} < 1000$$

$$\frac{h_m D_0}{k_f} (\mathrm{Pr}_f)^{-0.3} = 0.35 + 0.56\left(\frac{D_0 G}{\mu_f}\right)^{0.52} \tag{8.50}$$

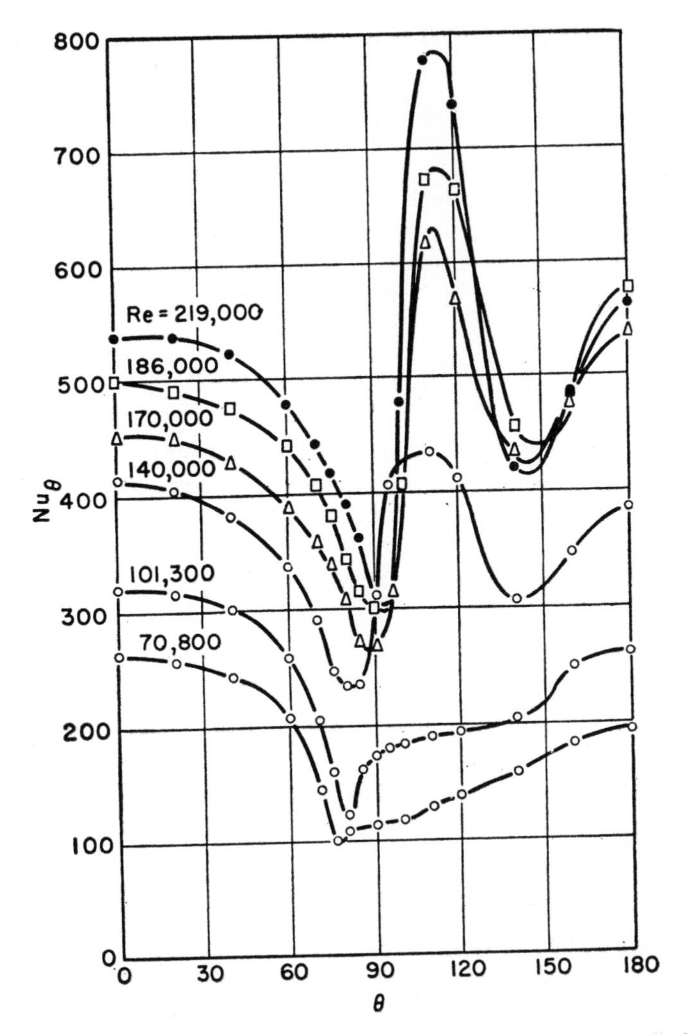

Fig. 8.17. Circumferential variation of h for crossflow of air over a cylinder [Giedt (33)].

Air data agreed well with this equation in the specified range of Re. For $1000 < D_0G/\mu_f < 50{,}000$ air data were correlated by

$$\frac{h_m D_0}{k_f} (\mathrm{Pr}_f)^{-0.3} = 0.26\left(\frac{D_0 G}{\mu_f}\right)^{0.60} \tag{8.51}$$

For flow of water and oil ($7.3 < \mathrm{Pr} < 380$) past single spheres (0.279 in. $< D_0 < 0.496$ in.) data of Kramers was correlated by McAdams in the range $1 < D_0G/\mu_f < 1000$ as follows:

$$\frac{h_m D_0}{k_f} (\text{Pr}_f)^{-0.3} = 0.97 + 0.68\left(\frac{D_0G}{\mu_f}\right)^{0.50} \tag{8.52}$$

As D_0G/μ_f gets smaller and approaches unity, the magnitude of Nu should approach 2, which is the calculated result for pure conduction of a sphere at uniform temperature in an infinite stagnant medium.

For flow across banks of tubes that are 10 or more rows deep, the data of Colburn (34) for $D_0G_{\max}/\mu_f > 6000$ can be correlated by the following equation:

$$\frac{h_m D_0}{k_f} = C\left(\frac{D_0G_{\max}}{\mu_f}\right)^{0.6} \text{Pr}_f^{1/3} \tag{8.53}$$

where $C = 0.33$ for staggered tubes and $C = 0.26$ for in-line tubes.

Using their test data, Kays and Lo (35) suggest the magnitudes in the following table for banks of tubes with less than 10 rows.

RATIO h_m FOR N ROWS TO h_m FOR 10 ROWS DEEP

N	1	2	4	6	8	10
Staggered tubes..............	0.68	0.75	0.89	0.95	0.98	1.0
In-line tubes..................	0.64	0.80	0.90	0.94	0.98	1.0

Data (36) for mercury ($\text{Pr} = 0.022$) flowing over a 10-row-deep bank of staggered $\frac{1}{2}$ in. tubes, 1.375 pitch-to-diameter ratio, was correlated as follows:

$$\text{Nu} = 4.03 + 0.228\,(\text{Re}_{\max}\,\text{Pr})_f^{0.67} \tag{8.54}$$

for the range $20{,}000 < \text{Re}_{\max} < 80{,}000$.

For flow through packed beds of spheres as discussed in Sec. 4.10, Eckert (2) suggests the following equation based on data of Denton:

$$\frac{hD_p}{k} = 0.80\left(\frac{D_pG_0}{\mu}\right)^{0.7} \text{Pr}^{1/3} \tag{8.55}$$

where the symbols are defined in Sec. 4.10 and the properties are those of the fluid. Correlations of other data are found in McAdams (15).

8.12 Turbulent Free Convection over a Vertical Flat Plate

We now illustrate the integral method of solution applied to a turbulent boundary layer for natural convection over a vertical flat plate at uniform temperature T_0 in a fluid of uniform temperature T_∞. Eckert and Jackson

(37) solved this problem assuming the thicknesses, δ, of the velocity and temperature boundary layers to be equal. Experimental data for velocity and temperature distribution for air showed reasonable agreement with the following approximations:

$$\frac{T - T_\infty}{T_0 - T_\infty} = \left[1 - \left(\frac{y}{\delta}\right)^{1/7}\right]$$
$$v_x = \Gamma\left(\frac{y}{\delta}\right)^{1/7}\left(1 - \frac{y}{\delta}\right)^4 \tag{8.56}$$

As in Sec. 4.8 for forced flow, these equations cannot be expected to give τ_0 and $(q/A)_0$ by their gradients at the wall. In Eqs. (7.62) and (7.63), the momentum and the energy integral equations, we replace* $\nu(dv_x/dy)_0$ by τ_0/ρ and $-\alpha(dT/dy)_0$ by $(q/A)_0/\rho c_p$ and use experimental values for these quantities. For convenient reference, these equations are reproduced here:

$$\frac{d}{dx}\int_0^\delta v_x^2\,dy = -\frac{\tau_0}{\rho} + g\beta\int_0^\delta (T - T_\infty)\,dy \tag{7.62}$$

$$\frac{d}{dx}\int_0^\delta (T - T_\infty)v_x\,dy = \frac{(q/A)_0}{\rho c_p} \tag{7.63}$$

The expressions for τ_0 and $(q/A)_0$ at the wall might be expected to be the same for the forced convection and the free convection boundary layers. For τ_0, we use Eq. (4.41):

$$\tau_0 = 0.0225\rho V_\infty^2\left(\frac{\nu}{V_\infty\delta}\right)^{1/4} \tag{4.41}$$

For $(q/A)_0$ we first combine Eqs. (8.46) and (4.42):

$$\frac{h_x}{c_p\rho V_\infty}\,\mathrm{Pr}^{2/3} = 0.0296\left(\frac{\nu}{V_\infty x}\right)^{1/5} \tag{8.46b}$$

$$\frac{\delta}{x} = 0.37\left(\frac{V_\infty x}{\nu}\right)^{-1/5} \tag{4.42}$$

In the resulting expression, we may replace V_∞ by Γ, since comparison of Eq. (8.56) with Eq. (4.40) shows that Γ and V_∞ are similar as $y/\delta \to 0$. Then the following equation results:

$$\frac{(q/A)_0}{\rho c_p} = 0.0225(T_0 - T_\infty)\,\Gamma\left(\frac{\nu}{\Gamma\delta}\right)^{1/4}\mathrm{Pr}^{-2/3} \tag{8.57}$$

* See footnote on page 185.

Substituting all of these expressions into the integral equations (7.62) and (7.63), we get

$$0.0523 \frac{d}{dx}(\Gamma^2\delta) = 0.125 g\beta(T_0 - T_\infty)\delta - 0.0225\Gamma^2\left(\frac{\nu}{\Gamma\delta}\right)^{1/4}$$

$$0.0366 \frac{d}{dx}(\Gamma\delta) = 0.0225\Gamma\left(\frac{\nu}{\Gamma\delta}\right)^{1/4} \mathrm{Pr}^{-2/3} \tag{8.58}$$

As before, we assume the solutions of the form $\Gamma = Ax^m$ and $\delta = Bx^n$, Eq. (7.60).

Substitute these in Eqs. (8.58) and perform the indicated operations. The magnitudes of A, B, m, and n are then determined by equating the coefficients and exponents of x in the two equations. The results are:

$$m = \tfrac{1}{2}, \quad n = 0.7$$

$$A = 0.0689\nu B^{-5} \mathrm{Pr}^{-8/3} \tag{8.59}$$

$$B^{10} = 0.00338\frac{\nu^2}{g\beta(T_0 - T_\infty)}[1 + 0.494\ \mathrm{Pr}^{2/3}]\ \mathrm{Pr}^{-16/3}$$

Then since $\Gamma = Ax^m$ and $\mathrm{Gr}_x = g\beta(T_0 - T_\infty)x^3/\nu^2$,

$$\Gamma = 1.185\frac{\nu}{x}\ (\mathrm{Gr}_x)^{1/2}[1 + 0.494\ \mathrm{Pr}^{2/3}]^{-1/2} \tag{8.60}$$

and since $\delta = Bx^n$,

$$\frac{\delta}{x} = 0.565(\mathrm{Gr}_x)^{-0.1}\ \mathrm{Pr}^{-8/15}[1 + 0.494\ \mathrm{Pr}^{2/3}]^{0.1} \tag{8.61}$$

Substituting these expressions of Γ and δ in Eq. (8.57) gives the following result:

$$\mathrm{Nu}_x = 0.0295(\mathrm{Gr}_x)^{2/5}\ \mathrm{Pr}^{7/15}[1 + 0.494\ \mathrm{Pr}^{2/3}]^{-2/5} \tag{8.62a}$$

or the average value is

$$\mathrm{Nu} = 0.0246(\mathrm{Gr}_L)^{2/5}\ \mathrm{Pr}^{7/15}[1 + 0.494\ \mathrm{Pr}^{2/3}]^{-2/5} \tag{8.62b}$$

This equation agrees well with experimental data for water and for air. It has not been checked for wide range of Pr. Because of the assumption of equal thermal and velocity boundary-layer thickness, it probably is not valid for liquid metals.

8.13 Experimental Results for Natural Convection

Figure 8.18 shows representative data for air and water in natural convection over vertical plates. The transition between laminar and

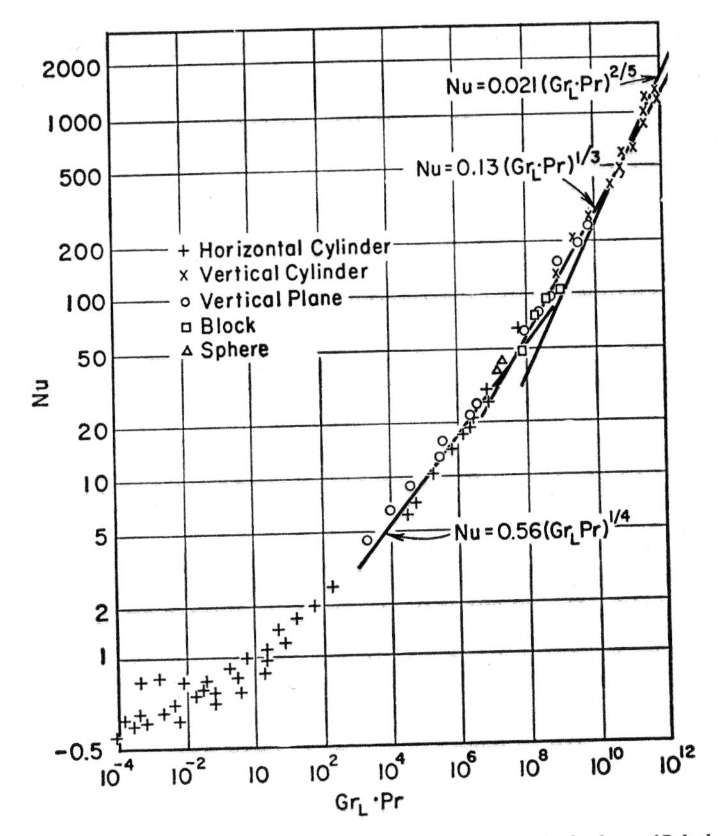

Fig. 8.18. Free convection of air and water over vertical plates [Jakob (40)].

turbulent flows occurs in the neighborhood of $\mathrm{Gr}\cdot\mathrm{Pr} = 10^8$ to 10^{10}, usually taken as 10^9.

In the laminar flow range $10^4 < \mathrm{Gr}\cdot\mathrm{Pr} < 10^9$, Eq. (7.59b) correlates data over a wide range of Pr, Fig. 7.15:

$$\frac{\mathrm{Nu}}{\sqrt[4]{\mathrm{Gr}_L/4}} = \frac{0.902\ \mathrm{Pr}^{1/2}}{(0.861 + \mathrm{Pr})^{1/4}} \tag{7.59b}$$

In the range of Pr near unity, data have been correlated by the following equation:

$$\mathrm{Nu} = C_L(\mathrm{Gr}_L\cdot\mathrm{Pr})^{1/4} \tag{8.63}$$

where $C_L = 0.56$ for Pr in the range 1 to 10.

In the turbulent range $10^9 < \mathrm{Gr}\cdot\mathrm{Pr} < 10^{12}$, Eq. (8.62) correlates air and water data very well and may be valid for a much wider range of Pr, but this has not been verified experimentally. In the Prandtl number range from 1 to 10 the data have been correlated by the following equation:

$$\mathrm{Nu} = C_T(\mathrm{Gr}_L \cdot \mathrm{Pr})^{1/3} \tag{8.64}$$

where $C_T = 0.13$.

Equation (8.62) may be written as follows:

$$\mathrm{Nu} = K_T(\mathrm{Gr}_L \cdot \mathrm{Pr})^{2/5} \tag{8.65}$$

where K_T is a function of Pr as follows:

Pr	1	5	10	100
K_T	0.0210	0.0192	0.0178	0.013

These two equations are shown plotted in Fig. 8.18.

Note that in turbulent flow since $\mathrm{Gr} \sim L^3$ and $\mathrm{Nu} \sim L$ the magnitude of h is independent of L in Eq. (8.64) and only weakly dependent in Eq. (8.65).

Data for horizontal cylinders in air and water with $10^3 < \mathrm{Gr} \cdot \mathrm{Pr} < 10^9$ are correlated by Eq. (8.63) if the length L is replaced by $\pi D/2$. Data for liquid metals (Na, NaK, Pb, Pb-Bi, and Hg) are correlated (38) in the laminar range by the following equation:

$$\mathrm{Nu} = 0.53\left[\frac{\mathrm{Pr}}{0.952 + \mathrm{Pr}}\right]^{1/4}(\mathrm{Gr}_D \cdot \mathrm{Pr})^{1/4} \tag{8.66}$$

The effect of the variation of properties with temperature has been correlated with the above equations by evaluating the properties at $T_f = \frac{1}{2}(T_0 + T_\infty)$.

The correlation equations may be simplified for any specific fluid over a limited temperature range. The following simplified equations result from those above using properties of air:

Vertical Plates:

$$10^{-2} < L^3\,\Delta T < 10^3, \quad h = 0.29(\Delta T/L)^{1/4} \tag{8.67a}$$

$$10^3 < L^3\,\Delta T < 10^6, \quad h = 0.21(\Delta T)^{1/3} \tag{8.67b}$$

Horizontal pipes:

$$10^{-2} < D^3\,\Delta T < 10^3, \quad h = 0.25(\Delta T/D)^{1/4} \tag{8.67c}$$

$$10^3 < D^3\,\Delta T < 10^6, \quad h = 0.18(\Delta T)^{1/3} \tag{8.67d}$$

Horizontal square, hot plate facing up (cold plate facing down):

$$0.1 < L^3\,\Delta T < 20, \quad h = 0.27(\Delta T/L)^{1/4} \tag{8.67e}$$

$$20 < L^3\,\Delta T < 30{,}000, \quad h = 0.22(\Delta T)^{1/3} \tag{8.67f}$$

Horizontal square, hot plate facing down (cold plate facing up):

$$0.3 < L^3 \, \Delta T < 30{,}000, \quad h = 0.12(\Delta T/L)^{1/4} \qquad (8.67g)$$

These simplified equations apply reasonably well to air in the temperature range of 100 to 1500 F and also to CO, N_2, O_2, and flue gases.

REFERENCES

1. Reichardt, H., *Z. ang. Math. u. Mech.*, **20**:6, 297–328 (Dec. 1940); *Arch. d. Wärmetech.*, 2 Jahrgang 1951, Heft 6/7, 129–42.
2. Eckert, E. R. G., *Introduction to Heat and Mass Transfer*, McGraw-Hill, New York, 1950.
3. Reynolds, O., *Proc. Manchester Lit. Phil. Soc.*, **8** (1874), reprinted in *Scientific Papers of Osborne Reynolds*, Vol. II, Cambridge, London, 1901.
4. Prandtl, L., *Z. Physik*, **11,** 1072 (1910); **29,** 487 (1928).
5. Taylor, G. I., *Brit. Adv. Comm. Aero. Rept. Mem.*, **272**:31, 423–29 (1916).
6. Von Kármán, T., *Trans. ASME*, **61,** 705 (1939).
7. Martinelli, R. C., *Trans. ASME*, **69,** 947 (1947).
8. Gazely, Carl, Discussion of ref. 7.
9. Jenkins, R., *Fl. Mech. and Ht. Trans. Inst.*, 1951, Stanford U. Press, pp. 147–58.
10. Deissler, R. G., *Nat. Advisory Comm. Aeronaut. Res. Mem.* E52F05 (1952).
11. Lykoudis, P. S., and Y. Touloukian, *Trans. ASME*, Apr. 1958, pp. 653–66.
12. Rohsenow, W. M., and L. S. Cohen, M.I.T. Heat Transfer Laboratory Report June, 1960.
13. Desmon, L. G., and E. W. Sams, *Nat. Advisory Comm. Aeronaut. Res. Mem.* E50H23 (1950).
14. Viskanta, R., and Y. Touloukian, "Heat Transfer to Liquid Metals with Variable Properties," ASME Paper No. 59-A-148.
15. McAdams, W. H., *Heat Transmission*, 3rd Ed., McGraw-Hill, New York, 1954.
16. Colburn, A. P., *Trans. Am. Inst. Chem. Engrs.*, **29,** 174 (1933).
17. Sieder, E. N., and G. E. Tate, *Ind. Eng. Chem.*, **28,** 1429–36 (1936).
18. Lyons, R. W., *Chem. Eng. Prog.*, **47,** 75–79 (1951).
19. Seban, R. A., and T. T. Shimazaki, *Trans. ASME*, **73,** 803 (1951).
20. Hartnett, J. P., *Trans. ASME*, **77**:7, 1211 (Nov. 1955).
21. Latzko, H., *Z. angew. Math. u. Mech.*, **1,** 268–90 (1921).

22. Deissler, R. G., *Trans. ASME*, **77**:7, (Nov. 1955); also *Nat. Advisory Comm. Aeronaut. Tech. Note* 3016.

23. Boelter, L. M. K., G. Young, and H. W. Iversen, *Nat. Advisory Comm. Aeronaut. Tech. Note* 1451 (1948).

24. Jeschke, D., *Z. ver. deut. Ing.*, **69,** 1526 (1925).

25. Nunner, W., *Z. ver. deut. Ing., Forschungsheft*, **455** (1956).

26. Martinelli, R. C., and L. M. K. Boelter, *U. of Cal. Publ.*, **5**:2, 23–58 (1942).

27. Pigford, R. L., publ. in ref. 15, p. 236.

28. Eubank, O. C., and W. S. Proctor, S.M. Thesis, Chem. Engrg. Dept., M.I.T., 1951.

29. Eckert, E. R. G., and A. J. Diagula, *Trans. ASME*, **75,** 497–504 (1954).

30. Eckert, ref. 2, p. 225.

31. Seban, R. A., and D. L. Doughty, *Trans. ASME*, **78**:1, 217 (1956).

32. Johnson, H. A., and M. W. Rubesin, *Trans. ASME*, **71,** 447 (1949).

33. Giedt, W. H., *Trans. ASME*, **71** (1949).

34. Colburn, A. P., *Trans. Am. Inst. Chem. Engrs.*, **29,** 174–210 (1933).

35. Kays, W. M., and R. K. Lo, TR-15, NR-035-104, Mech. Engrg. Dept., Stanford U., Aug. 1952.

36a. Hoe, R. J., D. Dropkin, and O. E. Dwyer, *Trans. ASME*, **79,** 899 (1957).

36b. Richards, C. L., O. E. Dwyer, and D. Dropkin, ASME Paper 57-HT-11, 1957.

37. Eckert, E. R. G., and T. W. Jackson, *Nat. Advisory Comm. Aeronaut. Tech. Note* 2207 (Oct. 1950).

38. *Reactor Handbook*, Vol. 2, *Engineering*, AECD-3646 (May 1955), p. 283.

39. Reynolds, W. C., W. M. Kays, and S. J. Kline, "Effect of Location of Transition on Heat Transfer in the Turbulent Incompressible Boundary Layer." Mech. Engrg. Dept. Rept. under Contract NAW-6494, Stanford U., July 1957.

40. Jakob, M., *Heat Transfer*, Vol. I, Wiley, 1949, p. 529.

PROBLEMS

8.1 Obtain Eqs. (8.7) and (8.8) by time-averaging term by term Eqs. (8.4c) and (8.5c).

8.2 Write Eq. (8.7) in cylindrical coordinates for turbulent flow in a round tube. Show that for steady, fully developed turbulent flow in a tube, τ_{app} varies linearly with the radius. State all assumptions leading to the final result.

8.3 Using the Reynolds analogy, $E = 1$, and the approximate expression for the friction factor of smooth pipes, Eq. (4.7), plot Nu_x vs. Re_D for $Pr = 1$ in the turbulent flow range $10^4 < Re_D < 10^6$.

8.4 Start with Eqs. (8.21) and (8.22) and derive an expression for the temperature distribution in steady, fully developed turbulent flow between parallel plates, assuming $\epsilon_m = \epsilon_h$ and the velocity profile, τ_0, and $(q/A)_0$ are known from measurements.

8.5 Assume the fluid in fully developed flow in a pipe may be divided into two zones—a laminar region (ϵ_h and ϵ_m negligibly small compared with α and ν) near the wall ($0 < y < \delta$) and a fully turbulent region (α and ν negligibly small compared with ϵ_h and ϵ_m) in the core $\delta < y < r_0$. Divide Eq. (8.21) by Eq. (8.22) and integrate first between $y = 0$ and δ and then between $y = \delta$ and y_m, the location where $v = V$ and $T = T_m$. Combine the results to eliminate T_δ and obtain the following equation:

$$\frac{h}{c_p G} = \frac{Ef/2}{1 + (v_\delta/V)(E\,\mathrm{Pr} - 1)}$$

With $E = 1$, this is the Prandtl (1910) analogy modified later by Taylor (1916). Hoffman [*Z. ges. Kälte-Ind.*, **44.** 99 (1937)] showed that with $E = 1$, v_δ/V in the above equation could be replaced by $7.54\ \mathrm{Pr}^{-1/6}\sqrt{f/2}$ and results agreed well with experimental data for $Pr > 0.5$ (gases, water, oils, etc.).

8.6 Oil flows in a long 2-in. pipe at an average velocity of 10 ft/sec. If the pipe wall is at 200 F and the bulk temperature of oil is 100 F, compare the heat transfer coefficients calculated by Eqs. (8.30), (8.34), (8.35), and (8.36).

8.7 Plot the temperature profile for turbulent flow of air in a smooth pipe, using Eqs. (8.25a, b, c) for $Re = 100{,}000$. Assume $E = 1$ and evaluate properties of air at 100 F. Use the profile to find the Nusselt number defined by Eq. (8.29). Also calculate the Nusselt number from Eq. (8.30).

8.8 Water flows in a heat exchanger tube 3 ft long and 1 in. in diameter at a velocity of 15 ft/sec. The tube wall temperature is kept constant at 210 F by condensing steam. If the inlet temperature is 60 F, calculate the exit temperature. Neglect the entrance effects and use properties of water evaluated at the average of the inlet and exit bulk temperatures.

8.9 Mercury at a mean temperature of 200 F flows through a $\frac{1}{2}$-in. diameter tube at the rate of 10,000 lb_m/hr. Using the value of E from Fig. 8.5, calculate the average heat transfer coefficient from Eq. (8.30). Check your answer by calculating h directly from Eq. (8.38).

8.10 Power generation in a nuclear reactor is limited principally by the ability to transfer heat in the reactor. A solid-fuel reactor is cooled by fluid flowing inside 0.25-in. diameter stainless steel tubes. If the tube wall temperature is 600 F, compare the relative merits of using water or liquid sodium as the coolant. In each case, the velocity is 15 ft/sec and the fluid inlet temperature is 400 F.

8.11 Air at 60 F flows over a 1-ft-square plate maintained at 340 F. For an air velocity of 150 ft/sec:

(a) Locate the transition point, assuming $Re_{tr} = 3.2 \times 10^5$.

(b) Plot the local heat transfer coefficient in the laminar boundary-layer region.

(c) Plot the local heat transfer coefficient in the turbulent boundary-layer region.

(d) Calculate the average heat transfer coefficient over the whole plate.

(e) Calculate the rate of cooling of the plate.

8.12 A tube bank consists of 1-in. O.D. tubes on 2-in. centers in square in-line arrangements of 12 rows in each direction. The tube wall is at 200 F and air at 60 F flows across the tube bank at mass velocity of 6500 $lb_m/hr\ ft^2$ (based on minimum area). Determine the heat transfer coefficient for this system. How does this value compare with the value calculated for a single cylinder using the maximum velocity?

8.13 Derive Eq. (8.58) by the procedure briefly outlined in Sec. 8.12. List and explain the major assumptions underlying the derivation.

8.14 A furnace 5 by 5 by 5 ft is installed on a concrete floor. If the sides and top are at 150 F and the ambient air temperature is 50 F, calculate the heat loss from the furnace.

8.15 An electrical transmission line of $\frac{1}{2}$-in. O.D. carries 200 amps and has a resistance of 1×10^{-4} ohm/ft of length. If still air around the line is at 60 F, what will be the surface temperature of the line at steady state? What will be the surface temperature on a windy day, assuming wind blows across the line at 20 mph?

CHAPTER 9

Heat Transfer with Boiling

Heat transfer associated with boiling had as its primary purpose, until recently, the conversion of liquid into vapor. With the development of the nuclear reactor, the rocket nozzle, and spacecraft, great interest has developed in the boiling process as a method of increasing heat transfer rates at modest temperature differences. Heat transfer rates in modern boilers are in the neighborhood of 20,000 to 90,000 Btu/hr ft^2, but in nuclear reactors, rocket nozzles, and spacecraft may be of the order of 10^6 to 10^7 Btu/hr ft^2. In a boiling process, the average liquid temperature may remain well below the saturation temperature with the wall temperature above saturation producing "local" boiling at the wall with subsequent condensation in the colder bulk of the fluid. This is known as *subcooled* or *local boiling*. Boiling in a liquid at saturation temperature is known as *saturated* or *bulk boiling*.

In some designs of evaporators the heating surface is submerged beneath a free surface of liquid. This is known as *pool boiling*. When the liquid flows through a tube with subcooled or saturated boiling, this bounded convection process is called *forced-convection boiling* even though the circulation may occur either by density differences or by a pump in a fluid circuit.

We begin with a discussion of the different regimes of boiling, without and with superimposed forced convection. Although boiling is a familiar phenomenon, we shall find that it is an extremely complicated process. We then focus our attention on nucleate boiling, including peak heat flux, which encompasses areas of greatest engineering interest. We shall examine in some detail the physical mechanisms of nucleate boiling and explain semiempirical procedures that have led to a number of successful correlations. We conclude the chapter with a brief discussion of the film boiling process.

9.1 Regimes of Boiling

To illustrate the regimes of boiling, consider an electrically heated horizontal wire submerged in a pool of liquid at saturation temperature T_{sat}. Figure 9.1 represents the type of heat transfer data obtained. As the wire surface temperature is raised above the saturation temperature, convection currents circulate the superheated liquid and steam is produced by evaporation at the free liquid surface. Further increase in wire surface temperature is accompanied by the formation of vapor bubbles which rise at favored spots on the metal surface and condense before reaching free liquid surface. In regime III larger and more numerous bubbles are formed and rise all the way to the free liquid surface. This is called *nucleate boiling*. Beyond the peak of the curve is the transition boiling regime IV; an unstable film forms around the wire and large bubbles originate at the outer upper surface of the film. This vapor film is not stable

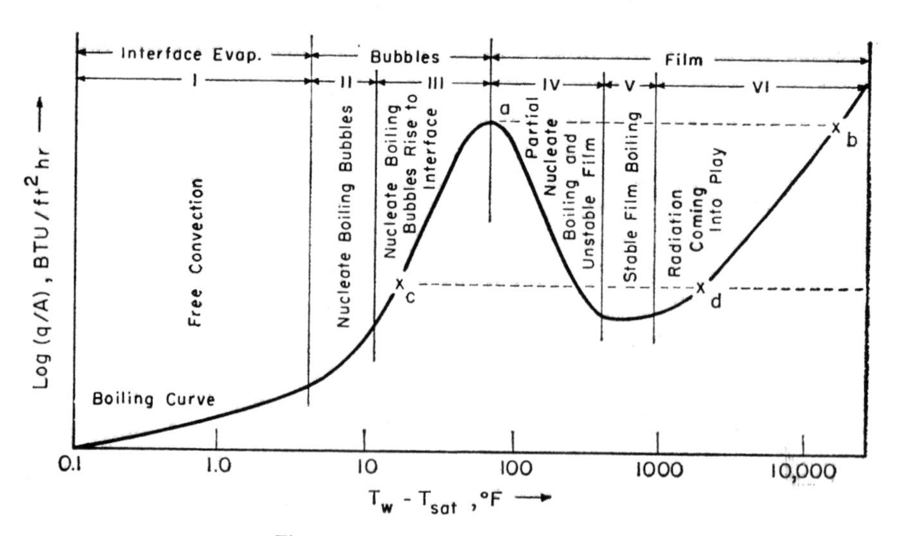

Fig. 9.1. Typical pool boiling data.

but, under the action of circulation currents, collapses and reforms rapidly. The presence of this film provides additional resistance to heat transfer and reduces the heat transfer rate. For values of ΔT in the range 400–1000 F, the film around the wire is stable in the sense that it does not collapse and reform repeatedly, but the shape of the outer film surface varies continuously. For values of ΔT beyond 1000 F, the influence of radiation becomes pronounced. In this regime the vapor film is very stable, and the orderly discharge of bubbles suggests that the frequency and location of bubble origination is controlled by factors operating at the outer surface of the film and that favored spots along the wire are without effect. This regime is called *stable film boiling*. See frontispiece for pictures.

The characteristic curve of Fig. 9.1 is readily obtained if the surface is heated by a condensing-vapor, but when electrical heating is used, the regime IV is difficult to obtain. As the electric energy, and hence q/A is increased, the resulting ΔT increases in regime III. When the peak value of q/A is reached and exceeded slightly, the boiling process cannot remove heat equal to the electrical energy input. The difference between these two quantities causes a rise in internal energy and hence temperature of the wire, which in turn is accompanied by a further decrease in q/A. In short, the system is unstable and unless the electrical input is reduced, the system will proceed toward point b of Fig. 9.1. where the wire temperature is very high. Generally, this temperature is above the melting point for most metals, and the wire melts before reaching point b. For this reason, point a, the peak heat flux in nucleate boiling, is sometimes called the "burnout point."

With an electrically heated wire, both nucleate and film boiling can exist simultaneously on different portions of the wire, corresponding to, say, points c and d of Fig. 9.1.

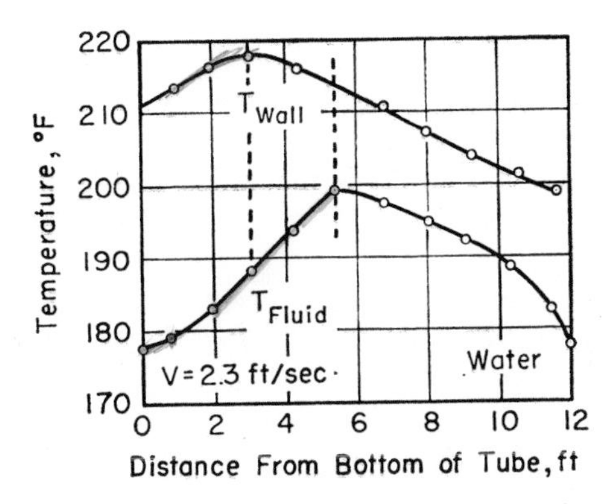

Fig. 9.2. Typical forced convection boiling data for water in a vertical tube [Boarts *et al.* (2)].

The process of boiling inside a tube with a flow of fluid through the tube is more complicated than the pool boiling processes. Figure 9.2 shows curves of temperature distribution in a vertical tube (2). One observes three distinct regions separated by the vertical dotted lines. In the first part of the tube where both wall and fluid temperatures are rising, no boiling takes place; the heat is transferred as in ordinary bounded convection. In the central section the fluid temperature is not at the local saturation temperature, but local boiling takes place at the fluid solid interface without net generation of vapor. Beyond the point where the fluid has reached, and perhaps slightly exceeded, the local saturation temperature, nucleate boiling with net vapor generation exists. Here the fluid temperature (saturation) decreases upwards because of the hydrostatic head effect.

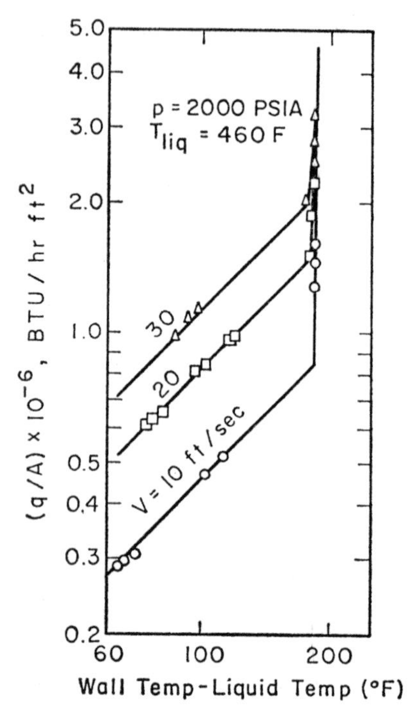

Fig. 9.3. Forced convection boiling of water in a 0.180 in. diameter nickel tube [Rohsenow and Clark (3)].

Forced-convection boiling data have been gathered for a number of different fluids and tube geometries. Figure 9.3 shows data (3) for water in an electrically heated nickel tube (0.18-in. diam., 9.4 in. long). All points are for a constant liquid temperature well below saturation. The lines with a slope of unity are in the nonboiling region and show a velocity effect of $(q/A) \sim V^{0.8}$ as expected in nonboiling. When ΔT is increased sufficiently to cause boiling to begin, the curve bends sharply upward.

The boiling region of Fig. 9.3 is magnified in Fig. 9.4 by plotting the data with $(T_w - T_{sat})$ instead of $(T_w - T_{liq})$. The effect of fluid velocity is clearly revealed.

If the $T_{liq} = T_{sat}$, then the curves of Fig. 9.4 tend asymptotically to $(q/A) = 0$ when $T_w \to T_{sat}$. If $T_{liq} < T_{sat}$, these curves approach the following asymptote as $T_w \to T_{sat}$:

$$\frac{q}{A} = h(T_{sat} - T_{liq}) \tag{9.1}$$

where h is the forced convection coefficient for the liquid [Eq. (8.34)] and

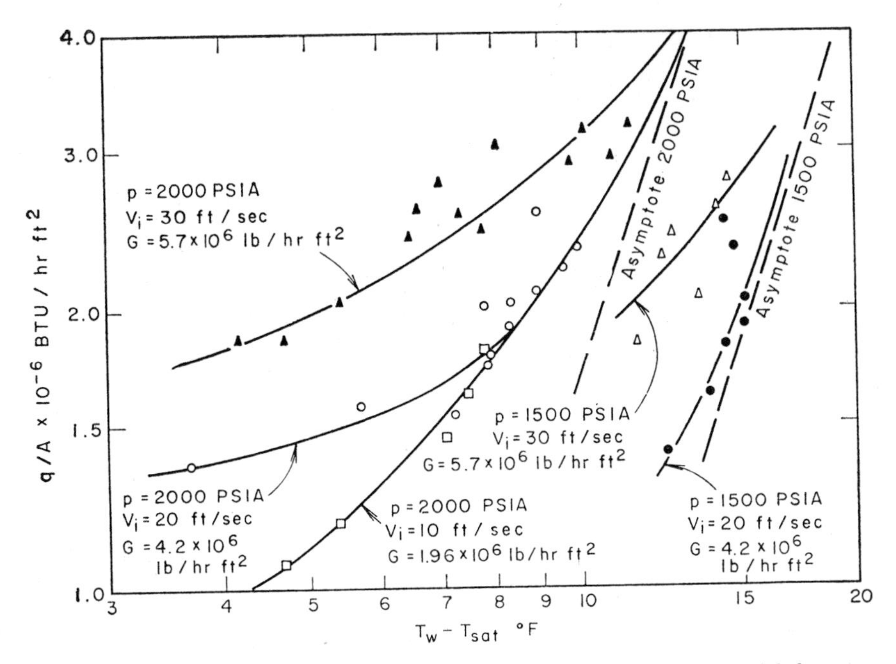

Fig. 9.4. Forced convection boiling of water in a 0.180 in. diameter nickel tube [Rohsenow and Clark (3)].

depends upon velocity, geometry, and fluid properties. Figures 9.5 and 9.6 show the type of results expected at low $(T_w - T_{sat})$ when $T_{liq} < T_{sat}$. Except near the low end of the curve, the effect of liquid subcooling disappears rapidly indicating that $(T_w - T_{sat})$ is the effective driving force for heat transfer with forced-convection boiling.

By counting and measuring bubbles in high-speed motion pictures, it has been shown (14) that the sudden upturn in the boiling curves of Fig. 9.3 could not be explained in terms of latent heat quantities. The increased heat transfer can only be explained by the increased agitation of the fluid by the bubble motion which alternately "pumps" liquid to and from the heating surface.

9.2 Nucleate Boiling

(a) *Nucleation*

Nucleate boiling involves two separate processes—the formation of bubbles (nucleation) and the subsequent growth and motion of these bubbles. Nucleation theories based on equilibrium or nonequilibrium thermodynamics of pure liquids all predict superheats required for nuclea-

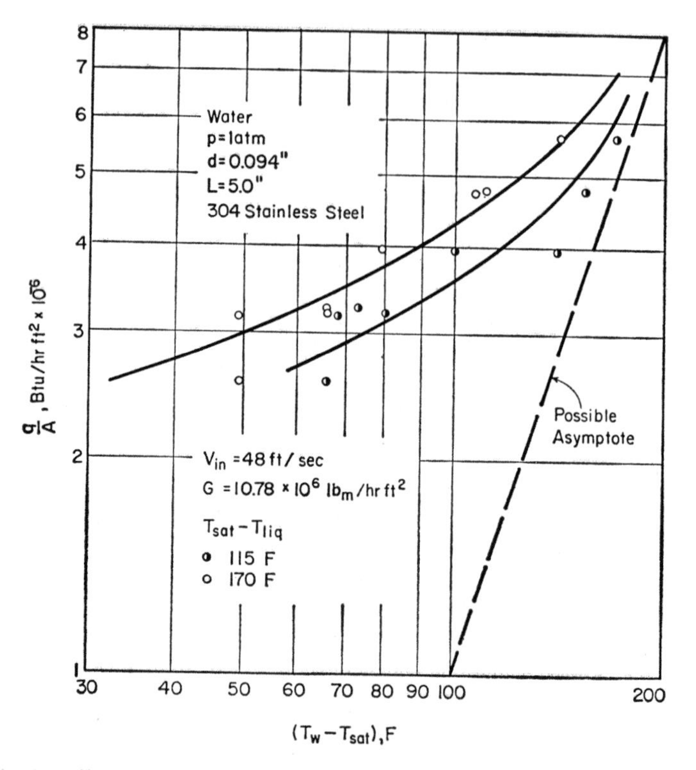

Fig. 9.5. Effect of liquid temperature on boiling heat transfer when $T_{liq} < T_{sat}$ [Bergles (41)].

tion which are far in excess of those experimentally observed in boiling except with de-aerated, pure fluids in clean, carefully made glass containers. For most fluids with commercially obtained heating surfaces, nucleation starts when T_w is only a few degrees above T_{sat}. The more commonly accepted theory of nucleation in boiling systems suggests that bubbles originate at small conelike cavities in the heating surface (4).

Imagine a spherical vapor bubble in a liquid; at equilibrium

$$\pi r^2(p_v - p_l) = 2\pi r\sigma$$

or

$$\Delta p = p_v - p_l = \frac{2\sigma}{r} \tag{9.2}$$

where σ is the surface tension and r the bubble radius. If T_v is the saturation temperature of the vapor, then at equilibrium, since $p_l < p_v$ and $T_{liq} = T_v$, the liquid temperature must be superheated with respect to the liquid pressure.

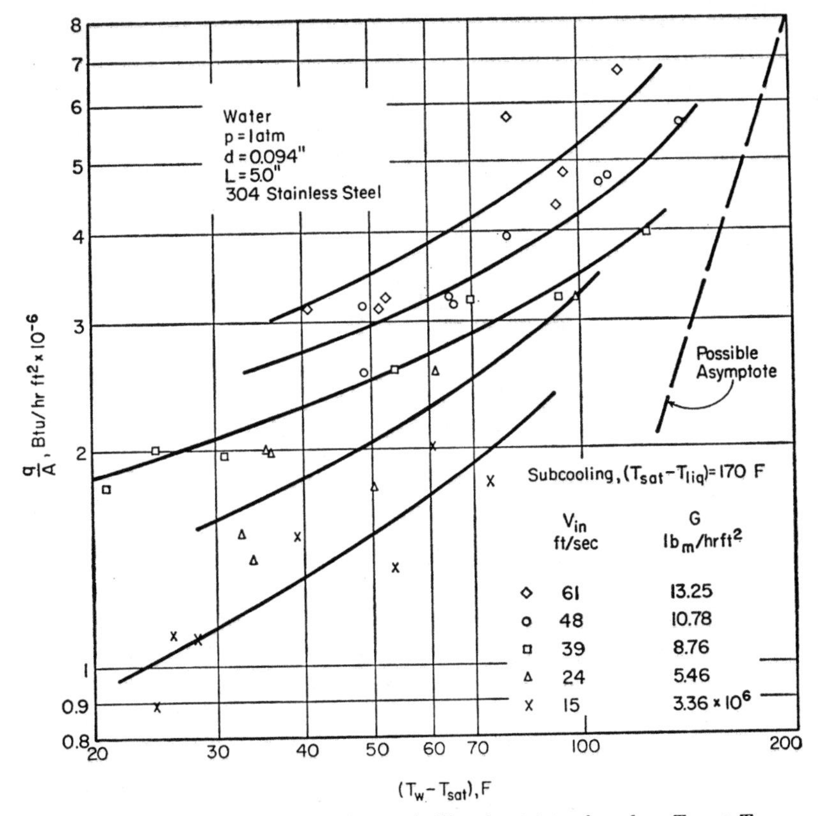

Fig. 9.6. Effect of inlet velocity on boiling heat transfer when $T_{liq} < T_{sat}$ [Bergles (41)].

The Clausius-Clapeyron equation with the perfect gas approximation relates T and p along the saturation line:

$$\frac{dp}{dT} \cong \frac{h_{fg}\rho_v}{T_v} \cong \frac{h_{fg}p_v}{R_v T_v^2} \tag{9.3}$$

where R_v is the gas constant of the vapor.

Since $(p_v - p_l) \cong (T_v - T_{sat})(dp/dt)$, we can combine Eqs. (9.2) and (9.3) to get

$$T_v - T_{sat} \cong \frac{2R_v T_{sat}^2 \sigma}{h_{fg} p_l r} \tag{9.4}$$

If the $(T_{liq} - T_{sat})$ is greater than $(T_v - T_{sat})$ calculated from Eq. (9.4), a bubble of radius r will grow; if smaller, the bubble will collapse.

Griffith and Wallis (5) carried out nucleation experiments by immersing in water, at various pressures at and below atmospheric, copper sheets with a variety of cavities. The water was raised in temperature and the superheat measured when bubbles formed repeatedly at a cavity. The superheat calculated with Eq. (9.4), using cavity radius, agreed well with their data.

Unless special care is taken, inert gas molecules are usually present in a liquid either as single molecules dispersed throughout the liquid or as a colloidal suspension of submicroscopic gas bubbles. If, in addition to the vapor at p_v, a gas at partial pressure p_g is present in a bubble at equilibrium, Eqs. (9.2) and (9.3) are modified as follows:

$$p_v - p_l = \frac{2\sigma}{r} - p_g$$
$$T_v - T_{\text{sat}} \cong \frac{R_v T_{\text{sat}}^2}{h_{fg} p_l}\left(\frac{2\sigma}{r} - p_g\right) \tag{9.5}$$

Equation (9.5) indicates that less liquid superheat is required when non-condensable gas is present in the bubble.

(*b*) *Nucleation at a Solid Surface*

Using the evidence just outlined above, we postulate that a surface contains many cavities, and bubbles form at a heated surface from cavities which already have some gas or vapor present—so-called *active cavities.* When heat is added, the vapor pocket in an active cavity grows by evaporation at the liquid-vapor interface near the heated wall, as shown in Fig. 9.7(a). In a liquid at or near the saturation temperature, the bubble grows and detaches, trapping vapor in the cavity as shown in Fig. 9.7(b). This trapped vapor is the nucleus for the next bubble. Usually a bubble encompasses many cavities before it detaches; thus, this vapor-trapping process can induce an inactive cavity (one entirely filled with the pure liquid) into activity.

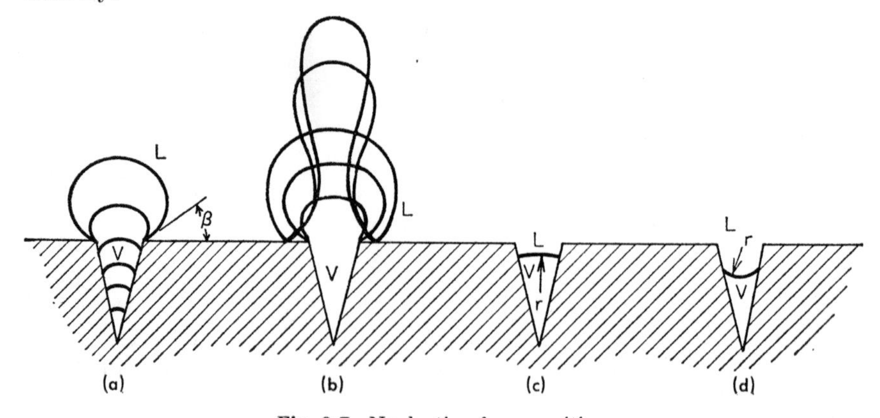

Fig. 9.7. Nucleation from cavities.

When the liquid is highly subcooled, the bubble grows and collapses while attached to the surface. Even in this case, the vapor pocket remains in the cavity, ready for the next bubble to grow.

It is not necessary that inert gas be present. Some cavities can contain pure vapor even at very highly subcooled temperatures. Two cases of different contact angle β and opposite curvature of the interface may be considered: $p_v > p_l$ in Fig. 9.7(c) and $p_v < p_l$ in Fig. 9.7(d). As the surface is allowed to cool below the saturation temperature of the liquid, the cavity in Fig. 9.7(c), if it contained pure vapor, would collapse and be completely filled with liquid, thus becoming inactive. As cavity (d) cools down, however, the interface recedes into the cavity, decreasing the radius of curvature and reducing p_v. Since p_v decreases, T_{sat} of the vapor decreases; the cavity does not collapse and is ready as an active nucleus when the surface is subsequently heated.

Griffith and Wallis (5) obtained the curves in Fig. 9.8(a) by counting nucleation spots n, number per sq ft, as $(T_w - T_{sat})$ was varied on a clean copper surface (finished with 3/0 emery paper) at the bottom of a pool of liquid. Cavity radius r calculated from Eq. (9.4) brings all three of these curves together as shown in Fig. 9.8(b). This curve is essentially an integration of a cavity distribution function such as that in Fig. 9.8(c)—integration being performed from $r = \infty$ to r for the last cavity to be activated and n_r being the number of cavities between size range r and $r + \Delta r$. The curve characterizes the particular surface for boiling heat transfer.

Motion pictures with magnification, taken by Westwater (6) of boiling on a surface tend to substantiate further the idea that bubbles form at cavities in a surface.

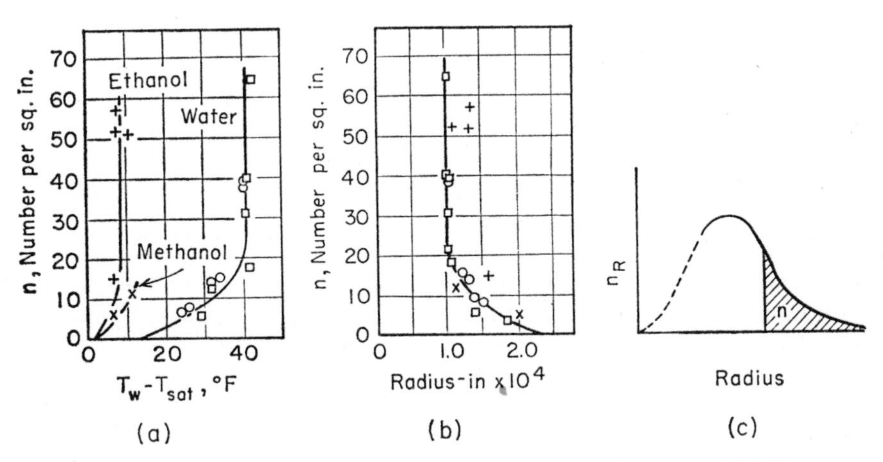

Fig. 9.8. Nucleation on a horizontal heated surface [Griffith and Wallis (5)].

(c) *Bubble Growth and Motion*

In a superheated liquid a bubble grows by heat conduction to the liquid-vapor interface, with resulting evaporation into the bubble. The differential equation for the growth of a bubble in a superheated liquid of infinite extent was solved (8), (9), (10) and the results compared well with experimental data. The following are the resulting expressions:

$$r = (\sqrt{\pi t})\sqrt{(k\rho c_p)_l}\left(\frac{T_l - T_v}{h_{fg}\rho_v}\right) \tag{9.6}$$

$$r\dot{r} = \frac{\pi}{2}\left(\frac{\sqrt{c_{pl}\rho_l k_l}(T_l - T_v)}{\rho_v h_{fg}}\right)^2 \tag{9.7}$$

Earlier, on the basis of a balance of buoyant and surface-tension forces, Fritz (10) found empirically that the bubble diameter, D_b, at detachment from the heating surface could be expressed as follows:

$$D_b = C_d\beta\sqrt{\frac{2g_0\sigma}{g(\rho_l - \rho_v)}} \tag{9.8}$$

where C_d was found to be 0.0148 for H_2O and H_2 bubbles and β is the bubble contact angle in degrees measured through the liquid, see Fig. 9.7(a).

Bubbles issue at active sites in a steady stream. A bubble grows while attached to the surface for time t_c; then a delay time t_d elapses before the next bubble forms. Zuber (11) proposed the following relation for frequency f times D_b:

$$fD_b = 1.18\left(\frac{t_c}{t_c + t_d}\right)\left[\frac{g_0 g\sigma(\rho_l - \rho_v)}{\rho_l^2}\right]^{1/4} \tag{9.9}$$

Usually $t_c \cong t_d$. In this case, the bubble center travels a distance D_b in time $(t_c + t_d)$ or $(1/f)$; hence, bubble velocity is $D_b/(1/f)$ or

$$V_b \cong fD_b \tag{9.10}$$

For any particular fluid, fD_b is a function of pressure, but not of q/A over most of the range of nucleate boiling. At low heat flux, Jakob (12) observed from motion pictures that $q/A \sim n$, the number of active sites on a surface. Staniszewski (13) found

$$\frac{q}{A} \sim n^m \tag{9.11}$$

where $m = 1$ at low q/A, but decreases to around $\frac{1}{2}$ at high q/A.

(d) Pool Boiling Heat Transfer

The slope of the q/A-vs.-$(T_w - T_{sat})$ curve depends on the cavity size distribution, Fig. 9.8(c). As q/A increases, the total number of active sites must increase, Eq. (9.11). This requires, Fig. 9.8(b)(c), that smaller-sized cavities be activated which, from Eq. (9.4), require larger superheat at the heating surface, $(T_w - T_{sat})$. Very few data are available regarding cavity size distribution, but direct measurements of q/A vs. $(T_w - T_{sat})$ indicate that for most commercially available surfaces—machined, rolled, or drawn—q/A is proportional to $(T_w - T_{sat})^3$ as shown in Fig. 9.9.

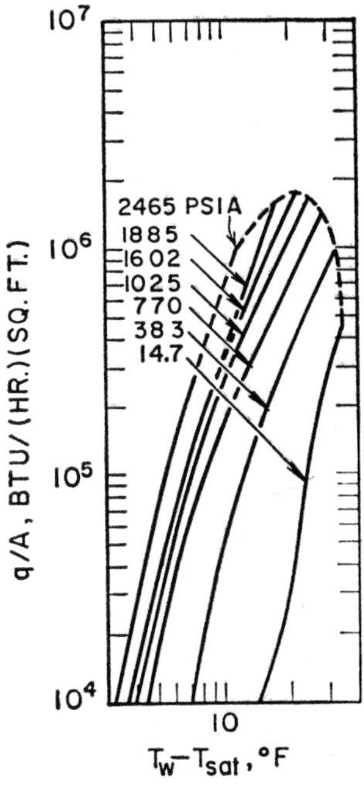

Fig. 9.9. Pool boiling of water on a horizontal platinum wire, 0.024 in. diameter [Addoms (19)].

Effect of pressure. The curves of Fig. 9.9 show the decreasing required wall superheat $(T_w - T_{sat})$ for higher pressures. This is explained by Eq. (9.4) as follows: a particular surface has a given cavity size distribution, Fig. 9.8(b) (c), and at a particular q/A a definite n is required, Eq. (9.11); then, the superheat required must activate a particular size of cavity, r. The right side of Eq. (9.4) with r constant decreases with increasing pressure; hence, less superheat is required, showing that the higher pressure curves should lie to the left as they do in Figs. 9.9.

Effect of roughness. Artificially changing the surface (cavity size distribution) with emery cloth can change the slope of the q/A-vs.-$(T_w - T_{sat})$ curve and move the curve to the right for the finer surfaces, Fig. 9.10.

Effect of gases. A liquid containing dissolved gas releases the gas at the hot surface. The resulting gas bubbles, which come out of solution well below the normal saturation temperature, agitate the liquid in the same way as vapor bubbles. In Fig. 9.11, McAdams' (16) data show the heat transfer curve to lie well above the curve for a degassed liquid. This further substantiates the theory that the heat transfer is increased by the bubble agitation and not by the latent heat effects.

The effect of gases discussed here refers to gas dissolved in the liquid. It has been observed that inert gas in a closed evaporator-condenser system containing Freon can cause the curves to move to higher $(T_w - T_{sat})$ magnitudes than those obtained in the same system purged of all inert gas.

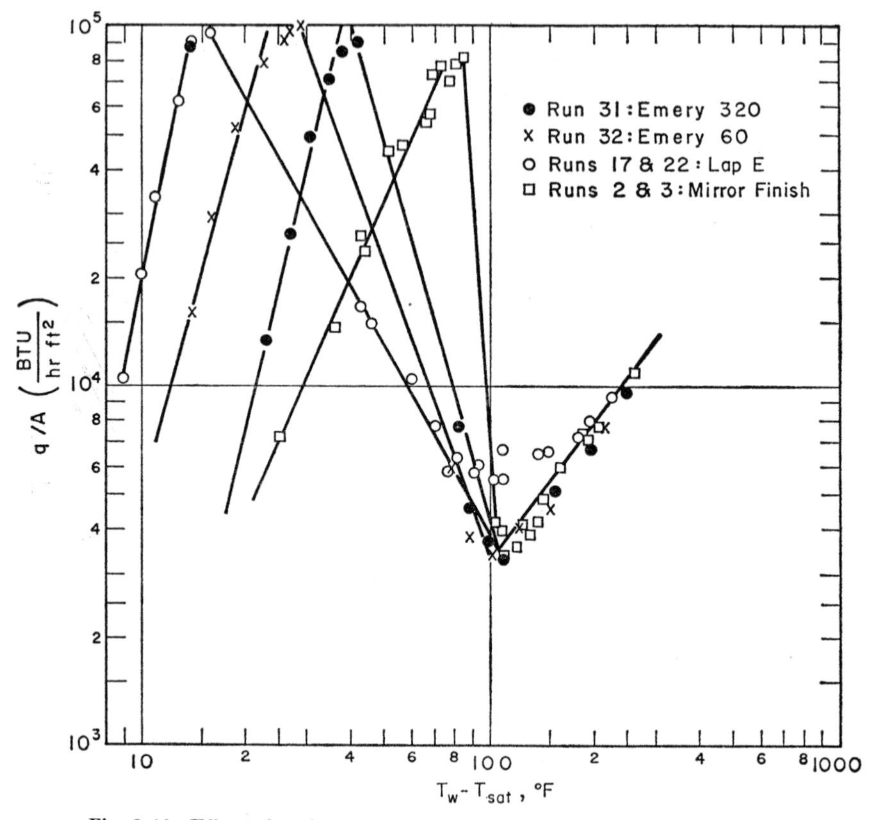

Fig. 9.10. Effect of surface finish on pool boiling of pentane on horizontal copper surface [Berenson (15)].

Aging. Surfaces that have been in service for long periods of time often require higher $(T_w - T_{sat})$ for the same (q/A). This is due either to the added resistance of an oxide or deposited layer, or to the decrease in cavity size resulting from mild oxidation.

Surface geometry. Pool boiling heat transfer data seem to be relatively insensitive to geometry. The same heat transfer data in the nucleate boiling regime are applicable to horizontal, vertical, or inclined flat surfaces and to wires above 0.004 in. in diameter. At the lower magnitudes of ΔT, where convection effects, natural or forced, are important, the results differ.

Effect of gravity. Siegel and Usiskin (37) showed photographically that at zero gravity in free fall, nucleate boiling stopped and a large bubble of vapor surrounded a heating surface. Merte and Clark (38) showed that in region III of Fig. 9.1, there is very little change in the position of the (q/A)-vs.-$(T_w - T_{sat})$ curve, but in regions I and II the curves move to

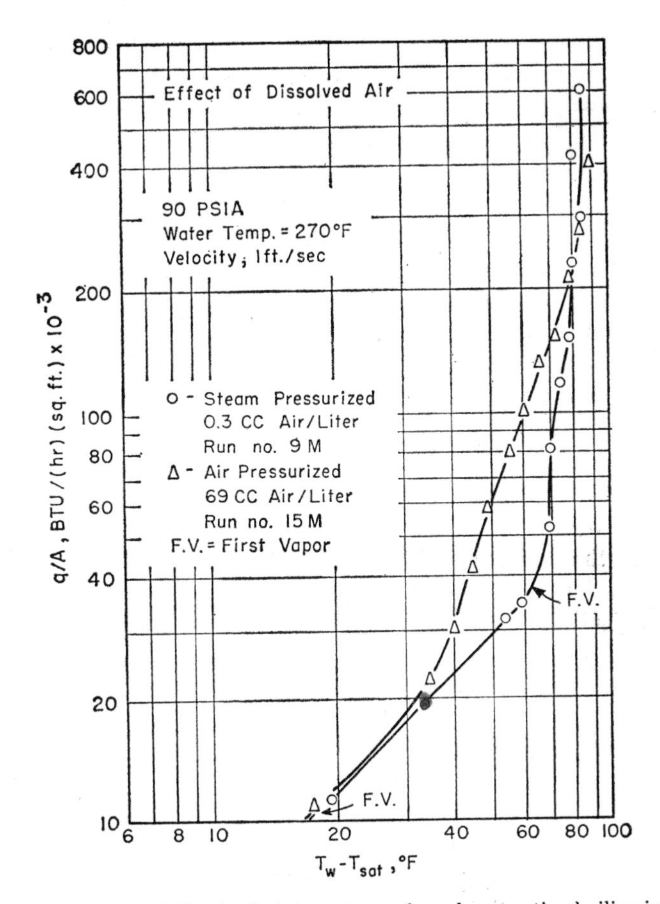

Fig. 9.11. Effect of dissolved air in water on forced convection boiling in an annulus [McAdams et al. (16)].

the left owing to greatly increased natural convection as gravity increases from 1 to 20 *G*'s. The effect on peak heat flux is discussed in Sec. 9.3(c).

9.3 Correlation of Boiling Heat Transfer Data

The phenomena of boiling heat transfer are more complex than those associated with single-phase convection. Since we are unable to describe the fluid motions, we cannot analyze the associated heat transfer process, particularly in nucleate boiling.

A single equation could not possibly correlate the data over the entire range of ΔT represented in Fig. 9.1 since the fluid flow patterns differ so radically in the various regimes. In region I, the data correlate with the normal natural convection relations, Sec. 8.13. In the nucleate boiling region (III), data have been correlated with equations developed from dimensional analysis. Region II is essentially a superposition of regions I and III. In the film boiling regions, V and VI, fluid motion in the vapor film has been described and analysis has resulted in correlations which agree with data. Here heat transfer data are independent of solid surface properties. In region IV no correlation exists based on descriptions of fluid motion. Figure 9.10 shows that surface properties influence the data in this region; so it is concluded that the liquid must touch the surfaces at least part of the time, suggesting that the film alternately forms and collapses. Other points of interest to be correlated are the heat flux at the upper limit of nucleate boiling and the heat flux at the lower limit of film boiling.

(*a*) *Nucleate Pool Boiling Data*

Since the increased heat transfer rate in boiling probably results from the increased agitation of the liquid (14) and the heat is transferred primarily from the surface directly to the liquid, we postulate that there should exist a bubble Reynolds number which measures the effect of this increased agitation on the heat transfer. Then by similarity with single-

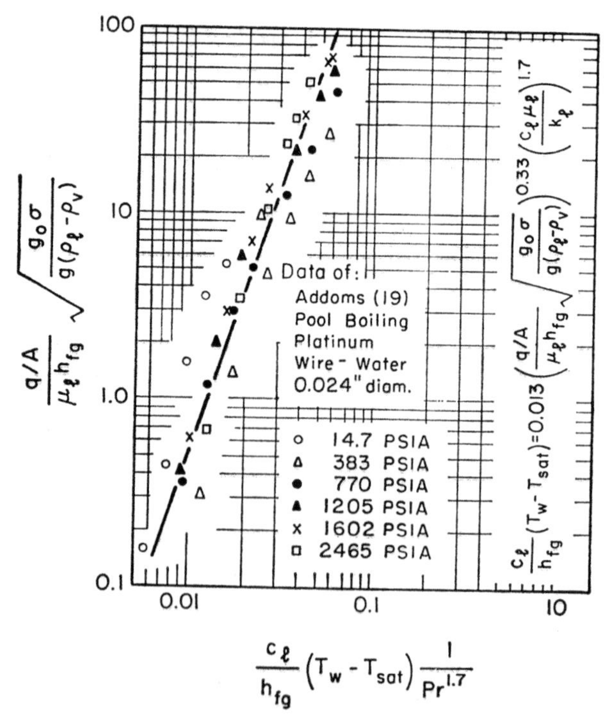

Fig. 9.12. A correlation of pool boiling data [Rohsenow (18)].

phase forced-convection correlations

$$\mathrm{Nu}_b = \phi(\mathrm{Re}_b, \mathrm{Pr}_l) \tag{9.12}$$

Much of the available pool boiling data was correlated (18) by defining the following groups:

$$\mathrm{Nu}_b \equiv \frac{(q/A)_b D_b}{(T_w - T_{\mathrm{sat}})k_l} \qquad \mathrm{Re}_b \equiv \frac{G_b D_b}{\mu_l} \tag{9.13}$$

where D_b is maximum bubble diameter at departure or just before collapse given by Eq. (9.8) and G_b is the average mass velocity of vapor receding from the surface; $G_b = (q/A)/\rho_v h_{fg}$ for saturated liquid. Details of the analysis are omitted here, but the final correlating equation is as follows:

$$\frac{c_l(T_w - T_{\mathrm{sat}})}{h_{fg}} = C_{sf}\left[\frac{(q/A)_b}{\mu_l h_{fg}}\sqrt{\frac{g_0\sigma}{g(\rho_l - \rho_v)}}\right]^{1/3} \mathrm{Pr}_l^{1.7} \tag{9.14}$$

where each group of terms is dimensionless. In this equation C_{sf} depends on the nature of the heating surface-fluid combination. The correlation of Addoms' data of Fig. 9.9 is shown in Fig. 9.12 where $C_{sf} = 0.013$. Table 9.1 gives values of C_{sf} for various other surface-fluid combinations as deduced from experimental data.

TABLE 9.1 VALUES OF C_{sf} FOR EQ. (9.14)

Surface-Fluid Combination	C_{sf}
Water-Nickel (3)*	0.006
Water-Platinum (19)	0.013
Water-Copper (21)	0.013
Water-Brass (22)	0.006
CCl_4-Copper (21)	0.013
Benzene-Chromium (20)	0.010
n-Pentane-Chromium (20)	0.015
Ethyl Alcohol-Chromium (20)	0.0027
Isopropyl Alcohol-Copper (21)	0.0025
35% K_2CO_3-Copper (21)	0.0054
50% K_2CO_3-Copper (21)	0.0027
n-Butyl Alcohol-Copper (21)	0.0030

* Numbers in parentheses refer to source of data.

Using Eq. (9.12), Forster and Greif (23) defined the following groups:

$$\mathrm{Nu}_b \equiv \frac{(q/A)_b}{\rho_v h_{fg}}\left(\frac{2\sigma}{\alpha_l\,\Delta p}\right)^{1/2}\left(\frac{\rho_l}{\Delta p}\right)^{1/4} \qquad \mathrm{Re}_b \equiv \frac{\rho_l \dot{r}r}{\mu_l} \tag{9.15}$$

where $2\sigma/\Delta p$ from Eq. (9.2) is r, $r\dot{r}$ is given by Eq. (9.7) and $\alpha = k/\rho c$. The following is the final result:

$$\left(\frac{q}{A}\right)_b = K_{sf}\left(\frac{\alpha c_l \rho_l T_{\text{sat}}}{778 h_{fg}\rho_v \sqrt{\sigma}}\right)\left(\frac{c_l T_{\text{sat}} \sqrt{\alpha_l}}{778(h_{fg}\rho_v)^2}\right)^{1/4}\left(\frac{\rho_l}{\mu_l}\right)^{5/8}(\Delta p)^2 \operatorname{Pr}_l^{1/3} \qquad (9.16)$$

This equation is identical in form with the equation proposed earlier by Forster and Zuber (42), differing only by a constant factor 0.817. In these equations $q/A \sim (\Delta T)^2$ rather than $(\Delta T)^3$. With $K_{sf} = 0.0012$, Eq. (9.16) was shown (23) to agree with data for a very limited number of fluids and pressures. When the exponents of Eq. (9.16) were determined to provide $q/A \sim (\Delta T)^3$, it was shown (43) that the resulting coefficient varied considerably for various sets of data, being 0.25 for n-pentane-chromium and 250 for ethanol-chromium. Since Eq. (9.16) involves only fluid properties and since the boiling curve can be shifted by changing only surface finish, Fig. 9.10, it is to be expected that the coefficient K_{sf} would change with surface condition.

Gilmour (24) reports the following correlation equation which satisfactorily represents the data of a number of investigators:

$$\frac{(q/A)_b}{(T_w - T_{\text{sat}})c_l G} = 0.001\left(\frac{DG}{\mu_l}\right)^{-0.3}\left(\frac{c_l\mu_l}{k_l}\right)^{-0.6}\left(\frac{p^2}{\rho_l \sigma}\right)^{0.425} \qquad (9.17)$$

where

$$G = \frac{q/A}{h_{fg}\rho_v}\rho_l$$

and p is the absolute pressure.

(b) Nucleate Boiling with Convection

The convection effect, either forced in a tube or unbounded natural for surfaces in a pool, has been shown, Rohsenow (25), to be superimposed on the bubble motion effect:

$$\frac{q}{A} = \left(\frac{q}{A}\right)_c + \left(\frac{q}{A}\right)_b \qquad (9.18)$$

Here $(q/A)_c$ is the heat transfer associated with either forced or natural convection in the absence of boiling and is calculated by $h(T_w - T_{\text{liq}})$ where h is evaluated by equations such as (7.41), (7.59), (8.34), or (8.63) as applicable. The term $(q/A)_b$ is the heat transfer associated with bubble motion alone in the absence of convection and is calculated by Eqs. (9.14), (9.16), or (9.17). Figure 9.13 shows the data of Fig. 9.4 plotted with $[(q/A) - (q/A)_c]$ in Eq. (9.14). Here (q/A) was the measured flux and $(q/A)_c$ was calculated by Eq. (8.34) with 0.023 replaced by 0.019 at the high pressures. A number of other data have been correlated in this manner.

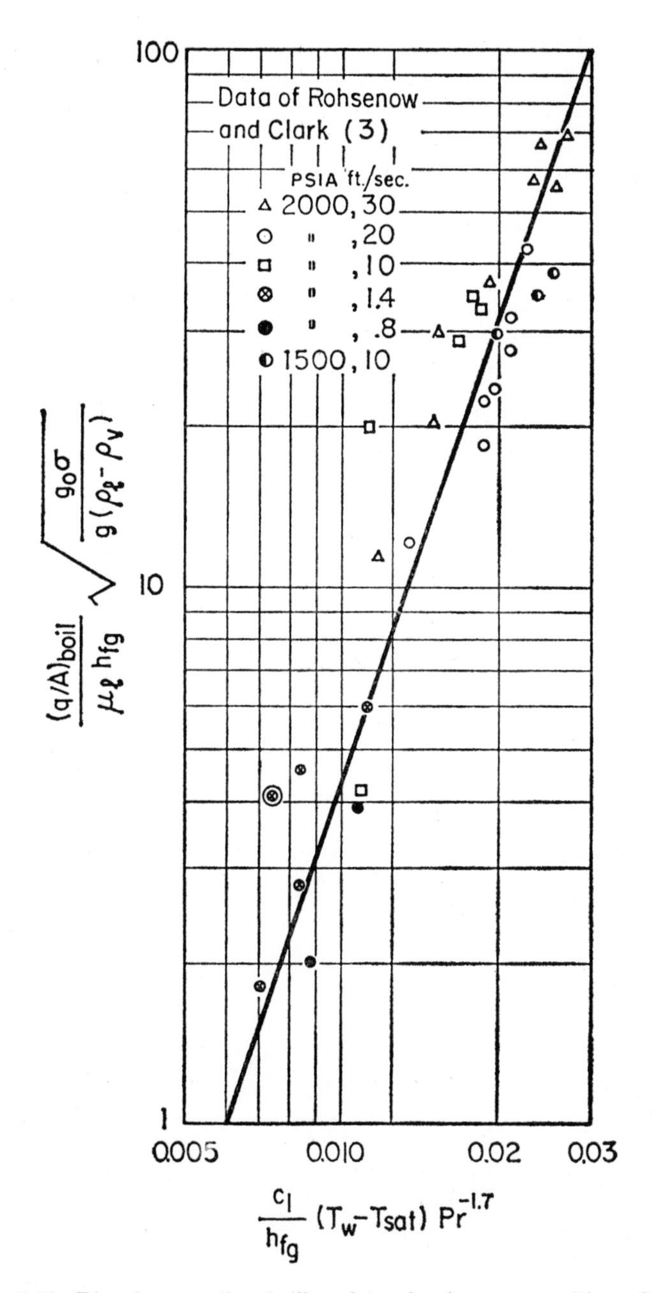

Fig. 9.13. Forced convection boiling data showing superposition effect [Rohsenow (25)].

For forced convection boiling in a tube or channel, the q/A vs. $(T_w - T_{sat})$ relationship seems to be relatively insensitive to the quality of the mixture. With uniform q/A the $(T_w - T_{sat})$ is very nearly constant along the tube as quality increases up to the limiting quality for nucleate boiling, burnout condition, as given, for example, by the curve of Fig. 9.14.

(c) *Peak Heat Flux Data*

The maximum heat flux in nucleate boiling is of great interest to the designer since at a given (q/A) near the maximum, surface temperatures in film boiling are orders of magnitude higher than in nucleate boiling. The magnitude of the peak heat flux seems to be relatively insensitive to the character or finish of reasonably smooth surfaces, as illustrated by the data plotted in Fig. 9.10. Knurled or threaded surfaces have been found to have slightly higher peak apparent heat fluxes.

One description of the maximum heat flux condition visualizes that at this condition the active points on the surface become so numerous that bubbles at the surface coalesce to form a vapor blanketing layer. Imagining bubbles packed together on a surface such that there are $1/D_b$ bubbles per foot, then the number per square foot would be $(1/D_b)^2$. We then assume the actual number of bubbles per square foot at peak heat flux to be

$$n = \frac{C_{vb}}{D_b^2} \tag{9.19}$$

where C_{vb} is at most a function of pressure. We also assume $(q/A)_b$ is proportional to the heat flux to the bubbles at any pressure or

$$\left(\frac{q}{A}\right)_b = C_q\left(\frac{\pi}{6}D_b^3\right)\rho_v f \cdot n h_{fg} \tag{9.20}$$

where C_q is the proportionality factor and is at most a function of pressure.

Substituting (9.19) into (9.20),

$$\frac{(q/A)_{\max}}{h_{fg}\rho_v} = \frac{\pi}{6}C_q C_{vb}(f \cdot D_b) \tag{9.21}$$

From Eq. (9.9), $(f \cdot D_b)$ is a function of $g^{1/4}$ and other quantities which are functions of pressure alone. Rohsenow and Griffith (26) have correlated pool boiling peak heat flux data with the following equation, where $(\rho_l - \rho_v)/\rho_v$ is taken as the function of pressure for purposes of correlation.

$$\frac{(q/A)_{\max}}{\rho_v h_{fg}} = 143 g^{1/4}\left(\frac{\rho_l - \rho_v}{\rho_v}\right)^{0.6} \tag{9.22}$$

where g is acceleration of gravity in G's. Here the $g^{1/4}$ is added to encompass the data of Usiskin and Siegel (27) taken in the range $0.06 \leq g \leq 1.0$.

Zuber (28) (29) developed a quantitative expression for peak heat flux based on observations of photographs taken by Westwater and Santangelo (6). At the peak heat flux point, the pictures show a regular spacing of "bubbles" or columns of liquid rising from the liquid-vapor interface of the vapor film. This is a kind of hydrodynamic instability which is perhaps related to the kind associated with the stability of an interface between a more dense and a less dense fluid. It looks like a wave motion.

Zuber imagines that at the peak heat flux the vapor coming off the vapor film is in the form of pulsating round jets, vortex sheets. Between these jets the volume rate of liquid flow toward the heating surface equals the volume rate of vapor flow away from the surfaces. The relative velocity at the interface produces an instability causing the jet to break up into spheres.

The resulting expression for peak heat flux in pool boiling is as follows:

$$\frac{(q/A)_{\max}}{\rho_v h_{fg}} = 0.18\left[\frac{\sigma(\rho_l - \rho_v)gg_o}{\rho_v^2}\right]^{1/4}\left[\frac{\rho_l}{\rho_l + \rho_v}\right]^{1/2}, \quad \text{ft/hr} \tag{9.23}$$

where 0.18 is determined empirically.

Equations (9.22) and (9.23) are of the same form, e.g., $(q/A)_{\max}/\rho_v h_{fg}$ equals $g^{1/4}$ times a function of the pressure; however, the quantities which are functions of pressure differ in the two equations.

The equations above apply to pool boiling in a saturated liquid. The peak heat flux increases significantly as subcooling increases. Zuber (29) presents a correlation for this effect.

Forced convection increases the peak heat flux in nucleate boiling. Many totally empirical equations have been proposed to correlate data of this type. The following is one: Mirshak, et al. (30) correlated a wide variety of data for water from flow in tubes to flow across heated metal strips with the following equation:

$$\left(\frac{q}{A}\right)_{\max} = 480{,}000(1 + 0.0365V)(1 + 0.00508\,\Delta T_{sc})(1 + 0.0131p) \tag{9.24}$$

where the units and ranges of test data covered were as follows: $V = 5$ to 45 ft/sec, subcooling $\Delta T_{sc} = 9$ to 135 F, $p = 25$ to 85 psia, and equivalent diameter of 0.21 to 0.46 in. The equation agreed with the data within ± 16 per cent.

Griffith (31) developed an empirical burnout correlation for forced convection covering the range from subcooled liquid to as high as 70 per cent quality. There is correlated a variety of data, including data on round tubes and rectangular channels, covering a wide range of conditions shown in Fig. 9.14. The quantities are all defined on the graph. The data correlate within ± 33 per cent. For the ordinate quantity h_g is the saturated vapor enthalpy and h_b is the bulk enthalpy of the fluid which may be subcooled

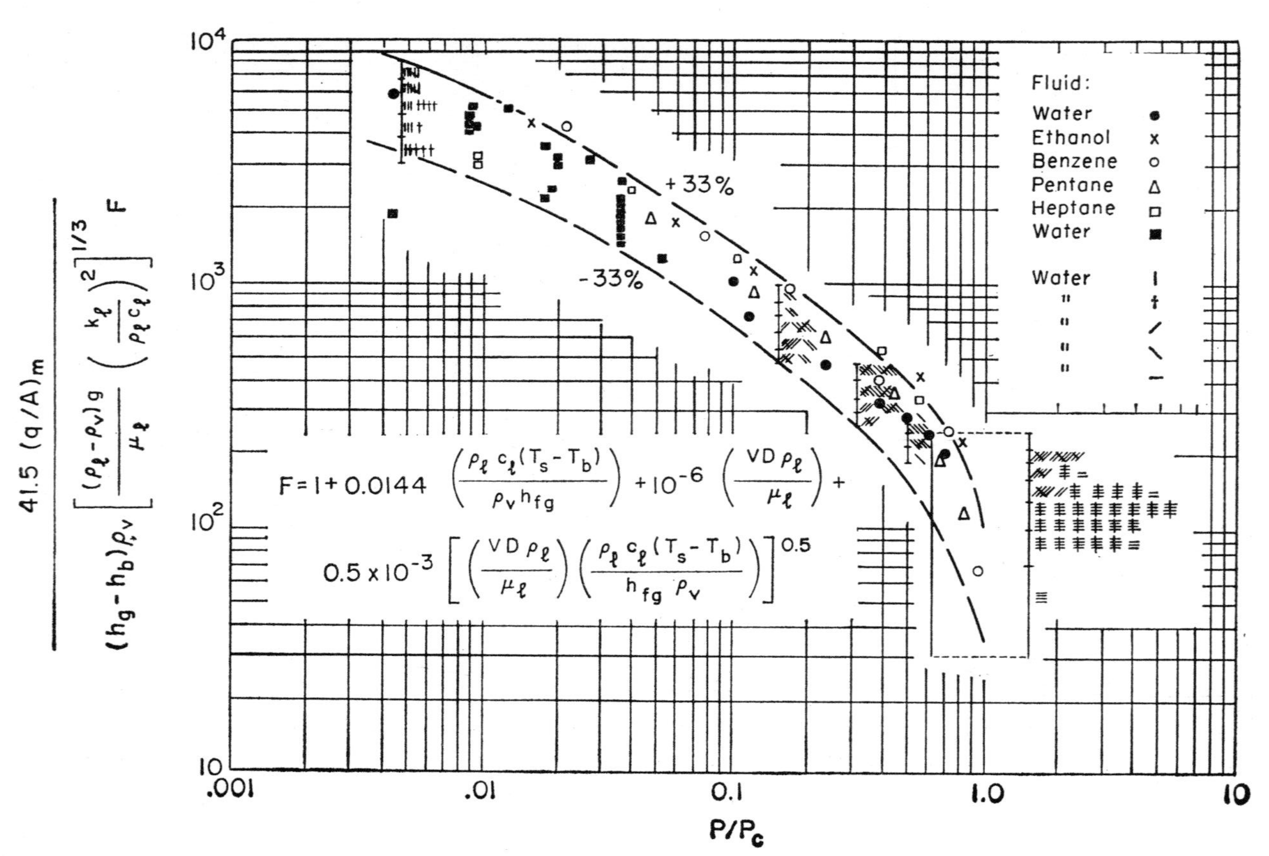

Fig. 9.14. Peak heat flux correlation for forced convection [Griffith (31)].

liquid, saturated liquid or a two-phase mixture at some quality (less than 70 per cent).

Lowdermilk, Lanzo, and Siegel (32) report measurements of burnout heat flux for water at pressures from atmospheric to 100 psia in tubes of 0.051-in. to 0.188-in. diameter, $L/D = 25$ to 250, $V = 0.1$ to 98 ft/sec, and inlet subcooling from 0 to 140 F. Burnout heat fluxes ranged from 0.9×10^6 to 13.2×10^6 Btu/hr ft^2 with net steam generated from near zero to near 100 per cent quality. In the low-velocity, high-quality region, the data were correlated by

$$1 < \frac{G}{(L/D)^2} < 150: \left(\frac{q}{A}\right)_{\max} = \frac{270G^{0.85}}{D^{0.2}(L/D)^{0.85}} \tag{9.25a}$$

and in the high-velocity, low-quality region by

$$150 < \frac{G}{(L/D)^2} < 10{,}000: \left(\frac{q}{A}\right)_{\max} = \frac{1400G^{0.50}}{D^{0.2}(L/D)^{0.15}} \tag{9.25b}$$

At the transitional value of $G/(L/D)^2 = 150$ the quality at burnout ranged from 40 to 60 per cent.

Other phenomena which increase the peak flux are the presence of an ultrasonic field in the liquid (33), the presence of an electrostatic field in the liquid (34) (see frontispiece), and forced vortex effects (35) (36).

9.4 Film Boiling

Film boiling as a cooling process has not had wide commercial applications because of the accompanying high surface temperatures; it may find use in the future as better materials become available. It is, however, often encountered in chemical process equipment and in cryogenic systems.

In film boiling, vapor is generated at the liquid-vapor interface by conduction and radiation from the heating surface through the vapor film. Using the hydrodynamic instability of the liquid-vapor boundary, Zuber (29) arrived at the following equation representing the minimum heat flux for film boiling:

$$\left(\frac{q}{A}\right)_{\min} = 0.09\rho_{v_f}h_{fg}\left[\frac{g(\rho_l - \rho_v)}{\rho_l + \rho_v}\right]^{1/2}\left[\frac{g_0\sigma}{g(\rho_l - \rho_v)}\right]^{1/4} \tag{9.26}$$

where ρ_{vf} is the vapor density at average film temperature and the other properties are evaluated at saturation temperature. The coefficient 0.09 was determined empirically by Berenson (15) from data such as those shown in Fig. 9.10. It is significant that the data for any surface finish converge on this point, Fig. 9.10, and in the film boiling region are independent of surface finish. In the transition boiling region the data are influenced by the same factors that influence nucleate boiling data, suggesting that the film in regime IV, Fig. 9.1, does occasionally touch the heating surface.

Extending Zuber's analysis, Berenson arrives at the following expression for ΔT at the minimum heat flux point:

$$\Delta T_{\min} = 0.127 \frac{\rho_{v_f} h_{fg}}{k_{v_f}} \left[\frac{g(\rho_l - \rho_v)}{\rho_l + \rho_v} \right]^{2/3} \left[\frac{g_0 \sigma}{g(\rho_l - \rho_v)} \right]^{1/2} \left[\frac{g_0 \mu_f}{g(\rho_l - \rho_v)} \right]^{1/3} \tag{9.27}$$

The stable film regime was studied experimentally and analytically by Bromley (39) (40) for horizontal tubes and vertical plates. By balancing the buoyant and frictional forces on the vapor flowing in the film on the outside of a horizontal tube, Bromley arrived at the following equation representing the h_c associated with conduction alone:

$$h_c = 0.62 \left[\frac{k_v^3 \rho_v (\rho_l - \rho_v) g (h_{fg} + 0.4 c_{p_v} \cdot \Delta T)}{D_0 \mu_v (T_w - T_{\text{sat}})} \right]^{1/4} \tag{9.28}$$

Radiation contributes to the heat transfer and increases the vapor film thickness, reducing the effective contribution of the conduction. The total heat transfer coefficient is given by

$$h = h_c \left(\frac{h_c}{h} \right)^{1/3} + h_r \tag{9.29}$$

where h_r is calculated for radiation between two parallel planes taking the emmissivity of the liquid as unity, Chap. 13.

For forced convection flow of the liquid across the tube, Bromley (40) suggests the following equation when $V_\infty \geq 2\sqrt{gD_0}$:

$$h_c = 2.7 \sqrt{\frac{V_\infty k_v \rho_v (h_{fg} + 0.4 c_{p_v} \Delta T)}{D_0 (\Delta T)}} \tag{9.30}$$

and

$$h = h_c + \tfrac{7}{8} h_r \tag{9.31}$$

Film boiling on a horizontal surface was analyzed by Berenson (15) and compared with data for pentane (Fig. 9.10), CCl_4, benzene, and ethyl alcohol. The following equation results:

$$h = 0.425 \left[\frac{k_{v_f}^3 \rho_{v_f} (\rho_l - \rho_v) g (h_{fg} + 0.4 c_{p_v} \Delta T)}{\mu_f (\Delta T) \sqrt{\dfrac{g_0 \sigma}{g(\rho_l - \rho_v)}}} \right]^{1/4} \tag{9.32}$$

Note this is identical in form with Eq. (9.28) with D_0 replaced by $\sqrt{g_0\sigma / g(\rho_l - \rho_v)}$ which is proportional to bubble diameter, Eq. (9.8).

REFERENCES

1. Farber, E. A., and R. L. Scorah, *Trans. ASME*, **70,** 4, 369–84 (May 1948).

2. Boarts, R. M., W. L. Badger, and S. J. Meisenburg, "Temperature Drop and Liquid-Film Coefficients in Vertical Tubes," *Ind. Eng. Chem.*, August 1937, p. 912.

3. Rohsenow, W. M., and J. A. Clark, "Heat Transfer and Pressure Drop Data for High Heat Flux Densities to Water at High Sub-Critical Pressures," *Fluid Mech. Heat Transfer Inst.*, Stanford Univ. Press, 1951.

4. Bankoff, S. G., "Ebullition from Solid Surfaces in Absence of Pre-existing Gaseous Phase," *Heat Transfer and Fluid Mech. Inst.*, Stanford Univ. Press, 1956.

5. Griffith, P., and J. D. Wallis, "The Role of Surface Conditions in Nucleate Boiling," presented at 3rd Nat. Ht. Trans. Conf. ASME-AIChE, Aug. 1959, to be published in *AIChE Jour.*

6. Westwater, J. W., and J. G. Santangelo, "Photographic Study of Boiling," *Ind. Eng. Chem.*, August 1955, p. 1605.

7. Forster, H. K., and N. Zuber, *Jour. Appl. Phys.*, **25,** 474 (1954); *AIChE Jour.*, **1,** 531 (1955).

8. Plesset, M. S., and S. A. Zwick, *Jour. Appl. Phys.*, **23,** 95 (1952) and **24,** 493 (1954).

9. Dergarabedian, P., *Jour. Appl. Mech.*, **20,** 537 (1953).

10. Fritz, W., *Z. Physik,* **36,** 379 (1935).

11. Zuber, N., "Hydrodynamic Aspects of Boiling Heat Transfer," Ph.D. dissertation, U.C.L.A., June 1959.

12. Jakob, M., "Local Temperature Differences as Occurring in Evaporation, Condensation, and Catalytic Reaction," in *Temperature, Its Measurement and Control in Science and Industry*, pp. 834, Reinhold, New York, Vol. 1, 1941.

13. Staniszewski, B. E., "Nucleate Boiling Bubble Growth and Departure," Technical Report No. 16, DSR 7673, Office of Naval Research Contract Nonr-1841(39), Heat Trans. Lab., Mass. Inst. of Tech., August 1959.

14. Rohsenow, W. M., and J. A. Clark, "A Study of the Mechanism of Boiling Heat Transfer," *Trans. ASME*, July 1951.

15. Berenson, P., "Transition Boiling Heat Transfer from a Horizontal Surface," AIChE Paper No. 18, ASME-AIChE Heat Transfer Conference, Buffalo, August 14–17, 1960; also, *ASME Jrl. Ht. Trans.*, Aug. 1961, p. 351.

16. McAdams, W. H., W. E. Kennel, C. S. Minden, C. Rudolf, C. Picornell, and J. E. Dow, "Heat Transfer at High Rates to Water with Surface Boiling," *Ind. Eng. Chem.*, **41,** 1945 (Sept. 1949).

17. Jakob, M., *Mech. Eng.*, **58,** 643 (1936).

18. Rohsenow, W. M., "A Method of Correlating Heat Transfer Data for Surface Boiling of Liquids," *Trans. ASME*, July 1952.

19. Addoms, J. N., "Heat Transfer at High Rates to Water Boiling Outside Cylinders," D.Sc. Thesis, Chem. Engrg. Dept., M.I.T., June 1948.

20. Cichelli, M. T., and C. F. Bonilla, *Trans. Am. Inst. Chem. Engrs.*, **41,** 755–87 (1945).

21. Piret, E. L., and H. S. Isbin, "Two-Phase Heat Transfer in Natural Circulation Evaporators," AIChE Heat Transfer Symposium, St. Louis, December 13–16, 1953.

22. Cryder, D. S., and A. C. Finalborgo, *Trans. Am. Inst. Chem. Engrs.*, **33,** 346 (1937).

23. Forster, K., and R. Greif, *Trans. ASME, Jour. Ht. Trans.* **81,** C, 1, pp. 43–53 (Feb. 1959).

24. Gilmour, C. H., *Chem. Eng. Prog.*, **54:**10, 77–79 (1958).

25. Rohsenow, W. M., "Heat Transfer, A Symposium 1952," Engrg. Res. Inst., Univ. of Mich.

26. Rohsenow, W. M., and P. Griffith, "Correlation of Maximum Heat Flux Data for Boiling of Saturated Liquids," AIChE-ASME Heat Trans. Symposium, Louisville, Ky., March 1955.

27. Usiskin, C. M., and R. Siegel, ASME Paper 60-HT-10, 1960.

28. Zuber, N., *Trans. ASME*, **80,** 711–20 (Apr. 1958).

29. Zuber, N., and M. Tribus, "Further Remarks on the Stability of Boiling Heat Transfer," U.C.L.A. Report No. 58–5, January 1958.

30. Mirshak, S., W. S. Durant, and R. H. Towell, "Heat Flux at Burnout," DP-355, Feb. 1959 (Und. AEC Rept.).

31. Griffith, Peter, "Correlation of Nucleate Boiling Burnout Data," ASME Paper 57-HT-21.

32. Lowdermilk, W. H., C. D. Lanzo, and B. L. Siegel, *Nat. Advisory Comm. Aeronaut. Tech. Note* 4382 (Sept. 1958).

33. Isakoff, S. E., "Effect of an Ultrasonic Field on Boiling Heat Transfer," *Heat Transfer and Fluid Mech. Inst.*, Stanford Univ. Press, 1956.

34. Choi, H. Y., Tufts University Mechanical Engineering Report No. 60–2, May 1960.

35. Gambill, W. R., and N. D. Greene, "Study of Burnout Heat Fluxes Associated With Forced-Convection Sub-cooled, and Bulk-Nucleate Boiling of Water in Source-Vortex Flow," preprint No. 29, 2nd Nat. Ht. Trans. Conf., Chicago, Aug. 1958, AIChE-ASME; also *Chem. Eng. Prog.*, Oct. 1958, p. 68.

36. Gambill, W. R., R. D. Bundy, and R. W. Wansbrough, "Heat Transfer, Burnout and Pressure Drop for Water in Swirl Flow Through Tubes with Internal Twisted Tapes," AIChE Paper No. 22, ASME-AIChE Ht. Trans. Conference, Buffalo, August 14–17, 1960.

37. Siegel, R., and C. Usiskin, "Photographic Study of Boiling in Absence of Gravity," *Trans. ASME, Jour. Ht. Trans.* **81**, 3 (Aug. 1959).

38. Merte, J., Jr., and J. A. Clark, "Study of Pool Boiling in an Accelerating System," ASME Paper 60-HT-22.

39. Bromley, L. A., *Chem. Eng. Prog.*, **46:**5, 221–27 (May 1950).

40. Bromley, L. A., *et al.*, *Ind. Eng. Chem.*, **45,** 2639–46 (1953).

41. Bergles, A. E., Memo. 8767-1, Proj. DSR, Heat Trans. Lab., Mass. Inst. of Tech., Apr. 10, 1961.

42. Forster, H. K., and N. Zuber, *AIChE Trans.* **1,** 531, (1955).

43. Rohsenow, W. M., P. Griffith, and P. J. Berenson, Discussion, *Trans. ASME*, **80,** 716–17, (Apr. 1958).

PROBLEMS

9.1 Calculate the size of cavities activated with heating-surface superheats $(T_w - T_{sat})$ of 1, 5, 10, 20 F in water at 14.7 psia; in water at 1000 psia.

9.2 Calculate the q/A vs. $(T_w - T_{sat})$ curve for saturated water at 11.5 psia boiling on a nickel wire surface in a pool.

(a) Use Eq. (9.14) and Table 9.1.

(b) Use Eq. (9.17).

(c) Calculate the maximum heat flux using Eqs. (9.22) and (9.23).

9.3 An electric current passing through a horizontal brass tube (0.375 in. diam.) immersed in a pool of water at atmospheric pressure provides the following data: $q/A = 61{,}000$ Btu/hr ft^2 at $T_w = 233$ F and $T_l = T_{sat} = 212$ F.

(a) Calculate C_{sf} in Eq. (9.14).

(b) For the measured temperatures, calculate $(q/A)_b$ from Eq. (9.17).

(c) Calculate the maximum heat flux for nucleate boiling from Eqs. (9.22) and (9.23).

(d) Calculate the minimum (q/A) for film boiling from Eqs. (9.26).

(e) Using Eqs. (9.28) through (9.31), calculate (q/A) vs. $(T_w - T_{sat})$ for the film boiling region.

(f) Plot the entire boiling curve of q/A vs. $T_w - T_{sat}$ from the calculations above.

9.4 Since electric power demand on weekends is low, there is excess hydroelectric power available in Maine at these times; this excess power is offered free of charge to small manufacturers. One shoe manufacturing plant decided to install electric coil heaters in a few tubes of its currently operating oversized fire-tube steam generator. The blanked-off tubes are 2 in. O.D. and 10 ft long. How many tubes must be used if 1000 kw of electrical energy are available for generating steam at 200 psia from feed water of 150 F? Design the heaters to keep the q/A below 1/2 of $(q/A)_{max}$ for nucleate boiling.

9.5 A nuclear reactor has cylindrical cooling passages 0.20 in. in diameter and 4 ft long. Water at 680 psia and 400 F enters the tube at a velocity of 20 ft/sec and the heat flux is uniform along the length and equal to q/A = 700,000 Btu/hr ft^2.

(a) Calculate the wall temperature distribution along the tube, assuming $C_{sf} = 0.013$ and Eqs. (9.14) and (9.18) apply along the entire tube.

(b) For the same flow conditions and inlet conditions, calculate the maximum heat flux and exit conditions when film boiling sets in (burnout point) using Fig. 9.14.

CHAPTER 10

Condensation

When a cold surface at temperature T_w below saturation temperature, T_{sat}, is exposed to vapor, either saturated or superheated, liquid condensate forms on the surface. If the liquid wets the surface, it spreads out and establishes a stable film; consequently the process is called *film condensation*. The vapor condenses on the liquid at the interface because of the heat transferred through the liquid film. With pure saturated vapor the liquid-vapor interface temperature is also essentially at the saturation temperature.

If the liquid does not wet the surface, droplets form, Fig. 10.1, and run down the surface, coalescing as they travel downward. Little is understood about the mechanism of dropwise condensation or the associated heat transfer process except that heat transfer rates are from 2 to 20 times those for film condensation with the same ΔT. Commercially, it is difficult to maintain dropwise condensation for long periods of time, so practically all condensing equipment is designed on the assumption that film condensation will exist.

10.1 Elementary Analysis of Laminar Film Condensation on Vertical Plate

An early film condensation analysis for a vertical plate, by Nusselt (1), assumed the temperature distribution in the liquid film to be linear, pure

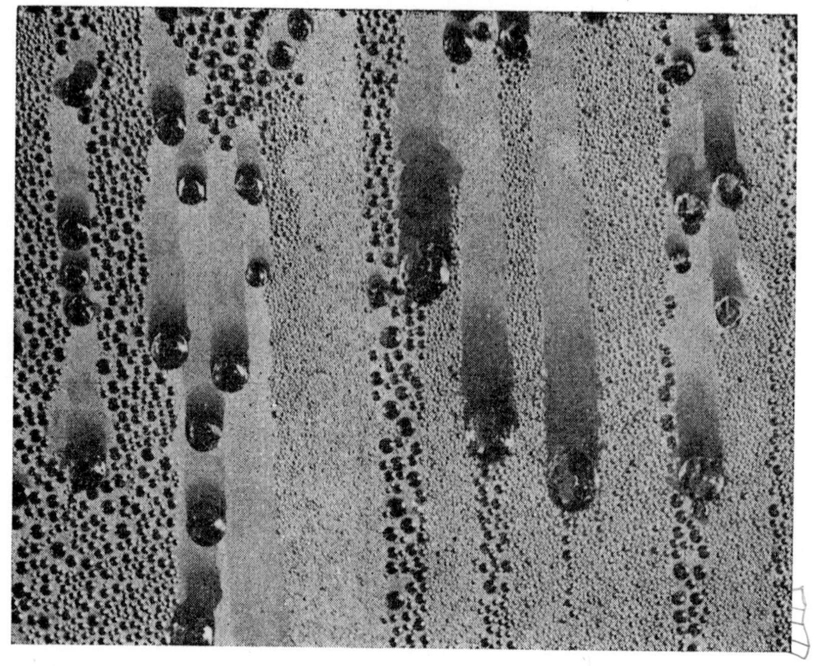

Fig. 10.1. Dropwise condensation. Steam condensing on a vertical surface of vinyl plastic.

vapor at T_{sat}, T_w uniform, no shear at liquid-vapor interface. The following is a modified Nusselt analysis. Neglecting momentum effects, a balance of the shear, gravity, and pressure forces ($dp/dz = g\rho_v$) on the element of liquid, Fig. 10.2, between y and δ, the liquid film thickness, results in the following equation:

$$\mu \frac{dv_z}{dy} = g\rho(\delta - y) - g\rho_v(\delta - y)$$

Then integrate from $v_z = 0$ at $y = 0$:

$$v_z = \frac{g(\rho - \rho_v)}{\mu}\left(\delta y - \frac{y^2}{2}\right) \tag{10.1}$$

The flow rate per unit width, Γ, is then calculated as follows:

$$\Gamma = \rho \int_0^\delta v_z \, dy = \frac{g\rho(\rho - \rho_v)}{\mu} \frac{\delta^3}{3} \tag{10.2}$$

Then the change of Γ with δ is given by

$$\frac{d\Gamma}{d\delta} = \frac{g\rho(\rho - \rho_v)}{\mu} \delta^2 \tag{10.3}$$

With the assumed linear temperature gradient, the average enthalpy change of the vapor in condensing to liquid and subcooling to the average liquid temperature is

$$h_{fg} + \frac{1}{\Gamma} \int_0^\delta \rho v_z c(T_{\text{sat}} - T)\, dy = h_{fg} + \frac{3}{8} c(T_{\text{sat}} - T_w)$$

and the heat flux to the wall is $(k/\delta)(T_{\text{sat}} - T_w)$. Then equating heat flux with enthalpy change

$$\frac{q}{A} = k\,\frac{\Delta T}{\delta} = \left(h_{fg} + \frac{3}{8} c\, \Delta T\right) \frac{d\Gamma}{dz} \tag{10.4}$$

Equating expressions for $d\Gamma$ from (10.3) and (10.4),

$$\delta^3\, d\delta = \frac{k\mu\, \Delta T}{g\rho(\rho - \rho_v) h'_{fg}}\, dz$$

where $h'_{fg} \equiv (h_{fg} + \frac{3}{8} c\, \Delta T)$. Integrating with $\delta = 0$ at $z = 0$, the following expression for δ is obtained:

$$\delta = \sqrt[4]{\frac{4k\mu\, \Delta T z}{g\rho(\rho - \rho_v) h'_{fg}}} \tag{10.5}$$

Since $h_z = (q/A)/(T_{\text{sat}} - T_w) = k/\delta$, the local heat transfer coefficient from Eq. (10.5) is

$$h_z = \sqrt[4]{\frac{g\rho(\rho - \rho_v) k^3 h'_{fg}}{4z\mu\, \Delta T}} \tag{10.6}$$

or the average coefficient is

$$h = (1/L) \int_0^L h_z\, dz$$

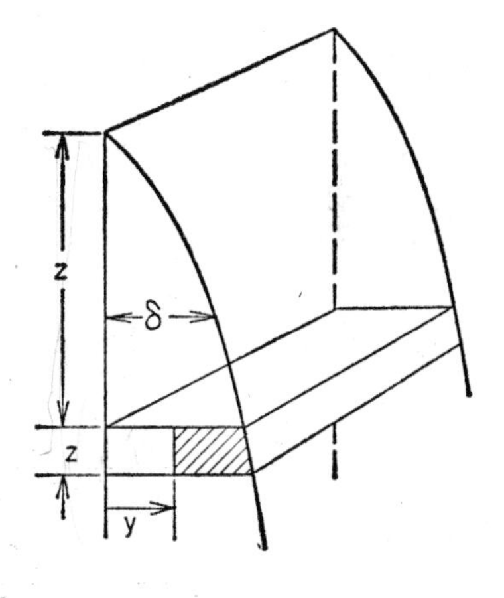

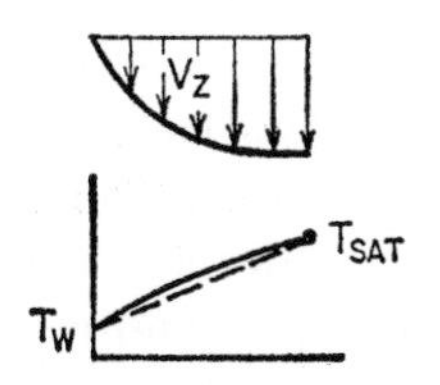

Fig. 10.2. Sketch of condensate film.

or

$$h = 0.943 \sqrt[4]{\frac{g\rho(\rho - \rho_v) k^3 h'_{fg}}{L\mu\, \Delta T}} \tag{10.7a}$$

Since $h_z \sim z^{-1/4}$, the average h between $z = 0$ and L is $\frac{4}{3}$ times the local h_L at $z = L$.

A modified integral analysis was developed (2) for this problem, applied to the fluid element, and integrated to determine the actual temperature distribution. The actual distribution is slightly curved, the film thickness slightly greater, but the heat transfer rate is slightly larger than that of the above analysis, Fig. 10.2. The result could be closely approximated by Eq. (10.7a) with h'_{fg} replaced by $(h_{fg} + 0.68c\ \Delta T)$ or

$$h = 0.943 \sqrt[4]{\frac{g\rho(\rho - \rho_v)k^3 h_{fg}(1 + 0.68c\ \Delta T/h_{fg})}{L\mu\ \Delta T}} \tag{10.7b}$$

This equation is valid for $\mathrm{Pr} > 0.5$ and $c\ \Delta T/h_{fg} \leq 1.0$. Evaluate μ at $T_f = T_w + \frac{1}{4}\Delta T$.

The above analysis also applies to a flat plate inclined at angle ϕ to the horizontal plane if g in the preceding equations is replaced by $g \sin \phi$, the component of gravity acting along the plate.

10.2 Boundary-Layer Analysis of Condensation on Vertical Plates

Sparrow and Gregg (3) first analyzed the liquid condensate film as a boundary layer. The boundary-layer equation (3.18) requires modification to include the gravity effect as follows:

$$\begin{gathered}
\frac{\partial v_z}{\partial z} + \frac{\partial v_y}{\partial y} = 0 \\
v_z \frac{\partial v_z}{\partial z} + v_y \frac{\partial v_z}{\partial y} = g\left(1 - \frac{\rho_v}{\rho}\right) + \nu \frac{\partial^2 v_z}{\partial y^2} \\
v_z \frac{\partial T}{\partial z} + v_y \frac{\partial T}{\partial y} = \frac{k}{\rho c_p} \frac{\partial^2 T}{\partial y^2}
\end{gathered} \tag{10.8}$$

These equations include the momentum effects omitted from the simple analysis in Sec. 10.1. The similarity transformation can be shown to be $y/z^{1/4}$. As in the development in Sec. 7.6, we now introduce the following definitions of θ, ψ, η, and F:

$$\begin{gathered}
\theta \equiv \frac{T - T_{\text{sat}}}{T_w - T_{\text{sat}}} \\
v_z \equiv \frac{\partial \psi}{\partial y}; \quad v_y \equiv -\frac{\partial \psi}{\partial z} \\
\eta \equiv \left[\frac{g(\rho - \rho_v)}{4\nu^2 \rho}\right]^{1/4} \frac{y}{z^{1/4}} \\
F(\eta) \equiv \frac{\psi}{4\nu z^{3/4}} \left[\frac{g(\rho - \rho_v)}{4\nu^2 \rho}\right]^{-1/4}
\end{gathered} \tag{10.9}$$

With these quantities the momentum and energy equations of Eq. (10.8) become

$$F''' + 3FF'' - 2(F')^2 + 1 = 0$$
$$\theta'' + 3 \, \mathrm{Pr} \, F\theta' = 0 \tag{10.10}$$

where the prime indicates differentiation with respect to η. For uniform T_w and zero shear at the liquid-vapor interface, the boundary conditions are as follows:

$$\begin{aligned} y &= 0 \qquad v_z = v_y = 0 \qquad T = T_w \\ y &= \delta \qquad \partial v_z/\partial y = 0 \qquad T = T_{sat} \end{aligned} \tag{10.11}$$

or

$$\begin{aligned} \eta &= 0 \qquad F = F' = 0 \qquad \theta = 1.0 \\ \eta &= \eta_\delta \qquad F'' = 0 \qquad \theta = 0 \end{aligned}$$

where η_δ is η at $y = \delta$.

To obtain a relation involving η_δ we write the energy balance equation (10.4) in integral form as follows:

$$\int_0^z k\left(\frac{\partial T}{\partial y}\right)_{y=0} dz = \int_0^\delta h_{fg}\rho v_z \, dy + \int_0^\delta \rho v_z c(T_{sat} - T) \, dy \tag{10.12a}$$

which in terms of the new variables, Eq. (10.9), becomes (3)

$$\frac{c \, \Delta T}{h_{fg}} = -3\frac{F(\eta_\delta)}{\theta'(\eta_\delta)} \, \mathrm{Pr} \tag{10.12b}$$

Equations (10.10) with Eqs. (10.11) and (10.12b) were solved numerically to determine $F(\eta)$ and $\theta(\eta)$ which are functions of Pr and $c_p \, \Delta T/h_{fg}$. Then, since $h_z = (k/\Delta T)(\partial T/\partial y)_{y=0}$,

$$\mathrm{Nu}_z \left[\frac{g(\rho - \rho_v)z^3}{4\nu^2\rho}\right]^{-1/4} = \left(\frac{d\theta}{d\eta}\right)_{\eta=0} \tag{10.13}$$

The results of the calculation in terms of Eq. (10.13) are shown plotted in Fig. 10.3. These results agree well with Eq. (10.7b), calculated with omission of the momentum terms, when $\mathrm{Pr} \geq 1$ and $c \, \Delta T/h_{fg} \leq 1.0$. It should be noted that since $h_z \sim z^{-1/4}$, then $h = \frac{4}{3}h_L$.

In the preceding analyses, the shear stress at the liquid-vapor interface was neglected. Calculations were made by Chen (4) and Koh *et al.* (5) for the case in which the vapor is assumed stagnant, thus producing a slight upward shear stress on the liquid surface. The results are shown dotted in Fig. 10.3.

The preceding two sections discuss condensation on vertical plates, but actually the results apply equally well to condensation inside or outside of

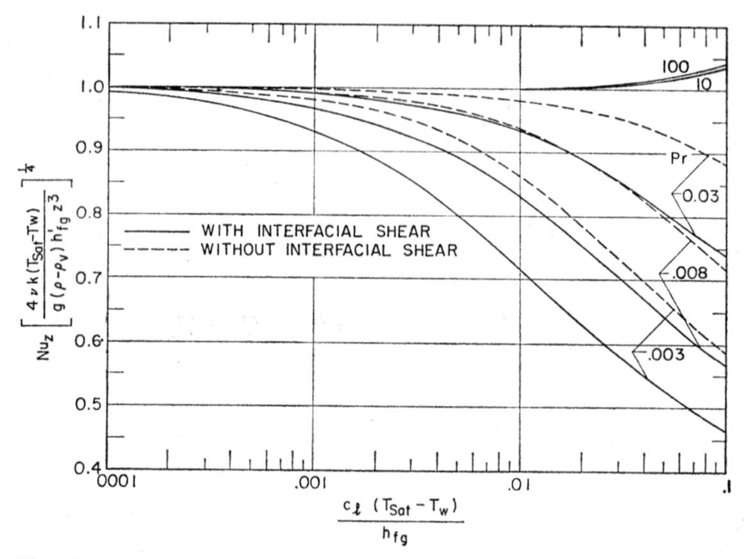

Fig. 10.3. Local Nusselt number for condensation on vertical plate [Koh, Sparrow, Hartnett (5) and Sparrow, Gregg (3)]

most vertical tubes since the liquid films are usually quite thin (0.001 to 0.010 in.).

10.3 Effect of Vapor Shear Stress on Laminar Film Condensation

For condensation in a confined space such as in a tube or between parallel plates, all of the condensate that forms enters the space as vapor, usually at one end of the tube. Because of the large density difference between liquid and vapor, the incoming vapor may have a rather high velocity, particularly for the larger length-to-diameter ratios. This produces large shear stress τ_v at the liquid-vapor interface.

The analysis in Sec. 10.1 is readily modified to include the effect of vapor shear stress, τ_v, at the interface. Then the force balance and Eqs. (10.1) and (10.2) become

$$-\tau = \mu \frac{dv_z}{dy} = g(\rho - \rho_v)(\delta - y) \pm \tau_v$$

$$v_z = \frac{g(\rho - \rho_v)}{\mu}\left(\delta y - \frac{y^2}{2}\right) \pm \frac{\tau_v}{\mu} y \qquad (10.14)$$

$$\Gamma = \frac{g\rho(\rho - \rho_v)}{\mu}\frac{\delta^3}{3} \pm \frac{\tau_v}{\mu}\frac{\delta^2}{2}$$

where the plus sign is for downward flow of vapor and the minus sign for upward flow of vapor.

Again assuming a linear temperature distribution and Eq. (10.4), the following results (6) are obtained if τ_v is assumed uniform at some average value between $z = 0$ and $z = z_L$, the plate height:

$$z^* = (\delta^*)^4 \pm \frac{4}{3}\tau_v^*(\delta^*)^3$$

$$\frac{4\Gamma}{\mu}\frac{1}{(1 - \rho_v/\rho)} = \frac{4}{3}(\delta_L^*)^3 \pm 2\tau_v^*(\delta_L^*)^2 \tag{10.15}$$

$$h^* = \frac{h}{k}\left(\frac{\nu^2}{g}\right)^{1/3} = \frac{4}{3}\frac{(\delta_L^*)^3}{z_L^*} \pm 2\tau_v^*\frac{(\delta_L^*)^2}{z_L^*}$$

where

$$\delta^* = \delta\left(\frac{g}{\nu^2}\right)^{1/3}$$

$$z^* = \frac{4z\,\Delta T}{\mathrm{Pr}}\frac{c}{h'_{fg}}\left(\frac{g}{\nu^2}\right)^{1/3}\frac{1}{(1 - \rho_v/\rho)} \tag{10.16}$$

$$\tau_v^* = \frac{\tau_v}{g(\rho - \rho_v)(\nu^2/g)^{1/3}}$$

Any two of the dimensionless quantities may be eliminated since there are three equations in Eq. (10.15). The results for downward vapor flow are plotted in Fig. 10.5 to the left of the dotted curve and are the same on each graph, e.g., independent of Pr. In these curves, $\rho_v \ll \rho$; hence, $(1 - \rho_v/\rho) \approx 1.0$.

10.4 Turbulent Film Condensation on a Vertical Plate

When the vertical plate or tube is long enough or the vapor flow high enough, the laminar liquid film changes to turbulent flow. Figure 10.4 represents the estimated film Reynolds number $4\Gamma/\mu$ at transition. A freely falling film ($\tau_v = 0$) is observed (7) to change to turbulent flow at $4\Gamma/\mu = 1800$. Then for this condition the shear stress at the wall is calculated from Eqs. (10.14) with $\tau_v = 0$. The steep curve up to $\tau_v^* \cong 11$, Fig. 10.4, results from assuming that the wall shear stress remains the same at transition, independent of the magnitude of τ_v.

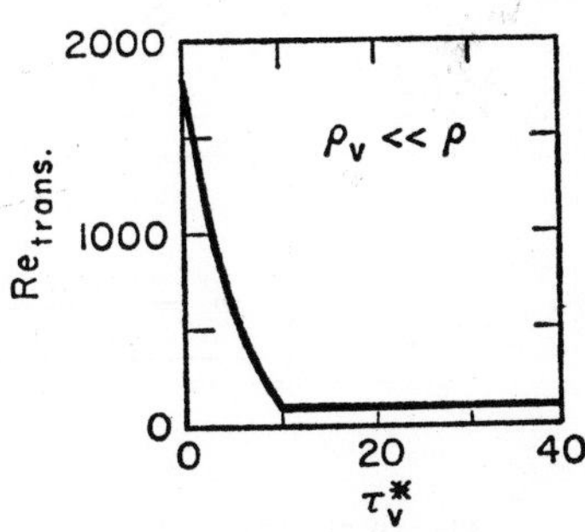

Fig. 10.4. Transition Reynolds number for liquid film (6).

For $\tau_v^* > 11$ the film thickness is very

small. Here we assume, after Carpenter (8), that $\delta^+ \equiv (\delta/\nu)\sqrt{\tau_0/\rho} = 6$. Experimental data are limited, but agreement with Carpenter's ethanol data was within 25 per cent for $(4\Gamma/\mu)_{tr}$ at $\tau_v^* = 110$.

The heat transfer associated with turbulent film condensation with significant vapor shear was analyzed [Rohsenow, Webber, and Ling (6)] by assuming the universal velocity distribution, Fig. 4.14, to apply and then using the methods of Sec. 8.4 based on the analogy between momentum and heat transfer. The details of the analysis are omitted here but the results for downward vapor flow are shown in Fig. 10.5 for Pr = 1 and 10.

In these graphs a uniform τ_v was assumed. Actually τ_v varies along the length because of the mass removed by condensation. Lehtinen (9) performed a stepwise analysis and determined that τ_v should be calculated for the following average mass velocity of the vapor:

$$G_{v_m} = 0.4(G_{\text{top}} + 1.5G_{\text{bottom}}) \tag{10.17a}$$

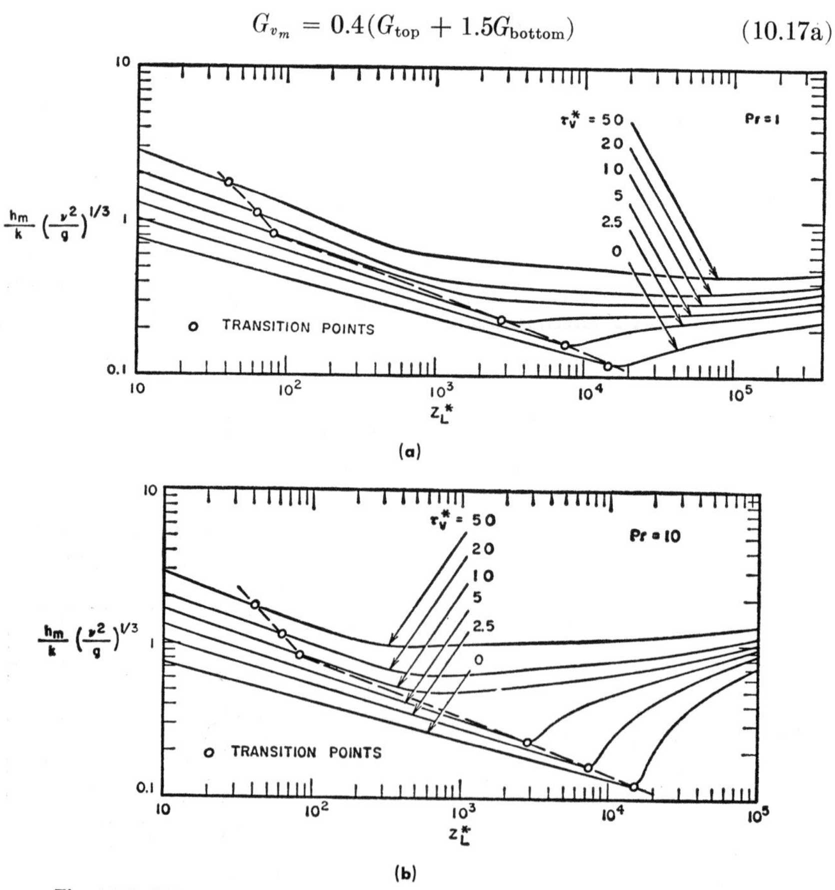

Fig. 10.5. Effect of turbulence and vapor shear stress on condensation [Rohsenow, Webber, Ling (6)].

If all of the vapor condenses ($G_{\text{bottom}} = 0$), this reduces to

$$G_{v_m} = 0.4G_{\text{top}} \tag{10.17b}$$

Then the proper average τ_v is calculated as follows:

$$\tau_v = \frac{fG_{vm}^2}{2\rho_v} \tag{10.18}$$

where f is taken from the data of Bergelin *et al.* (10) shown plotted in Fig. 10.6, where σ_w/σ is the ratio of surface tensions of water and the particular fluid being condensed, Γ_L corresponds to mean liquid flow, and ρ is the liquid density.

For condensation of a particular fluid at a given pressure on a vertical surface of a given geometry and magnitude of τ_v and T_w, the value of Pr

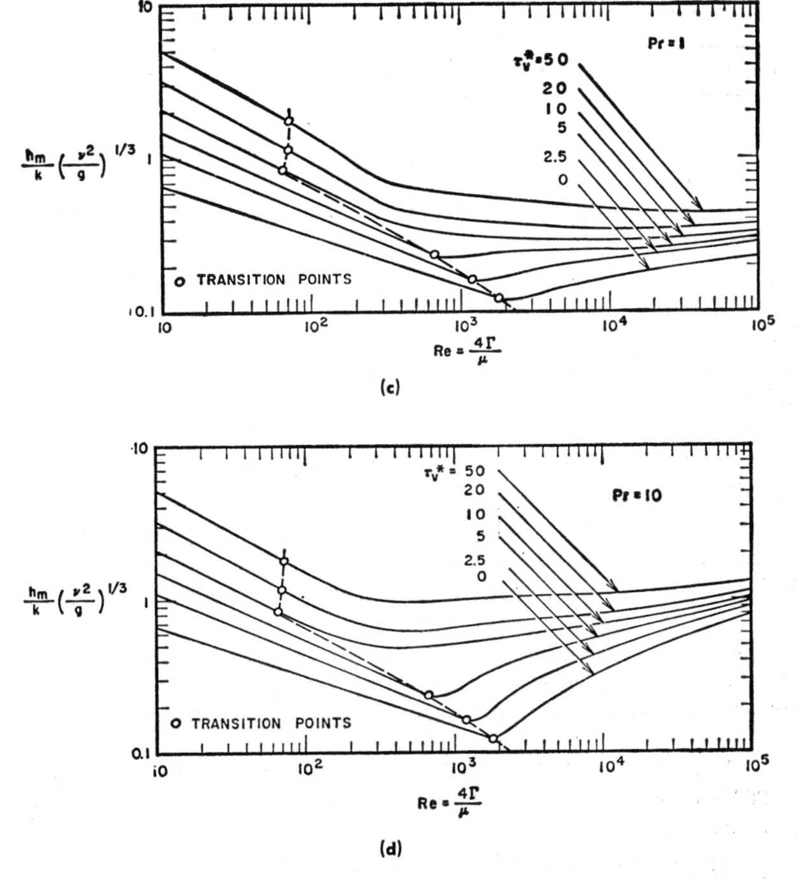

Fig. 10.5. (Con.) Effect of turbulence and vapor shear stress on condensation.

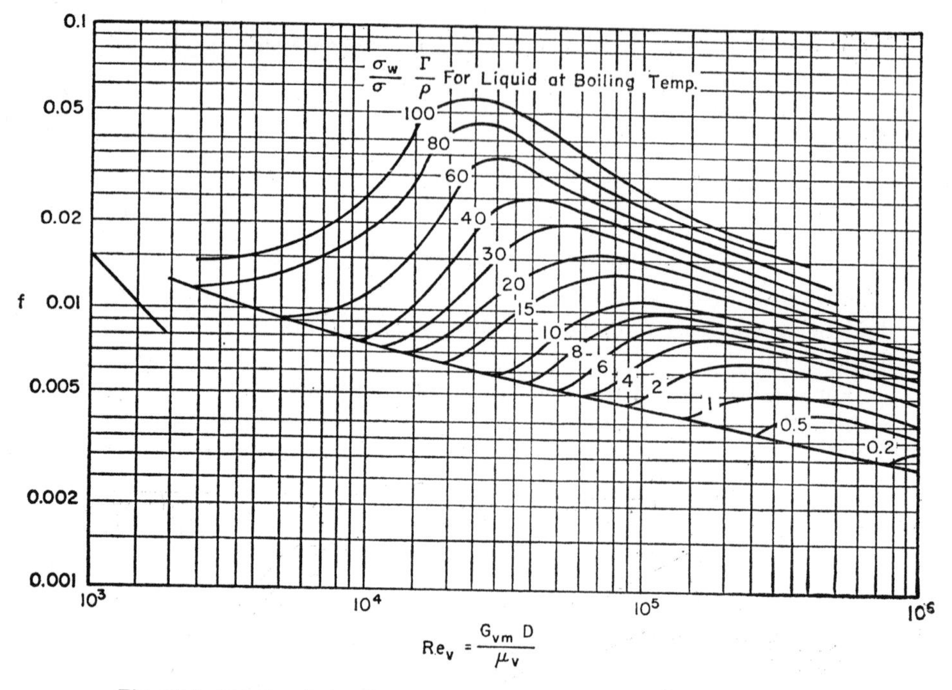

Fig. 10.6. Friction factor for gas (air) flowing in a tube with a liquid layer on the wall [Bergelin *et al.* (10)].

and z_L^* are readily calculated. Interpolation in Fig. 10.5 will give a value of $(h_m/k)(\nu^2/g)^{1/3}$ from which h_m is obtained and $(q/A)_{\text{avg}}$ is calculated. This value of h_m includes the effect of a laminar film on the upper part of the surface and, if present, the effect of a turbulent film on the remainder of the surface.

In determining the condensation rate for a surface of a given size, the procedure requires a trial-and-error type calculation with these curves, calculating the performance for several magnitudes of τ_v^* and interpolating.

In analyzing test data, the condensation rate is often measured. Then $(h_m/k)(\nu^2/g)^{1/3}$ is read directly from the curve at the particular $4\Gamma/\mu$ corresponding to the condensation rate.

The results of the analysis leading to the graphs of Fig. 10.5 were shown (6) to agree well with the experimental data of Carpenter and Colburn (12) which was taken for fluids whose Pr was in a limited range of 2 to 5.

Example 10.1: Determine the condensation rate for saturated steam at 262 F on a vertical plate 1 ft wide and 12 ft high which is maintained at a temperature of 162 F. Neglect the effect of vapor shear stress, $\tau_v = 0$.

The mean film temperature is 212 F; so $k = 0.412$ Btu/hr ft F, $c = 1.0$ Btu/lb F, $\rho = 59.9$ lb/ft³, and $\rho_v = 0.04$ lb/ft . At 262 F, $h_{fg} = 937.3$ Btu/lb.

It is suspected that this condensate film will be in turbulent flow, so μ is evaluated at 212 F, $\mu = 0.687$ lb/hr ft, and $h'_{fg} \cong h_{fg} = 937$.

For the chart quantities of Fig. 10.5,

$$\mathrm{Pr} = \frac{(1.0)(0.687)}{(0.412)} = 1.67$$

$$h = h^* k\left(\frac{g}{\nu^2}\right)^{1/3} = h^*(0.412)\left[\frac{(4.16 \times 10^8)(59.9)^2}{(0.687)^2}\right]^{1/3} = 6030h^*$$

$$z_L^* = \frac{4(12)(100)}{(1.67)(937)}(14{,}620) = 45{,}000$$

From Figs. 10.5(a), (b) at $\tau_v^* = 0$ read $h^* = 0.48$ at Pr = 10 and $h^* = 0.15$ at Pr = 1. Plot these two points on log-log graph paper and join by a straight line; then read $h^* = 0.18$ at Pr = 1.67.

So

$$h = 6030(0.18) = 1085 \text{ Btu/hr ft}^2 \text{ F}$$

$$q = Ah(\Delta T) = 12(1085)(100) = 1{,}300{,}000 \text{ Btu/hr}$$

and

$$w = \frac{q}{h_{fg}} = \frac{1{,}300{,}000}{970} = 1340 \text{ lb/hr}$$

Example 10.2: What length of 2-in. I.D. vertical tube is required to condense 1450 lb/hr of saturated steam at 262 F if the tube wall temperature is 162 F? The fluid properties are those listed in Example 10.1. For vapor at 262 F, $\mu_v = 0.032$ lb/hr ft.

If there is no vapor flow at the bottom (e.g., all vapor is condensed)

$$G_{vm} = 0.4(1450) \div (\pi/4)(2/12)^2 = 26{,}700 \text{ lb/hr ft}^2$$

Then for the vapor

$$\mathrm{Re}_v = \frac{(26{,}700)(2/12)}{(0.032)} = 139{,}000$$

The relative surface tension $\sigma_w/\sigma = 1.0$ and average Γ is

$$\Gamma = \frac{725}{\pi}\frac{12}{2} = 1380 \text{ lb/hr ft}; \quad \frac{\sigma_w \Gamma}{\sigma \rho} = \frac{1380}{(59.9)(1.0)} = 22.0$$

Then from Fig. 10.6, $f = 0.014$.

The film Reynolds number at exit is

$$\mathrm{Re}_L = \frac{4\Gamma_e}{\mu} = \frac{4(2760)}{(0.687)} = 16{,}200$$

Therefore, the condensate film is turbulent.

From steam tables $\rho_v = 0.0877\ \text{lb/ft}^3$. Then from Eq. (10.18)

$$\tau_v = 0.014\frac{(26{,}700)^2}{(2)(32.2)(3600)^2(0.0877)} = 0.136\ \text{lb}_f/\text{ft}^2$$

and from Eq. (10.16), where $(g/\nu^2)^{1/3} = 14{,}630\ \text{ft}^{-1}$,

$$\tau_v^* = \frac{(0.136)(14{,}630)}{(59.9\text{–}0.0877)} = 33.3$$

From Figs. 10.5(c), (d), $h^* = 0.90$ at Pr $= 10$ and $h^* = 0.38$ at Pr $= 1$. Interpolating linearly on a plot of log h vs. log Pr, $h^* = 0.41$ at Pr $= 1.67$. Then as calculated in Example 10.1,

$$h = 6030h^* = 2470\ \text{Btu/hr ft}^2\ \text{F}$$

$$q = wh'_{fg} = (1450)(937) = 1{,}360{,}000\ \text{Btu/hr}$$

and $$A = \frac{q}{h(\Delta T)} = \frac{(1{,}360{,}000)}{(2470)(100)} = 5.5\ \text{ft}^2$$

so $$L = \frac{A}{\pi D} = \frac{(5.5)}{\pi(\frac{2}{12})} = 10.5\ \text{ft}$$

10.5 Condensation on Horizontal Tubes

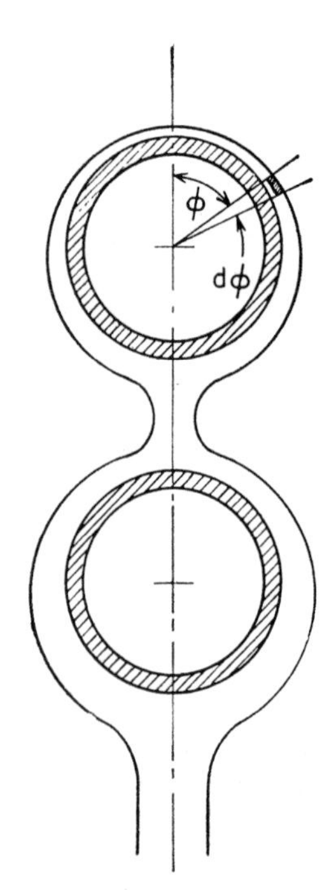

Fig. 10.7. Condensate film on horizontal tubes.

The condensate film on the outside of horizontal tubes flows around the tube and off the bottom in a sheet, shown exaggerated in Fig. 10.7. The liquid film actually is very thin so that the type of analyses of Secs. 10.1 and 10.2 applies here except that g is replaced by $g \sin \phi$ and the average value of h follows from integration over the range of ϕ values from 0 to 180°.

The details of the analysis are omitted here, but the result equivalent to Eq. (10.7a) is as follows for the average h on the top tube:

$$h = 0.728\sqrt[4]{\frac{g\rho(\rho - \rho_v)k^3h'_{fg}}{D\mu\ \Delta T}} \qquad (10.19)$$

The results of the boundary-layer analysis assuming zero shear stress at the liquid-vapor interface [Sparrow and Gregg (11)] and of the integral equation analysis assuming stagnant vapor [Chen (4)] are shown in Fig. 10.8. The same graph also applies to the average h on a vertical flat

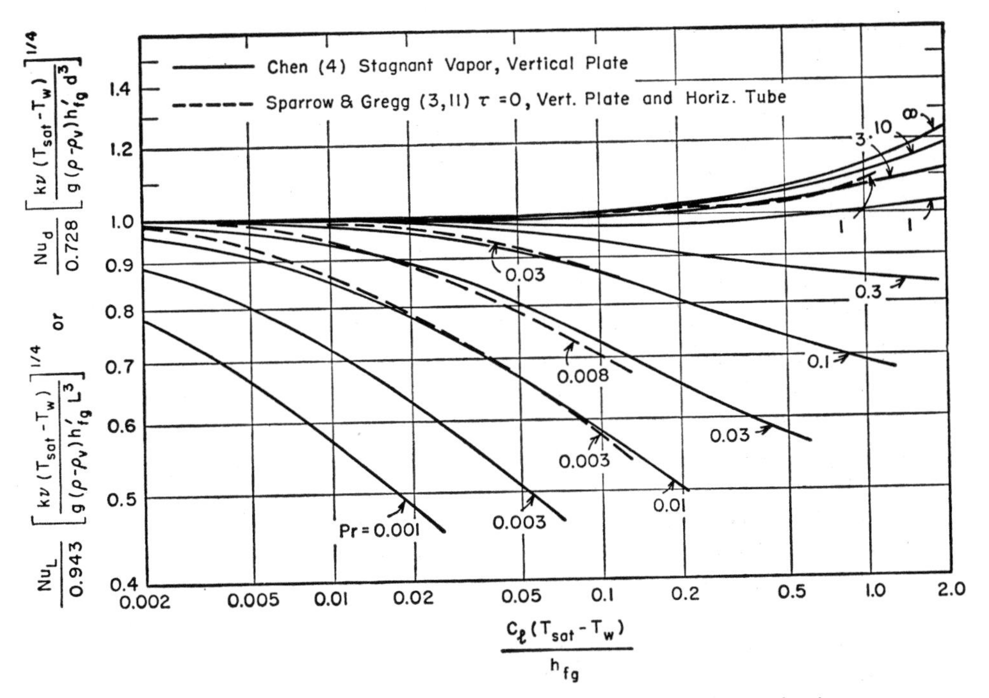

Fig. 10.8. Average Nusselt number for condensation on horizontal tubes and vertical plates [Sparrow and Gregg (3), Chen (4)].

plate with the ordinate interpreted as indicated. Inspection of Fig. 10.8 shows Eq. (10.19) is adequate for $\Pr \geq 1$ for $c\,\Delta T/h_{fg} < 1$.

The heat transfer coefficient on a horizontal tube decreases from a maximum value at $\phi = 0$ to essentially zero for $\phi = 180°$. The condensing rate on the upper half of the tube is 46 per cent greater than on the lower half.

Comparing Eqs. (10.7a) with (10.19), the same average heat transfer coefficient exists on vertical plates and horizontal tubes for identical fluid conditions and ΔT when $0.728/D^{1/4} = 0.943/L^{1/4}$ or

$$L = 2.78\,D \tag{10.20}$$

Tube bank. Nusselt (1) analyzed condensation on vertical banks of horizontal tubes, Fig. 10.7, by the method of Sec. 10.1. Here all of the condensate dropping from any tube is assumed to fall on the next lower tube. The result for the heat transfer coefficient averaged over all n tubes in a vertical bank is

$$h = 0.728\sqrt[4]{\frac{g\rho(\rho - \rho_v)k^3h'_{fg}}{nD\mu\,\Delta T}} \tag{10.21}$$

This result indicates that the average h for n tubes equals the h for the top tube divided by $\sqrt[4]{n}$. Actually, experimental data for banks of tubes fall well above the results of Eq. (10.21).

Chen (4) suggested that since the liquid film was subcooled an average of $\frac{3}{8}(T_{sat} - T_w)$, it is possible that additional condensation occurs on the liquid layer between tubes. Assuming all of this subcooling is removed, his analysis provided the following result:

$$h = 0.728\left[1 + 0.2\frac{c\,\Delta T}{h_{fg}}(n - 1)\right]\sqrt[4]{\frac{g\rho(\rho - \rho_v)k^3h'_{fg}}{nD\mu\,\Delta T}} \qquad (10.22)$$

which is a good approximation provided $[(n - 1)c\,\Delta T/h_{fg}] < 2$.

Equation (10.22) agrees well with most of the available data for condensation on banks of tubes. Most experimental data for pure vapors agree with or are slightly higher than the equations presented in this chapter. Departures are usually attributed to such phenomena as external vibrations causing ripples in the liquid surface, splashing off tubes in a bank, or uneven run-off due to bowing or inclination of tubes.

Experimental data for condensation of liquid metals are very scarce and inconsistent. Data for mercury (13) provide magnitudes of h which are 0.9 to 0.3 of the values predicted by Fig. 10.8. It was suggested that this might be due to an interfacial resistance at the liquid-vapor interface. This discrepancy remains unresolved.

Example 10.3: Determine the condensation rate and heat transfer rate for saturated steam at 216 F condensing on four 1-in. diameter, 2-ft long horizontal tubes arranged in a vertical bank. The tube wall temperature is 208 F. Properties of liquid water at 212 F are given in Example 10.1.

$$\frac{0.68c\,\Delta T}{h_{fg}} = \frac{(0.6)(1)(8)}{970.3} = 0.0056.$$

At $T_f = T_w + \Delta T/4 = 210$ F, $\mu = 0.695$ lb/ft hr. Then with $\tau_v = 0$,

$$B \equiv \left[\frac{h_{fg}\rho(\rho - \rho_v)k^3g}{\mu}\right]^{1/4} = \left[\frac{(970.3)(59.9)^2(0.412)^3(4.16 \times 10^8)}{0.695}\right]^{1/4} = 3500$$

$$C_n \equiv [1 + 0.2(c\,\Delta T/h_{fg})(n - 1)] = [1 + 0.2(0.0065)(4 - 1)] = 1.0039$$

Then

$$h_m = \frac{0.728C_nB(1 + 0.68c\,\Delta T/h_{fg})^{1/4}}{(nD\,\Delta T)^{1/4}}$$

$$= \frac{(0.728)(1.0039)(3500)(1.0056)^{1/4}}{(4 \times 0.0833 \times 8)^{1/4}} = 1990 \text{ Btu/hr ft}^2 \text{ F}$$

$$q = Ah_m(\Delta T) = 4\pi(0.0833)(2)(1990)(8) = 33{,}500 \text{ Btu/hr}$$

The condensate rate is

$$w = \frac{q}{h_{fg}} = \frac{33{,}500}{(973.3)} = 34.5 \text{ lb/hr}$$

and

$$\text{Re} = \frac{4\Gamma}{\mu} = \frac{4(34.5/2)}{0.695} = 100$$

which is less than 1800; so the flow is laminar, as assumed.

10.6 Effect of Superheated Vapor

When the vapor is superheated and the surface temperature of the cold surface is also above the saturation temperature, no condensation occurs; the vapor simply cools down, becoming less superheated. The heat transfer performance follows the pattern of any single phase fluid as discussed in Chapters 7 and 8.

Most experiments (15) show that when the surface temperature is below saturation and film condensation occurs, the condensation rate is only slightly increased by greater superheat. Assuming that the liquid-vapor interface remains very nearly at saturation temperature as superheat is increased, the analysis of Sec. 10.1 need be modified only by accounting for the larger enthalpy change from the superheated vapor to the slightly sub-cooled liquid. In other words, the enthalpy change h'_{fg} of Eqs. (10.4) through (10.7) should be replaced by

$$h''_{fg} = c_v(T_v - T_{sat}) + h_{fg} + \tfrac{3}{8}c(T_{sat} - T_w) \tag{10.23}$$

and the heat transfer rate calculated by

$$q/A = h(T_{sat} - T_w) \tag{10.24}$$

since $(T_{sat} - T_w)$ is still the driving force for heat transfer across the liquid condensate film. The condensate rate is also calculated from

$$\frac{w}{A} = \frac{1}{h''_{fg}}\left(\frac{q}{A}\right) \tag{10.25}$$

Although most experimental film-condensation evidence suggests a slight increase in (q/A) as superheat is increased (approximately 3 per cent increase with 200 F superheat for steam at atmospheric pressure), the experiments of Balekjian and Katz (16) show a decreasing heat transfer rate with increasing superheat of steam and Freon-114. More recent experiments (17) with Freon-11 disagreed with these results; they showed increasing (q/A) with increasing superheat.

In a heat exchanger, the wall temperature may be above saturation near the inlet and below saturation temperature near the outlet. Calculations for such a case should be divided into two parts—a desuperheating section and a condensing section.

10.7 Effect of Noncondensable Gas

When a condensable vapor is condensing in the presence of a noncondensable gas, the vapor must diffuse through the gas, requiring a decrease in vapor partial pressure toward the liquid-vapor interface. Thus, interface saturation temperature is significantly below the temperature of the main vapor-gas mixture Fig. 10.9a. This combined mass and heat trans-

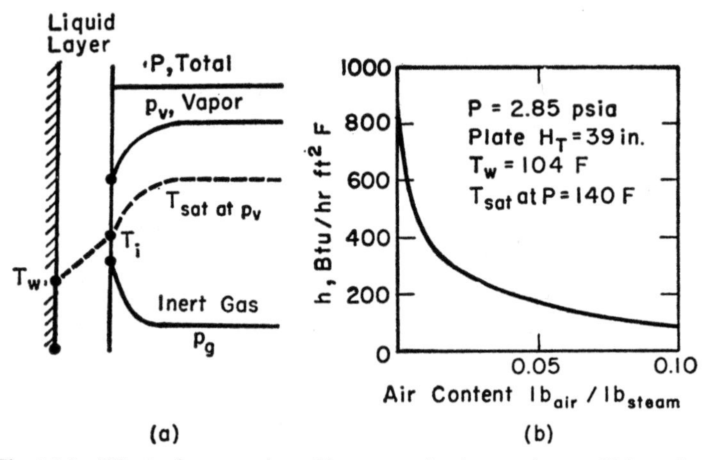

Fig. 10.9. Effect of noncondensable gas on heat transfer coefficient for condensing vapor [Langen (19)].

fer process requires a lengthy trial-and-error stepwise calculation to predict condenser performance accounting for this effect. Details are found elsewhere (18).

A significant decrease in h results from the presence of very small amounts of noncondensable gas. Figure 10.9b shows this effect.

In most condensing equipment, such as power plant steam condensers, provision is made to bleed off noncondensable gases that leak into the system. There are, however, cases in which noncondensable gases must be tolerated, such as in separating ammonia from air by condensation. Design of such condensers must allow for the presence of noncondensables.

10.8 Drop Condensation

Drop condensation has very large heat transfer coefficients, from 10,000 to 80,000 Btu/hr ft^2 F, depending on the tangential steam velocity.

This is an order of magnitude larger than the figure for film condensation, which has values from about 800 to 2000 Btu/hr ft^2 F. It has generally been supposed that most of the heat goes through the bare surfaces that exist between drops, an understandable line of thought, since one of the most obvious differences from film condensation is the existence of those apparently bare surfaces. Condensation on the bare surfaces has not been shown to occur. If this is what in fact occurs, the means by which the condensed water reaches existing drops or forms into drops is not yet known.

An alternate suggestion has been made by Emmons (21). Molecules reflected from the bare surfaces would have energies less than that of the saturated vapor and may lead to the formation of a thin layer of subcooled vapor next to the bare surfaces between drops. Emmons suggested that rapid condensation sets up violent eddy currents that move the subcooled layer to the drops.

Still another suggestion has been that conduction through the drops may be the mechanism for dissipation of the latent heat of condensation, with the condensation occuring on the surfaces of the drops. Because of the thermal conductivity of the liquid in static drops, this would require that the condensation occur very near the bottom of the drop, so that heat would have only to be conducted through the "edges" of the drops. Trefethen (22) has suggested that condensation need not occur at these edges, if the drops are assumed to be circulating internally, driven by surface tension differences between hot and cool regions of the drop surfaces, a condition that might occur, particularly if the bases of the drops are blanketed by noncondensable gases.

The mechanism of drop condensation is not understood, and no proposed explanation yet appears to have experimental confirmation. There is, however, one aspect that has considerable experimental backing. Drop condensation will occur only when the condensed liquid does not wet the solid surface. No known metallic surface, when clean, will produce drop condensation. To obtain drop condensation, metal surfaces must be coated with a film (two monomolecular layers are sufficient) of some substance that the liquid does not wet. The following are some promoters of dropwise condensation of steam: mercaptans or octyl thiocyanate on copper, and oleic acid on copper, brass, nickel and chromium. Hydrocarbon oils also promote dropwise condensation, but they wash off the surfaces rather quickly.

REFERENCES

1. Nusselt, W., *Z. ver. deut. Ing.*, **60**, 541, 569 (1916).
2. Rohsenow, W. M., *Trans. ASME*, **78**, 1645–48 (1956).

3. Sparrow, E. M., and J. L. Gregg, *Trans. ASME, Jour. Ht. Transfer*, Series C, 1959, pp. 13–18, including discussion by R. A. Seban.

4. Chen, M. M., *Trans ASME, Jour. Ht. Transfer*, Series C, **83,** 48–60, (Feb. 1961).

5. Koh, J. C. Y., E. M. Sparrow, and J. P. Hartnett, *Int. Jour. Ht. & Mass Trans.*, **2,** 69–82, March 1961.

6. Rohsenow, W. M., J. H. Webber, and A. T. Ling, *Trans. ASME*, **78,** 1637–44, (1956).

7. McAdams, W. H., *Heat Transmission*, 3rd Ed., p. 334, McGraw-Hill, New York, 1954.

8. Carpenter, F. G., Ph.D. Thesis, Chem. Engineering Dept., Univ. of Del., 1948.

9. Lehtinen, J. A., Sc.D. Thesis, Mech. Engineering Dept., M.I.T., Cambridge, Mass., June, 1957.

10. Bergelin, O. P., P. K. Kegel, F. G. Carpenter, C. Gazley, *Ht. Trans. and Fl. Mech. Inst.*, Berkeley, Calif., 1949.

11. Sparrow, E. M. and J. L. Gregg, *Trans. ASME, Jour. Ht. Transfer*, 81C, Nov. 1959, pp. 291–96.

12. Carpenter, F. G., and A. P. Colburn, *London Discussion of Heat Transfer*, ASME, July 1951, pp. 20–31.

13. Misra, B., and C. F. Bonilla, Heat Transfer Symposium AIChE, 1955, Louisville.

14. *Liquid Metals Handbook Na-NaK Supplement 1955*, AEC TID5277, p. 91.

15. Merkel, F., *Die Grundlagen der Wärmeübertragung*, T. Steinkopf, Leipzig, 1927.

16. Balekjian, G., and D. L. Katz, *AIChE Jour.*, **4,** 43 (March 1958).

17. Personal communication from J. K. Sparrell, Dynatech Corp., Cambridge, Mass, January, 1961.

18. Kern, D. Q., Process Heat Transfer, McGraw-Hill, New York, 1950.

19. Langen, E., *Forschung a.d. Geb. d. Ingenieurwes.* **2,** 359 (1931).

20. Euken, A., *Naturwissenschaften*, **25,** 209 (1937).

21. Emmons, H. W., *Trans. AIChE*, **35,** 109 (1939).

22. Trefethen, L., "Drop Condensation and the Possible Importance of Circulation within Drops Caused by Surface Tension Variation" GEL Rept. No. 58GL47, General Electric Co., Schenectady, N. Y., Feb. 3, 1958.

PROBLEMS

10.1 Film condensation takes place on a vertical plate whose surface is maintained at 208 F and is exposed to an atmosphere of saturated steam at 212 F. Calculate:

(a) the local h at 1 foot below the top

(b) the film thickness at 1 foot below the top

(c) condensation rate on a plate 1 ft x 1 ft.

10.2 Derive expressions for calculating h for laminar film condensation on the inside of a vertical circular tube with a uniform wall temperature. Neglect the effect of shear stress at the liquid surface. At what height does the liquid fill the tube?

10.3 Derive expressions and suggest in detail the calculation procedure for determining the heat transfer rate for a vertical flat plate of thickness x_w and of height L with a cooling fluid at a uniform temperature T_c on one side and a condensing vapor at saturation temperature T_{sat} on the other side. Assume the h on the coolant side is uniform.

10.4 A shallow pan is in an atmosphere of saturated steam at 212 F. The sides of the pan are insulators and the bottom is maintained at 208 F. Calculate the rate of condensate accumulation in the bottom of the pan. If the pan initially is empty, what will be the water depth after 5 sec? 1 minute? 5 minutes?

10.5 A horizontal tube ($\frac{5}{8}$ in. O.D.) is exposed to saturated steam at 212 F. The outer tube wall is maintained at 208 F by passing a liquid metal at high velocity through the tube.

(a) Calculate the condensing coefficient.

(b) How much steam condenses per foot of tube length in 1 minute? 5 minutes?

(c) Suppose this system were inside a satellite in space under zero-gravity conditions. How much steam condenses per foot in 1 minute? 5 minutes?

10.6 Calculate part (a) of Prob. 10.5 for a vertical bank of five horizontal tubes.

10.7 Perform an analysis for film condensation on a rotating disk situated in a large body of pure saturated vapor. The centrifugal field associated with the uniform rotation sweeps the condensate outward along the disk surface, and gravity and coriolis forces need not be considered. Your analysis should approximate the steps outlined in Nusselt's solution, Sec. 10.1.

(a) Arrive at expressions for the film thickness, δ, and the heat transfer coefficient, h.

10.8 A vertical tube 1 in. in diameter, 20 ft long is used to condense steam on the inside. The steam enters the top at 14.7 psia, 212 F, and the cooling water on the outside of the tube is at 60 F and is assumed to have a heat transfer coefficient $h = 2000$ Btu/hr ft^2 F between the water and the outside surface of the tube. The tube wall is 0.050 in. thick and is copper with $k = 200$ Btu/hr ft F. A scale coefficient on the water side is $h_{sc} = 1000$ Btu/hr ft^2 F.

Assume that the condensate film will be turbulent and determine the rate of condensation and the rate of heat transfer if the condensate is drained off at the bottom of the tube without permitting it to collect there. (Later check assumption that condensate film is turbulent.)

10.9 A tubular Freon-12 condenser is to be constructed with sixteen $\frac{5}{8}$-in. outside diameter, 10-ft long copper tubes (0.031 in. wall thickness) arranged horizontally in a bundle on triangular pitch. Cooling water flows at 6 ft/sec inside the tubes and enters at 80 F. Calculate the rate of condensation of

(a) saturated steam at 0.95 psia

(b) saturated Freon-12 at 171.8 psia.

10.10 A saturated vapor is condensed on the outside surface of a constant-temperature cone. Neglecting vapor shear stress at the liquid-vapor interface and subcooling, follow an analysis similar to the Nusselt theory and arrive at a single differential equation relating local film thickness to the temperature difference, fluid properties, and distance from apex of the cone.

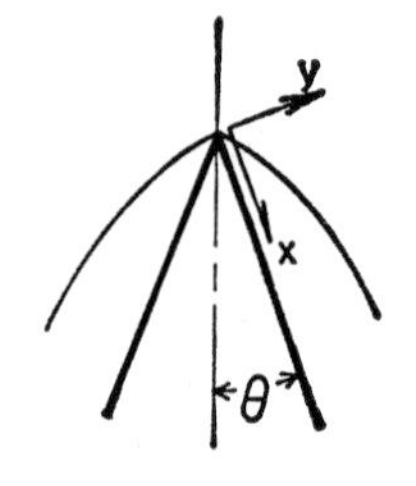

CHAPTER 11

Heat Transfer in High Velocity Flows and in Rarefied Gases

In fluid flow at high velocities over solid surfaces, temperature increases due to frictional effects, commonly known as *aerodynamic heating*, become important. Assumption of constant fluid properties may no longer be valid because of steep temperature gradients within the boundary layer. The presence of shock waves and their interaction with the boundary layer may be an additional complication. Furthermore, if the temperatures get high enough, gas molecules in the boundary layer may partially dissociate, thereby changing the components and the properties of the fluid. Gas may also become slightly ionized or electrically conducting, and its behavior influenced by the presence of a magnetic field (magneto-aerodynamics). At very high altitudes, gas rarefaction must be considered, necessitating a critical reappraisal of the underlying hypothesis of continuum flow (superaerodynamics). Clearly, problems of heat transfer at high velocities can be extremely complicated. We shall merely attempt here to outline the nature of the problems with a few simple examples and confine our attention to transfer processes occurring between a gas and a shock-free surface.

11.1 Adiabatic Wall Temperature and Recovery Factor

We shall first take a grossly simplified view that the compressibility effect may be neglected and that properties such as viscosity, thermal conductivity, and specific heat are constant. These assumptions take us back to Sec. 7.1 where we examined the effect of energy dissipation in a one-dimensional Couette flow system. It was then found that when the parameter $N = \mu V_1^2/2k(T_1 - T_0)$ exceeded unity, heat transferred not only to the lower plate, but also to the warmer upper plate. Let us again

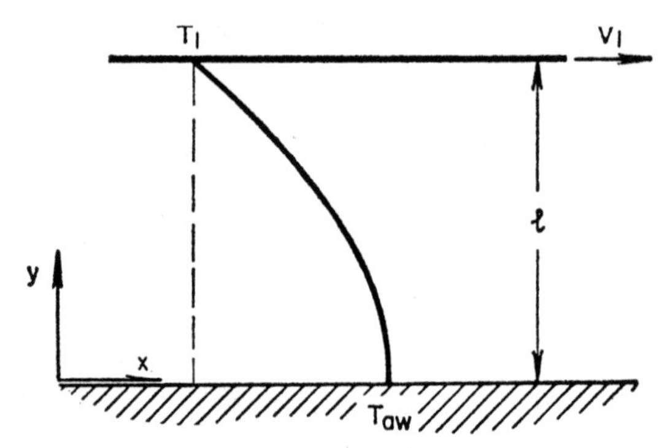

Fig. 11.1. Couette flow with insulated stationary plate.

examine the same system, Fig. 11.1, but with the lower plate perfectly insulated so that the new boundary conditions are

$$y = 0, \quad v_x = 0, \quad \partial T/\partial y = 0$$

$$y = l, \quad v_x = V_1, \quad T = T_1$$

If we again assume a linear velocity profile (zero dp/dx), Eq. (7.4) can be readily integrated to yield

$$T - T_1 = \frac{\mu V_1^2}{2k}\left(1 - \frac{y^2}{l^2}\right) \tag{11.1}$$

The temperature profile given by Eq. (11.1) is parabolic, with zero gradient at the insulated wall. The temperature excess of the lower plate over the upper plate is given by

$$T_0 - T_1 = \frac{\mu V_1^2}{2k} = (\mathrm{Pr})\frac{V_1^2}{2c_p} \tag{11.2}$$

This temperature of the insulated plate has a special significance in high-velocity systems and is called the *adiabatic wall temperature*, T_{aw}. It should be noted that although the velocity at the wall is zero, T_{aw} is not the stagnation temperature T_s, the latter being defined as the temperature that a gas stream will attain if brought to rest reversibly and adiabatically.

The stagnation enthalpy is related to the free-stream enthalpy by the steady-flow energy equation

$$i_s = i_\infty + \tfrac{1}{2}V_\infty^2$$

If the gas were ideal and if the specific heat were constant

$$T_s = T_\infty + \frac{V_\infty^2}{2c_p} \tag{11.3}$$

Also, $c_p = \dfrac{\gamma R}{\gamma - 1}$ and Mach number $M = \dfrac{V}{\sqrt{\gamma RT}}$

where $\gamma = c_p/c_v$. Therefore Eq. (11.3) can be written

$$T_s = T_\infty\left[1 + \frac{\gamma - 1}{2}M_\infty^2\right] \tag{11.4}$$

For a gas with variable specific heat, T_s may be determined with the help of the gas tables (31). The stagnation temperature is attained in the probe shown in Fig. 11.2a when the vent hole is very small and heat transfer is negligible. The stagnation temperature is not reached, in general, at any point on the blunt body shown in Fig. 11.2b. On the surface of the blunt body the gas stream is decelerated not only by pressure rise but also by momentum transfer (viscosity effects) within the boundary layer.

A parameter called the *recovery factor* and given the symbol r has been found useful and is defined as follows:

$$r = \frac{T_{aw} - T_\infty}{T_s - T_\infty} = \frac{T_{aw} - T_\infty}{V_\infty^2/2c_p} \tag{11.5}$$

Equation (11.5) expresses the ratio of the excess of the adiabatic wall temperature over the free-stream static temperature to the excess of the stagnation temperature over the static temperature. The adiabatic wall temperature may be expressed in terms of r by

$$T_{aw} = T_\infty + r\frac{V_\infty^2}{2c_p} \tag{11.6}$$

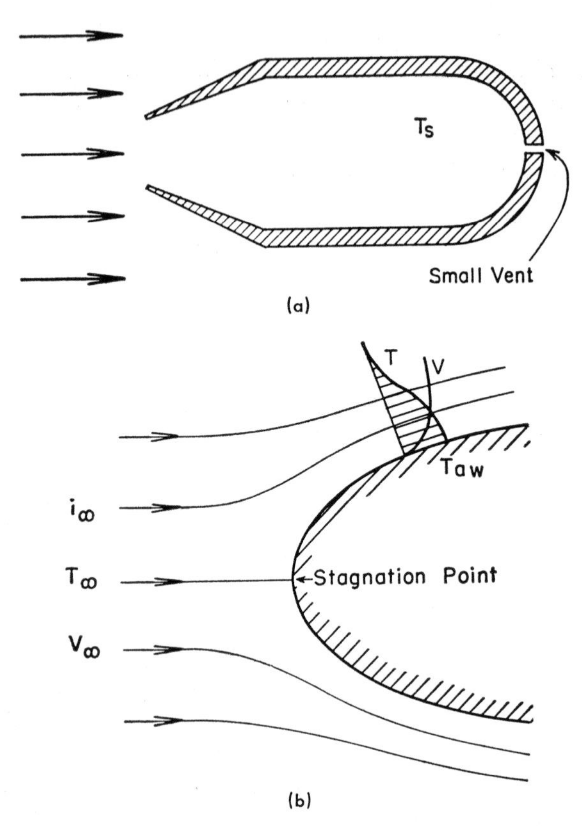

Fig. 11.2. Stagnation temperature and adiabatic wall temperature.

or, for an ideal gas,

$$T_{aw} = T_{\infty}\left[1 + r\frac{\gamma - 1}{2}M_{\infty}^{2}\right] \tag{11.7}$$

For the Couette flow of Fig. 11.1, replacing T_{∞} and V_{∞} in Eq. (11.5) by T_1 and V_1, the recovery factor from Eq. (11.2) is simply $r = \text{Pr}$. Other theoretical and experimental values of r and its use in heat transfer studies will be discussed in the following sections.

11.2 High Velocity Flow over Flat Plate

A flat-plate analysis is relatively simple and the geometry is basic to an understanding of the aerodynamic heating problems of high-speed vehicles. We shall therefore examine the system in some detail. If we neglect compressibility effects but assume properties as being variable, the usual boundary-layer assumptions lead to the following equations of continuity, momentum, and energy for a two-dimensional boundary layer over a flat plate:

$$\frac{\partial}{\partial x}(\rho v_x) + \frac{\partial}{\partial y}(\rho v_y) = 0 \tag{11.8}$$

$$\rho v_x \frac{\partial v_x}{\partial x} + \rho v_y \frac{\partial v_x}{\partial y} = \frac{\partial}{\partial y}\left(\mu \frac{\partial v_x}{\partial y}\right) \tag{11.9}$$

$$\frac{\partial p}{\partial y} = 0$$

$$\rho c_p\left(v_x \frac{\partial T}{\partial x} + v_y \frac{\partial T}{\partial y}\right) = \frac{\partial}{\partial y}\left(k \frac{\partial T}{\partial y}\right) + \mu \left(\frac{\partial v_x}{\partial y}\right)^2 \tag{11.10}$$

If we limit our solution to high subsonic and low supersonic flows ($M < 2$) and modest temperature differences, the effect of variable properties may be neglected.

Defining $\eta \equiv (y/\sqrt{x})\sqrt{V_\infty/\nu}$ and $f(\eta) = \psi/\sqrt{V_\infty \nu x}$ as before in Eqs. (3.22) and (3.23), Eq. (11.9) is transformed to Eq. (3.26) and Eq. (11.10) is transformed to

$$\frac{d^2\theta_r}{d\eta^2} + \frac{1}{2}(\text{Pr})(f)\frac{d\theta_r}{d\eta} + 2(\text{Pr})\left(\frac{d^2f}{d\eta^2}\right)^2 = 0 \tag{11.11}$$

where

$$\theta_r \equiv \frac{T - T_\infty}{V_\infty^2/2c_p} \tag{11.12}$$

(*a*) *Adiabatic Flat Plate*

We now solve Eq. (11.11) for an adiabatic flat plate with the following boundary conditions:

$$\begin{aligned} y &= 0, \quad v_x = v_y = 0, \quad \partial T/\partial y = 0 \\ y &= \infty, \quad v_x = V_\infty, \qquad T = T_\infty \end{aligned} \tag{11.13}$$

or equivalently,

$$\begin{aligned} \eta &= 0, \quad f(\eta) = f'(\eta) = 0, \quad \theta_r' = 0 \\ \eta &= \infty, \qquad f'(\eta) = 1, \quad \theta_r = 0 \end{aligned}$$

The velocity field for this case with constant properties is identical with that obtained in Sec. 3.3 and Fig. 3.5. This gives us $f(\eta)$ in Eq. (11.11). The temperature field, solution of Eq. (11.11), was obtained by Pohlhausen (1) by the method of variation of coefficients giving the following result:

$$(\theta_r)_{\text{ad}} = 2\,\text{Pr}\int_\eta^\infty \exp\left[-\frac{1}{2}\,\text{Pr}\int_0^\eta f\,d\eta\right]\left[\int_0^\eta \left(\frac{d^2f}{d\eta^2}\right)^2 \exp\left(\frac{1}{2}\,\text{Pr}\int_0^\eta f\,d\eta\right)d\eta\right]d\eta \tag{11.14}$$

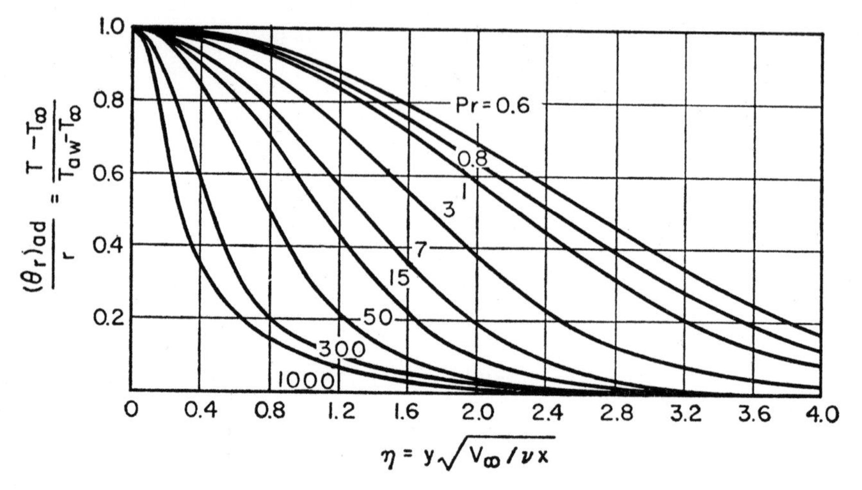

Fig. 11.3. Temperature distribution in a laminar boundary layer of a high velocity stream parallel to an adiabatic flat plate [Pohlhausen (1)].

Equation (11.14) gives θ_r as a function of η for the adiabatic wall and is shown plotted in Fig. 11.3.

By definition $(\theta_r)_{\text{ad}}$ evaluated from Eq. (11.14) at $\eta = 0$ is the recovery factor defined in Eq. (11.7). The result is shown plotted in Fig. 11.4. Over the range $0.5 < \text{Pr} < 5$, r may be approximated by the following simple relation:

$$r \cong \sqrt{\text{Pr}} \tag{11.15}$$

(b) *Nonadiabatic Flat Plate*

For the nonadiabatic flat plate the thermal boundary conditions of Eq. (11.13) change as follows:

$$\begin{aligned} y &= 0, \quad v_x = v_y = 0, \quad T = T_0 \\ y &= \infty, \quad v_x = V_\infty, \quad T = T_\infty \end{aligned} \tag{11.16}$$

where $T_0 \lessgtr T_{aw}$.

Since $f(\eta)$ is known from Eq. (3.28), Eq. (11.11) is linear in θ_r, and its solution with these boundary conditions is accomplished by superimposing the particular solution (adiabatic flat plate) and the general solution of the corresponding homogeneous equation which is identical with Eq. (7.34). Hence we solve Eq. (11.11) with the following boundary conditions:

$$\begin{aligned} y &= 0, \quad v_x = v_y = 0, \quad \partial T/\partial y = 0 \\ y &= \infty, \quad v_x = V_\infty, \quad T = T_\infty \end{aligned}$$

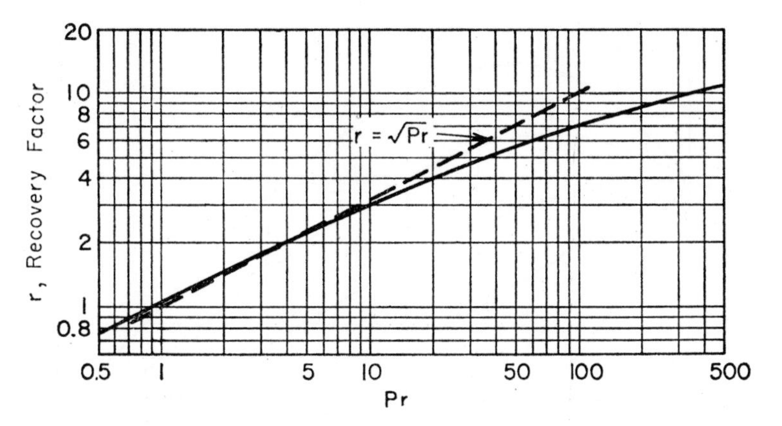

Fig. 11.4. Laminar recovery factor for flat plate.

and Eq. (7.34) with the following boundary conditions:

$$y = 0, \quad v_x = v_y = 0, \quad T = T_0$$

$$y = \infty, \quad v_x = V_\infty, \quad T = T_\infty$$

The final solution for T is obtained by adding the solutions from Eqs. (11.14) and (7.36a) to satisfy Eqs. (11.16)

$$T - T_\infty = (\theta_r)_{\text{ad}} \frac{V_\infty^2}{2c_p} + (T_0 - T_{aw})(1 - \theta) \tag{11.17}$$

Figure 11.5 shows this solution for the particular case in which Pr = 0.7 and $r = 0.84$. Note that when $(T_{aw} - T_\infty)/(T_0 - T_\infty) > 1$, the plate is heated by the air; when less than 1 it is cooled by the air.

From Eq. (11.17),

$$\frac{q}{A} = -k\left(\frac{\partial T}{\partial y}\right)_0 = k(T_0 - T_{aw})\left(\frac{\partial \theta}{\partial y}\right)_0 \tag{11.18}$$

since $[\partial(\theta_r)_{\text{ad}}/\partial y]_0 = 0$. Then with Eq. (7.37), Eq. (11.18) becomes

$$\frac{q}{A} = 0.332k \, \text{Pr}^{1/3} \sqrt{V_\infty/\nu x}\,(T_0 - T_{aw}) \tag{11.19a}$$

or

$$\frac{[(q/A)/(T_0 - T_{aw})]x}{k} = 0.332 \, \text{Pr}^{1/3} \sqrt{\text{Re}_x} \tag{11.19b}$$

The result of Eq. (11.19b) indicates the usual definitions of h and Nu based on the temperature difference $(T_0 - T_\infty)$ are no longer useful, since the heat flux is not proportional to this temperature difference. The definition of an alternate heat transfer coefficient is discussed in the next section.

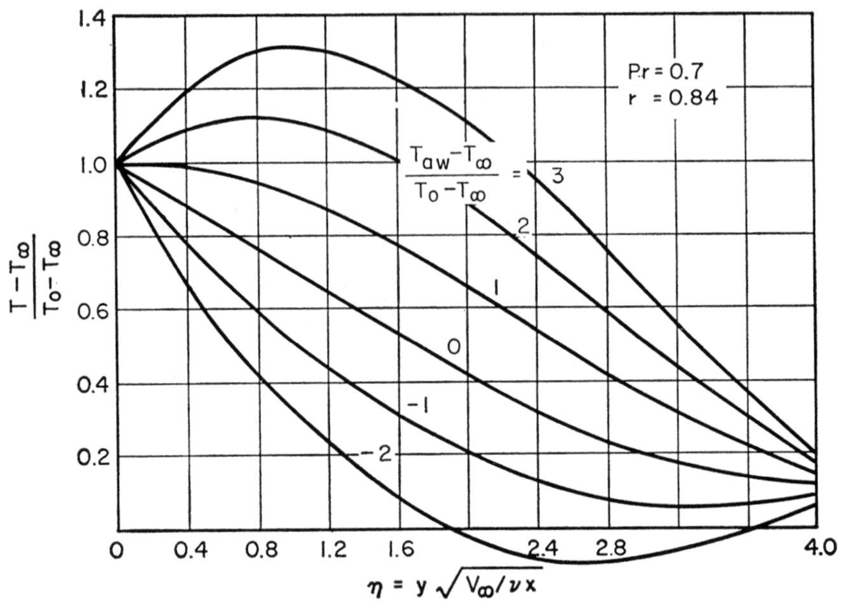

Fig. 11.5. Temperature distribution in a laminar boundary layer for air flowing at high velocity along a non-adiabatic flat plate (1).

Turbulent boundary layer in high velocity flow over a flat plate was studied theoretically by Ackerman (2). For Prandtl number in the neighborhood of unity ($0.5 < \mathrm{Pr} < 2$), his calculations of the recovery factor may be satisfactorily represented by the simple relation

$$r \cong (\mathrm{Pr})^{1/3} \tag{11.20}$$

Example 11.1. Air at 65 F flows over a plate 1 in. long at a velocity of 655 ft/sec. If the plate temperature is maintained at 95 F, calculate the adiabatic wall temperature. What do you conclude regarding heat transfer at the plate?

$$\mathrm{Re}_L = \frac{VL}{\nu} = \frac{(655)(3600)\left(\frac{1}{12}\right)}{0.60} = 3.27 \times 10^5, \text{ laminar}$$

From Eq. (11.15),

$$r = \sqrt{\mathrm{Pr}} = \sqrt{0.709} = 0.842$$

Eq. (11.7) gives

$$T_{aw} = T_\infty + r\frac{V_\infty^2}{2c_p} = 65 + 0.842\frac{(655)^2}{(2)(0.24)(778)(32.2)} = 95 \text{ F}$$

It is seen that the plate temperature is equal to the adiabatic wall temperature (obtained simply by insulating the plate), and heat transfer at the plate is negligible.

11.3 Theoretical and Experimental Heat Transfer Results

A realistic heat transfer analysis taking into account fluid compressibility is complicated not only by additional terms in the basic equations but also because the velocity field cannot be found independently of the temperature field. Property variations, especially the variation of viscosity with temperature, must be considered.

(*a*) *Theoretical Results*

For laminar compressible flow, Crocco (3) obtained a solution for a flat plate assuming constant free-stream velocity, plate temperature, and Prandtl number. The Crocco method was extended by van Driest to

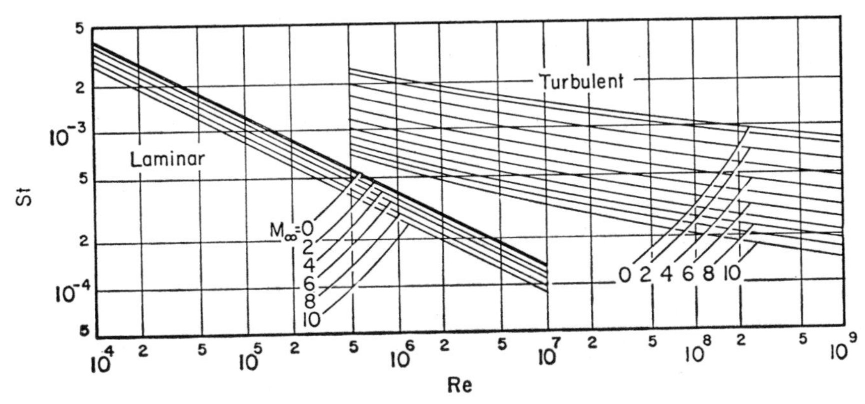

Fig. 11.6. Theoretical Stanton number for air as function of local Reynolds number and Mach number. It is based on the local effective coefficient h_e defined by Eq. (11.21) [van Driest (4)].

include the effect of variable Prandtl number. Solutions for the turbulent, compressible boundary layer on a flat plate have been obtained by many investigators using the integral method (4, 11), Fig. 11.6.

Variation of heat transfer coefficient over a body of revolution has received special attention because of its importance in modern rocketry. Figure 11.7 shows the solution obtained by Stine and Wanlass (5a) for laminar flow of air over an isothermal hemispherical nose at $M_\infty = 1.97$. It is seen that the maximum heat transfer occurs at the stagnation point (5b), and decreases over the hemisphere approaching the flat plate solution, Eq. (11.19).

The theoretical results indicate the trend of a solution, and in some instances agree reasonably well with existing experimental data. However, it is not yet possible to establish with certainty the validity of many of the underlying assumptions such as the property variation at high Mach numbers, since reliable heat transfer data in supersonic flows are scarce.

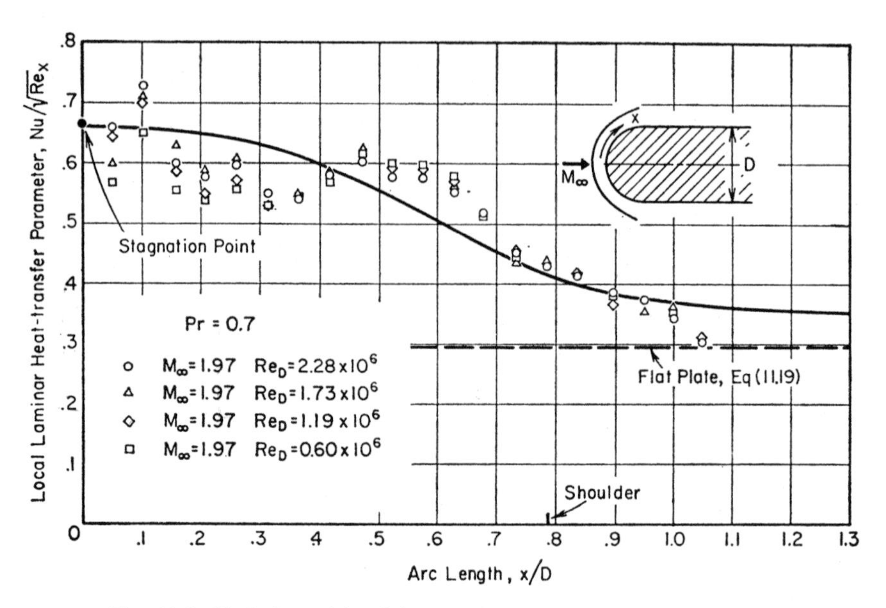

Fig. 11.7. Variation of local heat transfer coefficient along an isothermal hemispherical nose at $M_\infty = 1.97$

$$\mathrm{Nu} = h_e x/k_1 \qquad \mathrm{Re}_x = \rho_1 V_1 x/\mu_1 \qquad \mathrm{Re}_D = \rho_\infty V_\infty D/\mu_\infty$$

Subscript ∞ refers to the state upstream of shock, and subscript 1 refers to the state just outside the boundary layer downstream of shock. h_e is defined by Eq. (11.21) [Stine and Wanlass (5a)].

Therefore, for practical calculations, a simpler semi-empirical procedure is suggested.

(*b*) *Semi-empirical Results*

Figure 11.8 shows qualitatively the temperature distribution within the boundary layer at a solid surface during high velocity flow for the three possible cases—zero heat transfer at the wall (profile *a*), heat transfer from wall to fluid (profile *b*), and heat transfer from fluid to wall (profile *c*). In all cases, the wall is assumed to be at a temperature greater than the free-stream temperature. Note that heat may be transferred to the surface even though it is at a higher temperature than the free stream. Clearly, the heat transfer characteristics at the wall do not depend upon the difference between T_0 and T_∞, and the coefficient of heat transfer defined by Eq. (5.1) loses its meaning at high velocity.

The dilemma is solved by defining an alternate quantity, called the *effective heat transfer coefficient*, in which the free-stream temperature is replaced by the corresponding adiabatic wall temperature at the same free-

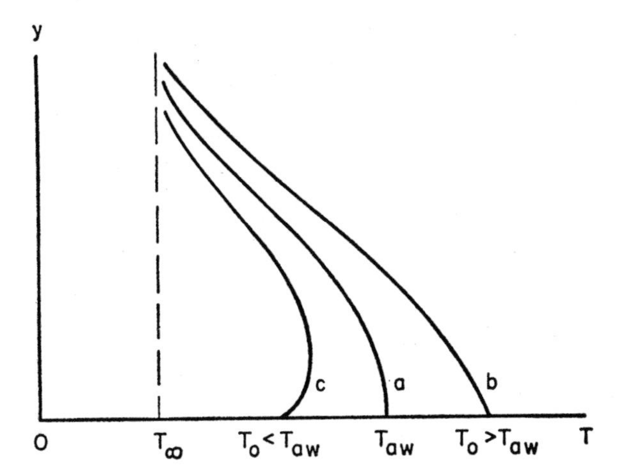

Fig. 11.8. Sketch of temperature profiles in boundary layer during high-velocity flow: Profile a—zero heat transfer at wall, Profile b—heat transfer from wall to fluid, Profile c—heat transfer from fluid to wall.

stream conditions:

$$\frac{q}{A} = h_e(T_0 - T_{aw}) \tag{11.21}$$

Before accepting the quantity h_e, we must test it for its usefulness. From Fig. 11.8 it is apparent that when $T_0 = T_{aw}$, the heat flux is zero as predicted by Eq. (11.21), and when $T_0 \neq T_{aw}$, the rate of heat transfer may be expressed in terms of a positive h_e regardless of the direction of transfer.

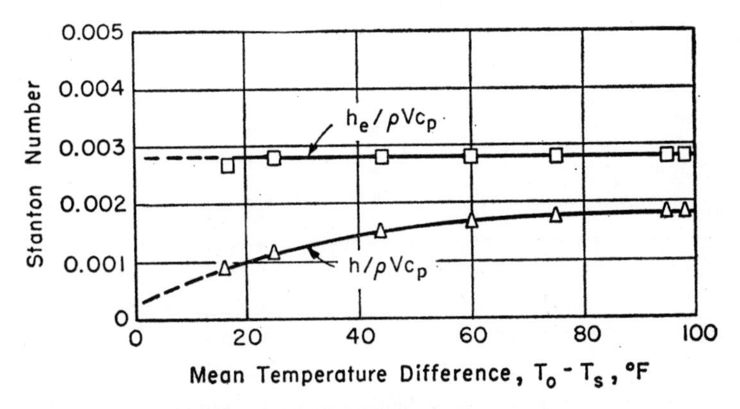

Fig. 11.9. Comparison of Stanton moduli defined in terms of different temperature potentials. T_0 is the temperature of the heated wall, and T_s is the stagnation temperature [McAdams, Nicolai, and Keenan (6)].

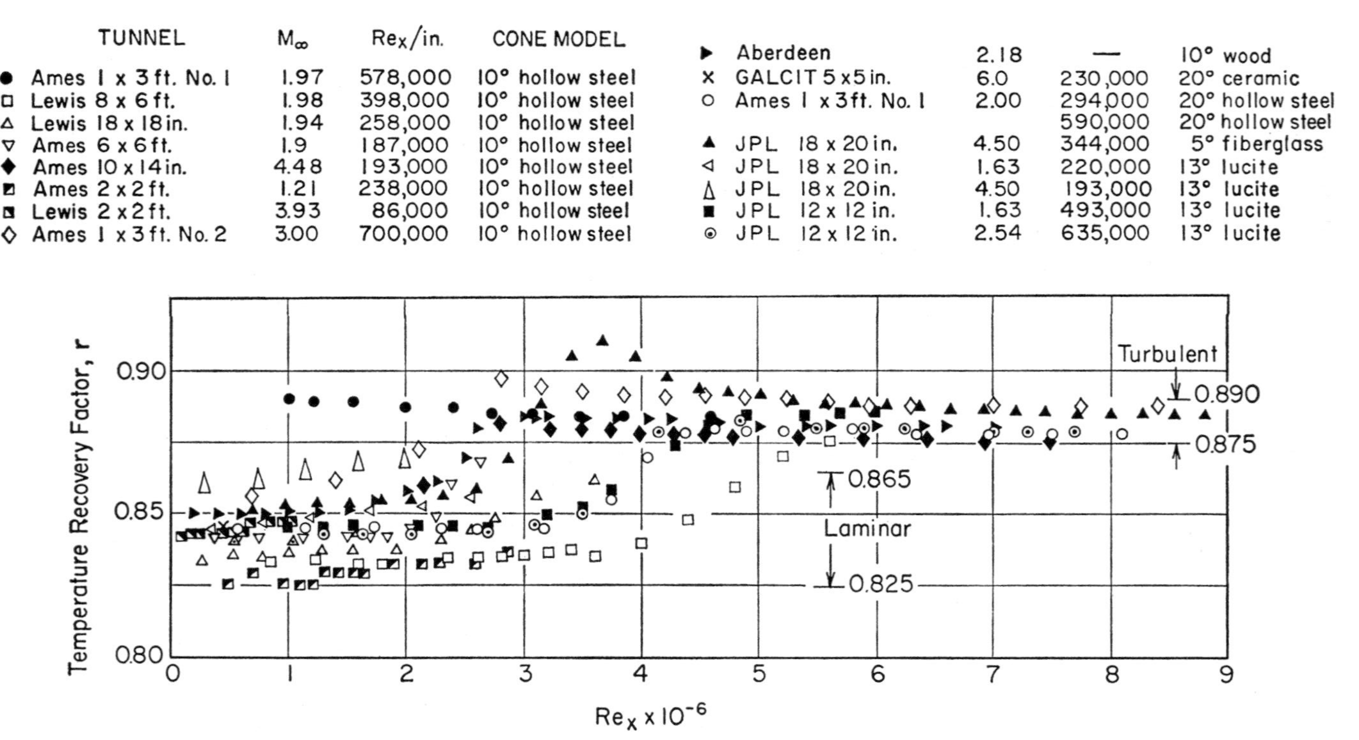

Fig. 11.10. Experimental recovery factors on cones. Re_x and r are based on the state outside the boundary layer downstream of shock, and M_∞ is based on the state upstream of shock [Mack (9)].

More significantly, McAdams, Nicolai, and Keenan (6) compared Stanton moduli defined in terms of different temperature potentials, using results from a number of high subsonic heat transfer experiments for flow in a tube, Fig. 11.9. They observed that the coefficient h_e was independent of the temperature difference, whereas the conventional coefficient h defined by Eq. (5.1) was not. Their experiments also showed that h_e was numerically identical with that for low velocities evaluated at the same Re and Pr. This latter conclusion has been verified theoretically for flat plates (7) and experimentally for other simple geometries such as cones.

With the usefulness of Eq. (11.21) thus established, we can state an important practical rule for calculating heat transfer rates in high velocity flows, namely that *the rate of heat transfer in a high velocity boundary layer may be calculated by the same relations as for constant-property, low velocity boundary layers except for the replacement of T_∞ by T_{aw} in the temperature potential determining the flux.* This rule holds for laminar as well as turbulent flows. A major source of error in its application is that T_{aw}, or equivalently the recovery factor, r, is not definitely known in the higher ranges of supersonic flows ($M > 4$). Also it is not clear what temperature should be used to evaluate the fluid properties in the determination of h_e.

The values of r given by Eq. (11.15) for laminar flow and Eq. (11.20) for turbulent flow over flat plates and for Prandtl number near unity may be applied as a first approximation to other shapes such as cones and cone-cylinders. Figure 11.10, compiled by Mack (9), shows some typical experimental data on the recovery factor of cones for laminar and turbulent boundary layers. The Reynolds number is based upon the distance from the cone tip and external stream conditions. Figure 11.11 shows the experimental values obtained by Stalder, Rubesin, and Tendeland (10a) on a

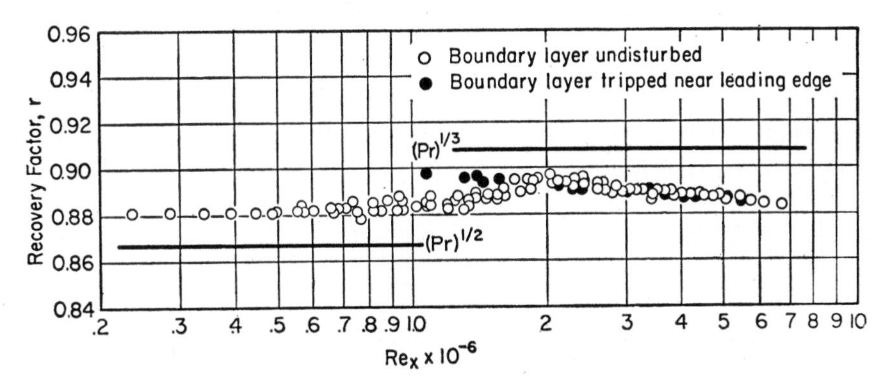

Fig. 11.11. Experimental recovery factors on flat plates at $M_\infty = 2.4$ [Stalder, Rubesin, and Tendeland (10a)].

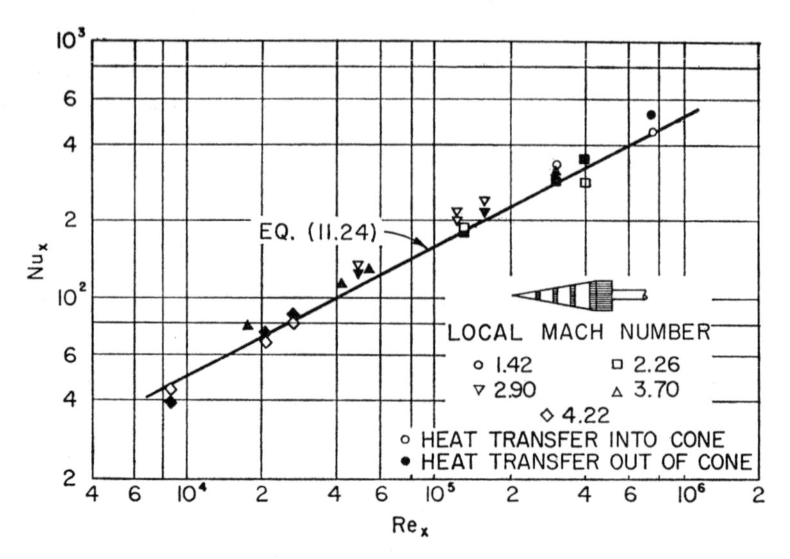

Fig. 11.12. Comparison between calculated and measured heat transfer rates on cones [Eber (12)].

flat plate, at $M = 2.4$. Their results are corroborated by Slack (10b). The approximate theoretical lines of $\sqrt{\text{Pr}}$ in the laminar region and $(\text{Pr})^{1/3}$ in the turbulent region, with air properties evaluated at the free-stream temperature, are also shown. In the transition region, the recovery factor may be expected to vary from the lower laminar value to the higher turbulent value.

The following heat transfer correlations derived for low velocity flow may be applied to high velocity flow, provided the specific heat c_p can be considered constant, and provided all fluid properties are evaluated at a reference temperature T^* determined empirically [Eckert (8a)] to be:

$$T^* = T_\infty + 0.50(T_0 - T_\infty) + 0.22(T_{aw} - T_\infty) \tag{11.22}$$

For laminar boundary layers, the local Nusselt number based upon the effective heat transfer coefficient can be calculated by

Flat plates: $$\text{Nu}_x = 0.332(\text{Re}_x)^{1/2}(\text{Pr})^{1/3} \tag{11.23}$$

Cones: $$\text{Nu}_x = 0.575(\text{Re}_x)^{1/2}(\text{Pr})^{1/3} \tag{11.24}$$

Figure 11.12 shows the data obtained by Eber (12) for cones in a wind tunnel with Mach numbers ranging from 1.42 to 4.22. Data show $(\text{Nu})_x$ is independent of M in this range, and generally corroborate Eq. (11.24) plotted as a solid line in the figure.

For turbulent boundary layers, the following equation for subsonic flows may be used as a first approximation for both flat plates and cones:

$$(\mathrm{Nu})_x = 0.029(\mathrm{Re}_x)^{0.8}(\mathrm{Pr})^{1/3} \tag{11.25}$$

When the temperature variation within the boundary layer is so large that the variation of c_p becomes important, then Eckert recommends basing heat transfer correlations on enthalpies rather than temperatures. This procedure will be illustrated in Sec. 11.4 in connection with a dissociated boundary layer flow.

It is emphasized that the correlations in high velocity flow are not in so satisfactory a state as the correlations in low velocity flow because of the scarcity of reliable data. This is particularly true in the hypersonic-speed ranges where the effects of dissociation, ionization and rarefaction are expected to be significant.

It is interesting to note also the effect of compressibility on the friction factor. Figure 11.13 shows the turbulent friction factor data obtained by a number of investigators for a flat plate plotted against the free-stream Mach number. C_f is the actual value measured, and $C_{f,i}$ is the corresponding value for incompressible flow. Results of analyses performed by van Driest

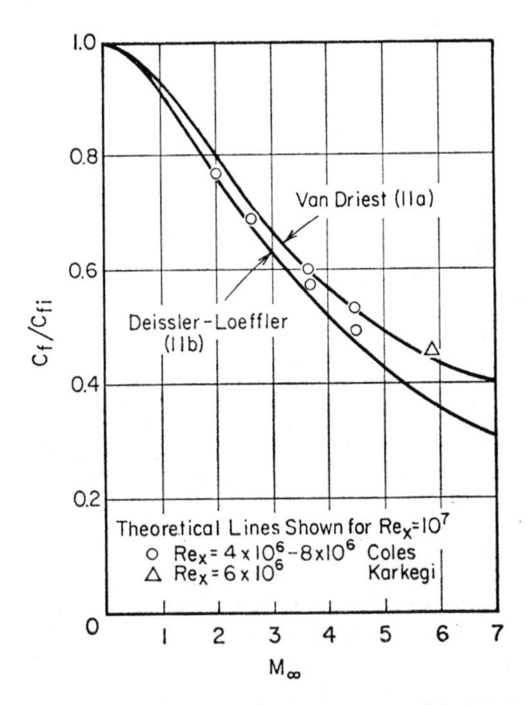

Fig. 11.13. Ratio of compressible to incompressible friction coefficients for turbulent flow over flat plate [van Driest (4)].

(11a) and Deissler and Loeffler (11b) are also shown. Both curves are seen to be excellent representations of the experimental data.

Example 11.2. In example 11.1, if the plate is maintained at 300 F, find the rate of heat transfer to or from the plate.

$$\mathrm{Re}_L = \frac{VL}{\nu} = \frac{(655)(3600)(\frac{1}{12})}{0.85} = 2.31 \times 10^5, \text{ laminar}$$

$$r = \sqrt{0.695} = 0.834$$

$$T_{aw} = 65 + 0.834\frac{(655)^2}{(2)(0.24)(778)(32.2)} = 94 \text{ F}$$

From Eq. (11.22)

$$T^* = 65 + 0.5(300 - 65) + 0.22(94 - 65) = 189 \text{ F}$$

It is seen that the reference temperature T^* is not appreciably different from the mean temperature of 180 F of the system.

Eq. (11.23) gives the local Nu_x. Using instead Eq. (7.41a) for the average Nusselt number,

$$\mathrm{Nu}_{\mathrm{avg}} = 0.664(2.31 \times 10^5)^{\frac{1}{2}}(0.695)^{\frac{1}{3}} = 283$$

$$h_{e,\mathrm{avg}} = \frac{\mathrm{Nu}_{\mathrm{avg}}k}{L} = \frac{(283)(0.018)}{(\frac{1}{12})} = 61 \text{ Btu/hr ft}^2\text{ F}$$

$$\frac{q}{A} = (61)(300 - 94) = 12{,}600 \text{ Btu/hr ft}^2$$

11.4 Dissociation and Ionization

At extremely high velocities, the thermal energy of a gas may be sufficiently great (due either to shock waves or frictional heating) to excite the

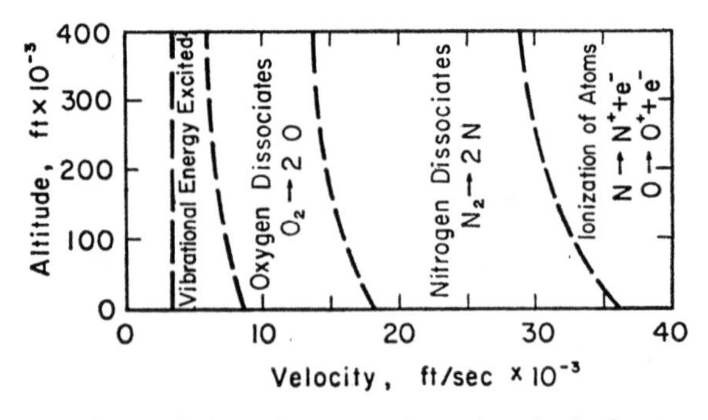

Fig. 11.14. State of air at the stagnation point of a body at various altitudes and velocities [Hansen and Heims (16)].

vibrational energy and subsequently to dissociate partially or even ionize the gas molecules. At 8000°K, more than 50 per cent of the air molecules may be in a dissociated state. Figure 11.14 (16) shows the approximate ranges in which such chemical changes are produced at a stagnation point of a body; for example, during re-entry into the earth's atmosphere. The pressure dependency of these reactions is reflected in the altitude effect.

For simplicity, we shall consider the phenomena of dissociation and ionization separately. In the study of a dissociated boundary layer, we might expect the transfer fluxes at a wall to be influenced by (1) the changes in the thermodynamic and transport properties of the atom-molecule mixture, and (2) the rate of dissociation reactions. Figure 11.15a shows the variation of a thermodynamic property, the specific heat; Fig. 11.15b, c show the variation of transport properties, μ and k, and Fig. 11.15d shows the variation of a property parameter, the Prandtl number, all as functions of absolute temperatures. The values are given for an equilibrium molecule-atom mixture of air.

The peaks in the specific heat curve correspond to the dissociation and ionization reactions. The curve is shown for a pressure of 0.01 atm; as pressure decreases, the peaks become sharper and shift to lower tempera-

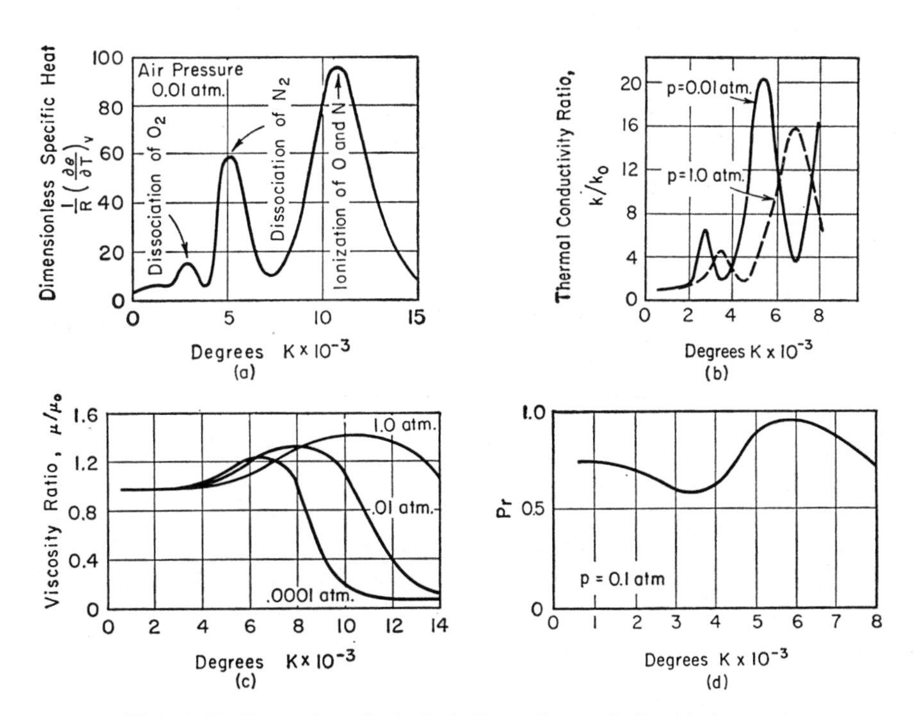

Fig. 11.15. Properties of air including effects of dissociation and ionization (16).

tures. In the graphs, μ_0 and k_0 are given by Sutherland-type formulas (Eq. 20.25):

$$\mu_0 = 1.46 \times 10^{-5} T^{1/2} \left(1 + \frac{112}{T}\right)^{-1} \text{ gm/cm-sec}$$

and

$$k_0 = 4.76 \times 10^{-6} T^{1/2} \left(1 + \frac{112}{T}\right)^{-1} \text{ cal/cm-sec K}$$

The Prandtl number, interestingly enough, does not vary appreciably from its low temperature value of 0.71.

The heat transfer problem of finite dissociation rates is extremely difficult to analyze, particularly since very little is known about these reaction rates. Two limiting cases, however, may be considered:

(a) *The equilibrium boundary layer.* The reaction rates are sufficiently rapid compared with the rates of diffusion of molecules and atoms that the boundary-layer mixture may be assumed to be in a state of thermochemical equilibrium. The composition of the mixture at every point within the boundary layer may be calculated as a function of temperature and pressure at that point. However, unlike an inert mixture of ideal gases, which also behaves like an ideal gas, the equation of state relating p, ρ, and T of a dissociating mixture generally cannot be approximated by the ideal gas equation, even though each component of the mixture may be ideal. Properties such as the specific heat and enthalpy are no longer functions of temperature only.

(b) *The "frozen" boundary layer.* The reaction rates are slow enough so that the transfer processes at a wall are governed mainly by molecular and atomic diffusion. This becomes a problem in mass transfer, in which a diffusion equation must be written for each diffusing species in addition to the appropriate equations of continuity, momentum, and energy. Some simplification is possible in the case of air by treating it as a mixture of "air" molecules and "air" atoms (a reasonable approximation since nitrogen and oxygen have similar atomic weights and transport properties); this means dissociated air can be considered as a binary mixture with molecules and atoms as species, and methods of analysis discussed in Chap. 15 will apply. Another important question in connection with a frozen boundary layer is whether or not appreciable recombination occurs at the wall. If the atoms do not recombine, their concentration will build up at the wall. If they do recombine, either because the wall temperature is below the dissociation temperature or because the wall is made of a material that catalyzes recombination, there is a high rate of chemical reaction at the wall; in essence, there is a heat "source" at the wall.

Even for these limiting cases, solution of the laminar boundary layer equations is not easy. We shall instead consider the Couette flow problem of Fig. 11.1 in conjunction with dissociating air in thermochemical equilibrium. The example serves to show why enthalpy is more convenient than temperature as the driving potential in heat transfer analyses involving dissociating gases.

The solution to the Couette problem in Secs. 7.1 and 11.1 was obtained assuming constant properties. However, in a dissociating gas, the properties can vary greatly, as Fig. 11.15 shows, and Eq. (7.4) derived for constant properties does not apply. Instead of retracing the derivation in Sec. 7.1, it is simple enough in this case to derive the appropriate energy equation directly. If we assume, as in Sec. 7.1 and 11.1, that the pressure gradient and the heat conduction in the axial direction can be neglected and that the lower plate is insulated, an energy balance on an element $\Delta x \Delta y$ and unity in depth, yields

$$\frac{d}{dy}\left(\frac{q}{A}\right)_y + \frac{d}{dy}(v_x \tau) = 0$$

$$-\frac{d}{dy}\left(k\frac{dT}{dy} + \mu v_x \frac{dv_x}{dy}\right) = 0$$

or

$$k\frac{dT}{dy} + \mu v_x \frac{dv_x}{dy} = 0 \tag{11.26a}$$

where the constant of integration is zero since the boundary conditions of the problem are

$$y = 0, \qquad v_x = 0, \qquad \frac{dT}{dy} = 0$$

$$y = l, \qquad v_x = V_1, \qquad T = T_1$$

Further integration of Eq. (11.26a) is not easy since both k and μ are not simple functions of the temperature, Fig. 11.15(b) and (c). We note, however, that although the enthalpy of a dissociating mixture is a function of both pressure and temperature, we can still write $di = c_p\, dT$ since the pressure is constant throughout the system. Substitution of this in Eq. (11.26a) gives

$$\frac{1}{\Pr}\frac{di}{dy} + v_x \frac{dv_x}{dy} = 0 \tag{11.26b}$$

Figure 11.15(d) shows that the Prandtl number of dissociating air may be taken to be approximately constant. Integration of Eq. (11.26b) with the boundary conditions above then yields

$$\frac{i}{\text{Pr}} + \frac{v_x^2}{2} = \text{constant}$$

or

$$i_0 - i_1 = (\text{Pr})\frac{V_1^2}{2} \tag{11.26c}$$

Equation (11.26c) gives the enthalpy difference between the lower and the upper plates and may be compared with Eq. (11.2).

The heat transfer problem near a stagnation point of bodies of revolution in hypersonic flight was considered by Fay and Riddell (14), Lees (17), Eggers *et al.* (18), and others. For an equilibrium boundary layer, Fay and Riddell obtained the following expression:

$$\frac{\text{Nu}}{\sqrt{\text{Re}}} = 0.76(\text{Pr})^{0.4}\left(\frac{\rho_s\mu_s}{\rho_0\mu_0}\right)^{0.4}\left[1 + (\text{Le}^{0.52} - 1)\frac{i_D}{i_s}\right] \tag{11.27}$$

in which the local Nusselt and Reynolds numbers are defined by

$$\text{Nu} \equiv \frac{q}{A}\frac{c_{p_0}x}{k_0(i_s - i_0)}$$

$$\text{Re} \equiv \frac{V_e x}{\nu_0}$$

x is the distance along a meridian profile, c_p is the constant pressure specific heat of dissociated air, Pr and Le are the Prandtl and Lewis numbers of dissociated air, i_D is the dissociation enthalpy; and among the subscripts,

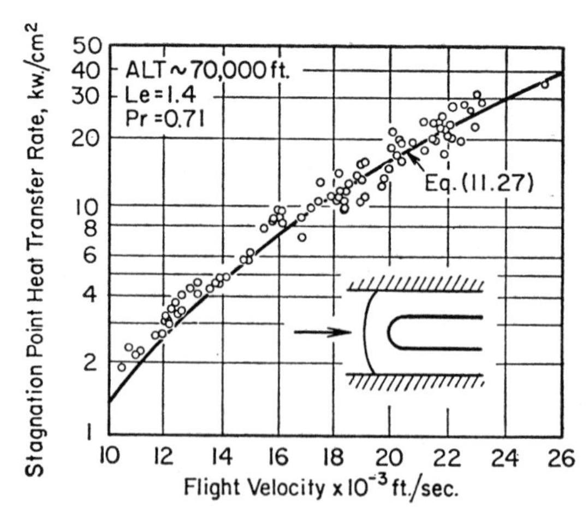

Fig. 11.16. Heat transfer at the stagnation point with dissociated air [Rose and Stark (13)].

s refers to the stagnation state in external flow, 0 refers to the wall, e refers to the external flow.

Note that the definition of the local Nusselt number differs from the usual definition in that the "driving force" for heat transfer is the enthalpy difference of the fluid at the wall and in the external flow, not the temperature difference. Equation (11.27) shows the heat transfer parameter $\text{Nu}/\sqrt{\text{Re}}$ depends only on the total variation in $\rho\mu$ across the boundary layer for a given Pr and for Le = 1. The numerical solution for Pr = 0.71 and Le = 1.4 is compared in Fig. 11.16 with the stagnation-point heat transfer data measured in a shock tube by Rose and Stark (13). The close agreement is quite remarkable when we consider the difficulties of measurement in a shock tube and the uncertainties in the property values used to calculate the theoretical curve.

Eckert (8a) suggests a very simple empirical procedure for calculating the heat transfer rates when the specific heat of the gas in the boundary layer varies significantly, as in the case of a dissociating gas. Paralleling the treatment of Sec. 11.3(b), an effective heat-transfer coefficient and the corresponding Nusselt and Stanton numbers may be defined in terms of the enthalpies of the gas as follows:

$$\frac{q}{A} = h_{ei}(i_0 - i_{aw})$$

$$(\text{Nu}_i)_x = \frac{c_p h_{ei} x}{k} \tag{11.28a}$$

$$\text{St}_i = \frac{h_{ei}}{\rho V_\infty}$$

where i_{aw} is the adiabatic wall enthalpy identical in concept with the adiabatic wall temperature and the subscript ∞ refers to the state in the external flow at the outer edge of the boundary layer. (Note that the state at the outer edge of the boundary layer is not the free-stream state when a shock wave is present; see for example, Fig. 11.16.)

An enthalpy recovery factor may also be defined as follows

$$r_i \equiv \frac{i_{aw} - i_\infty}{i_s - i_\infty} = \frac{i_{aw} - i_\infty}{V_\infty^2/2} \tag{11.28b}$$

For the Couette flow, Eq. (11.26c) gives r_i = Pr.

Based on the foregoing definitions, the heat-transfer conditions derived for the low-velocity laminar and turbulent flows may be used to evaluate the parameters Nu_i and St_i, provided the properties in these correlations are introduced at a reference enthalpy given by

$$i^* = i_\infty + 0.5(i_0 - i_\infty) + 0.22(i_{aw} - i_\infty) \tag{11.28c}$$

For example, for laminar flow over a flat plate of a dissociating gas in local equilibrium, the following correlation may be used:

$$(\text{Nu}_i)_x = 0.332\ \text{Re}_x^{1/2}\text{Pr}^{1/3}$$

This is indeed a remarkable conclusion, since we might expect the dissociation in the boundary layer to influence strongly the heat transfer rates. This conclusion is confirmed by theoretical calculations performed by Freedman (8b) on laminar flow of a dissociating iodine gas. Eckert further suggests that the method may be extended to non-equilibrium states of the boundary layer provided the wall is catalytic.

When gas in the boundary layer becomes electrically conducting because of either extreme temperatures or ablation from a solid metallic surface, it should be possible to influence the flow field by an electrical or a magnetic field since electrical or magnetic body force would affect fluid motion the same way as the more familiar gravitational or centrifugal body forces.

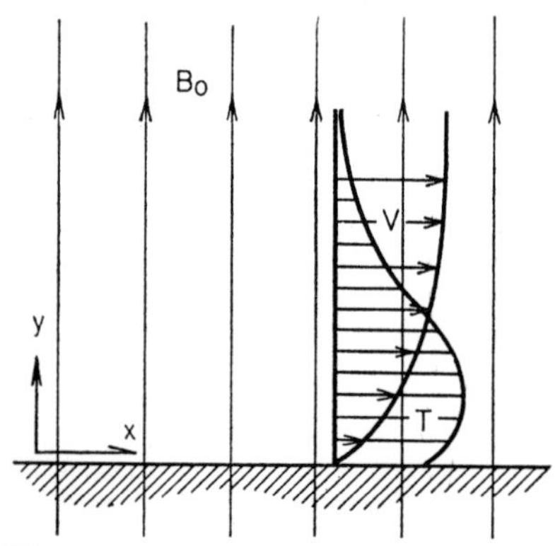

Fig. 11.17. Sketch of ionized boundary layer over a flat plate with a transverse magnetic field.

Consider a flat plate with a magnetic field applied transverse to the flow field as shown in Fig. 11.17. Consider a control volume $\Delta x\ \Delta y$ within the boundary layer which is assumed to be ionized. We shall assume that (1) the flow is laminar, steady, and incompressible, (2) there is no excess charge (approximately equal number of positively and negatively charged particles in the volume) and no imposed electric field, and (3) the magnetic induction resulting from current flow in the fluid is negligibly small compared with the applied magnetic field. The usual momentum equation of the boundary layer will then contain an additional electromagnetic body-force term given in vector notation by

$$\mathbf{F} = \mathbf{j} \times \mathbf{B}_0 \tag{11.29a}$$

where $\mathbf{j}$ is the current density vector, amp/unit area, and $\mathbf{B}_0$ is the applied magnetic lines of force/unit area. The usual energy equation of the boundary layer will contain an additional Joulean heating term given by

$$Q_{\text{elect}} = j^2/\sigma \tag{11.29b}$$

where σ is the electrical conductivity of the fluid, mho/unit length. From the electromagnetic theory, when a conductor moves at velocity $\mathbf{V}$ through magnetic lines of force $\mathbf{B}_0$, the current in the conductor is given by (19)

$$\mathbf{j} = \sigma \mathbf{V} \times \mathbf{B}_0$$

neglecting the induced magnetic field caused by current flow. Substitution of $\mathbf{j}$ into Eq. (11.29a) gives

$$\mathbf{F} = \sigma(\mathbf{V} \times \mathbf{B}_0) \times \mathbf{B}_0 = \sigma[\mathbf{B}_0(\mathbf{B}_0 \cdot \mathbf{V}) - \mathbf{V}(\mathbf{B}_0 \cdot \mathbf{B}_0)]$$

In Fig. 11.17, $\mathbf{B}_0$ is applied perpendicular to the free-stream velocity $\mathbf{V}_\infty$. However, since the direction of the local velocity $\mathbf{V}$ in the boundary layer differs negligibly from the free-stream direction, we may assume $\mathbf{B}_0$ is perpendicular to $\mathbf{V}$. Then, $(\mathbf{B}_0 \cdot \mathbf{V}) = 0$ and the electromagnetic body-force is given by

$$\mathbf{F} = -\sigma B_0^2 \mathbf{V} \tag{11.30a}$$

The negative sign indicates that $\mathbf{F}$ acts in an opposite direction to that of motion $\mathbf{V}$.

Similarly, the Joulean heating term may be rewritten by substituting the expression above for $\mathbf{j}$ into Eq. (11.29b) and noting that $\mathbf{B}_0$ is normal to $\mathbf{V}$:

$$Q_{\text{elect}} = \sigma(\mathbf{V} \times \mathbf{B}_0)^2 = \sigma V^2 B_0^2 \tag{11.30b}$$

The addition of the force term, Eq. (11.30a), and the energy term, Eq. (11.30b), to the respective incompressible boundary-layer equations of momentum (x-direction) and energy yields

$$v_x \frac{\partial v_x}{\partial x} + v_y \frac{\partial v_x}{\partial y} = \nu \frac{\partial^2 v_x}{\partial y^2} - \frac{\sigma B_0^2 v_x}{\rho} \tag{11.31}$$

$$\rho c_p \left(v_x \frac{\partial T}{\partial x} + v_y \frac{\partial T}{\partial y} \right) = k \frac{\partial^2 T}{\partial y^2} + \mu \left(\frac{\partial v_x}{\partial y} \right)^2 + \sigma B_0^2 v_x^2 \tag{11.32}$$

Equations (11.31) and (11.32) were solved by Rossow (15a) by the same general procedure as in the case of the incompressible nonconducting boundary layer. He used the transformation $\eta = y\sqrt{V_\infty/\nu x}$ and assumed the following expressions for the stream function ψ and the temperature parameter θ:

$$\psi = \sqrt{V_\infty \nu x}\,(f_0(\eta) + \sqrt{mx}\, f_1(\eta) + mx\, f_2(\eta) + \cdots)$$

$$\theta = \theta_0(\eta) + \sqrt{mx}\,\theta_1(\eta) + mx\,\theta_2(\eta) + \cdots$$

where $f_0(\eta)$ and $\theta_0(\eta)$ are the Blasius and the Pohlhausen solutions for the nonconducting boundary layers, the additional solutions $f_1(\eta)$, $f_2(\eta) \cdots$ and $\theta_1(\eta), \theta_2(\eta) \cdots$ are assumed to account for the electromagnetic effects, and $m = \sigma B_0^2/\rho V_\infty$ is a magnetic parameter. The problems of ionized boundary layer flows under re-entry conditions are not easy to interpret in terms of a simple analytical model. These are discussed by Hess (15b). Integral solutions for uniform velocity (not pressure) over a flat plate at uniform temperature with both laminar and turbulent boundary layers were performed by Moffatt (15c).

11.5 Rarefied Gases

In the usual macroscopic analysis of transfer processes, fluid media are treated as continua, and macroscopic properties such as density, velocity, and temperature are assumed to vary continuously in time and space. When rarefaction effects are present, a critical reappraisal of concepts regarding fluid systems becomes necessary. Our purpose in studying rarefied systems is (1) to extend analysis of transfer processes to high-speed, high-altitude flight, and (2) to broaden the conceptual basis of transfer phenomena and delineate clearly the limitations of continuum theories.

Let us consider a system comprised of air flowing over a heated flat plate (Fig. 11.18). If the mean free path λ of air were small compared with any characteristic dimension L of the system (for air at sea-level conditions $\lambda \approx 6 \times 10^{-6}$ cm), continuum theory postulates existence of hydrodynamic and thermal boundary layers at the plate surface, and rates of transfer of momentum and energy within the boundary layers are governed by a series of random molecular collisions, or diffusional mechanism. Velocity and temperature profiles within the boundary layers are given by curves labeled *a* in Fig. 11.18. Phenomenologically, the rate equations of Chap. 1 relate the transfer rates to the appropriate potential gradients at the wall. When air is rarefied so that λ is not negligibly small compared with L, and in fact may be of the same order of magnitude as L, intermolecular collisions becomes less frequent and molecules arriving at the solid surface are unable to come into equilibrium with the surface. As a result, significant velocity and temperature discontinuities may develop at the gas-solid

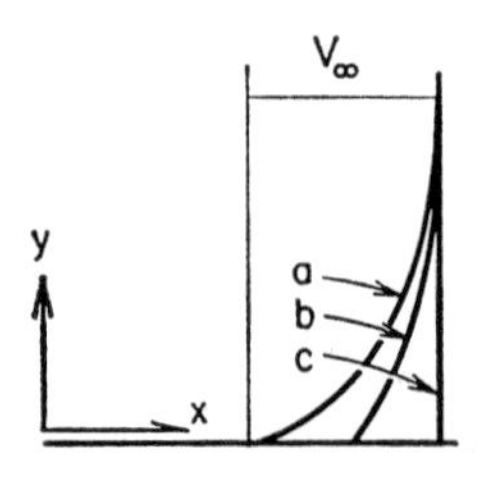

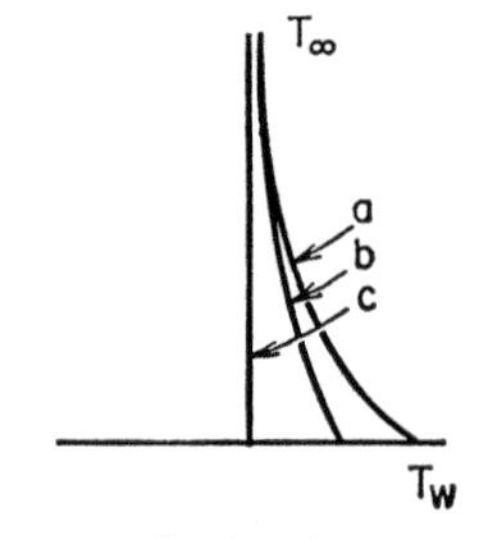

Fig. 11.18. Sketch of velocity and temperature distributions over a heated plate: a—continuum profiles, b—slip profiles, c—free-molecule profiles.

boundary shown by curves labeled *b*. Rates of transfer are no longer governed solely by intermolecular collisions within the boundary layers, but depend also upon effectiveness of property exchange between the gas molecules and the wall. The mechanism of interaction between molecules and wall is extremely complicated, and empirical parameters called *reflection coefficient* (Eq. 11.37) in the case of momentum transfer and *accommodation coefficient* (Eq. 11.40) in the case of energy transfer may be used to account for their effect. If air is highly rarefied so that $\lambda \gg L$, frequency of intermolecular collisions may be totally negligible. At the solid surface, the impinging and the re-emitted streams of molecules do not interact to any significant degree; the boundary layers to all practical purposes disappear as shown by curves *c*, molecules adjacent to the surface maintaining their free-stream identity. The physical model becomes simple enough to permit calculation of transfer rates by the kinetic theory. It is important to note that the mean free path is not the distance between molecules so that even in a highly rarefied gas, there are still a sufficient number of molecules in a unit element of the gas to give meaning to gas properties such as density, velocity, and temperature. The variation of mean free path with altitude is shown in Fig. 11.19.

The dimensionless parameter formed by the ratio of λ to a characteristic system dimension L is known as the *Knudsen number*:

$$\text{Kn} \equiv \frac{\lambda}{L}$$

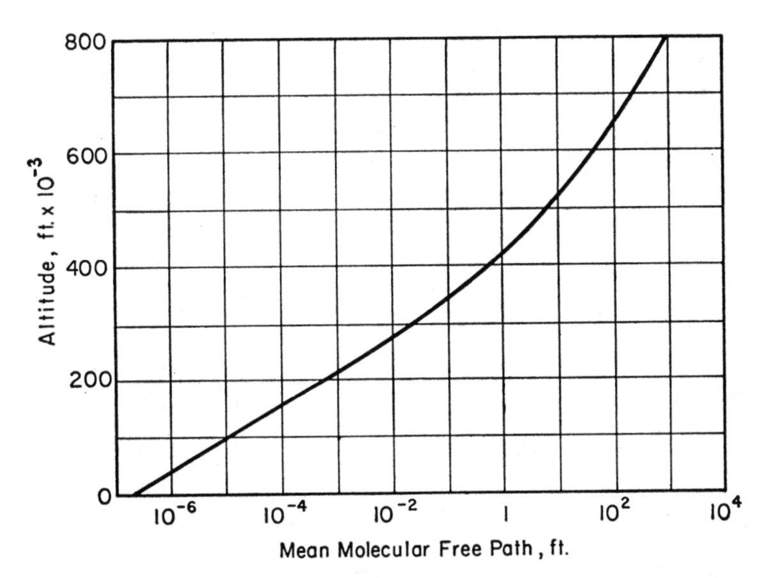

Fig. 11.19. Altitude variation of mean molecular free path (25).

As in the case of the Reynolds number, the concept of the Knudsen number must be applied in context, in connection with a particular scale (L) and particular phenomena of interest. In aerodynamic applications, it has been found useful to express the Knudsen number in terms of the Mach and the Reynolds numbers as follows.

The elementary kinetic theory gives the viscosity of a gas as

$$\mu = \tfrac{1}{3}\rho\bar{V}\lambda \tag{20.12}$$

where $\bar{V}$ is the mean molecular velocity and may be expressed in terms of the velocity of sound V_s and the ratio of the specific heats γ by

$$\bar{V} = \sqrt{\frac{8}{\pi\gamma}}\,V_s$$

More rigorous assumptions regarding the behavior of the elastic, spherical molecules, in particular taking into account the persistence of velocities after collision (an average tendency of molecules to continue moving in their original direction after a collision), lead to modification of the numerical factor in Eq. (20.12) from $\frac{1}{3}$ to 0.499, or for all practical purposes $\frac{1}{2}$. This value of the viscosity combined with the equation for $\bar{V}$ gives the following expression for the Knudsen number:

$$\text{Kn} = \frac{\lambda}{L} = \sqrt{\frac{\pi\gamma}{2}}\left(\frac{\mu}{\rho V_s L}\right) \tag{11.33}$$

The quantity within brackets, multiplied and divided by a free-stream velocity V_∞, is readily seen to be the ratio of a free-stream Mach number to a Reynolds number referred to a characteristic dimension L of a system. Equation (11.33) indicates that the continuum concept is likely to fail either at very high Mach numbers or at very low Reynolds numbers.

For flows at very low Reynolds numbers, the characteristic system dimension is usually a body dimension l, and from Eq. (11.33) we can write

$$\text{Kn}_l = \frac{\lambda}{l} \sim \frac{M}{\text{Re}_l}, \quad \text{for low Re}_l \tag{11.34}$$

For flows at large Reynolds numbers, the significant characteristic dimension is of the order of the boundary-layer thickness δ. Kn for such systems may therefore be defined in terms of δ, but since a Reynolds number defined in terms of δ is not in common usage, we need additional information in order to relate Kn to Re_l. If flow is laminar, from Eq. (3.29),

$$\frac{\delta}{l} \sim \frac{1}{\sqrt{\text{Re}_l}}$$

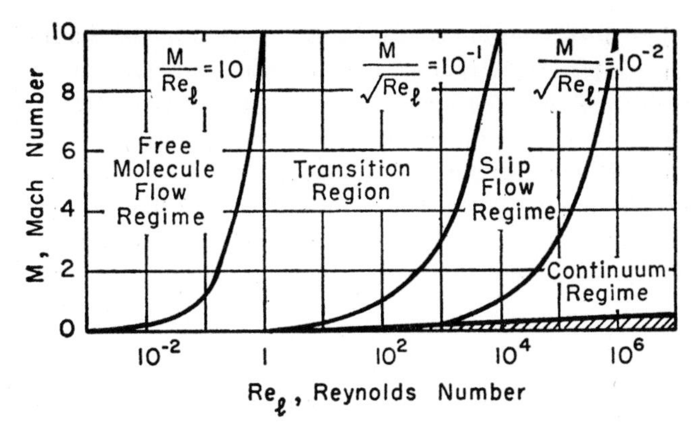

Fig. 11.20. Flow regimes in gas dynamics [Schaaf and Chambré (25)].

Therefore,

$$\text{Kn}_\delta = \frac{\lambda}{\delta} = \frac{\lambda l}{l\delta} \sim \frac{M}{\sqrt{\text{Re}_l}}, \quad \text{for high Re}_l \tag{11.35}$$

Tsien (29) first proposed the classification of flow regimes in terms of the characteristic ranges of M and Re_l, as shown in Fig. 11.20.

The lines of demarcation between the various flow regimes are not known with precision for all Reynolds numbers. The following criteria are suggested (25):

Regime boundary between:

Free-molecule and Transition $\qquad \dfrac{M}{\text{Re}_l} = 3 \text{ to } 10$

Transition and Slip
$$\begin{cases} \dfrac{M}{\text{Re}_l} = 0.1 & \text{Re}_l < 1 \\[2ex] \dfrac{M}{\sqrt{\text{Re}_l}} = 0.1 & \text{Re}_l > 1 \end{cases}$$

Slip and Continuum
$$\begin{cases} \dfrac{M}{\text{Re}_l} = 0.01 & Re_l < 1 \\[2ex] \dfrac{M}{\sqrt{\text{Re}_l}} = 0.01 & \text{Re}_l > 1 \end{cases}$$

The lack of precision in the definition of these boundaries is apparent, since the same order of magnitude is used for the regime boundaries defined by both M/Re_l and $M/\sqrt{\text{Re}_l}$.

Note that M/Re_l is the sole criterion for the edge of the free-molecule regime since high rarefaction is always associated with low Re flow. A system in this region no longer possesses the usual attributes of a fluid and analysis of transfer phenomena must be based on molecular theories.

In the slip flow regime, M will usually be large and/or Re_l small, indicating the presence of compressibility or viscosity effects with rarefaction. The transfer phenomena in the slip and transition regimes are extremely complex, and analysis is possible, at best, by semi-empirical procedures.

11.6 Momentum and Energy Transfer in the Slip Regime

The basic conservation equations of momentum and energy as well as the continuity equation might be expected to hold for rarefied systems. However, allowance must be made for the following facts. In the range usually designated as the slip flow regime, either the Mach number must be large (compressibility effect) or the Reynolds number small (viscosity effect), or both. Re may be too low for the boundary layer theory to be completely valid and not low enough for the inertia terms to be neglected. In addition, the rate equations of transfer must be able to account for the non-diffusional aspects of the transfer mechanism at the bounding surface. It is seen that the analysis of transfer processes involving slip flow can be quite complex and that no solution as yet exists that satisfactorily accounts for the rarefaction, compressibility, and viscosity effects by general formulation of the differential equations and boundary conditions. It has been argued that some modified form of the Navier-Stokes equations, such as the Burnett equations (20a) or the Thirteen Moment equations (20b) could account for some of the experimentally observed phenomena associated with slip flow. However, the validity of these equations has yet to be rigorously established on both the theoretical and the experimental grounds (25). In view of the complexity and the uncertain validity of these equations, we shall not discuss them here, but instead present an approximate analysis that has yielded results fairly well corroborated by experiment in the slip regime. In essence, in the approximate analysis, the simple rate equations of Chapter 1 are assumed valid in slip flow, and solutions are obtained as in the conventional continuum analysis, using the Navier-Stokes equations, with the stipulation that at the flow boundary they match the rarefaction velocity and temperature discontinuities.

Slip boundary conditions. An expression for the velocity discontinuity at a wall in slip flow, Fig. 11.18, may be obtained as follows. An infinitesimal gas layer adjacent to the wall is acted upon on the gas side by a shear force proportional to the velocity gradient in the gas and on the solid side by a shear force which may be assumed to be proportional to the velocity difference between the gas and the wall. Equating these two forces, we can write

$$\mu \left(\frac{\partial v_x}{\partial y}\right) = \beta(V_0 - V_w) \tag{11.36}$$

where μ is the coefficient of viscosity of the gas, β is a proportionality constant, V_0 is the gas velocity at the wall, and V_w is the wall velocity. The ratio of μ/β measures the effectiveness of tangential momentum exchange at the wall.

We may interpret this slip effect in terms of a simple kinetic theory picture. In Eq. (20.9), it is shown that molecules near the wall have their last collision, before striking the wall, at an average distance of $2\lambda/3$. From Eqs. (20.8) and (20.2), the number of molecules striking a surface per unit area and time is $\bar{V}n/4$ where n is the number of molecules per unit volume and $\bar{V}$ is the average molecular velocity. Then for the velocity distribution shown in curve (b) of Fig. 11.18, the tangential momentum of the molecules that strike the surface is

$$m(V_0 + \tfrac{2}{3}\lambda\, dv_x/dy)$$

If these molecules in striking the surface are assumed to transfer the fraction F of their tangential momentum, then

$$F\frac{m\bar{V}n}{4}\left(V_0 + \tfrac{2}{3}\lambda \frac{dv_x}{dy} - V_w\right) = \mu \frac{dv_x}{dy}$$

Substituting for μ its equivalent from Eq. (20.12), $nm\bar{V}\lambda/3$, this reduces to

$$V_0 - V_w = \frac{2}{3}\frac{2 - F}{F}\lambda \frac{dv_x}{dy}$$

or from Eq. (11.36)

$$\frac{\mu}{\beta} = \frac{2}{3}\frac{2 - F}{F}\lambda$$

These are the results of the elementary kinetic theory picture. A more refined analysis (21) originally made by Maxwell, in 1879, leads to similar results with the $\frac{2}{3}$ replaced by 1.0 in the last two equations:

$$\frac{\mu}{\beta} = \frac{2 - F}{F}\lambda \tag{11.37}$$

$$V_0 - V_w = \frac{2 - F}{F}\lambda \left(\frac{dv_x}{dy}\right)_{y=0} \tag{11.38}$$

TABLE 11.1 SPECULAR REFLECTION COEFFICIENT, F

Gas	Surface	F	Reference
Air	Oil	0.90	(21)
He	Oil	0.87	(21)
H_2	Oil	0.93	(21)
CO_2	Oil	0.92	(21)
Air	Hg	1.00	(21)
Air	Machined brass	1.00	(21)
He	Polished Ag_2O	1.00	(22)
H_2	Polished Ag_2O	1.00	(22)
O_2	Polished Ag_2O	0.99	(22)
Air	Polished Ag_2O	0.98	(22)

F may be interpreted as the fraction of the incident molecules that are reflected diffusely from the surface, the rest being reflected specularly,* and is usually determined empirically. Typical values are listed in Table 11.1. The fact that its value approximates unity in all cases is not surprising, since even an apparently perfectly smooth surface must appear extremely jagged on a molecular scale, and therefore the dominance of diffuse reflection is perhaps to be expected. Note that when λ is negligibly small compared with the scale of the system under study, Eq. (11.38) reduces to the usual boundary condition of a continuum system, v_x at the wall (V_0) equals V_w.

By a similar procedure, the temperature discontinuity at the wall may be expressed as

$$k\left(\frac{\partial T}{\partial y}\right) = K(T_0 - T_w) \tag{11.39}$$

where k is the thermal conductivity of the gas and K is a proportionality constant. The ratio k/K can be expressed by the kinetic theory in terms of a parameter A called the *thermal accommodation coefficient*.

$$\frac{k}{K} = \frac{2}{\mathrm{Pr}}\left(\frac{2 - A}{A}\right)\left(\frac{\gamma}{\gamma + 1}\right)\lambda \tag{11.40}$$

The coefficient A measures the effectiveness of energy exchange at the wall; it is given to a good approximation by

$$A = \frac{T_r - T_i}{T_w - T_i}$$

* *Specular reflection*: Molecules reflect in perfectly elastic fashion as balls bounding from a smooth, hard surface. The tangential velocity component is conserved and the normal velocity component undergoes a reversal in direction but no change in magnitude.

Diffuse reflection: Reflection of molecules is completely random with respect to both magnitude and direction.

TABLE 11.2 THERMAL ACCOMMODATION COEFFICIENT, A

Gas	Surface	A	Reference
H_2	Bright paint	0.32	(23)
H_2	Black paint	0.74	(23)
O_2	Bright paint	0.81	(23)
O_2	Black paint	0.93	(23)
CO_2	Bright paint	0.84	(23)
CO_2	Black paint	0.96	(23)
Air	Flat lacquer on bronze	0.88–0.89	(24)
Air	Polished bronze	0.91–0.94	(24)
Air	Machined bronze	0.89–0.93	(24)
Air	Polished cast iron	0.87–0.93	(24)
Air	Machined cast iron	0.87–0.88	(24)
Air	Polished aluminum	0.87–0.95	(24)
Air	Machined aluminum	0.95–0.97	(24)
Air	Etched aluminum	0.89–0.97	(24)

when T_i is the temperature of the incident molecular stream, T_r is the temperature of the reflected (or re-emitted) stream, and T_w is the temperature of the wall.

Unlike F, A varies over a rather wide range, depending upon the nature of the gas and the condition of the surface. Some typical values of A are listed in Table 11.2. Values of A less than unity indicate molecular interactions at the wall are not adequate for the molecular stream leaving the wall to adjust to, or become "accommodated" to the temperature of the wall.

A striking similarity in concept between A and the absorptivity in radiant energy transfer (Chap. 13) should be noted. By combining Eqs. (11.39) and (11.40), the temperature boundary condition in the slip regime may be written as follows:

$$\text{at } y = 0, \quad T_0 - T_w = \frac{2}{\mathrm{Pr}}\left(\frac{2 - A}{A}\right)\left(\frac{\gamma}{\gamma + 1}\right)\lambda\left(\frac{\partial T}{\partial y}\right) \tag{11.41}$$

There follow two examples illustrating the application of the continuum equations to the slip flow regime.

Slip flow in tube at low Mach numbers. The simplest illustration of the approximate analysis is provided by fully developed laminar slip flow of air in a tube. Assuming the Navier-Stokes equations applicable, we saw in Chap. 3 that they reduce for incompressible fluid (very low M) to

$$\frac{dp}{dx} = \mu\left[\frac{1}{r}\frac{d}{dr}\left(r\frac{dv_x}{dr}\right)\right] \tag{3.12}$$

The rarefaction effect is introduced into the problem as a boundary condition at the wall. The velocity discontinuity is given by Eq. (11.38), which can be simplified by assuming F equals unity.

The boundary conditions are therefore

$$r = 0, \quad \frac{dv_x}{dr} = 0$$

$$r = r_0, \quad v_x = +\lambda \frac{dv_x}{dy} = -\lambda \frac{dv_x}{dr}$$

Integration of Eq. (3.12) gives the following expression for the velocity distribution:

$$v_x = \frac{1}{4\mu} \frac{dp}{dx}(r^2 - r_0^2 - 2\lambda r_0) \tag{11.42}$$

and

$$\frac{v_c}{V} = \frac{1 + 4\ Kn_D}{\frac{1}{2} + 4\ Kn_D} \tag{11.43}$$

where v_c is the centerline or maximum velocity, V is the average velocity and

$$\mathrm{Kn}_D = \frac{\lambda}{D} = \sqrt{\frac{\pi\gamma}{2}} \frac{M}{\mathrm{Re}_D} \tag{11.33}$$

Equation (11.43) shows V approaches v_c as Kn_D increases, or the velocity distribution in the tube becomes more uniform owing to increasing "slip" at the wall. On the other hand, as Kn_D approaches zero, the ratio v_c/V approaches the well-known value of 2 for Hagen-Poiseuille flow of a continuum.

The friction factor defined by Eq. (4.3) is found to be

$$f = \frac{16}{\mathrm{Re}_D}\left(\frac{1}{1 + 8\ \mathrm{Kn}_D}\right) \tag{11.44}$$

Figure (11.21) is a plot of f as a function of M and Re_D, assuming $F = 1$ and $\gamma = 1.4$. Note that here M is not an indicator of high velocity but rather of rarefaction in conjunction with Re_D. For a given M the solution approaches the continuum solution as Re_D is increased.

Flat plate. We shall consider momentum and heat transfer in steady laminar slip flow over a flat plate aligned in the direction of motion. The boundary-layer equation for a flat plate, assuming constant properties, is given by

$$v_x \frac{\partial v_x}{\partial x} + v_y \frac{\partial v_x}{\partial y} = \nu \frac{\partial^2 v_x}{\partial y^2} \tag{3.18}$$

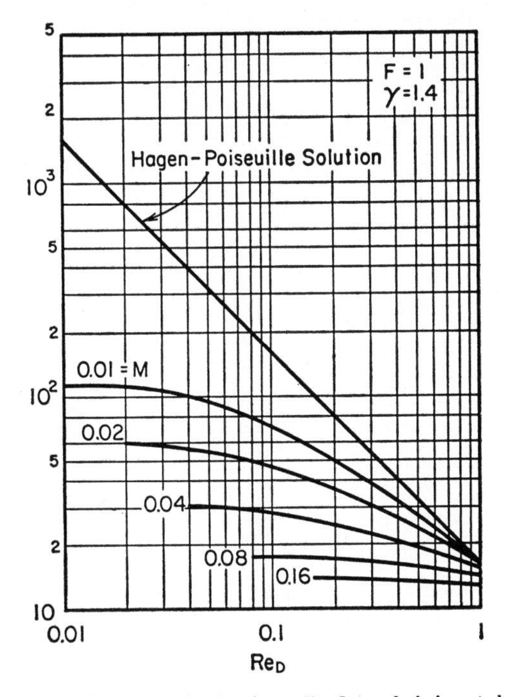

Fig. 11.21. Friction factor for laminar slip flow of air in a tube at very low Mach numbers [Tsien (29)].

For small deviations from the free-stream flow, Oseen (30) suggested that the velocity components can be represented as the sum of a constant and a perturbation term:

$$v_x = V_\infty + v_x'$$

$$v_y = v_y'$$

where v_x' and v_y' are the perturbation terms which are small with respect to V_∞. The boundary-layer equation, Eq. (3.18) reduces to a linear form known as the *Rayleigh equation*:

$$V_\infty \frac{\partial v_x}{\partial x} = \nu \frac{\partial^2 v_x}{\partial y^2} \tag{11.45}$$

Note that this linearized equation is a poor approximation for a continuum boundary layer, but under conditions of increasing slip, the assumption of small deviation from the free-stream flow becomes progressively better. The boundary conditions are (assuming $F = 1$)

$$x = 0, \quad y > 0, \quad v_x = V_\infty$$

$$y = 0, \quad x > 0, \quad v_x = \lambda \frac{\partial v_x}{\partial y}$$

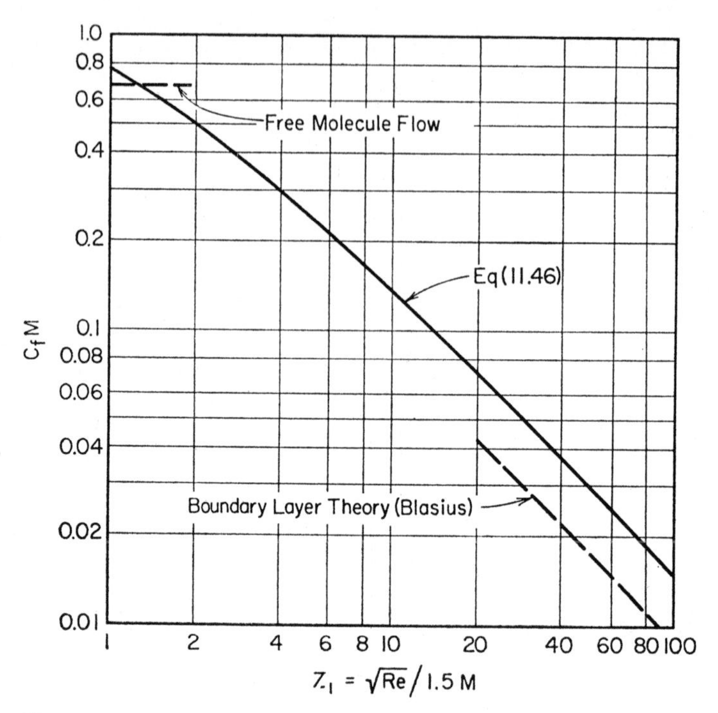

Fig. 11.22. Friction coefficient for slip flow over a flat plate [Schaaf and Sherman (26)].

Equation (11.45) may be solved by the method of Laplace transform. The results are given here for air in terms of the total drag on one side of the plate:

$$C_f = \frac{1.33}{Z_1^2 M}\left[\exp(Z_1^2)\operatorname{erfc} Z_1 - 1 + \frac{2Z_1}{\sqrt{\pi}}\right] \tag{11.46}$$

C_f is the average friction coefficient defined by Eq. (3.34) and

$$Z_1 = \frac{\sqrt{Re}}{1.5M}$$

Figure 11.22 shows a curve representing Eq. (11.46). Agreement of experimental data (26) with this equation is satisfactory in the slip region.

The boundary-layer energy equation, (7.32) may be similarly reduced to the following linear form:

$$V_\infty \frac{\partial T}{\partial x} = \alpha \frac{\partial^2 T}{\partial y^2} \tag{11.47}$$

The boundary conditions are

$$x = 0, \quad y > 0, \qquad T = T_\infty$$

$$y = 0, \quad x > 0, \quad T - T_w = \frac{1.75\lambda}{\mathrm{Pr}} \frac{\partial T}{\partial y}$$

The second boundary condition is written from Eq. (11.41) for a diatomic gas, $\gamma = 1.4$, and for $A = 0.8$.

Equation (11.47) may be solved for the average Stanton number over the plate length:

$$\mathrm{St} = \frac{0.38}{Z_2^2 M}\left[\exp(Z_2^2)\ \mathrm{erfc}\ Z_2 - 1 + \frac{2}{\sqrt{\pi}} Z_2\right] \tag{11.48}$$

where $$Z_2 = \frac{\sqrt{\mathrm{Re\ Pr}}}{2.63M} \quad \text{and} \quad \mathrm{St} = \frac{h}{\rho c_p V_\infty}$$

Heat transfer from spheres and cylinders under slip-flow conditions was analyzed by Drake and Kane (27a). Figure 11.24 compares their result with experimental data.

11.7 Momentum and Energy Transfer in Free-Molecule Regime

In the free-molecule regime, a flow medium loses its identity as a fluid and analysis of transfer processes must be based entirely upon the molecular theories. As a basic postulate, the molecular stream incident on a solid surface is assumed to be effectively undisturbed by the reflected stream;

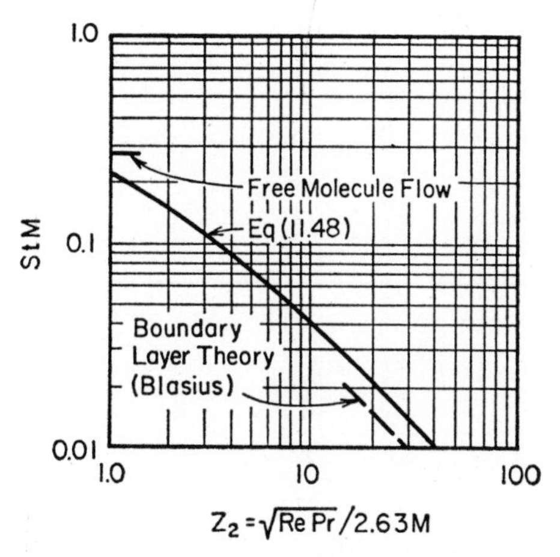

Fig. 11.23. Stanton number for slip flow over a flat plate [Schaaf (34)].

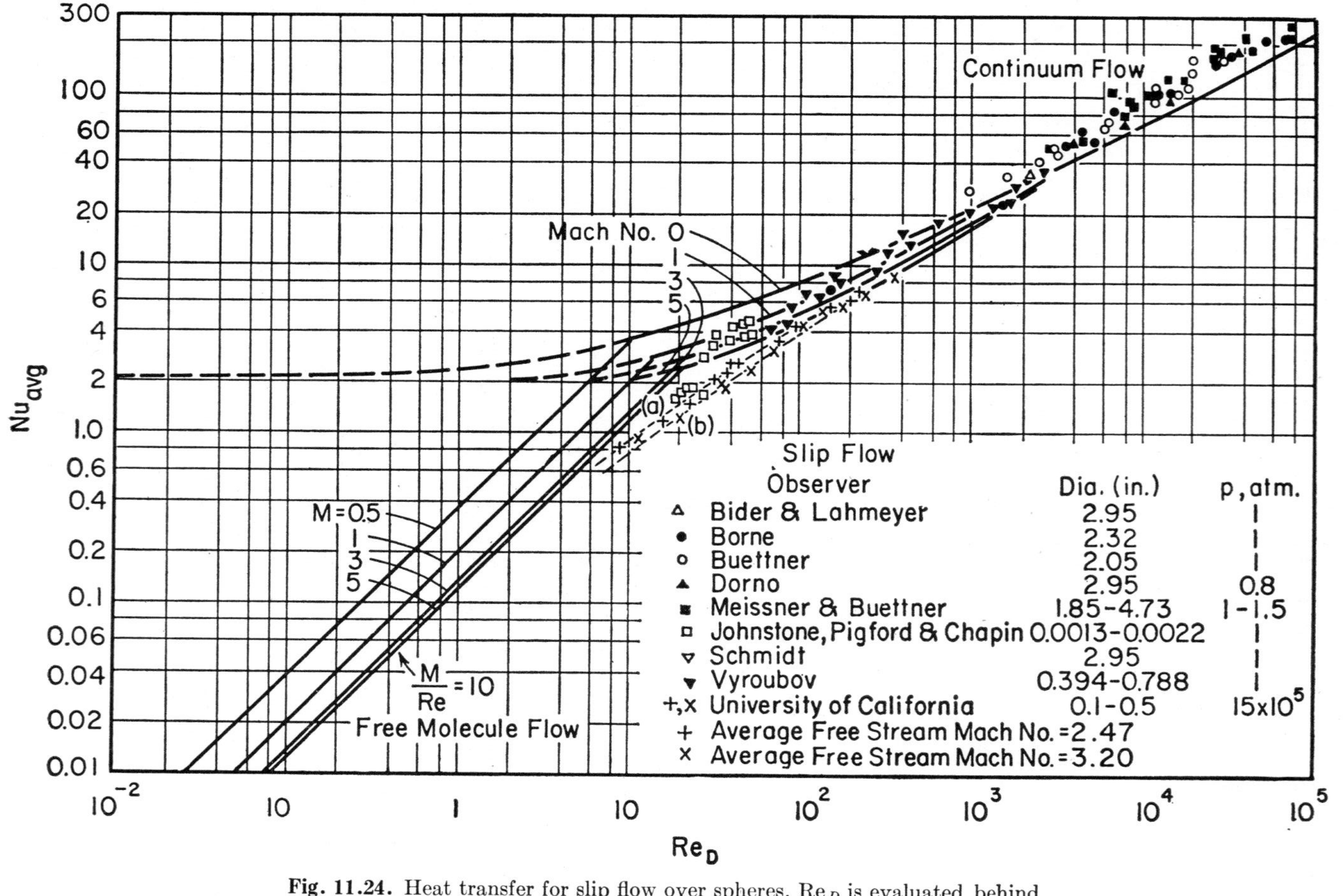

Fig. 11.24. Heat transfer for slip flow over spheres. Re_D is evaluated behind hypothetical normal shock for $M > 1$ [Drake and Backer (27b)].

thus, it may be represented by the classical Maxwellian velocity distribution superimposed on a uniform mass velocity. Drag and heat transfer at the surface are found by considering separately the contributions of the incident and the reflected molecules.

We shall investigate the transfer of momentum and energy by free molecules to and from a surface. For a frame of reference for which the gas is at rest, if v_x', v_y', v_z' are the Cartesian components of velocity of a molecule, the number of incident molecules per unit volume possessing velocity components in the range v_x' to $v_x' + dv_x'$, v_y' to $v_y' + dv_y'$, and v_z' to $v_z' + dv_z'$ is

$$n_i f \, dv_x' \, dv_y' \, dv_z'$$

n_i is the number of molecules per unit volume and f is the Maxwellian velocity distribution function which, for an ideal gas, is given by

$$f = \left(\frac{1}{2\pi RT}\right)^{3/2} \exp\left(-\frac{v_x'^2 + v_y'^2 + v_z'^2}{2RT}\right)$$

where R is the gas constant for the particular gas.

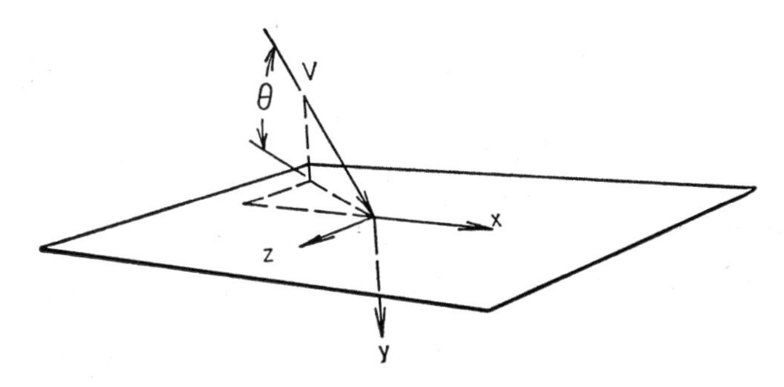

Figure 11.25.

Now, if the gas has a mass motion with velocity components $-V_x$, $-V_y$, $-V_z$, relative to the surface, Fig. 11.25, then the velocity components of the molecule relative to the surface are

$$v_x = v_x' - V_x$$

$$v_y = v_y' - V_y$$

$$v_z = v_z' - V_z$$

Therefore, in this new coordinate system, the number of molecules per unit volume having velocity components in the range dv_x, dv_y, and dv_z is

$$n_i\left(\frac{1}{2\pi RT_i}\right)^{3/2} \exp\left(-\frac{(v_x + V_x)^2 + (v_y + V_y)^2 + (v_z + V_z)^2}{2RT_i}\right) dv_x \, dv_y \, dv_z$$

Among this number, the number that will strike a unit area of the surface in unit time is found by multiplying the expression above by v_y. The total number N_i of molecules striking this unit area per unit time is then obtained by integrating the resulting expression over all possible values of v_x, v_y, and v_z that may be possessed by the molecules:

$$N_i = \int_{-\infty}^{\infty}\int_{-\infty}^{0}\int_{-\infty}^{\infty} n_i\left(\frac{1}{2\pi RT_i}\right)^{3/2} v_y \times \exp\left(-\frac{(v_x + V_x)^2 + (v_y + V_y)^2 + (v_z + V_z)^2}{2RT_i}\right) dv_x\, dv_y\, dv_z$$

The upper limit for v_y is zero, since no molecule with negative velocity in the y-direction can strike the surface. Upon integration, we get

$$N_i = n_i\sqrt{\frac{RT_i}{2\pi}}\left[e^{-\eta^2} + \sqrt{\pi}\,\eta(1 + \operatorname{erf}\eta)\right]$$

where $\eta = V_y/V_m = \sqrt{\gamma/2}M \sin\theta$ is the ratio of the normal component of the mass velocity to the most probable speed and $V_m = \sqrt{2RT_i}$. By multiplying the equation above by the mass of the molecule, the molecular mass flux incident on the surface is given by

$$m_i = \rho_i\sqrt{\frac{RT_i}{2\pi}}\left[e^{-\eta^2} + \sqrt{\pi}\,\eta(1 + \operatorname{erf}\eta)\right] \tag{11.49}$$

where ρ_i is the density.

We shall assume all molecules are reflected at the surface in a diffuse manner. This means that the incident molecules give up to the surface all their tangential momentum; the reflected molecules, having no preferred direction of motion, contribute no net momentum to the surface. Consequently, to find the shear stress at the wall, we need only calculate the total tangential momentum of the incident molecules.

The shearing stress is

$$\tau_i = \int_{-\infty}^{\infty}\int_{-\infty}^{0}\int_{-\infty}^{\infty} \rho_i\left(\frac{1}{2\pi RT_i}\right)^{3/2} v_x v_y \times \exp\left(-\frac{(v_x + V_x)^2 + (v_y + V_y)^2 + (v_z + V_z)^2}{2RT_i}\right) dv_x\, dv_y\, dv_z$$

The result of integration may be expressed in terms of the friction factor C_f as follows:

$$C_f = \sqrt{\frac{2}{\pi\gamma}}\left(\frac{1}{M}\right)\cos\theta e^{-\eta^2} + \sin\theta\cos\theta[1 + \operatorname{erf}\eta] \tag{11.50}$$

If the surface is aligned in the direction of motion, as is usual in the flat plate problems, $\theta = 0$, and $\eta = 0$, and Eq. (11.50) reduces to

$$C_f = \sqrt{\frac{2}{\pi\gamma}}\frac{1}{M}$$

for $\gamma = 1.4$, $C_f M = 0.674$. This value is plotted as a horizontal line in Fig. 11.22.

Net energy transfer to or from a surface in free-molecular flow, in the absence of radiant heat transfer, is given by the energy balance

$$\frac{q}{A} = E_i - E_r \tag{11.51}$$

where E_i is the incident energy flux and E_r is the re-emitted energy flux.

The incident and re-emitted fluxes of energy may be calculated by the same general method as the momentum fluxes. However, the following procedure is simpler [Sänger (32), Tsien (29)]. Let ϵ_{trans}, ϵ_{rot}, and ϵ_{vib} be the translational, the rotational, and the vibrational energies of the molecules per unit mass of the gas. Then, assuming completely diffuse reflection,

$$E_i = m_i(\tfrac{1}{2}V^2 + \epsilon_{i,\text{trans}} + \epsilon_{i,\text{rot}} + \epsilon_{i,\text{vib}}) \tag{11.52a}$$

$$E_r = m_i(\epsilon_{r,\text{trans}} + \epsilon_{r,\text{rot}} + \epsilon_{r,\text{vib}}) \tag{11.52b}$$

where m_i is the molecular mass flux incident and re-emitted from the surface, given by Eq. (11.49). The corresponding energies at T_w are $\epsilon_{w,\text{trans}}$, $\epsilon_{w,\text{rot}}$, and $\epsilon_{w,\text{vib}}$. Sänger then suggests that the accommodation coefficient for the translational and the rotational energies may be taken to be unity, while the accommodation coefficient for the vibrational energy is zero because of its slow rate of adjustment. Then,

$$\epsilon_{i,\text{trans}} - \epsilon_{r,\text{trans}} = \epsilon_{i,\text{trans}} - \epsilon_{w,\text{trans}}$$

$$\epsilon_{i,\text{rot}} - \epsilon_{r,\text{rot}} = \epsilon_{i,\text{rot}} - \epsilon_{w,\text{rot}}$$

$$\epsilon_{i,\text{vib}} - \epsilon_{r,\text{vib}} = 0$$

or

$$\epsilon_{r,\text{trans}} = \epsilon_{w,\text{trans}}; \quad \epsilon_{r,\text{rot}} = \epsilon_{w,\text{rot}}; \quad \epsilon_{r\text{vib}} = \epsilon_{i,\text{vib}}$$

By substituting these into Eqs. (11.52) and the resulting expressions for E_i and E_r in Eq. (11.51), we get

$$\frac{q}{A} = m_i[\tfrac{1}{2}V^2 + (\epsilon_{i,\text{trans}} - \epsilon_{w,\text{trans}}) + (\epsilon_{i,\text{rot}} - \epsilon_{w,\text{rot}})] \tag{11.53a}$$

In accordance with the principle of equipartition of energy, the internal energy of a gas molecule in a system in equilibrium at temperature T is given by $jRT/2$ where j is the number of degrees of freedom. For diatomic molecules such as the nitrogen and oxygen, we can write

$$\epsilon_{\text{trans}} = \tfrac{3}{2}RT, \qquad \epsilon_{\text{rot}} = \tfrac{2}{2}RT$$

Therefore, Eq. (11.53a) may be rewritten as

$$\frac{q}{A} = m_i\left[\frac{1}{2}V^2 + \frac{5}{2}R(T_i - T_w)\right] \tag{11.53b}$$

where m_i is the molecular mass flux given by Eq. (11.49). For the case $\theta = 0$ (plate aligned in the direction of motion), $\eta = 0$

$$\frac{q}{A} = \frac{\rho V_m}{2\sqrt{\pi}}\left[\frac{1}{2}V^2 + \frac{5}{2}R(T_i - T_w)\right] \tag{11.54}$$

The result may be more simply expressed in terms of the Stanton number, noting that $V/V_m = \sqrt{\gamma/2}\,M$ and assuming $\gamma = 1.4$:

$$\text{St}\,M = 0.240 \tag{11.55}$$

The solution is shown as a horizontal line in Fig. 11.23. The result is of the right form; the constant differs by only 20 per cent from that obtained by Oppenheim by a more generalized treatment in terms of the molecular structure. Figures 11.26 and 11.27 reproduced from Oppenheim's paper present heat transfer information in the free-molecule regime for a few fundamental shapes such as flat plates, spheres, and cylinders.

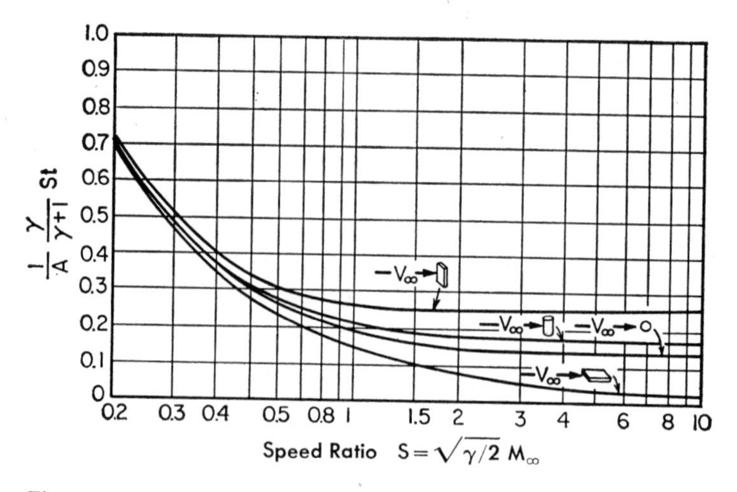

Fig. 11.26. Heat transfer from a flat plate, a sphere, and a transverse circular cylinder in free-molecule flow [Oppenheim (28)].

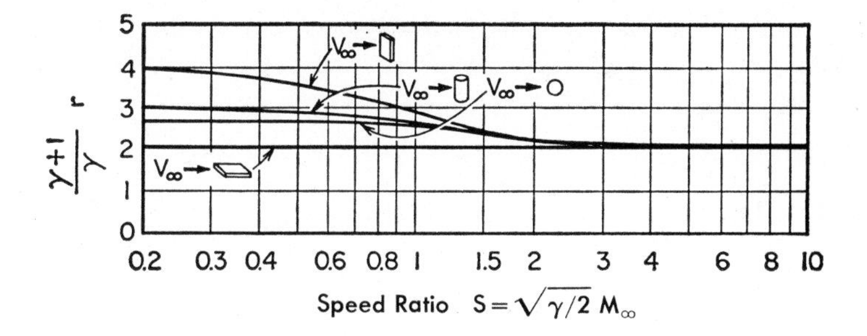

Fig. 11.27. Recovery factor for a flat plate, a sphere, and a transverse circular cylinder in free-molecule flow [Oppenheim (28)].

Example 11.3. An aluminum sphere 2 ft in diameter travels at $M_\infty = 10$ at an altitude of 500,000 ft. At that altitude the temperature is 450 F and the density is 3×10^{-9} lb_m/ft^3. Calculate (*a*) the adiabatic wall temperature and (b) the average heat transfer coefficient on the surface of the sphere.

(a) $V_\infty = M_\infty \sqrt{\gamma R T_\infty} = 10\sqrt{(1.4)(53.34)(32.2)(910)} = 14{,}800$ ft/sec

From Fig. 11.19, at 500,000 ft, $\lambda \cong 10$ ft

Therefore, $\lambda/d = 10/2 = 5$, free molecule regime

The speed ratio $S = M_\infty\sqrt{\gamma/2} = 10\sqrt{(1.4)/2} = 8.36$

From Fig. 11.27, $r(\gamma + 1)/\gamma = 2$ or $r = 1.17$

From Eq. 11.5

$$r = \frac{T_{aw} - T_\infty}{V_\infty^2/2c_p}$$

or

$$T_{aw} = \frac{(1.17)(14{,}800)^2}{(2)(32.2)(778)(0.248)} + 910 = 21{,}500R$$

(b) From Fig. 11.26, at $S = 8.36$,

$$\frac{1}{A}\frac{\gamma}{\gamma + 1}\text{St} \cong 0.12$$

For an aluminum surface in air, from Table 11.2, we can assume $A = 0.95$

$$\text{St} = \frac{(0.12)(0.95)(2.4)}{1.4} = 0.195$$

$$h_{\text{avg}} = \text{St}\,\rho V c_p = (0.195)(3 \times 10^{-9})(14{,}800)(3600)(0.24)$$

$$= 0.075 \text{ Btu/hr ft}^2 \text{ F}$$

REFERENCES

1. Pohlhausen, E., *Z. Angew. Math. u. Mech.*, **1,** 115–21 (1921).

2. Ackermann, G., *Forsch. Gebiete Ingenieurw.*, **13,** 226 (1942).

3. Crocco, L., *Monographie Scientifiche di Aeronautica*, **3** (1946), reviewed by van Driest, *Nat. Advisory Comm. Aeronaut. Tech. Note* 2597 (1952).

4. van Driest, E. R., "Convective Heat Transfer in Gases," Sec. F of Vol. 5, *High Speed Aerodynamics and Jet Propulsion*, ed. C. C. Lin, Princeton Univ. Press, 1959.

5. (a) Stine, H. A., and K. Wanlass, *Nat. Advisory Comm. Aeronaut. Tech. Note* 3344 (1954).
(b) Sibulkin, M., J. Aero. Sc. **19**:8, 570 (1952).

6. McAdams, W. H., L. A. Nicolai, and J. H. Keenan, *Trans. AIChE*, **42,** 907 (1946), also *Nat. Advisory Comm. Aeronaut. Tech. Note* 985 (1945).

7. Brown, W. B., and P. L. Donoughe, *Nat. Advisory Comm. Aeronaut. Tech. Note* 2479 (1951).

8. (a) Eckert, E. R. G., *Trans. ASME*, **78,** 1273 (1956); also Wright Air Dev. Ctr. Tech. Rpt. 59-624 (1960).
(b) Freedman, S. I., Ph.D. Thesis, Mech. Engrg. Dept., Mass. Inst. of Tech., Jan. 1961.

9. Mack, L. M., Cal. Inst. Tech. Jet Propul. Lab. Report 20-80, (1954).

10. (a) Stalder, J. R., M. W. Rubesin, and T. Tendeland, *Nat. Advisory Comm. Aeronaut. Tech. Note* 2077 (1950).
(b) Slack, E. G., *Nat. Advisory Comm. Aeronaut. Tech. Note* 2686 (1952).

11. (a) van Driest, E. R., J. Aero Sc., **18,** 145 (1951).
(b) Deissler, R. G., and A. L. Loeffler, Jr., *Nat. Advisory Comm. Aeronaut. Tech. Note* 4262 (1958).

12. Eber, G. R., *J. Aero. Sc.*, **19**:1, 1 (1952).

13. Rose, P. H., and W. I. Stark, *J. Aero. Sc.*, **25**:2, 86 (1958).

14. Fay, J. A., and F. R. Riddell, *J. Aero. Sc.*, **25**:2, 73 (1958).

15. (a) Rossow, V. J., *Nat. Advisory Comm. Aeronaut. Tech. Note* 3971 (1957).
(b) Hess, R. V., Nat. Aeronaut. and Space Admin. Memo 4-9-59L (1959).
(c) Moffatt, W. C., Sc.D. Thesis, M.E. Dept., M.I.T., June 1961.

16. Hansen, C. F., and S. P. Heims, *Nat. Advisory Comm. Aeronaut. Tech. Note* 4359 (1958).

17. Lees, L., *Jet Propulsion*, **26**:4, 259 (1956).

18. Eggers, A. J., Jr., C. F. Hansen, and B. E. Cunningham, *Nat. Advisory Comm. Aeronaut. Tech. Note* 4229 (1958).

19. Spitzer, L., Jr., "Physics of Fully Ionized Gases," *Interscience*, 1956.

20. (a) Burnett, D., *Proc. London Math. Soc.* **40,** 382 (1935).
(b) Grad, H., Commun. on Pure and Applied Math. **2,** 331 (1949).

21. Millikan, R. A., *Phys. Rev.*, **21,** 238 (1923).

22. Blankenstein, E., *Phys. Rev.* **22,** 582 (1923).

23. Knudsen, M., *Kinetic Theory of Gases*, Methuen, London, 1934.

24. Wiedmann, M. L., and P. R. Trumpler, *Trans. ASME*, **68,** 57 (1946).

25. Schaaf, S. A., and P. L. Chambré, "Flow of Rarefied Gases," Sec. H, Vol. 3, *High Speed Aerodynamics and Jet Propulsion*, Princeton Univ. Press, 1958.

26. Schaaf, S. A., and F. S. Sherman, *J. Aero. Sc.*, **21**:2, 85 (1954).

27. (a) Drake, R. M., and E. D. Kane, General Discussion on Heat Trans., London, Sept., 1951.
(b) Drake, R. M., and G. H. Backer, *Trans. ASME*, **74**:7, 1241 (1952).

28. Oppenheim, A. K., *J. Aero. Sc.*, **20**:1, 49 (1953).

29. Tsien, H. S., *J. Aero. Sc.*, **13**:1, 653 (1946).

30. Oseen, C. W., *Arkiv for mathematik, astronomi och fysik*, **6,** 29 (1910).

31. Keenan, J. H., and J. Kaye, *Gas Tables*, Wiley, New York, 1948.

32. Sanger, E., and J. Bredt, *Deutsche Luftfahrtforschung, Untersuchungen u. Mitteilunten*, No. 3538, 141 (1944).

33. Warfield, C. N., *Nat. Advisory Comm. Aeronaut. Tech. Note* 1200 (1947).

34. Schaaf, S. A., "Heat Transfer, A Symposium 1952", *Engr. Res. Inst.*, Univ. of Mich.

PROBLEMS

11.1 Consider the Couette flow system shown in Fig. 11.1. The upper plate is maintained at 200 F and moves at a velocity of 800 ft/sec and the plates are spaced 0.01 in. apart. Calculate the recovery factor, the adiabatic wall temperature, and the stagnation temperature, if the fluid is (a) air and (b) hydrogen. Neglect any pressure gradient in the flow field and assume properties do not change significantly.

11.2 Air at 60 F flows past an insulated plate at $M_\infty = 3$. Plot the distribution of the stagnation temperature through the boundary layer (plot the ratio of the stagnation temperature to the stagnation temperature of the free stream vs. η) using the results of the Pohlhausen analysis as given by Figs. 11.3, 11.4, and 3.5. Interpret your result.

11.3 A supersonic plane flies at $M = 3$ at an altitude of 30,000 ft. Find the adiabatic surface temperature 6 in. behind the leading edge of a fin. Approximate the fin as a flat plate.

11.4 A thermocouple 0.025 in. in diameter is held normal to an airstream, and at $M = 0.7$, it reads 140 F. If the recovery factor is known to be $r = 0.9$, find the static temperature of the air.

11.5 Air at 100 F flows at (a) 400 ft/sec and (b) 2400 ft/sec over a 2-in. plate aligned in the direction of flow. Calculate the Mach number and the adiabatic wall temperature in each case. Assume transition occurs at $\text{Re} = 5 \times 10^5$.

11.6 If the plate of Prob. 11.5 is maintained at a temperature of 140 F,
(a) estimate the rate of heat transfer at the plate in each case using the appropriate semiempirical correlations.
(b) Repeat the calculations using Fig. 11.6. Compare your results with part (a).

11.7 In Prob. 11.5, estimate the drag force per unit area of the plate, using Fig. 11.13.

11.8 A wedge-shaped body (wedge angle $= 5°$) travels at $M = 2.45$ at an altitude of 50,000 ft. Using the Gas Tables (31), it is estimated that the external flow over the body behind an oblique shock wave is at a Mach number of 2.24, a pressure of 2.3 psia, and a density of 0.145 lb_m/ft^3. Calculate the adiabatic wall temperature and the local heat transfer coefficient 4 in. from the apex of the wedge.

11.9 Estimate the rate of cooling necessary to maintain the flat surface 1 ft square of a body at a maximum temperature of 400 F, if the free-stream flow past the body at an altitude of 100,000 ft is 20,000 ft/sec. Air under these conditions is in a dissociated state. Assume as a first approximation that the enthalpy recovery factors may be calculated by

$$r_i \cong (\text{Pr})^{1/2}, \qquad \text{laminar}$$

$$r_i \cong (\text{Pr})^{1/3}, \qquad \text{turbulent}$$

The enthalpy of air may be found from the Gas Tables (31).

11.10 The energy equation for continuum flow past a sphere is given by

$$\rho c_p\left(v_r \frac{\partial T}{\partial r} + \frac{v_\theta}{r}\frac{\partial T}{\partial \theta} + \frac{v_\phi}{r \sin\theta}\frac{\partial T}{\partial \phi}\right)$$

$$= k\left[\frac{1}{r^2}\frac{\partial}{\partial r}\left(r^2 \frac{\partial T}{\partial r}\right) + \frac{1}{r^2 \sin\theta}\frac{\partial}{\partial \theta}\left(\sin\theta \frac{\partial T}{\partial \theta}\right) + \frac{1}{r^2 \sin^2\theta}\frac{\partial^2 T}{\partial \phi^2}\right]$$

State all assumptions necessary to reduce the equation above to the following equation approximately valid for slip flow past a sphere:

$$\rho c_p \frac{V_\infty}{r} \frac{\partial T}{\partial \theta} = k \left[\frac{1}{r^2} \frac{\partial}{\partial r} \left(r^2 \frac{\partial T}{\partial r} \right) \right]$$

where V_∞ is the uniform free-stream velocity. Write also the necessary temperature boundary conditions. The solution to this problem is given in Ref. 27a and appears on Fig. 11.24 with the Mach number as a parameter.

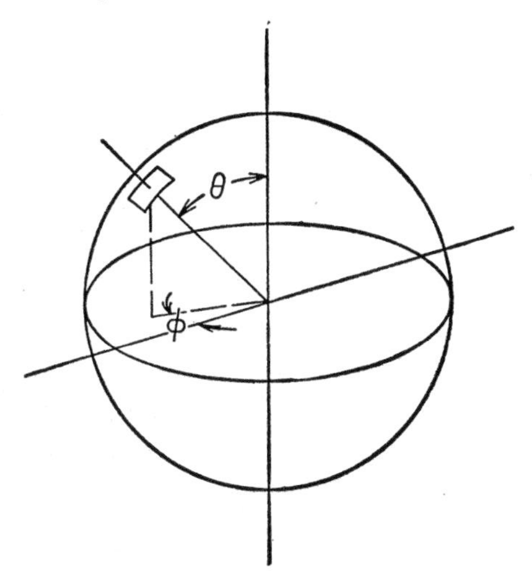

11.11 Using the usual continuum equation for the Couette flow and the slip velocity boundary condition, show that the skin friction coefficient on either plate, assuming zero pressure gradient and constant properties, is given by

$$C_f = \frac{\tau}{\frac{1}{2}\rho V^2} = \frac{2}{\text{Re}\left(1 + 2\frac{2 - F}{F}\frac{\lambda}{l}\right)}$$

where l is the distance between the plates and $\text{Re} = \rho V l/\mu$. Note that the solution reduces to the continuum solution when $\lambda/l \to 0$.

11.12 The two plates of Prob. 11.11 are 0.002 in. apart and the air between the plates is sufficiently rarefied so that $\lambda = 10^{-4}$ in. If the air temperature is 0 F and the velocity of the moving plate is 0.5 ft/sec, find the drag force per unit area of the plate when $F = 1$ and $F = 0.8$.

11.13 We shall consider the phenomenon of thermal transpiration. Consider two large chambers separated by a diaphragm containing a small orifice. The states of the gas in the two chambers are indicated by subscripts 1 and 2. If the orifice is extremely small (size of orifice $< \lambda^2$), the occasional loss of molecules from either chamber will have no appreciable effect on the states of the gas which can be assumed to be at equilibrium at all times. The number of molecules which escapes from either chamber per unit area of the orifice per unit time may be found from the kinetic theory to be

$$N = \tfrac{1}{4} n\bar{V}$$

where n is the molecular density and $\bar{V}$ is the average speed.

(a) Show that the mass flux is given as follows:

$$mN = \frac{p}{\sqrt{2\pi RT}}$$

where m is the mass of a molecule and R is the gas constant.

(b) If initially the two chambers have the same pressure but different temperatures, show that when the flow of molecules ceases, $p_2/p_1 = \sqrt{T_2/T_1}$. Note that if the orifice were large, the condition of zero flow would be simply $p_1 = p_2$.

11.14 An aluminum plate 1 ft square moves at $M_\infty = 2$ at an altitude of 500,000 ft where the mean free path is roughly 10 ft. Estimate the rate of heat transfer necessary to keep the plate surface temperature below 300 F

(a) if the plate is aligned in the direction of flow and
(b) if the plate is positioned perpendicular to flow.

Neglect radiant heat transfer to and from the plate. The answers would therefore provide conservative estimates.

CHAPTER 12

Heat Exchangers

A heat exchanger is a piece of equipment in which heat is transferred from a hot fluid to a colder fluid. In its simplest form, the two fluids mix and leave at an intermediate temperature determined from conservation of energy. This device is not truly a heat exchanger but rather a mixer. In most applications the fluids do not mix but transfer heat through a separating wall which takes on a wide variety of geometries. The sketches of Fig. 12.1 represent a few of the many types of heat exchangers in existence.

Figure 12.1(a) shows a concentric tube and (b) a shell-and-tube (multiple tubes) exchanger with counterflow. Parallel flow exists if the flow direction of either fluid in (a) and (b) is reversed. Near the entrance and exit pipes on the shell side, some crossflow occurs. In (c) a double-pass tube side unit is shown with a single-pass shell side, and the two most common baffle arrangements are drawn in (d) and (e). Figure 12.1(f) shows the core of a crossflow tubular exchanger and (g) the core of a plate-fin exchanger. Fins also are often used on tubes on the gas side of liquid-to-gas heat exchangers.

Heat exchanger design presents many stress and corrosion problems as well as problems of fluid mechanics and heat transfer. In most cases design aims at minimum cost, balancing the cost of pumping the fluids and initial

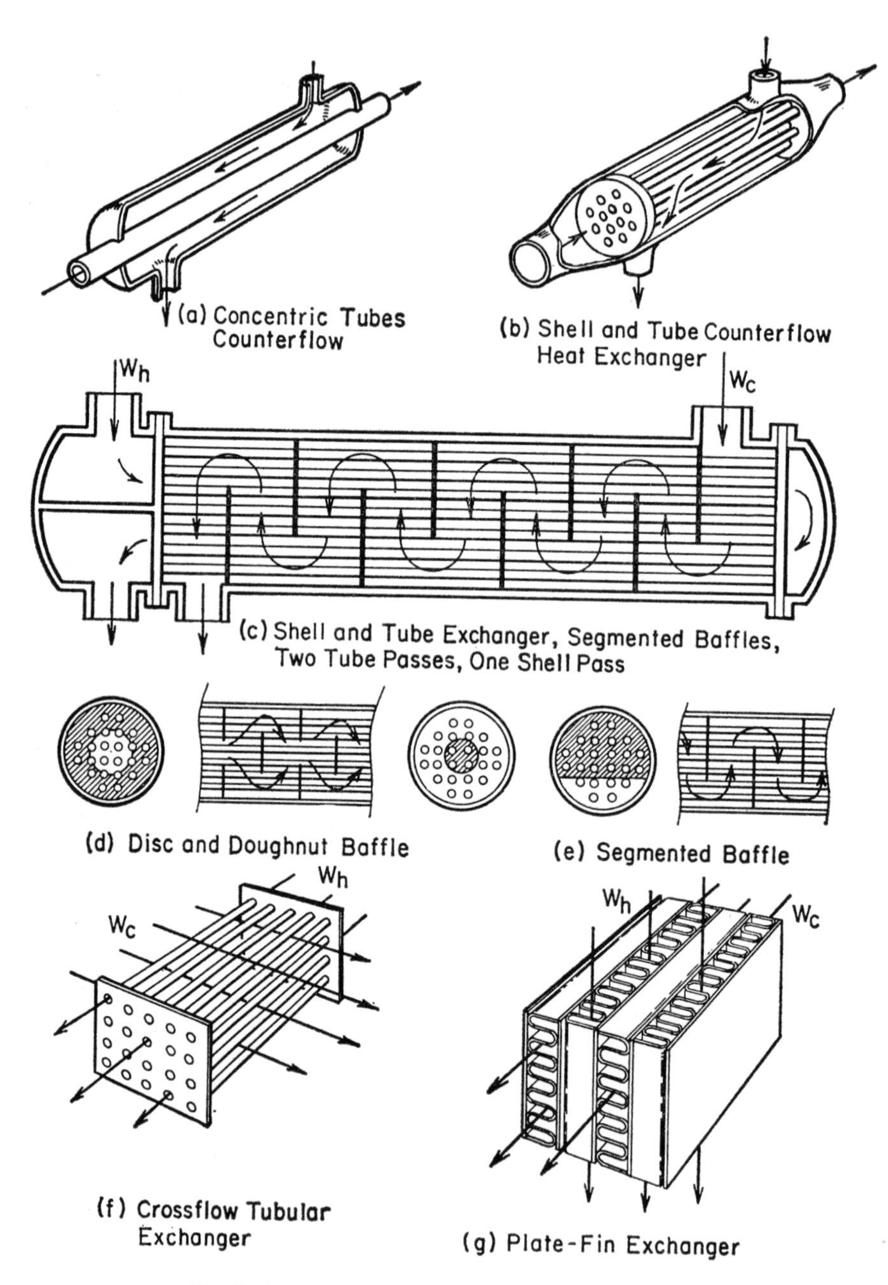

Fig. 12.1. Representative types of heat exchangers.

cost of the exchanger against the savings resulting from the heat transfer. In some cases—such as in missile, aircraft or shipboard applications—design may be governed by the necessity of minimizing either volume or weight.

In this chapter we discuss two alternative methods of evaluating heat exchanger performance or size, discuss heat transfer in electrically heated units and in nuclear reactors, and conclude with a simple illustrative example of heat exchanger design.

12.1 Over-all Heat Transfer Coefficient and Fouling Factors

Heat exchanger walls are usually single materials, although sometimes as a protection against corrosion the wall may be bimetallic (steel with aluminum cladding) or coated with a plastic. Most heat exchanger surfaces tend to acquire an additional heat transfer resistance which increases with time. This may be either a very thin surface oxidation layer or, at the other extreme, may be a thick crust deposit, such as that which results from salt water coolant in steam condensers. We define a scale coefficient of heat transfer h_s in terms of thermal resistance (Sec. 6.3) of this scale as follows:

$$R_s = \frac{1}{Ah_s} \tag{12.1}$$

where the area A is the original heat transfer area of the surface before scaling began.

The over-all coefficient of heat transfer (Sec. 6.5) for a single wall material is then expressed as follows:

$$\frac{T_h - T_c}{dq/dA_r} = \frac{1}{U_r} = \frac{1}{h_c}\frac{dA_r}{dA_c} + \frac{1}{h_{s_c}}\frac{dA_r}{dA_c} + R_w + \frac{1}{h_{s_h}}\frac{dA_r}{dA_h} + \frac{1}{h_h}\frac{dA_r}{dA_h} \tag{12.2}$$

where reference area A_r is usually selected either as A_c or A_h. For the flat plate with $A_c = A_h = A_r$, $R_w = x_w/k_w$, and for a cylinder,

$$R_w = \left[\frac{\ln (r_0/r_i)}{2\pi k_w}\right]\frac{dA_r}{dL}$$

The temperature drops and corresponding resistances across a wall are shown in Fig. 12.2. Table 12.1 gives magnitudes of $1/h_s$, called *fouling factor*, recommended for inclusion in U_r for calculating the required heat transfer surface of an exchanger. Looking at the order of magnitude of resistances (Sec. 6.6), it is seen that the scale resistance is usually significant only when there are liquids on both sides of the exchanger wall.

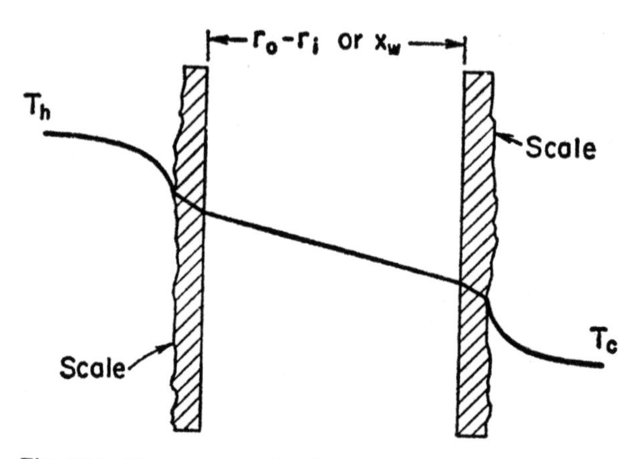

Fig. 12.2. Temperature distribution across exchanger wall.

A separating wall may be finned differently on each side, Fig. 12.3. On either side, the heat transfer takes place from the fins (subscript f in the equations that follow) as well as from the unfinned portion of the wall (subscript u); using Eq. (6.46), which introduces the fin efficiency, η, the total q is given by

$$q = (\eta A_f h_f + A_u h_u)\,\Delta T \tag{12.3}$$

where ΔT is either $(T_h - T_1)$ or $(T_2 - T_c)$. Taking $h_u = h_f = h$, an

TABLE 12.1 HEAT EXCHANGER FOULING FACTORS *

($r_s = 1/h_s$ in °F ft² hr/Btu)

Temperature of heating medium	up to 240 F		240–400 F	
Temperature of water	125 F or less		above 125 F	
Water velocity, ft/sec	3 and less	over 3	3 and less	over 3
Distilled water	0.0005	0.0005	0.0005	0.0005
Sea water	0.0005	0.0005	0.001	0.001
City or well water	0.001	0.001	0.002	0.002
Treated boiler feed water	0.001	0.0005	0.001	0.001
Mississippi River water	0.003	0.002	0.004	0.003
Liquid gasoline, organic vapors				0.0005
Refrigerating liquids, cooling brine, oil-bearing steam				0.001
Refrigerating vapors, distillate bottoms above 20° API, air				0.002
Fuel oil, salty crude oil, residual bottoms less than 20° API				0.005
Diesel exhaust gas, coke-oven gas, cracking unit residuum				0.010

* From *Standards of Tubular Exchanger Manufacturers' Association*, 3rd Ed., New York, 1952.

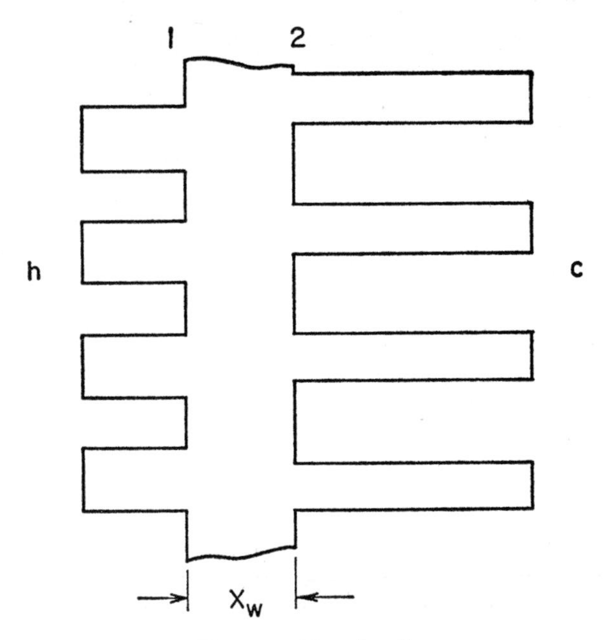

Fig. 12.3. Finned wall.

over-all surface efficiency $\mathcal{E}$ may be defined as follows:

$$\mathcal{E} \equiv \frac{q}{Ah\,\Delta T} = \eta\frac{A_f}{A} + \frac{A_u}{A} \tag{12.4}$$

where $A \equiv A_u + A_f$. An over-all heat transfer coefficient for the entire wall is then given by

$$\frac{T_h - T_c}{dq/dA_r} \equiv \frac{1}{U_r} = \frac{1}{\mathcal{E}_c h_{\text{eff}_c}}\frac{dA_r}{dA_c} + R_w\frac{dA_r}{dA_w} + \frac{1}{\mathcal{E}_h h_{\text{eff}_h}}\frac{dA_r}{dA_h} \tag{12.5}$$

where
$$\frac{1}{h_{\text{eff}}} \equiv \frac{1}{h_{\text{fluid}}} + \frac{1}{h_{\text{scale}}}$$

and η is computed as in Sec. 6.9 with h_{eff}. Here the reference area A_r is the total surface area $A = A_u + A_f$ for either the hot or cold sides, and A_w is the total primary wall area.

Example 12.1: Calculate the rate of heat transfer through a 12 in. × 12 in. section of wall (shown in Fig. 12.3) if

$$T_h = 200\text{ F},\quad T_c = 100\text{ F},\quad x_w = 0.25\text{ in.},\quad k_w = 24,$$

and

$$h_c = h_h = 2 = h_{\text{eff}_c} = h_{\text{eff}_h}$$

neglecting the scale resistance. Both sides of the wall have $\frac{1}{4}$-in. diameter × 2 in. long pin fins on $\frac{3}{4}$-in. centers in a square pitch arrangement, or 256 pins per sq ft.

$$A_f = \frac{\pi(0.25)(2.0625)(256)}{144} = 2.885 \text{ ft}^2$$

$$A_u = 1 - \frac{256(\pi)(0.25)^2}{4(144)} = 0.912 \text{ ft}^2$$

$$A_w = 1 \text{ ft}^2$$

$$A = A_u + A_f = 3.797 \text{ ft}^2$$

From the example in Sec. 6.9,

$$\eta = \frac{\tanh\eta\ [(4)(2.0625/12)]}{(4)(2.0625/12)} = 0.867$$

From Eq. (12.4),

$$\mathcal{E} = \frac{0.867(2.885) + 0.912}{3.797} = 0.90$$

Then, from Eq. (12.5)

$$\frac{T_h - T_c}{q} = \frac{1}{A_r U_r}$$

$$= 2\left(\frac{1}{(0.90)(3.797)(2)}\right) + \frac{0.25}{24(12)(1)} = 0.294$$

$$q = \frac{1}{0.294}(200 - 100) = 340 \text{ Btu/hr}$$

12.2 Temperature Distribution in Heat Exchangers

The preceding discussions of heat transfer from one fluid to another separated by a wall have been limited to the heat transfer processes occurring at a particular place in a heat exchange system where the fluid temperatures have particular values.

In a heat exchanger, such as the concentric tube arrangement, Fig. 12.1(a), the fluid temperatures change as the fluids flow along the heat exchanger length. If the fluid condenses or evaporates along the length, the temperature remains constant. Figure 12.4 shows typical temperature distributions which may be obtained in heat exchangers. For the element of length dx with associated heat transfer area dA, Fig. 12.4,

$$dq = U(T_h - T_c)\, dA \tag{12.6}$$

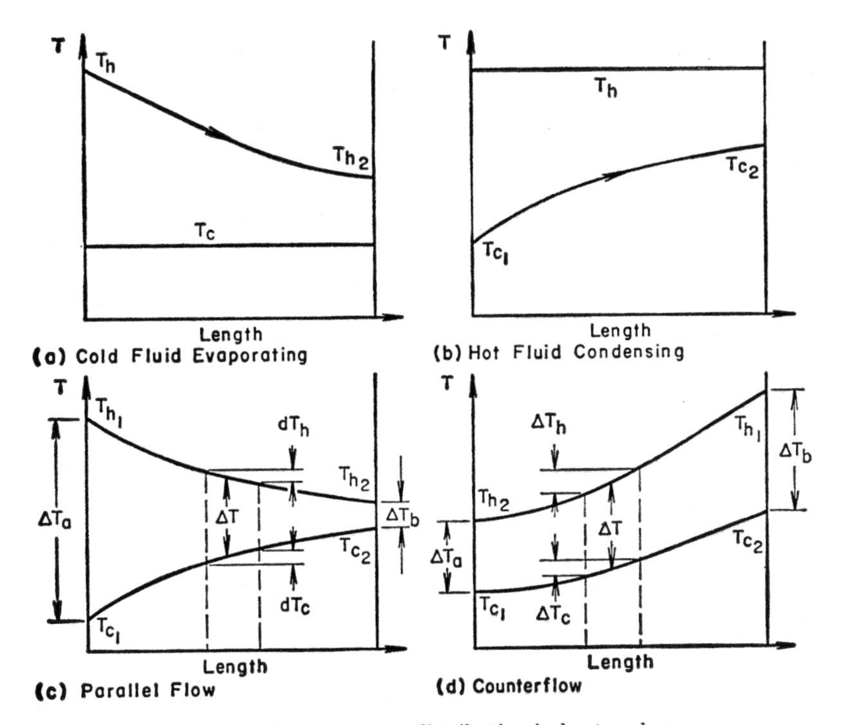

Fig. 12.4. Axial temperature distribution in heat exchangers.

Also from the steady flow energy equation for each fluid,

$$dq = w_c\, di_c = w_h\, di_h \tag{12.7}$$

where w is the flow rate, i is the enthalpy; the kinetic and potential energy changes of the fluids are neglected.

If the cold fluid evaporates and the hot fluid condenses and T_h and T_c are each constant along the exchanger length, Eq. (12.6) when integrated with U constant becomes

$$q = AU(T_h - T_c) \tag{12.8}$$

For the cases of parallel flow and counterflow of fluids without phase change, Figs. 12.4(c) and 12.4(d), the enthalpy change may be written as $di = c\, dt$; then

$$dq = w_c c_c\, dT_c = \pm w_h c_h\, dT_h \tag{12.9}$$

where the $(+)$ sign refers to counterflow since dT_h/dx is positive and the $(-)$ sign to parallel flow since dT_h/dx is negative. Then from Eq. (12.9),

$$d(T_h - T_c) = dT_h - dT_c = dq\left(\pm\frac{1}{w_h c_h} - \frac{1}{w_c c_c}\right) \tag{12.10}$$

Substituting for dq from Eq. (12.6)

$$\frac{d(T_h - T_c)}{(T_h - T_c)} = U\left(\pm\frac{1}{w_h c_h} - \frac{1}{w_c c_c}\right) dA \tag{12.11}$$

which, when integrated with constant values of U, $w_h c_h$, and $w_c c_c$ between limits ΔT_a and ΔT_b, results in

$$\ln\frac{\Delta T_b}{\Delta T_a} = UA\left(\pm\frac{1}{w_h c_h} - \frac{1}{w_c c_c}\right) \tag{12.12}$$

Similarly, integration of Eq. (12.9) results in

$$q = w_c c_c(T_{c_2} - T_{c_1}) = \pm w_h c_h(T_{h_1} - T_{h_2}) \tag{12.13}$$

Solving this for $w_c c_c$ and $w_h c_h$ and substituting in Eq. (12.12)

$$q = AU\frac{\Delta T_a - \Delta T_b}{\ln(\Delta T_a/\Delta T_b)} \tag{12.14}$$

for either parallel or counterflow. The quantity

$$\frac{\Delta T_a - \Delta T_b}{\ln(\Delta T_a/\Delta T_b)}$$

is the logarithmic mean value, ΔT_{lm}, of ΔT between ΔT_a and ΔT_b, so $q = AU\,\Delta T_{\text{lm}}$. It should be noted that ΔT, for example, is the difference in temperature of the fluids at a particular place in the heat exchanger; $\Delta T_a = (T_{h_1} - T_{c_1})$ for parallel flow, but $\Delta T_a = (T_{h_2} - T_{c_1})$ for counterflow.

In the case of counterflow with $w_c c_c = w_h c_h$, the quantity ΔT_{lm} is indeterminate since $(T_{h_1} - T_{h_2}) = (T_{c_2} - T_{c_1})$ and $\Delta T_a = \Delta T_b$. Equation (12.9) shows that for any increment of area dA, $dT_c = dT_h$ for this case; therefore, ΔT is uniform along the heat exchanger and Eq. (12.8) represents the over-all heat transfer performance with $(T_h - T_c) = \Delta T_a = \Delta T_b$. This may also be shown from Eq. (12.14) by applying the calculus of limits.

Schematic arrangements of other flow paths, such as multipass, are shown in Figs. 12.1 and 12.5. The integration of Eq. (12.6) for these other flow arrangements results in a form of an integrated mean temperature difference ΔT_m such that

$$q = AU\,\Delta T_m \tag{12.15}$$

where ΔT_m is a complex function of T_{h_1}, T_{h_2}, T_{c_1}, and T_{c_2}. Generally this function ΔT_m can be determined in terms of the following quantities:

$$\Delta T_{\text{lmc}} = \frac{(T_{h_2} - T_{c_1}) - (T_{h_1} - T_{c_2})}{\ln \dfrac{(T_{h_2} - T_{c_1})}{(T_{h_1} - T_{c_1})}}$$

$$\frac{C_c}{C_h} = \frac{w_c c_c}{w_h c_h} = \frac{T_{h_1} - T_{h_2}}{T_{c_2} - T_{c_1}} \tag{12.16}$$

$$\epsilon_c = \frac{T_{c_2} - T_{c_1}}{T_{h_1} - T_{c_1}}$$

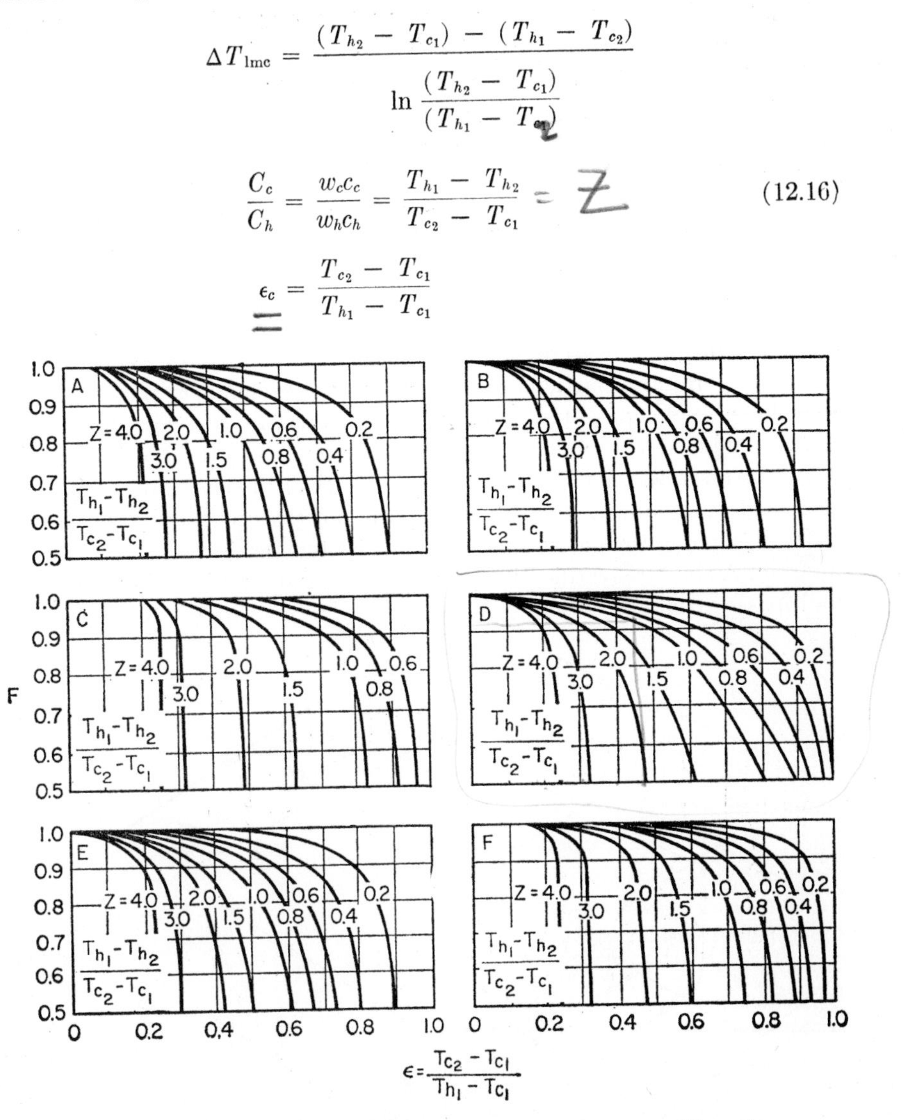

Fig. 12.5. Mean temperature difference in heat exchangers with various flow arrangements [Bowman, Mueller, and Nagle (1)]. (A) One shell pass and 2, 4, 6, etc., tube passes. (B) One shell pass and 3, 6, 9, etc., tube passes. (C) Four shell passes and 8, 16, 24, etc., tube passes. (D) Crossflow, both fluids unmixed, one tube pass. (E) Crossflow, shell fluid mixed, one tube pass. (F) Crossflow, shell fluid mixed, two tube passes, shell fluid flows over first and second passes in series.

These quantities may be interpreted as follows: ΔT_{lmc} is the logarithmic mean temperature difference for a counterflow arrangement with the same fluid inlet and outlet temperatures; C_h/C_c is the ratio of wc products of the two fluids; and ϵ_c is called the effectiveness of the heat exchanger on the cold fluid side because it is a measure of the ratio of the heat actually transferred to the cold fluid to the heat which would be transferred if the same fluid were to be raised to the temperature of the hot inlet fluid.

Charts (Fig. 12.5) for convenience of solution have been prepared for various flow arrangements with $F \equiv \Delta T_m/\Delta T_{lmc}$ as a function of ϵ_c and C_c/C_h. Then

$$q = AUF\,\Delta T_{lmc} \tag{12.17}$$

In a multipass or a crossflow arrangement, the fluid temperature may not be uniform at a particular distance into the exchanger unless the fluid is well mixed along the path length. For example, in crossflow (Fig. 12.6) the hot and cold fluids may enter at uniform temperatures, but if there are channels in the flow path to prevent mixing, the exit temperature distributions will be as shown. If such channels are not present, the fluids may be well mixed along the path length and the exit temperatures are more nearly uniform. A similar stratification of temperatures occurs in a shell-and-tube multipass exchanger. A series of baffles may be required if mixing of the

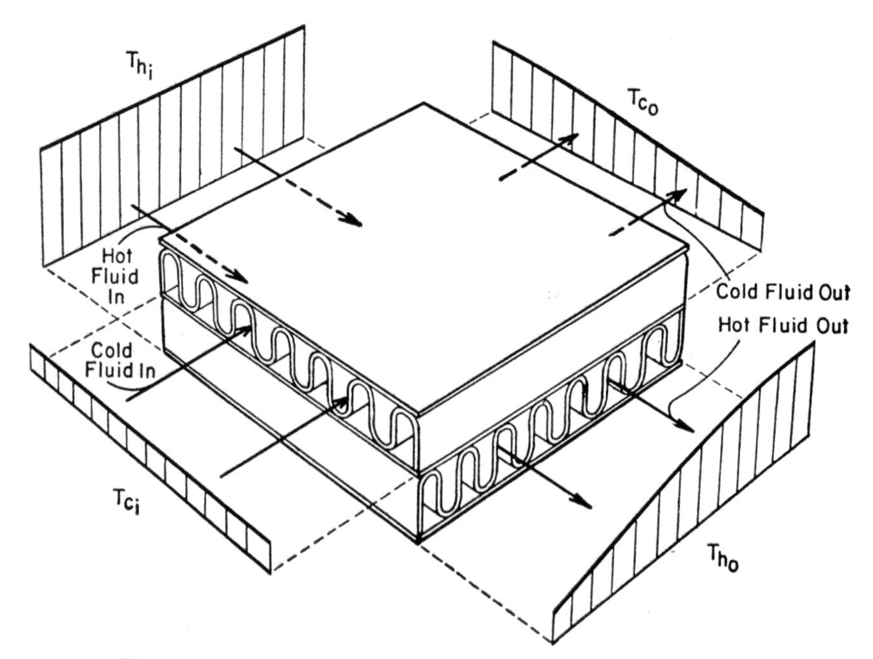

Fig. 12.6. Temperature distributions in crossflow heat exchanger.

shell fluid is to be obtained. Charts are presented for both mixed and unmixed fluid in Fig. 12.5.

The preceding analysis assumed U to be uniform throughout the heat exchanger. If U is not uniform the heat exchanger calculations may be made by subdividing the heat exchanger into sections over which U is nearly uniform and applying the previously developed relations to each subdivision.

Example 12.2: Calculate the required heat exchanger area for various types of flow arrangements if 10,000 lb/hr of water ($c = 1$ Btu/lb F) is cooled from $T_{h_1} = 200$ F to $T_{h_2} = 160$ F by means of an equal rate of flow of water at 100 F if $U = 400$ Btu/hr ft² F.

From Eq. (12.13),

$$q = 10{,}000(1)(200 - 160) = 400{,}000 \text{ Btu/hr} = 10{,}000(1)(T_{c2} - 100)$$

Then, $T_{c2} = 140$ F. From Eq. (12.15),

$$A = \frac{q}{U\,\Delta T_m} = \frac{400{,}000}{400\,\Delta T_m} = \frac{1000}{\Delta T_m}$$

(a) *Parallel flow.* From Fig. 12.4 and Eq. (12.14),

$$\Delta T_m = \Delta T_{\text{lm}} = \frac{(200 - 100) - (160 - 140)}{\ln\left[(200 - 100)/(160 - 140)\right]} = 49.6 \text{ F}$$

so

$$A = \frac{1000}{49.6} = 20.2 \text{ ft}^2$$

(b) *Counterflow.* With equal flow rates ΔT along the exchanger is uniform at $160 - 100 = 60$ F, so

$$A = \frac{1000}{60} = 16.67 \text{ ft}^2$$

(c) *Crossflow* with hot water mixed making one pass through the shell and cold water making one pass through tubes. From Eq. (12.16), $\Delta T_{\text{lmc}} = 60$ F, $C_h/C_c = 1.0$, and

$$\epsilon_c = \frac{140 - 100}{200 - 100} = 0.40$$

Therefore, from chart E, Fig. 12.5, $Y = 0.92$. Then

$$\Delta T_m = F\,\Delta T_{\text{lmc}} = 0.92(60) = 55.2 \text{ F}$$

and

$$A = \frac{1000}{55.2} = 18.1 \text{ ft}^2$$

(d) *Multipass* with hot water making one pass through a well-baffled shell and cold water making two passes through tubes. From chart A, Fig. 12.5, $Y = 0.91$; so

$$\Delta T_m = (0.91)(60) = 54.6\ \text{F} \quad \text{and} \quad A = \frac{1000}{54.6} = 18.3\ \text{ft}^2$$

12.3 NTU Design Method

The heat exchanger heat transfer equations such as Eqs. (12.12) and (12.13) may be written in dimensionless form (2) resulting in the following dimensionless groups:

(1) Capacity rate ratio,

$$C_{\min}/C_{\max} \tag{12.18a}$$

where $C_{\min}$ and $C_{\max}$ are respectively the smaller and larger of the two magnitudes C_h and C_c. Recall that C is the product of the fluid flow rate and its specific heat.

(2) Exchanger heat transfer effectiveness,

$$\epsilon \equiv \frac{C_h(T_{h_1} - T_{h_2})}{C_{\min}(T_{h_1} - T_{c_1})} = \frac{C_c(T_{c_2} - T_{c_1})}{C_{\min}(T_{h_1} - T_{c_1})} \tag{12.18b}$$

which is the ratio of the actual heat transfer rate in the exchanger to the thermodynamically limited maximum possible heat transfer rate which could be realized only in a counterflow heat exchanger of infinite heat transfer area. The first definition in Eq. (12.18b) is for $C_h = C_{\min}$ and the second for $C_c = C_{\min}$.

(3) Heat transfer area number,

$$N_A = \frac{AU}{C_{\min}} \tag{12.18c}$$

This dimensionless group has been called the *number of exchanger heat transfer units*, NTU.

In the equations of the preceding section assume that $C_c > C_h$; so $C_h = C_{\min}$ and $C_c = C_{\max}$. Equation (12.12) may then be written

$$\frac{\Delta T_a}{\Delta T_b} = \exp[-N_A(\pm 1 - c_{\min}/c_{\max})] \tag{12.19}$$

where the $(+)$ sign is for counterflow and $(-)$ for parallel flow. With Eqs. (12.18) and (12.13) and Fig. 12.4, the following identities are obtained:

Counterflow:

$$\frac{\Delta T_a}{\Delta T_b} = \frac{T_{h_2} - T_{c_1}}{T_{h_1} - T_{c_2}} = \frac{1 - \epsilon}{1 - (c_{\min}/c_{\max})\epsilon} \tag{12.20a}$$

Parallel flow:

$$\frac{\Delta T_a}{\Delta T_b} = \frac{T_{h_1} - T_{c_1}}{T_{h_2} - T_{c_2}} = \frac{1}{1 - \epsilon(1 + c_{min}/c_{max})} \tag{12.20b}$$

Combining Eqs. (12.19) and (12.20)

Counterflow:

$$\epsilon = \frac{1 - \exp\left[-N_A(1 - c_{min}/c_{max})\right]}{1 - (c_{min}/c_{max}) \exp\left[-N_A(1 - c_{min}/c_{max})\right]} \tag{12.21a}$$

Parallel flow:

$$\epsilon = \frac{1 - \exp\left[-N_A(1 + c_{min}/c_{max})\right]}{1 + c_{min}/c_{max}} \tag{12.21b}$$

These same results are obtained if it is assumed that $C_c < C_h$; then $C_c = C_{min}$ and $C_h = C_{max}$.

Two limiting cases are of interest—C_{min}/C_{max} equal to zero and unity. For $C_{min}/C_{max} = 1.0$, Eq. (12.21a) is indeterminate, but the following result may be obtained directly from Eqs. (12.8), (12.13), and (12.18) or by applying the calculus of limits to Eq. (12.21a).

For $(C_{min}/C_{max}) = 1.0$,

Counterflow:
$$\epsilon = \frac{N_A}{1 + N_A} \tag{12.22a}$$

Parallel flow:
$$\epsilon = \tfrac{1}{2}(1 - e^{-2N_A}) \tag{12.22b}$$

For $(C_{min}/C_{max}) = 0$, Figs. 12.4(a) and 12.4(b), for parallel or counterflow, Eqs. (12.21) become

$$\epsilon = 1 - e^{-N_A} \tag{12.22c}$$

These relations are shown graphically in Fig. 12.7(a) and 12.7(b). Similar relations and curves have been obtained for crossflow and for cross-counterflow exchangers and are shown in Figs. 12.7(c), (d), (e). In a multipass cross-counterflow arrangement if each fluid is "mixed" between passes and if N_A is equally distributed among the passes, the over-all effectiveness was derived (2) to be

$$\epsilon = \frac{\left[\dfrac{1 - \epsilon_p(c_{min}/c_{max})}{1 - \epsilon_p}\right]^n - 1}{\left[\dfrac{1 - \epsilon_p(c_{min}/c_{max})}{1 - \epsilon_p}\right]^n - \dfrac{c_{min}}{c_{max}}} \tag{12.23a}$$

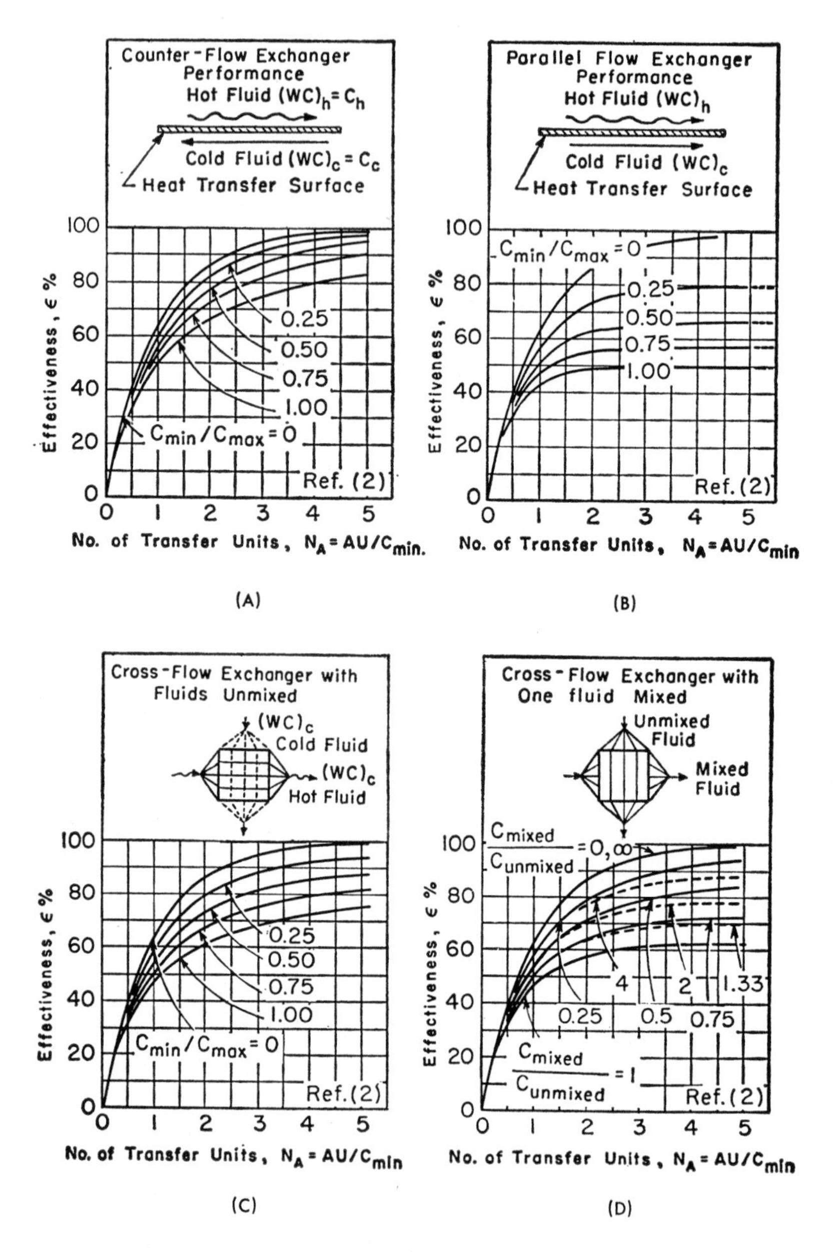

Fig. 12.7. Effectiveness vs. NTU for various types of heat exchanger flow arrangements [Kays, London and Johnson (2)].

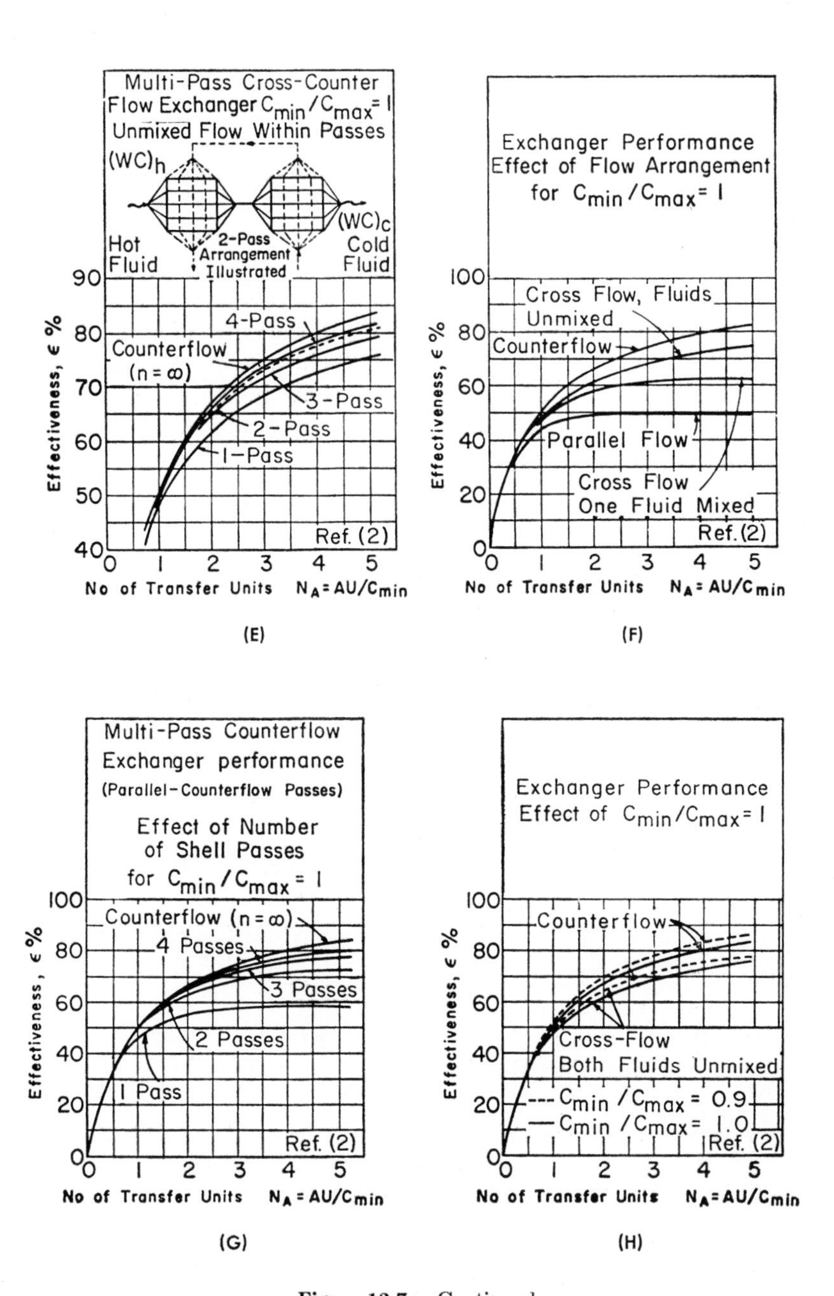

Figure 12.7.—Continued.

and if $C_{min}/C_{max} = 1.0$,

$$\epsilon = \frac{n\epsilon_p}{1 + (n - 1)\epsilon_p} \tag{12.23b}$$

where n is the number of identical passes and ϵ_p is the effectiveness of each pass, a function of N_A/n.

The curves of Fig. 12.7 show the asymptotic character of the ϵ-vs.-N_A relation. The curves become quite flat beyond $N_A \cong 3$; hence, it is usually found that the most economical exchanger of a particular flow arrangement and C_{min}/C_{max} will have a magnitude of N_A in the range of approximately 1 to 3. Figure 12.7 shows that a close approach to counterflow can be obtained by multipassing and that using more than three or four passes does not increase significantly the effectiveness. Figure 12.7(f) compares the performance of various flow arrangements for $C_{min}/C_{max} = 1.0$.

Example 12.3: Solve Example 12.2, parts (a), (b), and (c), using the charts of Fig. 12.7.

$C_{min} = C_{max} = 10{,}000$ Btu/hr F; so $C_{min}/C_{max} = 1.0$

From Eq. (12.18b) $\epsilon = 0.40$; and from Eq. (12.18c)

$$A = \frac{N_A(10{,}000)}{400} = 25N_A$$

(a) *Parallel flow.* From Fig. 12.7(b) or Eq. (12.22b), $N_A = 0.81$; so

$$A = 25(0.81) = 20.2 \text{ ft}^2$$

(b) *Counterflow.* From Fig. 12.7(a) or Eq. (12.22a), $N_A = 0.67$; so

$$A = 25(0.67) = 16.7 \text{ ft}^2$$

(c) *Crossflow, hot fluid mixed, cold fluid unmixed.* From Fig. 12.7(d), $N_A = 0.73$; so

$$A = 25(0.73) = 18.1 \text{ ft}^2$$

12.4 Prescribed Heat Flux Distribution

The heat flux at any point in the heat exchangers discussed in Secs. 12.2 and 12.3 is determined by the local temperature difference, which is itself determined by the temperature history of the two fluids at upstream points. Consider now the case in which there is an electrically heated rod along the centerline of an insulated tube through which a fluid flows, Fig. 12.8. In the figure, the lower-case a represents the surface area of the rod measured from section 1 to any section along the rod, and A represents the total surface area of the rod between sections 1 and 2. If rod diameter is uniform and the variation of electrical resistance R with temperature is neglected, the electric current I and hence, I^2R is uniform along the length; also, axial conduction is neglected. In the steady state, the heat transfer rate q is proportional to I^2R; therefore, q'', the heat transfer rate per unit surface area, is uniformly distributed.

For an incremental element of the rod with da surface area,

$$q''\,da = wc\,dT_c \tag{12.24}$$

Since q'' $(= dq/da)$, w, and c are uniform along the length, dT_c/da is a constant and hence T_c varies linearly as shown in Fig. 12.8. At any place along the rod $q'' = h(T_w - T_c)$. For a constant flow cross-sectional area, h is very

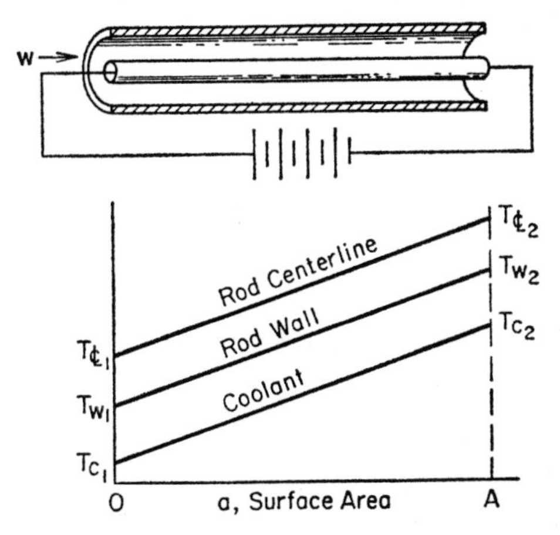

Fig. 12.8. Axial temperature distribution with uniform heat flux.

nearly uniform; then, since q'' is also uniform, $(T_w - T_c)$ is a constant along the length as shown in Fig. 12.8. The difference between centerline and surface temperature of a rod with uniformly distributed heat sources at any lengthwise position is given by Eqs. (6.34) and (6.36).

$$T_{\mathrm{¢}} - T_w = \frac{q''r_0}{2k} \tag{12.25}$$

where r_0 is the radius of the rod and k its conductivity. Again since r_0, k, and q'' are all constant, $(T_{\mathrm{¢}} - T_w)$ is constant along the length. The preceding analysis applies equally well to a nuclear fuel rod exposed to a uniform neutron flux distribution.

A simpler analysis may be performed for any prescribed variation of q'' with length. The heat transfer distribution in a nuclear reactor becomes another special case of this general problem.

A nuclear reactor with solid fuel may consist of an assemblage of fuel elements which may be in the form of solid rods, flat plates, cylinders, etc.

A coolant, which may be a gas or a liquid, passes through the reactor and receives heat transferred from these fuel elements. The neutron flux distribution in rectangular, unshielded reactors may be closely approximated by a sine curve distribution.

Consider the rod of Fig. 12.8 to be a nuclear fuel element of, say, uranium and the outer tube a duct which channels the coolant along the rod. The amount of heat generation in the fuel element is proportional to the neutron flux at any point; therefore, the heat transfer rate per unit area of fuel element ($q'' \equiv dq/da$) also has sine curve distribution or

$$q'' = q''_{\max} \sin \pi \frac{a}{A} \qquad (12.26)$$

as shown in Fig. 12.9. From the properties of a sine curve

$$q''_{\max} = 1.57 q''_{\text{avg}} \qquad (12.27)$$

Combining Eqs. (12.24) and (12.26) and integrating between $a = 0$ and a,

$$T_c - T_{c_1} = \int_0^a \frac{q''_{\max}}{wc} \sin \pi \frac{a}{A}\, da$$

$$= \frac{q''_{\max} A}{\pi wc}\left(1 - \cos \pi \frac{a}{A}\right) \qquad (12.28)$$

which is shown sketched in Fig. 12.9. At $a = A$, $T_c = T_{c_2}$; so,

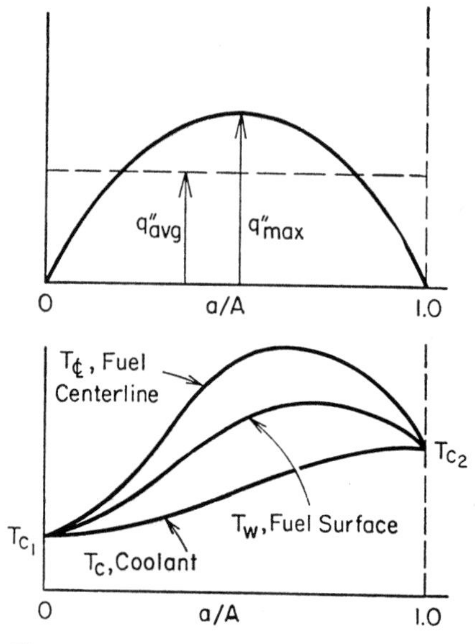

Fig. 12.9. Axial temperature distribution in nuclear reactor with cosine heat flux distribution.

$$(T_{c_2} - T_{c_1}) = \frac{2}{\pi} \frac{q''_{\max} A}{wc} \qquad (12.29)$$

Since $q'' = h(T_w - T_c)$ at any point and h is very nearly uniform along the length, the difference between the fuel surface temperature and the coolant temperature also varies sinusoidally along the length. Add $(T_w - T_c)$ to $(T_c - T_{c_1})$ from Eq. (12.28) to get

$$(T_w - T_{c_1}) = \frac{q''_{\max}}{h}\left[\frac{1}{\pi}\left(\frac{Ah}{wc}\right)\left(1 - \cos \pi \frac{a}{A}\right) + \sin \pi \frac{a}{A}\right] \qquad (12.30)$$

which is also sketched on Fig. 12.9.

Since r_0 and k are constant and q'' varies sinusoidally, Eq. (12.25) shows $(T_{℄} - T_w)$ also varies sinusoidally along the rod. Add Eqs. (12.30) and (12.25) with q'' given by Eq. (12.26). Then

$$(T_{℄} - T_{c_1}) = \frac{q''_{\max}}{h}\left[\frac{1}{\pi}\left(\frac{Ah}{wc}\right)\left(1 - \cos \pi\frac{a}{A}\right) + \left(1 + \frac{hr_0}{2k}\right)\sin \pi\frac{a}{A}\right] \quad (12.31)$$

These equations may be made dimensionless by dividing by Eq. (12.29). Then Eqs. (12.28), (12.30), and (12.31) become

$$\frac{T_c - T_{c_1}}{T_{c_2} - T_{c_1}} = \frac{1}{2}\left(1 - \cos \pi\frac{a}{A}\right) \quad (12.32)$$

$$\frac{T_w - T_{c_1}}{T_{c_2} - T_{c_1}} = \frac{1}{2}\left(1 - \cos \pi\frac{a}{A}\right) + \frac{\pi}{2}\left(\frac{wc}{Ah}\right)\sin \pi\frac{a}{A} \quad (12.33)$$

$$\frac{T_{℄} - T_{c_1}}{T_{c_2} - T_{c_1}} = \frac{1}{2}\left(1 - \cos \pi\frac{a}{A}\right) + \frac{\pi}{2}\frac{wc}{Ah}\left(1 + \frac{hr_0}{2k}\right)\sin \pi\frac{a}{A} \quad (12.34)$$

Inspection of these equations reveals the following common form:

$$Y = \frac{1}{2}\left(1 - \cos \pi\frac{a}{A}\right) + Z \sin \pi\frac{a}{A} \quad (12.35)$$

where related values of Y and Z are

Y	Z
$\dfrac{T_c - T_{c_1}}{T_{c_2} - T_{c_1}}$	0
$\dfrac{T_w - T_{c_1}}{T_{c_2} - T_{c_1}}$	$\dfrac{\pi}{2}\dfrac{wc}{Ah}$
$\dfrac{T_{℄} - T_{c_1}}{T_{c_2} - T_{c_1}}$	$\dfrac{\pi}{2}\dfrac{wc}{Ah}\left(\dfrac{1 + hr_0}{2k}\right)$

This equation is shown plotted in Fig. 12.10 and represents for $Z = 0$ the coolant temperature distribution, and for particular values of Z, the fuel element surface and centerline temperatures. The magnitude of Z for the centerline is greater than for the surface. The three temperature curves sketched in Fig. 12.9 become a part of the family of curves in Fig. 12.10.

The curves for centerline and surface temperature of the fuel element pass through maximum values. Differentiate Eq. (12.35) with respect to a/A and equate to zero to obtain $\pi a/A = \tan^{-1}(-2Z)$ at $Y_{\max}$. Then

$$Y_{\max} = \tfrac{1}{2}[1 + \cos \tan^{-1}(2Z)] + Z \sin \tan^{-1}(2Z) \quad (12.36)$$

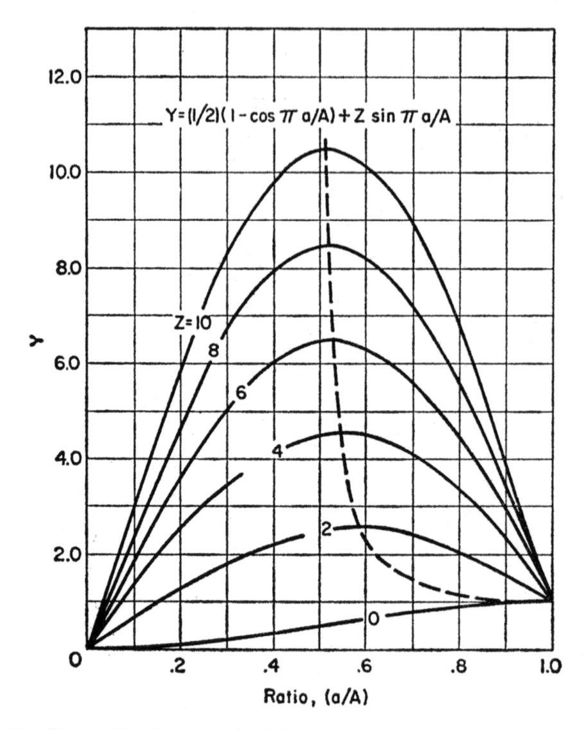

Fig. 12.10. Generalized graph of axial temperature distribution with cosine heat flux distribution.

These maxima impose limiting conditions on the power level of a nuclear reactor. For example, the maximum permissible temperature which the fuel element can withstand places a restriction on the maximum centerline temperature. For uranium this could be placed at around 1220 F where the material changes from the α to the β phase. The use of a liquid coolant may limit the maximum fuel surface temperature to the boiling-point temperature at the particular coolant pressure selected.

Example 12.4: A nuclear fuel element shown in Fig. 12.8 has $k = 12$, $r_0 = 0.125$ in., and is 4 ft long. The coolant is water ($c = 1$) and is flowing at a rate of 400 lb/hr per fuel element. It enters at 200 F, and the fuel element temperature should not exceed 900 F. At the fuel surface $h = 2000$ Btu/hr ft² F. Calculate the maximum exit coolant temperature, the total heat transferred from the fuel element, and the minimum allowable coolant pressure to prevent boiling of the coolant.

Solution: Area $A = \pi(0.25)\frac{4}{12} = 0.262 \text{ ft}^2$

$$\frac{hr_0}{2k} = \frac{2000(0.125)}{2(12)(12)} = 0.868; \frac{wc}{Ah} = \frac{400(1)}{(0.262)(2000)} = 0.762$$

$Z_{\mathbb{C}} = (\pi/2)(0.762)(1.868) = 2.24$. Then $Y_{\mathbb{C}_{max}} = 2.79$ from Eq. (12.36) or

$$\frac{900 - 200}{T_{c_2} - 200} = 2.79$$

so $T_{c_2} = 451$ F.

$$q = wc(T_{c_2} - T_{c_1}) = 400(251) = 100{,}400 \text{ Btu/hr per rod}$$

$Z_w = (\pi/2)(0.762) = 1.20$; $Y_{w_{max}} = 1.80$;

$$\frac{T_{w_{max}} - 200}{451 - 200} = 1.80$$

so

$$T_{w_{max}} = 651 \text{ F}$$

From the steam tables, saturation pressure at 651 F is 2210 psia.

12.5 Heat Exchanger Design

Heat exchanger design calculations are mainly of the following types: (i) for a given heat exchanger predict its performance (pressure drops and exit temperatures or total heat transfer rate) for specified flow rates and inlet temperatures of particular fluids; (ii) determine the geometry of a particular type heat exchanger for specified flow rates, temperatures, and pressure drops for particular fluids; or (iii) select from a series of these calculations the optimum design based on minimum cost, minimum weight, or minimum volume. In some cases design is restricted by a space requirement; often in aircraft applications one limiting space dimension governs the entire design. The relations needed include equations expressing heat transfer, pressure drop, and geometry.

The heat transfer equations involved are the equations of this chapter plus the correlation equations of Secs. 8.8, 8.11, 7.3, 7.7, 4.4, 4.5, and 4.9 for flow through and across tubes. In addition, test data (2) are available for a wide variety of plate-fin exchangers. Data for only one representative type are shown in Fig. 12.11, where the ordinate gives the Colburn j factor Eq. (16.17a) for heat transfer and the friction factor f.

Data for heat transfer and pressure drop in baffled shells [Fig. 12.1(c) (d)(e)] have been collected and correlated by Tinker (3). Details are omitted here.

Example 12.5: A flow rate of 50,000 lb/hr of air at 60 F and 1 atm is to be heated to 160 F in a shell-and-tube exchanger by condensing steam in the shell at 220 F, Fig. 12.1(b), with 2-in. I.D. steel tubes. Determine the size of the heat exchanger assuming $G_a = 7000$ lb/hr ft^2, and calculate the air side pressure drop.

For air at 110 F, $c_p = 0.24$, $\mu = 0.047$ lb/ft hr, $k = 0.016$, Pr $= 0.71$. The required number of tubes, n, is calculated from

$$n\frac{\pi D_i^2}{4} = \frac{w_a}{G_a} \quad \text{or} \quad n = \frac{50{,}000}{7000}\frac{4/\pi}{(\frac{2}{12})^2} = 328$$

$$\text{Re}_i = \frac{GD_i}{\mu} = \frac{7000(\frac{2}{12})}{(0.047)} = 24{,}800 \text{ (turbulent)}$$

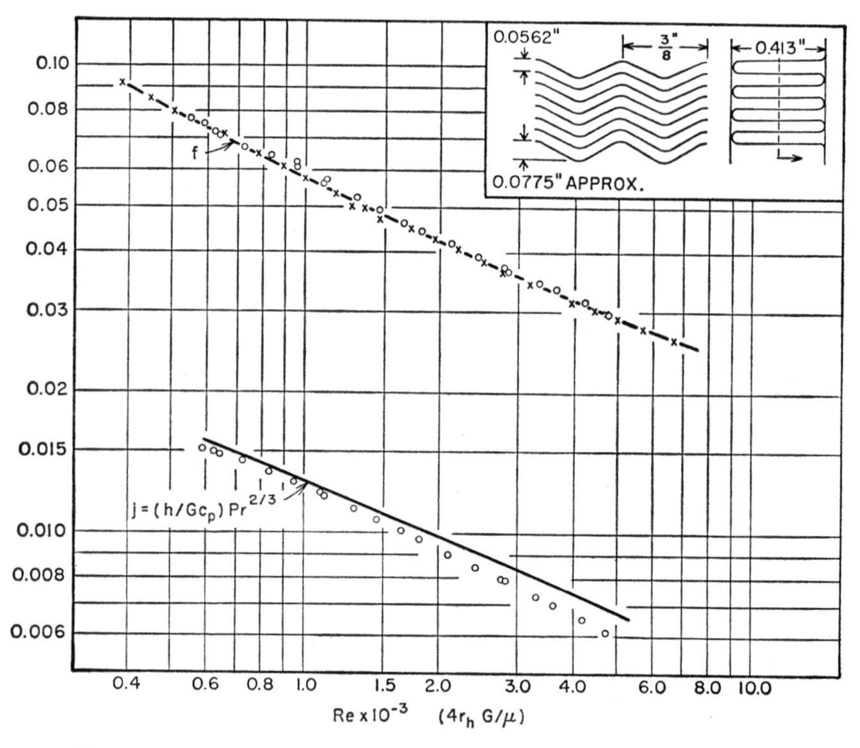

Fig. 12.11. f and j_{max} for compact heat exchanger surface [Kays *et al.* (2)].

Wavy-Fin Plate-Fin
Surface 17.8: 3/8*W*

Fin pitch: 17.8 per inch
Plate spacing: 0.413 in.
Flow passage hydraulic diameter: $4r_h$ 0.00696 ft
Fin metal thickness: 0.006 in.
Total heat transfer area/volume between plates: 514 ft²/ft³
Fin area/total area: 0.892

Note: Hydraulic diameter based on free-flow area normal to mean flow direction.

From Eq. (8.34),

$$h_a = 0.023\frac{0.016}{(\frac{2}{12})}(24{,}800)^{0.8}(0.71)^{0.4} = 6.15 \text{ Btu/hr ft}^2 \text{ F}$$

All other resistances are negligible compared with $(1/h_a)$ so $U_i \cong h_a$.

$$\Delta T_{lm} = \frac{(220 - 60) - (220 - 160)}{\ln (220 - 60)/(220 - 160)} = 102 \text{ F}$$

$$A_i = \frac{q}{U\,\Delta T_{lm}} = \frac{50{,}000(0.24)(160 - 60)}{(6.15)(102)} = 1910 \text{ ft}^2$$

$$A_i = \pi D_i L n; \qquad L = \frac{1910}{\pi(\frac{2}{12})(328)} = 11.1 \text{ ft}$$

From Eq. 4.7, $f = 0.046/(24{,}800)^{0.2} = 0.0059$. Then, as in Example 4.3,

$$\rho_1 = \frac{p}{RT} = \frac{14.7(144)}{(53.35)(520)} = 0.0763 \text{ lb/ft}^3$$

and

$$\rho_2 = 0.064 \text{ lb/ft}^3$$

then

$$\rho_m = 0.0696$$

From Eq. (4.24) for horizontal tubes, with $\alpha = 1$,

$$p_1 - p_2 = \left(\frac{7000}{3600}\right)^2 \frac{1}{32.2}\left[\left(\frac{1}{0.064} - \frac{1}{0.0763}\right) + 4(0.0059)\frac{11.1}{(\frac{2}{12})}\frac{1}{2(0.0696)}\right]$$

$$= 14.9 \text{ lb/ft}^2$$

Assume the area of the header is twice the flow area of the tubes,

$$\frac{A_1}{A_0} = \frac{A_2}{A_3} = 0.5$$

$V_1 = (7000/3600)(0.0763) = 25.5$ ft/sec; $V_2 = 30.4$, $V_0 = 12.7$, $V_3 = 15.2$ ft/sec. From Fig. 4.9, $K_e = 0.21$, $K_c = 0.31$.

Then from Eqs. (4.27) and (4.28),

$$p_0 - p_1 = \frac{0.0763}{2(32.2)}[0.31(25.5)^2 + (25.5)^2 - (12.75)^2] = 1.20 \text{ lb/ft}^2$$

$$p_2 - p_3 = \frac{0.064}{2(32.2)}[0.21(30.4)^2 + (15.2)^2 - (30.4)^2] = -0.50 \text{ lb/ft}^2$$

Therefore

$$p_0 - p_3 = 14.9 + 1.20 - 0.50 = 15.6 \text{ lb/ft}^2$$

In designing for a particular pressure drop limitation, the calculation usually follows the procedure above, selecting various magnitudes of G_i and interpolating to the desired pressure drop.

REFERENCES

1. Bowman, R. A., A. C. Mueller, and W. M. Nagle, *Trans. ASME*, **62,** 283–94 (1940).

2. Kays, W. M., and A. L. London and D. W. Johnson, *Gas Turbine Plant Heat Exchangers*, ASME, New York, 1951.

3. Tinker, T., *General Discussion of Heat Transfer*, p. 89, ASME, London, 1951; also *Trans. ASME*, **80,** 36–60 (Jan. 1940).

PROBLEMS

12.1 (a) Show that for a counterflow exchanger with $w_c c_c = w_h c_h$, $\Delta t_{lm} = \Delta t_a = \Delta t_b$.

(b) Show for this case that the temperature distribution curves of Fig. 12.4d are straight parallel lines.

12.2 (a) Derive Eqs. (12.21a) and (12.21b) for the case in which $C_c < C_h$.

(b) Derive Eq. (12.22a) from Eqs. (12.8), (12.13), and (12.18) and also from Eq. (12.21a) by the calculus of limits (L'Hopital's rule).

12.3 Water is evaporated continuously at 212 F in an evaporator by cooling 1000 lb per hour of air from 500 F to 300 F. The over-all heat transfer coefficient is 8. Calculate the heat transfer area required and the pounds of steam evaporated per hour if the liquid enters at 212 F.

12.4 Hot gas is used to heat cold water continuously at a rate of 10,000 lb per hr in an adiabatic heat exchanger. The gas enters at 1000 F and leaves at 400 F. The water enters at 45 F and leaves at 120 F. The over-all coefficient of heat transfer is 15 Btu/hr ft^2 F. Calculate the heat transfer area of the heat exchanger for:

(a) parallel flow, and

(b) counterflow.

12.5 A shell-and-tube heat exchanger is composed of 400 steel tubes (1.050 in. O.D., 0.824 in. I.D.). Steam condenses in the shell at 220 F (h = 900), and 250,000 lb/hr of water on the inside is heated from 100 F to 200 F (h = 300). A scale factor of h_{sc} = 1000 on the inside of the tubes is assumed.

(a) Calculate the over-all coefficient of heat transfer based on the inside area.

(b) Calculate the required length of the tube bundle.

12.6 A parallel-flow heat exchanger is operating at the conditions shown in the sketch. The properties for both streams (water) are: $\rho = 62.4$ lb/ft³, $c_p = 1$ Btu/lb F, $k = .34$ Btu/hr ft F, $\mu = 2.71$ lb/hr ft. If the heat transfer surface is made of thin-walled tubes, the heat transfer resistance of which is negligible, and the heat transfer coefficient on both sides is 100 Btu/hr ft² F, how much area is necessary in the heat exchanger?

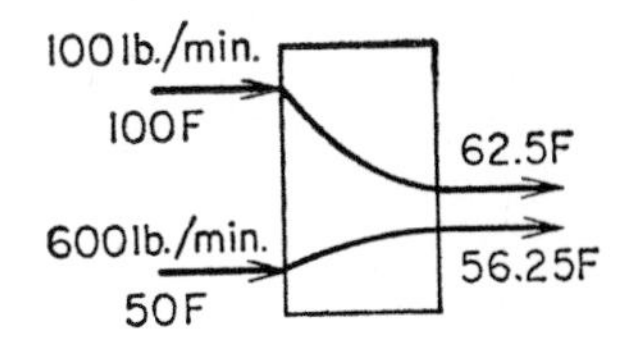

The heat transfer coefficients vary with the velocity (or flow-rate) according to Eq. (8.35) with $d = 0.05$ ft on each side, and the properties are constant with temperature. What are the exit temperatures

(a) if the hot-fluid flow rate is doubled?

(b) if the flow rates of both fluids are doubled?

12.7 An existing shell-and-tube exchanger cools 100,000 lb/hr of oil ($c = 0.6$) from 200 F to 100 F by a counterflow of 80,000 lb/hr of water entering at 70 F. It is proposed to build another heat exchanger of the same dimensions but longer, in order to cool the same flow of oil down to 80 F with the same water and water flow. Calculate the ratio of the lengths of the two tube bundles.

12.8 A heat exchanger for a gas turbine is to be designed to cool turbine exhaust air from 800 F to 520 F by an equal flow rate of air entering at 200 F. The gas velocities will be large enough to result in a $U = 4$. Calculate the heat transfer area required for each of the following flow arrangements: (a) parallel flow, (b) counterflow, (c) crossflow, hot gas making one pass through shell, cold gas making one pass through tubes ($w = 10$ lb/sec).

12.9 Oil with $c = 0.7$ Btu/lb F is cooled from 200 F to 120 F at a flow rate of 8000 lb/hr in a heat exchanger of Fig. 12.7 (E). The cooling water enters at 60 F and leaves at 140 F and makes two passes through the tubes. The oil makes one pass through the shell. Each tube bundle has 80 tubes of 0.50 in. I.D., and the over-all coefficient is $U = 1800$. Calculate (a) the cooling water flow rate and (b) the length of each bundle of tubes. Perform calculations by the method of Sec. 12.3.

12.10 A heat transfer test loop is composed of stainless steel ($k = 12$) tubing (0.375 in. O.D., 0.275 in. I.D.) with water circulating at 5 ft/sec at a pressure of 1500 psia. It is found that the water entering the test section is not hot enough. Someone suggests passing an electric current through the tube material itself to add heat to the water. If the length available for such an operation is 3 ft and all of the heat generated in the tube wall is transferred to the water, what is the maximum temperature rise of the water from 200 F initially if the stainless steel is to be limited in temperature to 800 F and the surface temperature is limited to the boiling point of the water? For this flow $h = 3000$.

12.11 Suppose a nuclear reactor has fuel elements in the form of large parallel flat plates of uranium with a thickness of x_0. Derive Eqs. (12.32), (12.33), and (12.34) for this case.

12.12 A nuclear reactor with heat flux distribution as shown in Fig. 12.9 is to be designed with uranium rods ($k = 12$) of diameter 0.375 in. and length 10 ft. The cooling water at a pressure of 1000 psia enters at 100 F with a flow rate of 1000 lb/hr and $h = 3000$. If the uranium temperature is not to exceed 1200 F and the rod surface temperature is not to exceed the boiling point, what maximum temperature rise of the water can be obtained for a flow of 1000 lb/hr per rod?

12.13 A single-pass shell-and-tube heater is to be designed in which 50,000 lb/hr of *n*-butyl alcohol will be heated from 70 to 120 F by steam condensing at 215 F in a well-baffled shell. The preliminary design will be based on 1-in. standard pipes. The following data apply:

Cost of pipes	\$0.075 per foot
Installation cost of pipes	\$1.00 per tube
Cost of shell	\$20.00 per ft of length of tube bundle
Fixed charges per year (interest, repairs, depreciation, insurance, taxes, etc.)	30% of first cost
No. of hours of operation per year	3500
Cost of power	\$0.02 per kw hr
Efficiency of pump and motor based on work required to pump the butyl alcohol reversibly and adiabatically	52%
Sufficient motor and pump capacity is available so that new pumping equipment will not have to be bought	
Specific gravity of the alcohol	0.81

For this initial design, use properties of the alcohol at the arithmetic mean temperature, assume the thermal resistance of the metal and of the film on the steam side to be negligible, and take the lost head due to end effects as 40% of the lost head due to friction in the pipes. On the basis of a cost analysis, recommend:

(a) number of tubes in parallel

(b) length of tubes

(c) velocity through tubes (ft/sec)

(d) mass velocity through tubes (lb/hr ft^2).

12.14 In a shell-and-tube heat exchanger it is desired to heat 10,000 ft^3/min of air (measured at 70 F and 1 atm) from 70 to 205 F by passing the air inside

1-in. std. steel pipes which are heated on the outside by steam condensing at 220 F. The fan available will deliver the cold air to the entrance header at 70 F and a static pressure of 1.5 in. of water gage. The hot air is to be discharged to the atmosphere. The cross-sectional areas of the entrance and exit headers will be twice the total internal cross-sectional area of the tubes. Assuming $U_i \simeq h_i$, calculate:

(a) the changes in static pressure for the entrance and exit headers

(b) the mass velocity of the air inside the tubes

(c) the heat transfer coefficient, h_i, from inside pipe wall to air

(d) the number of tubes and the length of each tube.

12.15 Design a heat exchanger for air-to-air use using surface 17.8–$\frac{3}{8}W$ on both sides. The heat transferred is 1000 Btu/min. One air stream enters at 500 F and leaves at 400 F. The other stream enters at 100 F. The pressure drop on the hot side is not to exceed 2 in. H_2O, this being the available pressure from a fan (whose head vs. flow curve is flat in this region of flow). The hot and cold air streams both exit at atmospheric pressure (14.7 psia).

(a) Calculate the size (volume) of the heat exchanger.

(b) How could the volume be reduced?

12.16 CO_2 from a gas-cooled reactor at Calder Hall in Great Britain is used to generate steam. A flow rate of 200,000 lb/hr of CO_2 at 60 psia enters the tubes of a shell- and-tube steam generator at 900 F. The CO_2 leaves the generator at 600 F, and the steam saturation temperature is 450 F. Using 1 in. I.D. copper tubes (0.08 in. wall thickness) and designing for a CO_2 mass flow rate of 80,000 lb/ft^2 hr, calculate

(a) the length and the number of tubes to be used, neglecting the steam side resistance

(b) CO_2 pressure drop in the tube bundle.

(c) If only half the number of tubes is used, calculate the ratio of the new tube length to the old length if the same amount of heat is to be transferred and the total CO_2 flow rate is held constant. Also, find the ratio of the new pressure drop to the old pressure drop. Properties of CO_2: $c_p = 0.28$ Btu/lb F, $k = 0.025$ Btu/hr ft F, $\mu = 0.072$ lb/hr ft, $\rho = 0.203$ lb/ft^3.

12.17 A shell-and-tube heater in an ammonia plant is preheating 40,000 cu ft of pure nitrogen per hour (measured at 1 atm abs and 70 F) from 70 to 150 F, using steam condensing at 20 psia. The mass velocity of the nitrogen through the 1.00-in. I.D. heated tubes is 10,000 lb/hr ft^2.

It has been decided to change over from ammonia synthesis to methanol synthesis. This heater would then be used to preheat CO_2 from 70 to 170 F, using process steam condensing at 35 psia. What output can be expected of the heater, in pounds of CO_2 per hour?

12.18 A large surface condenser in a power plant was tested at three water velocities, when new (with clean tubes) and again after considerable service. The results are tabulated below, as over-all coefficients of heat transfer, based on the outside surface of the tubes. The latter have outside and inside diameters of 1.00 and 0.902 in., respectively, and the metal of which they are made has a thermal conductivity of 63.

Condition of tubes	Clean			Dirty		
Water velocity, ft/sec	2.0	4.0	8.0	2.0	4.0	8.0
Over-all coefficient, U_i	357	550	795	293	410	534

Plot the data as $1/U_i$ vs. $1/V^{0.8}$ on log-log paper, and determine the following individual coefficients of heat transfer, expressed as Btu/hr ft² F. (This is known as the Wilson method.)

(a) Value of h on the steam side of the clean condenser, based on the *outside* surface.

(b) Value of h for the scale and slime deposited in the dirty tube, based on the *inside* surface.

(c) Value of h on water side at a water velocity of 4 ft/sec.

(d) What over-all coefficient would you expect with dirty tubes at a water velocity of 25 ft/sec?

12.19 Consider a shell-and-tube air heater with condensing saturated steam in the tube bundle. The following equations are applicable:

$$q = wc(T_2 - T_1)$$

$$q = U(\pi dLn)\,\Delta T_{lm}$$

$$U \simeq h_i$$

$$\frac{h_i d}{k} = 0.023\left(\frac{Gd}{\mu}\right)^{0.8} \mathrm{Pr}^{0.4}$$

$$G = \frac{4w}{\pi d^2 n}$$

$$\Delta p_t = 4f \frac{L}{d}\frac{G^2}{2\rho}$$

$$f = \frac{0.046}{\left(\dfrac{Gd}{\mu}\right)^{0.2}}$$

Below is a list of independent variables and some dependent variables. *When one independent variable is changed, all other independent variables are assumed to remain constant.*

Independent Variables	*Dependent Variables*
1. Inlet temperature of the air	A. Rate of heat transfer
2. Outlet temperature of the air	B. Total heat transfer area
3. Tube diameter	C. Mass velocity of air
4. Wall thickness of tubes	D. Over-all coefficient of heat transfer
5. Mass rate of flow of air	E. Pressure drop through tubes
6. Pressure of steam	F. Ratio of temperature rise of air to log mean temperature difference
7. Total cross-sectional area of tubes	G. Cost of installing the tubes at a fixed charge per tube
8. Surface coefficient on steam side	H. Cost of installing the heater at a fixed charge per foot of length
9. Thermal conductivity of tube material	I. Cost of the tube material at a fixed charge per pound
10. Density of tube material	

For constant values of 1, 3, 4, 5, 6, 7, 8, 9, 10		Increases	Remains the same	Decreases
If 2 is increased,	A	_____	_____	_____
	B	_____	_____	_____
	C	_____	_____	_____
	D	_____	_____	_____
	E	_____	_____	_____
	F	_____	_____	_____
	G	_____	_____	_____
	H	_____	_____	_____
	I	_____	_____	_____

For constant values of 1, 2, 3, 4, 5, 7, 8, 9, 10		Increases	Remains the same	Decreases
If 6 is increased,	A	_____	_____	_____
	B	_____	_____	_____
	C	_____	_____	_____
	D	_____	_____	_____
	E	_____	_____	_____
	F	_____	_____	_____
	G	_____	_____	_____
	H	_____	_____	_____
	I	_____	_____	_____

CHAPTER 13

Radiant Heat Transfer

Solid bodies, liquids, and some gases emit radiant energy which may be in many forms such as x-rays resulting from high-speed electron bombardment of a metal plate, gamma rays emitted from radioactive material, ultraviolet light emitted from an electric discharge in a gas, thermal radiation resulting from the thermal excitation of the molecules, and other forms. Modern theory describes the nature of this radiation in terms of electromagnetic waves which all travel at the velocity of light. The various forms of radiation differ only in wavelength, Fig. 13.1. In this chapter discussion will be confined to thermal radiation. The emission of this energy depends upon the temperature and character of the emitting body.

13.1 Emission and Absorption Characteristics

The emission and absorption characteristics of a body may be treated independently from each other but are related under certain circumstances.

For a surface exposed to a quantity of radiant energy e, Btu/hr ft^2, the absorptivity α, the transmissivity τ, and the reflectivity ρ are respectively the fractions of e which are absorbed, transmitted, and reflected. Then

$$\alpha + \tau + \rho = 1 \tag{13.1}$$

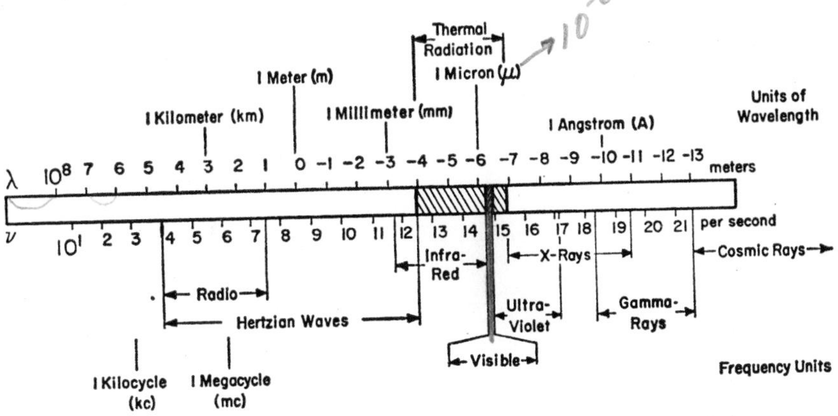

Fig. 13.1. Electromagnetic wave spectrum.

The part of the radiant energy which is absorbed or emitted by materials which are electric conductors is all completely absorbed or emitted in a layer of approximately 0.00005 in. below the surface. In electric nonconductors this layer is of the order of 0.05 in. thick. Because this region of influence is so small, the emissive and absorptive properties at a surface are not seriously influenced by strong temperature gradients at and normal to the surface. For most industrial applications liquids and solid bodies are almost always thick enough to be opaque to radiation ($\tau = 0$), so $\rho = 1 - \alpha$.

The reflection from a highly polished surface may be *specular*, Fig. 13.2(a); from a rough surface it is usually *diffuse*, Fig. 13.2(b), as illustrated.

In general, unoxidized surfaces of good electric conductors have low values of α, and surfaces of poor conductors and oxidized surfaces have high values. *A black body is defined as one which absorbs all oncoming (or incident) radiation,* $\alpha = 1$, $\rho = 0$. The concept of a black body is useful because the laws governing its radiation are simple and many real bodies may be idealized as black bodies. Although no bodies in nature are black bodies, many industrial materials have very high values of α, and hence approach black-body conditions. The term *black* is used because if a surface

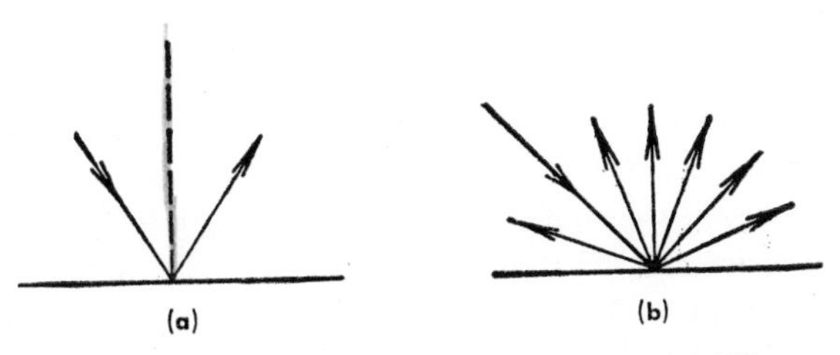

Fig. 13.2. Reflecting surfaces: (a) specular; (b) diffuse.

actually does absorb all radiant energy falling on it, the surface will appear black to the eye; this, of course, ignores the emitted radiation. Some surfaces absorb nearly all incident radiation yet do not appear black to the eye because they do not absorb all visible light rays. Indeed, freshly fallen snow, soot, and whitewashed walls all have absorptivities greater than 0.95 for thermal radiation.

Black bodies also emit radiation at all wavelengths, and the total intensity e_b (Btu/hr ft^2) depends only on the surface temperature. Then even without reference to any solid surface, black-body radiation of intensity e_b has an associated temperature. Further, since the radiation travels at finite velocity, an enclosed space may be imagined to be "filled" with radiation.

Imagine a region of space in which black-body radiation streams in all directions. For a non-black body 1 exposed to this black radiation, the difference between the energy emitted by the body, e_1, and the energy absorbed, $\alpha_1 e_b$, represents the net rate of heat transferred to the body by radiation or

$$\left(\frac{q}{A}\right)_{\text{net in}} = \alpha_1 e_b - e_1 \tag{13.2}$$

If the temperature of body 1 is the same as that of the black body from which the black radiation originated, the $q_{\text{net}} = 0$ or $e_1 = \alpha_1 e_b$. A similar relation exists for any other non-black body in thermal equilibrium with black radiation. Then

$$e_b = \frac{e_1}{\alpha_1} = \frac{e_2}{\alpha_2} = \frac{e_3}{\alpha_3} = \cdots \tag{13.3}$$

This generalization is known as *Kirchhoff's law*.

A similar type of reasoning can show that all black bodies at a particular temperature must emit radiant energy at the same rate. If body 1 in the previous discussion is a black body ($\alpha_1 = 1.0$), then at equilibrium $e_b = e_1$. If e_1 were to be greater than e_b, there would result a spontaneous cooling of body 1 by transfer of heat from the lower temperature to the higher black-body radiation temperature. Since the Second Law of thermodynamics shows this to be impossible, e_1 cannot be greater than e_b. In a similar way, it is readily shown that e_1 cannot be less than e_b at thermal equilibrium when $\alpha_1 = 1$. Therefore all black bodies have the same emissive power e_b at any particular temperature. A black body then has the maximum emissive power that any body can have at a particular temperature since α cannot be greater than unity; therefore, a black body, in addition to being a perfect absorber, may be considered to be a perfect emitter of radiant energy.

The emissive characteristics of a non-black body may be quite different from its absorptive characteristics. The emissive characteristic at a particular temperature may be represented by the *emissivity* ϵ defined as the ratio of actual rate of energy emission to the rate of energy emission of a black body *at the same temperature*. Hence

$$e = \epsilon e_b \tag{13.4}$$

The quantity ϵ is more properly called *total hemispherical emissivity* to distinguish it from monochromatic emissivity, ϵ_λ, and directional emissivity, ϵ_ϕ.

The reasoning which led to Eq. (13.3) also shows that at thermal equilibrium $e_1 = \epsilon_1 e_b = \alpha_1 e_b$ or $\epsilon_1 = \alpha_1$; similarly $\epsilon_2 = \alpha_2$, etc. This constitutes a restatement of Kirchhoff's law: at thermal equilibrium the absorptivity and emissivity of a body are the same.

Radiation from any surface actually streams from the surface in all directions of a hemisphere. The quantity e is known alternatively as *emissive power, emittance, total hemispherical radiation intensity*, or *radiant-flux density*. In discussing radiant transfers within enclosures, it is often convenient to employ the concept of fictitious rays which represent the total hemispherical radiation streaming to or from a surface or being absorbed, transmitted, or reflected at a surface. These rays aid in a pictorial representation of the process and hence in the analysis for determining the net rates of heat transfer.

13.2 Laboratory Black Bodies

A black body may be approximated by a hollow space with walls at a uniform temperature and a small hole, Fig. 13.3. Clearly, only a very small fraction of the energy entering the hole and being diffusely reflected inside at the nonblack walls will escape; hence the cavity with a small hole acts very nearly like a black body. This is perhaps easier to visualize with specular reflection.

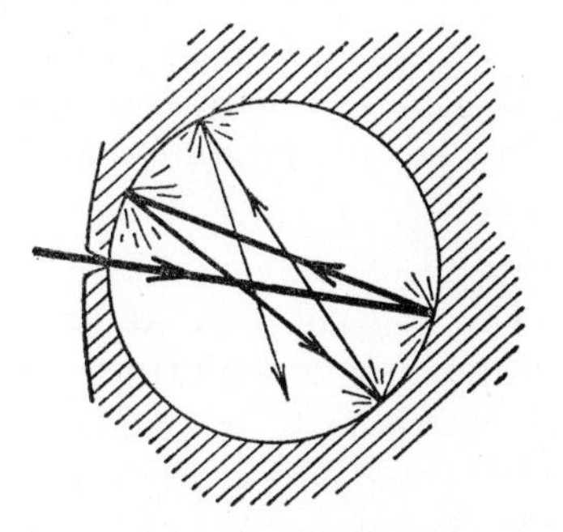

Fig. 13.3. Hohlraum.

The energy *emitted* from a portion of a surface in the cavity is ϵe_b. After one reflection this quantity is $\rho\epsilon e_b$, after two reflections, $\rho^2\epsilon e_b$, etc. Imagine the radiation leaving the hole of this cavity to be composed of rays which have been directly radiated, reflected once, reflected twice, etc. Then the energy emitted from the hole is

$$e = \epsilon e_b(1 + \rho + \rho^2 + \rho^3 + \cdots) = \epsilon e_b \frac{1}{1 - \rho} = e_b$$

since at a particular temperature $\epsilon = \alpha = 1 - \rho$. The energy emitted from the hole is black-body radiation. Other forms of black bodies are small-angle wedges, which are useful as laboratory black bodies, e.g., a 10-deg included-angle, wedge-shaped cavity has an emissivity of $\epsilon = 0.99$.

13.3 Spectral Energy Distribution

It is a characteristic of thermal radiation from a solid body that when it is dispersed by being passed through a prism, a continuous spectrum is formed with an energy distribution as shown in Fig. 13.4 (1). This is the distribution of black-body radiation. The area under any constant temperature curve is the total rate of energy emission per unit area given by

$$e_b = \int_0^\infty e_{b\lambda}\, d\lambda \tag{13.5}$$

which defines $e_{b,\lambda}$, the monochromatic emissive power. Planck's equation (2) expresses $e_{b,\lambda}$ as a function of temperature T and wavelength λ as (see Sec. 13.12)

$$e_{b,\lambda} = \frac{2\pi hc^2\lambda^{-5}}{e^{ch/\kappa\lambda T} - 1} = \frac{3.74041 \times 10^{-5}\lambda^{-5}}{e^{1.43868/\lambda T} - 1} \tag{13.6}$$

where c = velocity of light, 2.997902×10^{10} cm/sec; h = Planck's constant, 6.62377×10^{-27} erg-sec; κ = Boltzmann's constant, 1.38026×10^{-16} erg/K; T = temperature in K; λ = wavelength, cm; and $e_{b,\lambda}$, ergs/sec cm^3. The wavelength of maximum $e_{b,\lambda}$ is inversely proportional to the absolute temperature or $\lambda_{max}T = 0.2898$ cm K (Wien's displacement law). These values were all assembled by Snyder (3).

The emissive power e_b of a black body may be shown by the Second Law of thermodynamics to be proportional to the fourth power of the absolute temperature, and also by integrating Eq. (13.5) with Eq. (13.6),

$$e_b = \sigma T^4 \tag{13.7}$$

This is known as Stefan-Boltzmann's law, and the Stefan-Boltzmann constant is $\sigma = 0.1713 \times 10^{-8}$ Btu/ft^2 hr R^4.

Aschkinass (4) found that the emissive power e for bright metal surfaces was more nearly proportional to the fifth instead of the fourth power of absolute temperature. This results from the variation of ϵ_λ with wavelength. Actually, it is usually adequate to treat engineering problems in terms of total radiation expressed as the fourth power of the temperature, accounting for deviations from this true black-body relation by allowing the resulting total emissivity, absorptivity, or transmissivity to be a somewhat weak function of temperature.

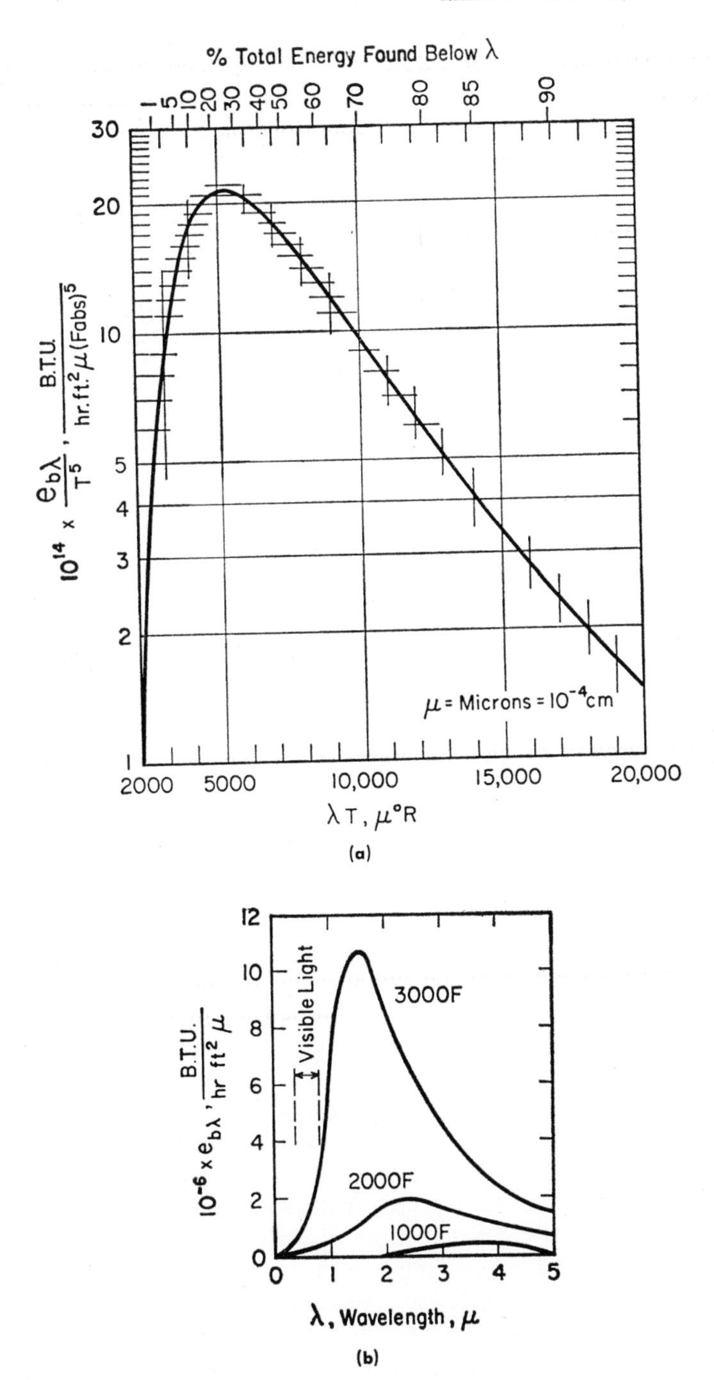

Fig. 13.4. Spectral energy distribution of a black body.

The total hemispherical emissivity ϵ varies with temperature, degree of roughness, and, if a metal, degree of oxidation. The directional emissivity ϵ_ϕ is nearly independent of direction ϕ except in the case of well-polished metals.

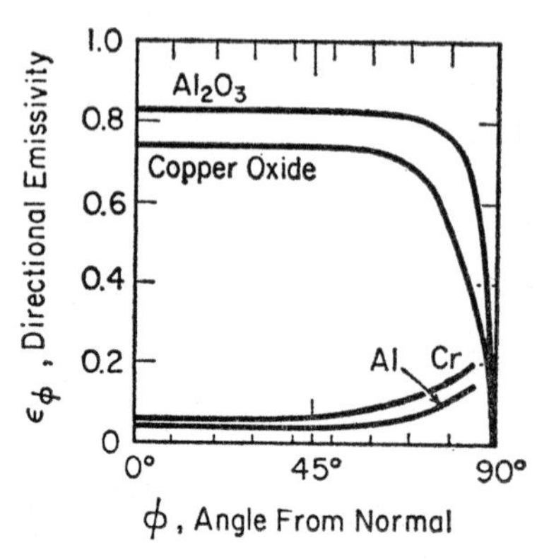

Fig. 13.5. Variation of directional emissivity of metals and oxides [Schmidt and Eckert (14)].

Figure 13.5 shows how ϵ_ϕ varies for some metals and metal oxides. For the well-polished metals, ϵ may be 15 to 20 per cent higher than ϵ_φ perpendicular to the surface.

Emissivities of pure metal surfaces are very low and nearly proportional to the absolute temperature, the proportionality constant varying nearly as the square root of the electrical resistance at a standard temperature (5). Even the slightest amounts of oxide on the metal surfaces cause the emissivities to rise to many times their value for polished pure metal surfaces. Emissivities of nonconductors are usually much higher than those of conductors and usually decrease as temperature increases. For most nonmetals emissivities at near room temperature are above 0.8, and for some materials may drop to as low as 0.3 at temperatures around 2000 F.

The absorptivity of a surface depends on those factors which affect the emissivity and, in addition, on the character of the oncoming radiation as determined by its spectral distribution. Hence two subscripts (6) should be assigned to α, the first to represent the temperature of the receiving surface and the second, that of the oncoming radiation. From Kirchhoff's law, for black incident radiation from a source at the same temperature as the surface, the emissivity ϵ equals the absorptivity $\alpha_{1,1}$. The Kirchhoff law is applicable also to the radiation at each wavelength.

The monochromatic absorptivity α_λ may vary considerably with wavelength; for example, ice is very transparent to radiation at wavelengths in the visible range but is almost a perfect absorber in the infrared zone of the spectrum.

Figure 13.6 shows how α varies with source temperature for black radiation. For nonmetals α_λ varies greatly with wavelength but not very much with temperature; then total absorptivity $\alpha_{1,2}$ varies more with T_2 than with T_1. Absorptivity $\alpha_{1,2}$ for metals has been found (7) to increase nearly linearly with $\sqrt{T_1 \cdot T_2}$.

If α_λ is a constant (independent of λ) the surface is said to be a *gray surface*. Then the total absorptivity α will be independent of the spectral

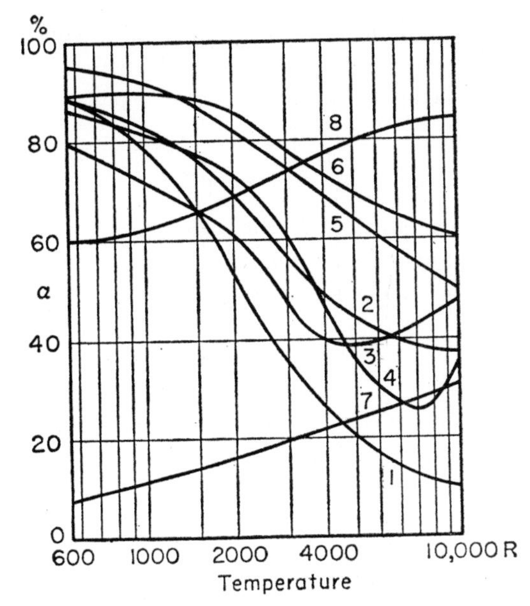

Fig. 13.6. Variation of total absorptivity of different materials for incident black radiation at the indicated temperatures [Sieber (15)]: (1) white fire clay; (2) asbestos; (3) cork; (4) wood; (5) porcelain; (6) concrete; (7) aluminum; (8) graphite.

distribution of the incident radiant energy and $\alpha_{1,2} = \alpha_{1,1} = \epsilon_1$. Then for gray surfaces $\epsilon = \alpha$ even though the temperatures of the incident radiation and of the receiving surface are not the same. The evaluation of radiant heat transfer between non-gray surfaces may be accomplished properly only by considering the radiation at each wavelength (monochromatic) and integrating over the entire spectrum. Fortunately, many industrial materials, especially those used in furnace construction, behave substantially as gray radiators.

Example 13.1: A glass is transparent to black-body radiation in the range 2.0 to 10.0 microns and may be considered opaque to the remainder of the radiation. What is its transmittance at 1200 F? Assume $\rho = 1.0$.

Solution: 1200 F = 1660 R. From Fig. 13.4, below $\lambda T = 2.0 \times 1660 = 3320$, 4.5 per cent of the total energy is absorbed. Above $\lambda T = 10 \times 1660 = 16{,}600$, 10.5 per cent of total energy is absorbed. So, transmittance is

$$1 - 0.045 - 0.105 = 0.85$$

13.4 Geometry Factor

Consider the radiant energy leaving black surface 1 and arriving at black surface 2 (Fig. 13.7) with a nonabsorbing medium between the surfaces. This will be proportional to the apparent areas ($dA \cos \phi$) of each surface as viewed from the other surface and inversely proportional to r^2 since the radiation is emitted throughout a hemisphere. Then

$$d^2E_{12} = I\frac{dA_1 \cos \phi_1 \, dA_2 \cos \phi_2}{r^2} \tag{13.8}$$

where I is the proportionality factor which is actually defined by this equation. The solid angle $d\omega_1$ subtended by surface 2 from the location of surface 1 is $dA_2 \cos \phi_2/r^2$. Then Eq. (13.8) may be written

$$\frac{d}{d\omega_1}\left(\frac{dE_{12}}{dA_1}\right) = I \cos \phi_1 \tag{13.9}$$

Since $dA_1 \cos \phi_1$ is the projected or apparent area 1 in any direction ϕ_1, the quantity I may be interpreted as an intensity of emitted radiation in that

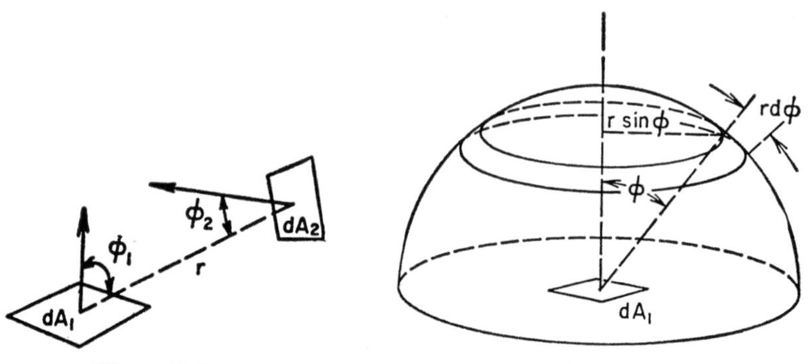

Figure 13.7. Figure 13.8.

direction or energy per unit time per unit apparent area per unit solid angle. For many surfaces I is very nearly independent of the direction ϕ_1. If I is independent of ϕ_1, Eq. (13.9) states that the energy emitted per unit of actual area dA_1 per unit solid angle $d\omega_1$ is proportional to the cosine of the angle ϕ_1. This has been called Lambert's cosine relation (1760).

The magnitude of I may be evaluated for black surface 1 by considering the radiation in the entire hemisphere, Fig. 13.8. Then

$$dA_2 = 2\pi r \sin \phi \, r \, d\phi \quad \text{and} \quad \cos \phi_2 = 1$$

Integration of Eq. (13.8) over the entire hemisphere results in

$$\frac{dE_{b_1}}{dA_1} = e_{b_1} = 2\pi I \int_0^{\pi/2} \sin\phi \cos \phi \, d\phi = \pi I \tag{13.10}$$

This states that the emissive power of a black body is π times the intensity of the emitted radiation.

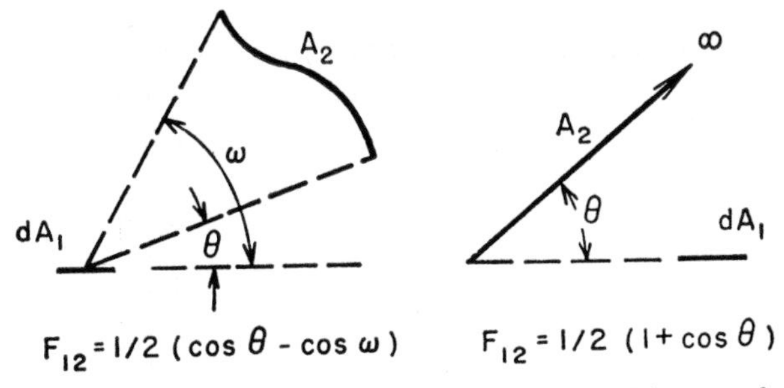

Fig. 13.9. Geometry factor from plane point surface dA_1 to surface A_2.

Then in Eq. (13.8), I may be replaced by e_{b_1}/π. For two finite-sized surfaces A_1 and A_2, Eq. (13.9) may be integrated with Eq. (13.10) in which I remains constant. Then

$$E_{12} = e_{b_1}\frac{1}{\pi}\int_{A_1}\int_{A_2}\frac{\cos\phi_1\cos\phi_2}{r^2}\,dA_2\,dA_1 \tag{13.11}$$

The double integral may be evaluated for any arrangement of surfaces and is a function of geometry alone. For convenience, a geometry factor F_{12} may be defined by the equation

$$A_1F_{12} \equiv \frac{1}{\pi}\int_{A_1}\int_{A_2}\frac{\cos\phi_1\cos\phi_2}{r^2}\,dA_2\,dA_1 \tag{13.12}$$

This equation may be evaluated for any desired geometry. Values of F_{12} have been calculated for various surface arrangements, some of which

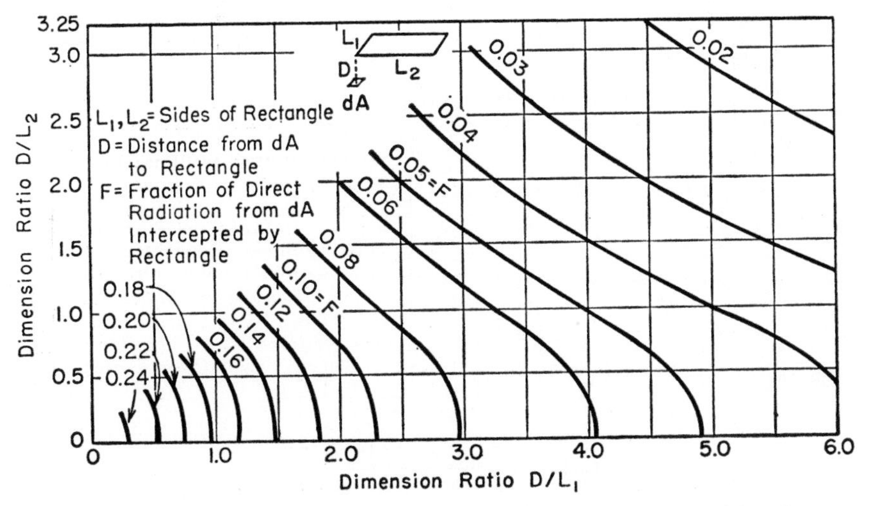

Fig. 13.10. View factor F for direct radiation between an element dA and a parallel rectangle with one corner opposite dA [Hottel (6)].

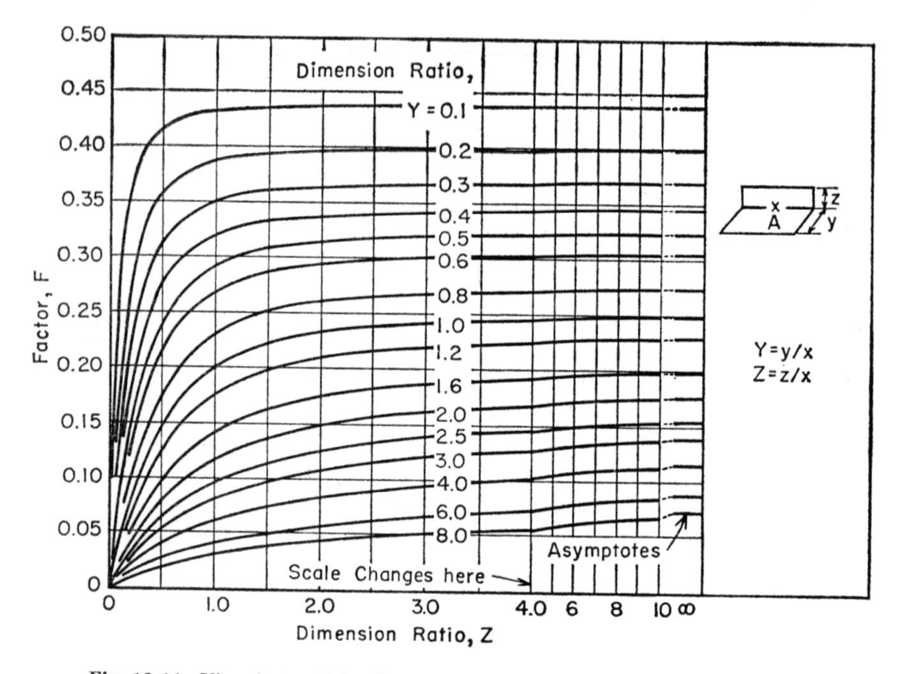

Fig. 13.11. View factor F for direct radiation between adjacent rectangles in perpendicular planes [Hottel (6)].

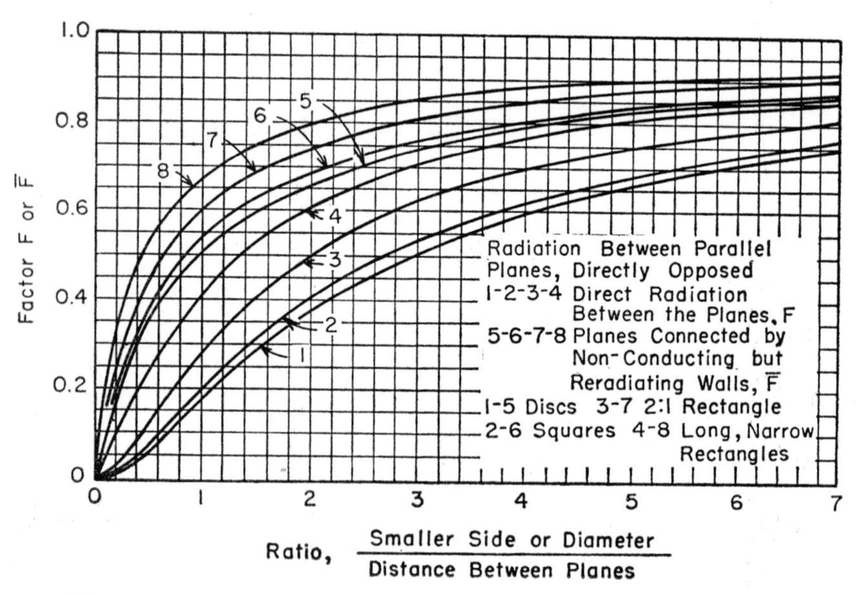

Fig. 13.12. View factor F and interchange factor $\bar{F}$ for radiation between parallel planes directly opposed (6).

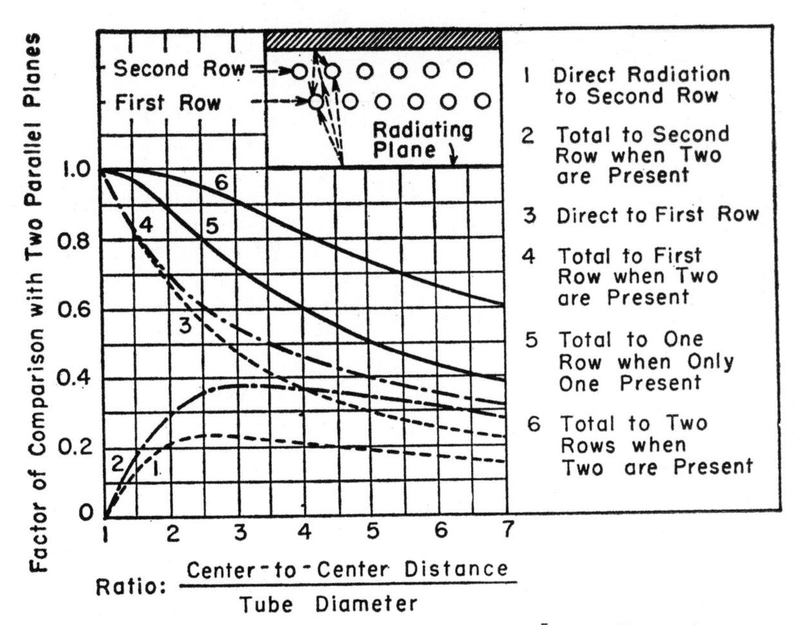

Fig. 13.13. View factor F and interchange factor $\bar{F}$ for radiation from a plane to one or two rows of tubes above and parallel to the plane (6).

are included here in Figs. 13.9 to 13.13. Other cases are found in the published literature, (6), (8). For two large parallel plates $F_{12} \cong 1.0$. For a small body, s, completely enclosed by a large surface, l, $F_{sl} \cong 1.0$.

13.5 Black Surfaces Separated by Nonabsorbing Medium

Equations (13.11) and (13.12) express the relation for the energy emitted by surface 1 and intercepted by surface 2. These become

$$E_{12} = e_{b_1} A_1 F_{12} \tag{13.13}$$

A similar expression for the energy originating at surface 2 and arriving at surface 1 is $E_{21} = e_{b_2} A_2 F_{21}$. Since the order of integration in Eq. (13.12) is immaterial,

$$A_1 F_{12} = A_2 F_{21} = AF \tag{13.14}$$

Then the difference between E_{12} and E_{21} expresses the net interchange between the two black surfaces.

$$q_{1 \leftrightarrows 2} = E_{12} - E_{21} = A_1 F_{12}(e_{b_1} - e_{b_2}) = A_1 F_{12} \sigma (T_1^4 - T_2^4) \tag{13.15}$$

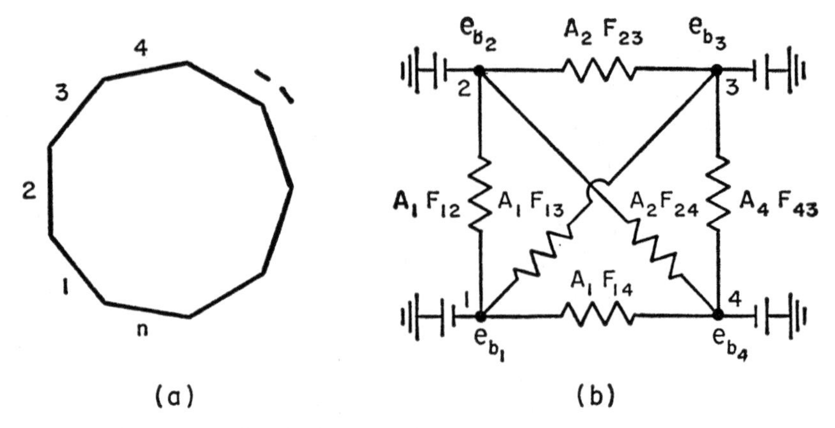

Fig. 13.14. Electrical analog of an enclosure of black surfaces.

Clearly, from Eq. (13.13), F_{12} may be interpreted as the *fraction of the radiation leaving a black surface in all directions which is intercepted by surface* A_2. Then from this interpretation it follows that in a black enclosure (Fig. 13.14) of surfaces 1, 2, 3, $\cdots$

$$F_{11} + F_{12} + F_{13} + \cdots = 1 \tag{13.16}$$

and the net heat flux from surface A_1 is, from Eqs. (13.15) and (13.16),

$$\begin{aligned} q_{1,\text{net}} &= A_1 F_{12} \sigma (T_1^4 - T_2^4) + A_1 F_{13} \sigma (T_1^4 - T_3^4) + \cdots \\ &= A_1 \sigma T_1^4 - (A_1 F_{11} \sigma T_1^4 + A_1 F_{12} \sigma T_2^4 + A_1 F_{13} \sigma T_3^4 + \cdots \end{aligned} \tag{13.17}$$

Oppenheim (9) shows that the black enclosure of Fig. 13.14a may be represented by an analogous electrical network, Fig. 13.14b (drawn here for four surfaces) where e_b (which equals σT^4) is the potential, the heat flow rate is analogous to the electrical current, and $A_i F_{ik}$ is the branch conductance between two nodes at potentials e_{b_i} and e_{b_k}. From an analogous Kirchhoff's law, the net heat flux leaving node point 1 (Fig. 13.14) is

$$q_{1,\text{net}} = \sum_{k=2}^{n} A_1 F_{1k} \sigma (T_1^4 - T_k^4) \tag{13.18}$$

which is identical with Eq. (13.17).

A common case in industrial furnace design involves the calculation of radiant heat transfer rates in an enclosure which consists of a heat source (fuel bed, electrical resistor bank, etc.), a heat sink (water tube bank of a boiler, surface of billets, etc.), and intermediate refractory surfaces which are essentially adiabatic surfaces except for a very small heat loss by conduction to the furnace exterior. Usually the heat transfer associated with convection flow of the gases inside will balance this conduction loss. In any event the radiant heat fluxes are usually so large that they predominate in furnace heat transfer calculation. These refractory surfaces may be considered to be "no-net-flux" surfaces for radiant transfer.

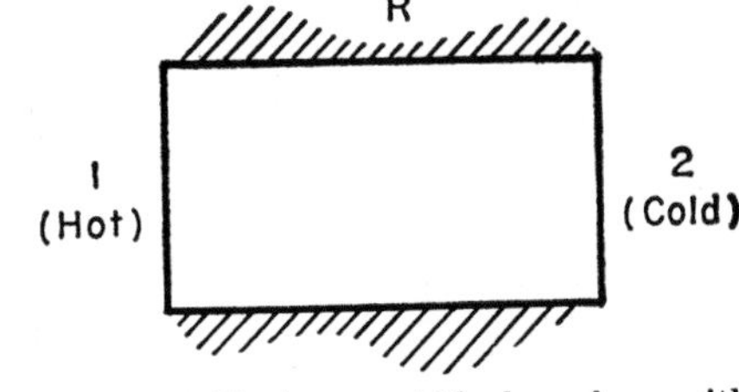

Fig. 13.15. Enclosure of black surfaces with a refractory (no net flux) surface.

Temporarily let the problem be restricted to an enclosure containing a single source surface 1, a single sink surface 2, and refractory surface R. Assume also that all of the surfaces are each at a uniform temperature T_1, T_2, and T_R, Fig. 13.15.

The net radiant heat transfer from hot surface 1:

$$q_{1,\text{net}} = A_1F_{12}\sigma(T_1^4 - T_2^4) + A_1F_{1R}\sigma(T_1^4 - T_R^4) \tag{13.19}$$

Since $q_{R,\text{net}} = 0$ and $A_2F_{2R} = A_RF_{R2}$,

$$A_1F_{1R}\sigma(T_1^4 - T_R^4) + A_2F_{2R}\sigma(T_2^4 - T_R^4) = 0 \tag{13.20}$$

When T_R is eliminated between Eqs. (13.19) and (13.20), then

$$q_{1,\text{net}} = \left(A_1F_{12} + \frac{1}{(1/A_1F_{1R}) + (1/A_2F_{2R})}\right)\sigma(T_1^4 - T_2^4) \tag{13.21}$$

A new factor $\bar{F}$ may be defined such that $\bar{F}_{12}$ represents the fraction of $A_1e_{b_1}$ radiating from surface A_1 which reaches A_2 directly and by detour via the refractory or "no-net-flux" surfaces. Then for the net interchange,

$$q_{1,\text{net}} = A_1\bar{F}_{12}\sigma(T_1^4 - T_2^4) \tag{13.22}$$

and by comparison with Eq. (13.21),

$$A_1\bar{F}_{12} = A_1F_{12} + \frac{1}{(1/A_1F_{1R}) + (1/A_2F_{2R})} \tag{13.23}$$

Note from Eq. 13.16 that $F_{1R} = (1 - F_{12})$ and $F_{2R} = (1 - F_{21})$. Figure 13.16 shows the electrical network equivalent to the furnace enclosure of Fig. 13.15. Equation (13.23) for the equivalent circuit conductance between points 1 and 2 may be written directly from Fig. 13.16, remembering that in parallel circuits conductances are directly additive and in series circuits the reciprocal conductances (resistances) are directly additive.

Values of $\bar{F}_{12}$ for various geometrical arrangements of surfaces 1, 2, and R may be determined from Eq. (13.23). Such values are plotted in Figs. 13.12 and 13.13.

Equations (13.22) and (13.23) solve many furnace problems and are in error by departures from uniformity of temperature over each of the surfaces. Hottel (6) extends this type of analysis to multisurface cases, each surface of which is at a uniform temperature.

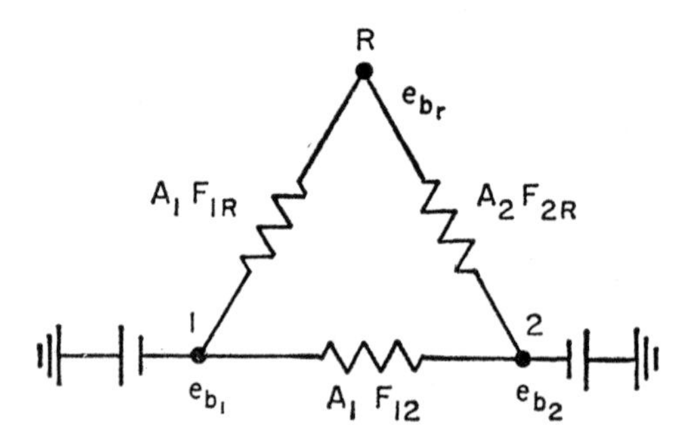

Fig. 13.16. Electrical analog of enclosure of black surfaces with a refractory surface.

13.6 Gray Enclosures

The radiant interchange between two gray opaque surfaces, each of which has an unobstructed view of the other (infinite parallel planes) involves no geometry factor since $F = 1.0$. Consider two gray planes as shown in Fig. 13.17. For opaque surfaces $\rho = 1 - \alpha$ and for gray surfaces $\alpha = \epsilon$. Surface 1 emits $\epsilon_1 e_{b_1}$ per unit time and area. Surface 2 absorbs the fraction α_2 (or ϵ_2) of this energy and reflects $(1 - \epsilon_2)$ back toward A_1. The resulting energy quantities are represented in Fig. 13.17 as rays. The net heat transferred per unit of surface 1 to 2 is the emission $\epsilon_1 e_{b_1}$ minus the fraction of $\epsilon_1 e_{b_1}$ and of $\epsilon_2 e_{b_2}$ which is ultimately absorbed by surface 1 after successive reflections. Then

$$q_{12,\text{net}} = A_1 \epsilon_1 e_{b_1}[1 - \epsilon_1(1 - \epsilon_2) - \epsilon_1(1 - \epsilon_1)(1 - \epsilon_2)^2 \cdots]$$
$$- A_2 \epsilon_2 e_{b_2}[\epsilon_1 + \epsilon_1(1 - \epsilon_1)(1 - \epsilon_2) + \epsilon_1(1 - \epsilon_1)^2(1 - \epsilon_2)^2 \cdots]$$

These infinite series reduce exactly to

$$q_{12,\text{net}} = A\sigma(T_1^4 - T_2^4)\frac{1}{1/\epsilon_1 + 1/\epsilon_2 - 1} \tag{13.24}$$

It will be convenient to define a quantity $\mathfrak{F}_{12}$ (introduced by Hottel (6)) which has the same definition as F_{12} and $\bar{F}_{12}$ but applies to the cases in which

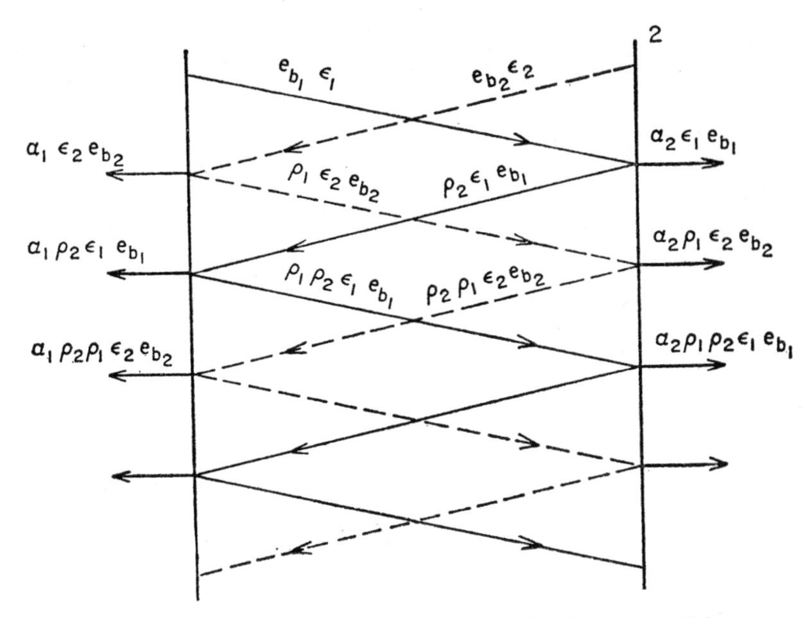

Fig. 13.17. Radiant energy quantities associated with two parallel gray surfaces of infinite extent.

the surfaces are gray. Then

$$q_{12,\mathrm{net}} = A_1\mathfrak{F}_{12}\sigma(T_1^4 - T_2^4) \tag{13.25}$$

and for the case of infinite parallel planes,

$$\mathfrak{F}_{12} = \frac{1}{1/\epsilon_1 + 1/\epsilon_2 - 1} \tag{13.26}$$

Now consider a small gray surface in a very large gray enclosure. By the reasoning associated with Fig. 13.2, the radiation in the large enclosure is essentially black radiation; therefore the interchange at the small gray surface is between a gray emitting surface at T_1 and black-body incident radiation at T_2. Then

$$q_{12,\mathrm{net}} = A_1\epsilon_1\sigma T_1^4 - A_1\alpha_{12}\sigma T_2^4$$

Since for gray surfaces $\alpha_{12} = \epsilon_1$, this may be written as Eq. (13.25) where

$$\mathfrak{F}_{12} = \epsilon_1 \tag{13.27}$$

Now consider the case of an enclosure composed of a gray heat source 1, a gray heat sink 2, and the remaining surfaces reradiating or refractory surfaces. The temperature of each type of surface is assumed to be uniform over the surface. The sketch in Fig. 13.15 represents this case also.

For the moment, direct attention to the radiation streaming in the vicinity of any surface. Consider g Btu/hr ft^2 of radiation streaming toward the surface from direct or reflected radiation from all surfaces in the enclosure. Then the amount of energy streaming away from the gray surface will be the sum of the amounts emitted and reflected from the surface:

$$r \equiv \rho g + \epsilon e_b \tag{13.28}$$

The net heat transferred from the surface per unit area is then the difference between this energy leaving the surface and the oncoming or incident radiation.

$$\left(\frac{q}{A}\right)_{\text{net}} = r - g \tag{13.29}$$

Eliminate g by combining the last two equations to get

$$q_{\text{net}} = \frac{\epsilon}{1 - \epsilon} A (e_b - r) \tag{13.30}$$

This equation provides a basis for a network representation of the gray enclosure. Oppenheim (9) showed that any of the three equivalent network circuits of Fig. 13.18 represents the gray enclosure of Fig. 13.15. Equation

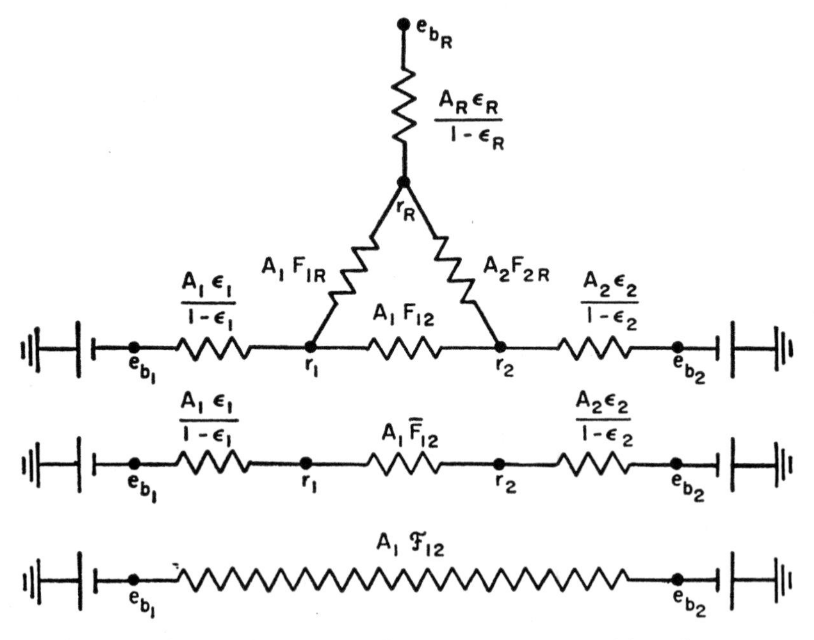

Fig. 13.18. Electrical analog of enclosure of gray surfaces with a refractory surface.

(13.30) indicates that a gray surface may be considered to have a potential of r with respect to the potential e_b of a black surface and a conductance (reciprocal of resistance) of $A\epsilon/(1 - \epsilon)$ between r and e_b. Then with $A_1\mathfrak{F}_{12}$ defined by Eq. (13.25) as the over-all effective conductance, the sum of the resistances in the circuit in Fig. 13.18 is

$$\frac{1}{A_1\mathfrak{F}_{12}} = \frac{1}{A_1}\frac{1 - \epsilon_1}{\epsilon_1} + \frac{1}{A_1\bar{F}_{12}} + \frac{1}{A_2}\frac{1 - \epsilon_2}{\epsilon_2} \tag{13.31}$$

or

$$\mathfrak{F}_{12} = \frac{1}{\dfrac{1}{\bar{F}_{12}} + \left(\dfrac{1}{\epsilon_1} - 1\right) + \dfrac{A_1}{A_2}\left(\dfrac{1}{\epsilon_2} - 1\right)} \tag{13.32}$$

If no reradiating surfaces exist, then $\bar{F}_{12} = F_{12}$ in this equation. This expression for $\mathfrak{F}_{12}$ along with Eq. (13.25) may be used to solve many furnace problems and other radiation problems involving gray surfaces. They are limited to cases involving transparent gases in cavities enclosed by a hot surface and a sink, each at a uniform temperature, and reradiating surfaces which may be assumed to be at a single uniform temperature. Equation (13.32) is readily shown to reduce to Eqs. (13.26) and (13.27) for the two special cases discussed.

From the network shown in Fig. 13.18 it is clear that the following equation is an alternative form of Eq. (13.25) for the case discussed:

$$q_{12,\text{net}} = A_1\bar{F}_{12}(r_1 - r_2) \tag{13.33}$$

In Fig. 13.18 it is clear that the node at R with potential r_R is a floating node. The actual magnitude of r_R is determined by the main $1R2$ circuit and is independent of the emissivity of the reradiating surface.

If the temperature of each type of enclosure surface departs significantly from uniformity, the problem must be solved by dividing the surfaces into a number of parts each of which may be considered to have a uniform temperature. Then there may be a multiplicity of source or sink surfaces and of no-net-flux (reradiating) surfaces. This more complex problem has been treated extensively by Hottel (6) with an energy analysis and by Oppenheim (9) with the network analysis.

Example 13.2: What is the geometry factor for radiation between two parallel disks of 6-in. diameter and separated by 3 in.?

The ratio of diameter to separation distance is $6/3 = 2$. Then from curve 1 of Fig. 13.12, $F_{12} = 0.38$.

Example 13.3: An oxidized steel tube ($\epsilon = 0.6$, O.D. = 3 in.) passes through a silica brick furnace ($\epsilon = 0.8$, inside dimensions 6 by 6 by 6 in.).

The temperature of the inside surface of the furnace wall is 1800 F and of the outer surface of the tube is 1000 F. What is the rate of heat transfer?

Tube Area:

$$A_t = \frac{\pi(3)(6)}{144} = 0.393 \text{ ft}^2$$

Furnace Area:

$$A_f = \frac{[(6 \times 6 \times 6) - 2(\pi/4)(3)^2]}{144} = 1.402 \text{ ft}^2$$

Here $F_{tf} = 1.0$, since every portion of tube sees the furnace through a hemisphere. From Eq. (13.32),

$$\mathfrak{F}_{tf} = \frac{1}{\frac{1}{1} + \left(\frac{1}{0.6} - 1\right) + \frac{0.393}{1.402}\left(\frac{1}{0.8} - 1\right)} = 0.575$$

From Eq. (13.25),

$$q = (0.393)(0.575)(0.171)[(22.6)^4 - (14.6)^4] = 8300 \text{ Btu/hr}$$

Example 13.4: A muffle furnace has a floor 4 by 6 ft at a temperature of 2000 F and an emissivity of 0.7. At the top, 5 ft away, are 2-in.-diameter tubes, 4 ft long on 4-in. centers backed by an insulated refractory roof. The tubes are oxidized steel ($\epsilon = 0.8$) and are at a surface temperature of 500 F. The side walls are insulated refractory, perfectly reradiating surfaces. Find the total rate of heat transferred to the tubes.

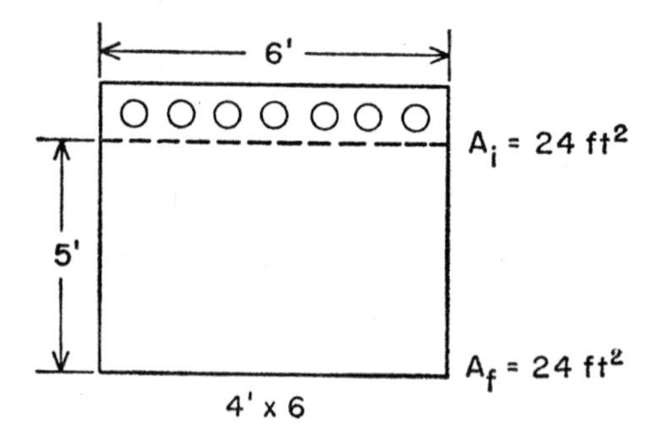

First consider a radiant interchange between the tubes of area A_t and an imaginary plane $A_i = 4$ by 6 ft just below the tubes. The tubes emit to the plane

$$A_t\mathfrak{F}_{ti}T_t^4 \quad \text{or} \quad A_i\mathfrak{F}_{it}T_t^4$$

From Fig. 13.13 $\bar{F}_{it} = 0.88$. Then from Eq. (13.32)

$$\mathfrak{F}_{it} = \frac{1}{\dfrac{1}{0.88} + \left(\dfrac{1}{1} - 1\right) + \dfrac{2}{\pi}\left(\dfrac{1}{0.8} - 1\right)} = 0.836$$

Then the tubes are equivalent to a plane A_i radiating with an emissivity of 0.836 at the same temperature of 500 F. We then treat the problem as one of radiation between planes i and f.

As an approximation, $\bar{F}_{if}$ for 4-by-6-ft rectangles separated by 5 ft may be evaluated as the geometric mean of the value for 4-by-4-ft squares and 3-by-6 ft rectangles separated by 5 ft. Then from Fig. 13.12 lines 6 and 7,

$$\bar{F}_{if} = \sqrt{0.48 \times 0.48} = 0.48$$

From Eq. (13–32)

$$\mathfrak{F}_{if} = \frac{1}{\dfrac{1}{0.48} + \left(\dfrac{1}{0.836} - 1\right) + 1\left(\dfrac{1}{0.7} - 1\right)} = 0.354$$

Then the net heat transfer rate is from Eq. (13.25),

$$q = (4 \times 6)(0.171)(0.354)[(24.6)^4 - (9.6)^4] = 520{,}000 \text{ Btu/hr}$$

13.7 Radiation From Nonluminous Gases

Gases absorb and emit radiation in rather narrow wavelength bands rather than in the continuous spectrum exhibited by solid surfaces. In general, the simpler molecules such as H_2, He, O_2, and N_2 are essentially transparent to radiation. Of the gases found in heat transfer equipment, those which emit a sufficient amount of energy to merit consideration are CO, CO_2, H_2O, SO_2, NH_3, HCl, the hydrocarbons, and the alcohols.

Figure 13.19 shows the emission bands for both CO_2 and H_2O vapor. The electromagnetic radiation (emitted in finite steps and called *quanta of energy*) is postulated to have its origin in the simultaneous changes in energy levels of rotation of the molecules and vibration of the atoms within the molecules. Each quantum absorbed or emitted is accompanied by a displacement of an electron between orbits of different energy levels within the molecule.

Radiation entering a gas volume is not absorbed within a small distance from the surface as is the case in most solids. Instead, the intensity decreases slowly as the radiation passes through the gas. At any position along the path the monochromatic radiation intensity decrease, dI_λ, in path length dx, is proportional to the intensity of the radiation and obeys the relation

$$-dI_\lambda = a_\lambda c I_\lambda \, dx \tag{13.34a}$$

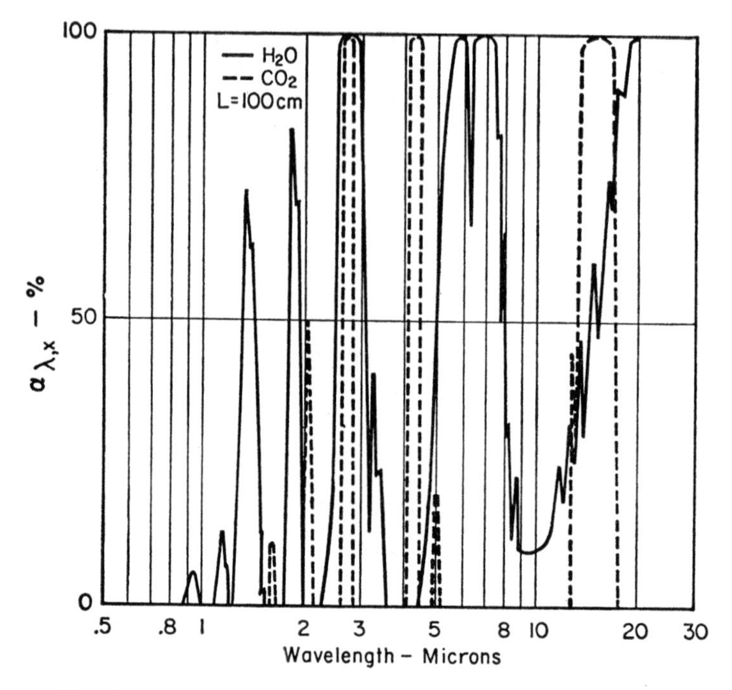

Fig. 13.19. Band emission of carbon dioxide and water vapor.

where a_λ is a function of both temperature and wavelength and c is the concentration of absorbing molecules. If c is uniform along the path and I_{λ_0} is the intensity at the entering surface, Eq. (13.34a) integrates to

$$I_\lambda = I_{\lambda_0} e^{-a_\lambda c L} \tag{13.34b}$$

The quantity $(I_{\lambda_0} - I_\lambda)/I_{\lambda_0}$ is α_λ, the monochromatic absorptivity. Equation (13.34) is known as *Beer's Law.* Intensity of monochromatic radiation emitted from within a gas mass follows similar equations. At a particular temperature c is porportional to p, the partial pressure of the radiating gas. Then summation over all significant wavelengths of the expression in Eq. (13.34) leads to the following equation for emissivity of a gas,

$$\epsilon_g = 1 - \frac{\sum_\lambda I_{\lambda_0} e^{-b_\lambda p L}}{\sum_\lambda I_{\lambda_0}} \tag{13.35}$$

where the summation is taken over all integer values of λ, and b_λ is a function of temperature.

Radiation analyses should be made for each wavelength of significance and the total effect obtained by summation. Actually, it is more convenient to treat the problem in terms of the total radiation absorption and emission characteristics of a gas. Hottel (6) has developed a series of charts based on experimental data for evaluating the total emissivity and absorptivity of various gases.

As a first approximation, the total number of radiating molecules present in a space determine emissivity or absorptivity of the gas. The approximation improves at lower pressures with accompanying large molecular spacings which minimize the molecular shielding effect, permitting each molecule to radiate as though it were alone in the volume. At a particular temperature the partial pressure, p_g, of the radiating gas measures the number of radiating molecules per unit volume; therefore, the product $p_g L$ measures the number of radiating or absorbing molecules in a particular

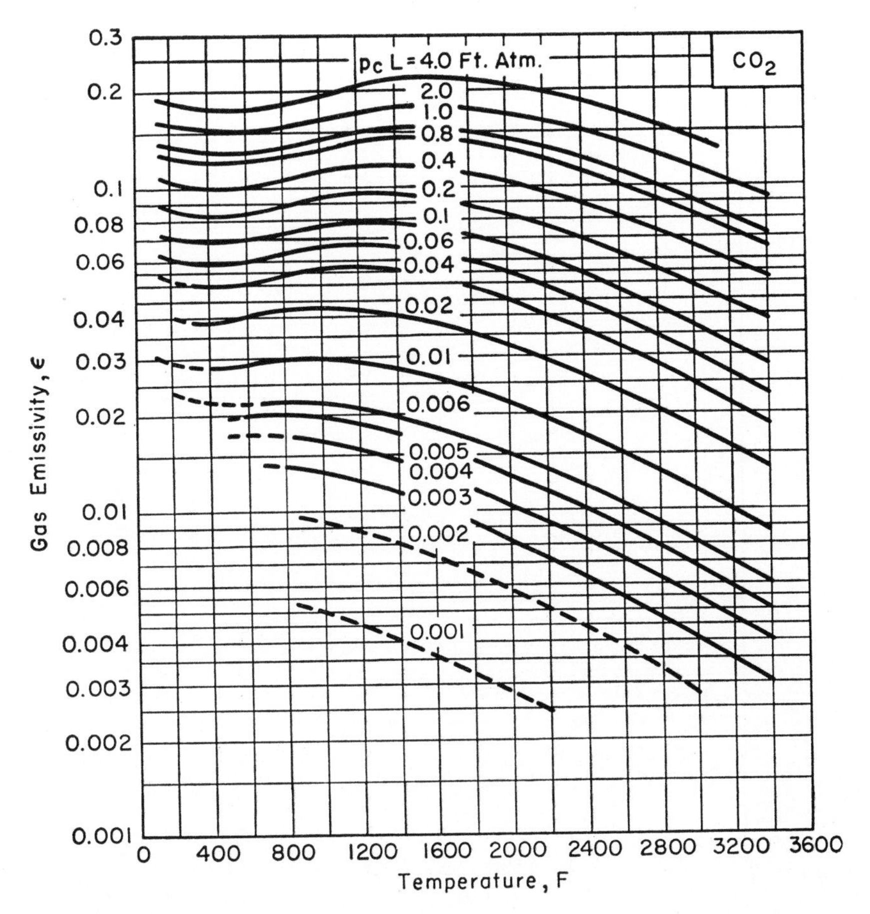

Fig. 13.20. Emissivity of carbon dioxide at 1 atmosphere total pressure and near zero partial pressure [Hottel and Egbert (17)].

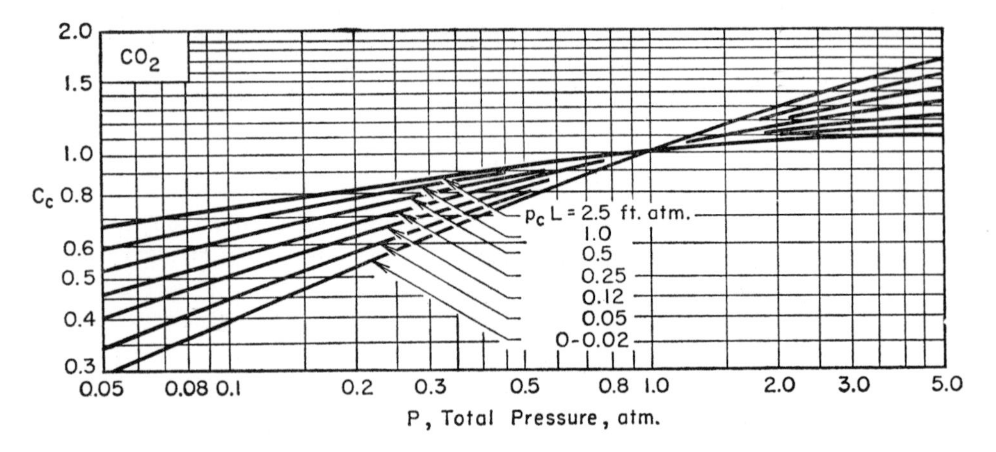

Fig. 13.21. Correction factor for converting emissivity of CO_2 at 1 atm total pressure to emissivity at P atm total pressure [Hottel (6)].

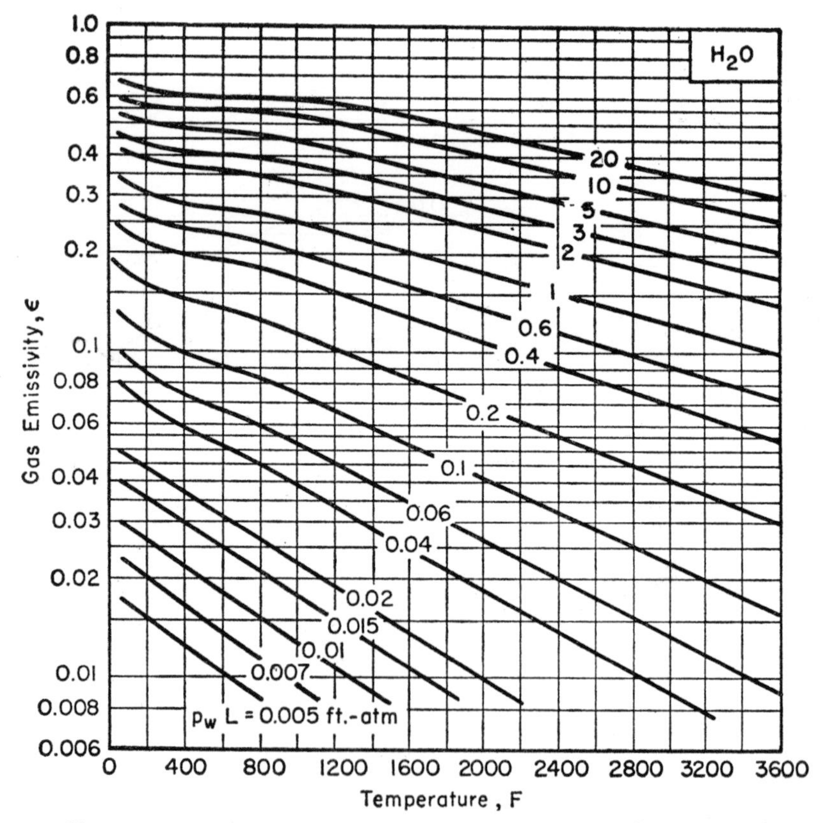

Fig. 13.22. Emissivity of water vapor at 1 atmosphere total pressure and near zero partial pressure [Hottel and Egbert (17)].

geometry and hence determines the emissivity or absorptivity of the gas, as was shown by Eq. (13.35).

Consider a hemisphere, as shown in Fig. 13.8, of radius L filled with gas at T_g exchanging heat radiantly with black surface element dA_1 at T_1. The emission from the gas per unit area of surface dA_1 is $\sigma T_g^4 \epsilon_g$ where ϵ_g is found from Figs. 13.20 and 13.21 for CO_2 and Figs. 13.22 and 13.23 for H_2O vapor. The value of ϵ_c from Fig. 13.20 obtained as a function of p_cL and T_g is multiplied by the correction factor C_c from Fig. 13.21 which allows for departures from 1 atm total pressure. These charts are for a mixture of CO_2 and any other nonradiating gases. Figures 13.22 and 13.23 are a similar set of charts for H_2O vapor.

The absorption by the gas of radiation from dA_1 is $\sigma T_1^4 \alpha_{g1}$ where the double subscript on α is necessary because the gas is not gray. The gas absorptivity equals its emissivity only when $T_1 = T_g$. When $T_1 \neq T_g$, the following equations empirically allow for the non-grayness effect (6).

$$CO_2: \quad \alpha_{g1} = C_c\left(\frac{T_g}{T_1}\right)^{0.65}\left[\epsilon_c\left(T_1,\ p_cL\frac{T_1}{T_g}\right)\right] \tag{13.36}$$

$$H_2O: \quad \alpha_{g1} = C_w\left(\frac{T_g}{T_1}\right)^{0.45}\left[\epsilon_w\left(T_1,\ p_wL\frac{T_1}{T_g}\right)\right] \tag{13.37}$$

where, for example, in the case of CO_2 the chart value of ϵ_c (Fig. 13.20) is evaluated at T_1 and an equivalent p_cL value obtained by multiplying the

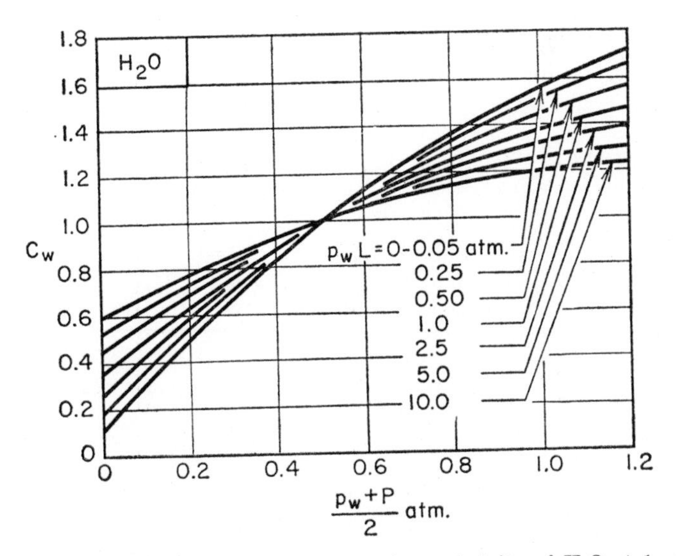

Fig. 13.23. Correction factor for converting emissivity of H_2O at 1 atm total pressure to emissivity at P atm total pressure [Hottel (6)].

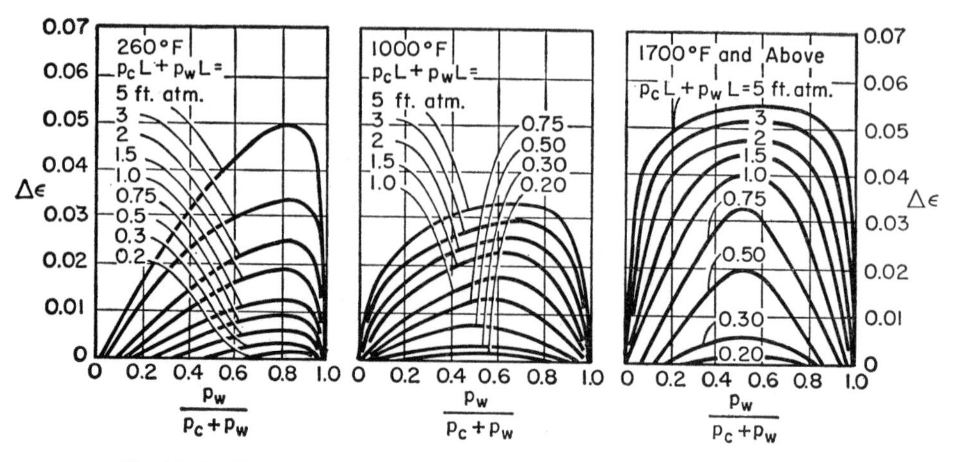

Fig. 13.24. Correction to gas emissivity due to spectral overlap of H_2O and CO_2 [Hottel (6)].

actual p_cL by T_1/T_g. This value of ϵ_c multiplied by $(T_g/T_1)^{0.65}$ and C_c from Fig. 13.21 results in α_{g1}.

When water vapor and CO_2 appear together in a mixture with other nonradiating gases the emissivity is less than the sum of ϵ_c and ϵ_w calculated for each gas alone because of the overlapping bands (Fig. 13.19). The correction term to be subtracted is shown in Fig. 13.24. Then for the mixture

$$\epsilon_g = \epsilon_c + \epsilon_w - \Delta\epsilon \tag{13.38}$$

The same correction factor applies in calculating α_{g1}.

Similar charts for other radiating gases—CO, NH_3, SO_2—are given by Hottel (6).

Example 13.5: Determine the emissivity and absorptivity of a gas composed of $2H_2O + 3CO_2 + 3N_2$, with $P = 2$ atm, $L = 2$ ft, $T_g = 2000R$, and $T_1 = 1200R$.

Then $p_c = (3/8)2 = 0.75$ atm, $p_w = (2/8)2 = 0.5$ atm, and $p_cL = 1.5$ ft atm, $p_wL = 1.0$ ft atm, $(p_w + p_c)/2 = 1.25$, and $p_w/(p_c + p_w) = 0.4$.

From Figs. 13.20 through 13.24, $\epsilon_c = 0.170$, $C_c = 1.10$, $\epsilon_w = 0.209$, $C_w = 1.5$, and $\Delta\epsilon = 0.048$. So

$$\epsilon_g = (0.170)(1.10) + (0.209)(1.5) - (0.048) = 0.453$$

For evaluation of absorptivity $p_cLT_1/T_g = 0.9$, $p_wLT_1/T_g = 0.6$,

$$(T_g/T_1)^{0.65} = 1.394, \text{ and } (T_g/T_1)^{0.45} = 1.258$$

Then from Figs. 13.20 through 13.24 and Eqs. (13.36) and (13.37),

$$\alpha_{g1} = (1.10)(1.394)(0.130) + (1.5)(1.258)(0.220) - (0.048) = 0.566$$

13.8 Geometry of Gas Radiation

The preceding gaseous radiation charts apply to hemispherical (of radius L) collections of gases radiating to an element of area at the center of the base. For other gas shapes there exists an equivalent mean beam length L, defined as the radius of a gas hemisphere which radiates to unit area at the center of its base the same as the average radiation over the area from the actual gas mass shape.

The following analysis (6) applies to a gas at uniform temperature at low values of p_gL, permitting the assumption that any radiation emitted by one portion of gas will not be reduced by absorption in intervening gas. Consider a hemisphere of gas radiating e_g or $\epsilon_g e_b$ to a small unit surface at the center of the base, A_1. The contribution to e_g resulting from the emission to A_1 from a hemispherical shell of the gas at radius x and of thickness dx is $e_b(d\epsilon_g/dx)\,dx$. From Eq. (13.10) the contribution to the intensity of radiation because of this amount of energy at A_1 is (Fig. 13.25)

$$dI = \frac{1}{\pi} e_b \frac{d\epsilon_g}{dx}\,dx$$

Now consider an element of shell at radius x of volume $dS\,dx$. From Eq. (13.9),

$$d^3E_{g1} = dI\,\cos\phi_1\,d\omega_1\,dA_1$$

Eliminating dI between these last two expressions and integrating over the three space coordinates—x, A_1, ω_1—results in

$$E_{g1} = \frac{e_b}{\pi}\int_{A_1}\int_0^{x_{\max}}\int_{\omega_1} \cos\phi_1\,d\omega_1\,\frac{d\epsilon_g}{dx}\,dx\,dA_1 \tag{13.39}$$

which applies to any shape enclosure.

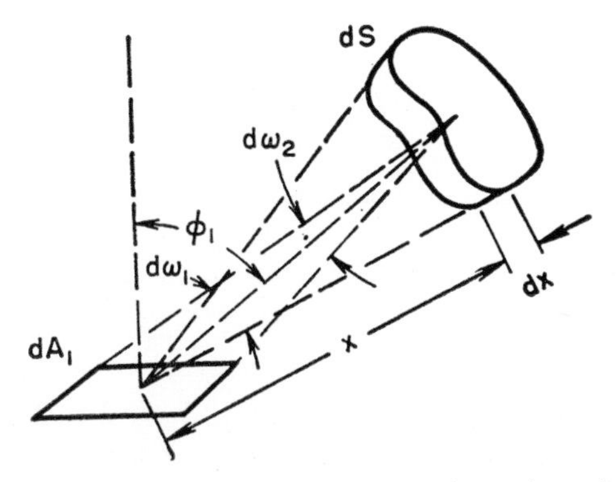

Fig. 13.25. Geometry for radiation from a gas volume element to a bounding surface element.

The emission to dA_1 per unit volume of gas $dS\,dx$ is obtained by dividing the integrand of Eq. (13.39) by $x^2\,d\omega_1\,dx$ or

$$\frac{e_b}{\pi}\frac{d\epsilon_g}{dx}\,d\omega_2$$

since $dS = x^2\,d\omega_1$ and $d\omega_2 = dA_1(\cos\phi_1)/x^2$.

This is the radiation per unit volume emitted through solid angle $d\omega_2$. To get the emission per unit volume in all directions (e_v), multiply this by $4\pi/d\omega_2$ since the solid angle of a sphere is 4π. Then

$$e_v = 4e_b\left(\frac{d\epsilon_g}{dx}\right)_{x\approx 0} = 4e_b p_g\left[\frac{d\epsilon_g}{d(p_gL)}\right]_{p_gL\approx 0} \tag{13.40}$$

Radiation charts such as in Figs. 13.20 and 13.22 provide magnitudes of $d\epsilon_g/d(p_gL)$ since at low p_gL values this derivative approaches (ϵ_g/p_gL).

For the gas hemisphere of radius L radiating to dA_1 at the center of the base consider a ringlike volume element of gas at radius x and angle ϕ_1 with the normal at dA_1. Then

$$dV = (2\pi x\sin\phi_1)x\,d\phi_1\,dx \quad\text{and}\quad d\omega_1 = \frac{(2\pi x\sin\phi_1)(x\,d\phi_1)}{x^2}$$

The emission per square foot of dA_1 becomes from Eqs. (13.39) and (13.40)

$$e_{g1} = \frac{e_v}{2}\int_0^{\pi/2}\int_0^L \cos\phi_1\sin\phi_1\,d\phi_1\,dx = \frac{e_vL}{4} \tag{13.41}$$

Radiation from an irregularly shaped gas mass to its bounding surface may be expressed in terms of an average emissive power $e_{g1_{avg}}$. Then

$$e_{g1_{avg}}A = e_vV \tag{13.42}$$

Now define an equivalent hemispherical mean beam length from Eq. (13.41) such that

$$L = \frac{4e_{g1\,avg}}{e_v} \tag{13.43}$$

Then from Eqs. (13.42) and (13.43),

$$L = \frac{4V}{A} \tag{13.44}$$

which is the value of L for an irregularly shaped gas mass and permits ϵ_g to be determined from charts such as Figs. 13.20 and 13.22. This magnitude of L applies to a uniform-temperature gas with (p_gL) very small. As p_gL increases, the actual L should be smaller than the value calculated with Eq. (13.43), decreasing as p_gL increases. For most practical values of p_gL,

TABLE 13.1 BEAM LENGTHS FOR GAS RADIATION (HOTTEL (6))

Shape	X Characterizing dimension	Factor by which X is multiplied to obtain mean beam length L: When $p_g L = 0$	For average values of $p_g L$
Sphere	Diameter	2/3	0.60
Infinite cylinder	Diameter	1	0.90
Semi-infinite cylinder, radiating to center of base	Diameter	...	0.90
Right-circular cylinder, height = diameter, radiating to center of base	Diameter	...	0.77
Same, radiating to whole surface	Diameter	2/3	0.60
Infinite cylinder of half-circular cross section. Radiating to spot on middle of flat side	Radius	...	1.26
Rectangular parallelepipeds:			
1:1:1 (cube)	Edge	2/3	...
1:1:4, radiating to 1 × 4 face	Shortest edge	0.90	...
radiating to 1 × 1 face	Shortest edge	0.86	...
radiating to all faces	Shortest edge	0.89	...
1:2:6, radiating to 2 × 6 face	Shortest edge	1.18	...
radiating to 1 × 6 face	Shortest edge	1.24	...
radiating to 1 × 2 face	Shortest edge	1.18	...
radiating to all faces	Shortest edge	1.20	...
1: ∞ : ∞ (infinite parallel planes)	Distance between planes	2	...
Space outside:			
Space outside infinite bank of tubes with centers on equilateral triangles; tube diameter = clearance	Clearance	3.4	2.8
Same as preceding, except tube diameter = one-half clearance	Clearance	4.45	3.8
Same, except tube centers on squares; diameter = clearance	Clearance	4.1	3.5

the actual L is about $0.85(4)(V/A)$. Table 13.1 presents values of L for various gas shapes.

13.9 Gas in Black Enclosure

With the mean beam length L giving ϵ_g and α_{g1}, the expression for the net interchange of radiant energy between a gas and its black bounding surface is the difference between the amount emitted by the gas (which is all absorbed by the black surface) and the fraction of the amount emitted by the surface which is absorbed by the gas.

$$\frac{q}{A} = \sigma(\epsilon_g T_g^4 - \alpha_{g_1} T_1^4) \tag{13.45}$$

When the bounding surfaces are grey of emissivity ϵ_1, part of the energy emitted by the gas is reflected and partially absorbed in successive passes through the gas. Equation (13.45), multiplied by a factor which lies between ϵ_1 and 1, expresses the net interchange for this case. For surfaces with ϵ_1 of the order of 0.7 or greater, the use of factor $(\epsilon_1 + 1)/2$ as a multiplying factor for Eq. (13.45) cannot lead to much error. For small values of ϵ_1 the methods of Sec. 13.10 must be used.

13.10 Gray Enclosure Filled with a Gray Gas

A gray gas is defined as one which has the same transmissivity for successive passages of reflected radiation, has ϵ_λ, α_λ and τ_λ independent of λ, and hence $\epsilon_\lambda = \alpha_\lambda = \epsilon = \alpha$. Therefore, for radiation between two surfaces 1 and 2, the following relation describes the role of the gas:

$$\tau_{1g2} + \alpha_{1g2} = \tau_{1g2} + \epsilon_{1g2} = 1 \tag{13.46}$$

Assuming Beer's law to apply to this gray gas, Eq. (13.35) becomes

$$\epsilon_g = \alpha_g = 1 - e^{-bpL} \tag{13.47a}$$

For low partial pressures of the radiating gas or small dimensions, $bpL \ll 1$. Then expansion of Eq. (13.47a) results in

$$e_g = \alpha_g = bpL \tag{13.47b}$$

neglecting higher order terms.

This linearity between α_g and L suggests that, for transfer between two finite surfaces 1 and 2, the absorptivity of the intervening gas may be approximated from Figs. 13.20 to 13.23 with a mean beam length defined as follows (11):

$$L_{\text{mean}} = \frac{1}{A_1F_{12}} \int r\, d(A_1F_{12}) = \frac{1}{A_1F_{12}} \frac{1}{\pi} \int_{A_1} \int_{A_2} r \frac{\cos\phi_1 \cos\phi_2\, dA_1 dA_2}{r^2} \tag{13.48}$$

The sketch in Fig. 13.26 represents a furnace consisting of a cold or a hot gray surface (1), a refractory (no-flux) surface R, and a gray gas, g, where

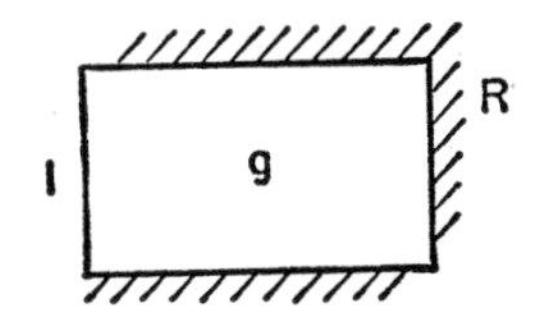

Fig. 13.26. Radiating gray gas in an enclosure of a gray hot or cold surface and a refractory surface.

each element is assumed to be at a uniform temperature—T_1, T_R, and T_g. Calculate ϵ_{g1} and ϵ_{gR} from Figs. 13.20 to 13.23 with L given by Eq. (13.44) and ϵ_{1gR} from Eq. (13.46) with L given by Eq. (13.48).

The radiative flux transmitted between surfaces 1 and R is $\tau_{1gR}A_1F_{1R}\ (r_1 - r_2)$ and between the gas and a surface is $A_1\epsilon_{g1}(r_1 - e_{b_g})$. The equivalent electrical network for the enclosure of Fig. 13.26 is shown in Fig. 13.27. With $\mathfrak{F}_{1g}$ defined by $q_{1g} = A_1\mathfrak{F}_{1g}\sigma(T_1^4 - T_2^4)$, the over-all network resistance $1/A_1\mathfrak{F}_{1g}$ is immediately written from Fig. 13.27 as follows:

$$\frac{1}{A_1\mathfrak{F}_{1g}} = \frac{1 - \epsilon_1}{A_1\epsilon_1} + \frac{1}{A_1\epsilon_{g1} + \dfrac{1}{(1/A_R\epsilon_{gR}) + (1/A_1F_{1R}\tau_{1gR})}} \tag{13.49}$$

These equations are approximate to the extent that the original assumptions are true. Actually the better solution involves subdividing both surfaces 1 and R into smaller elements to allow for variations in temperature over each surface. This more general problem is discussed by Hottel (6) and Oppenheim (9), resulting in a solution involving a determinant which reduces to Eq. (13.49) for the case discussed here.

If a gray gas is present in the gray enclosure shown in Fig. 13.15, the equivalent network of Fig. 13.18 must be modified as shown in Fig. 13.28. The basic $1R2$ circuit is similar to that in Fig. 13.18 but with the conductances multiplied by the transmissivity of the gas. The symbol τ_{1gR}, for example, represents the transmissivity of the gas for the radiant exchange between surfaces 1 and R. There are also exchanges between the gas and each of the solid bounding surfaces. The node at R is again a floating node, the emissivity of this reradiating surface being of no consequence in the solution.

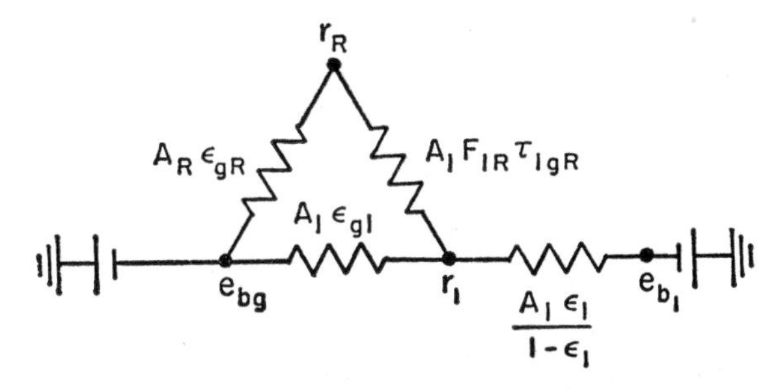

Fig. 13.27. Electrical analog of enclosure of Fig. 13.26.

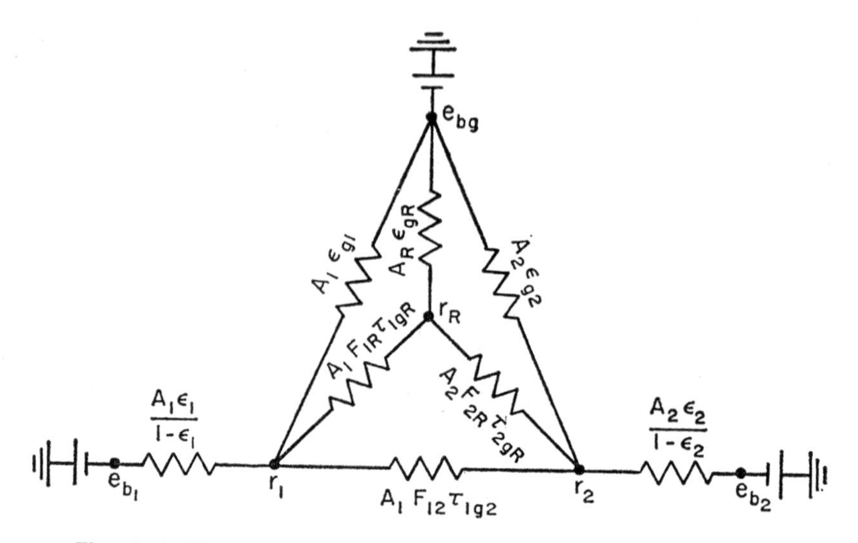

Fig. 13.28. Electrical analog of a radiating gray gas in an enclosure of a gray hot surface, a gray cold surface, and a refractory surface.

In furnaces such as in steam generators, the gas temperature may be unknown but is determined by the combustion process and the rate of heat transferred from the gas. For this case e_{b_g} in the preceding equation is treated as unknown. The energy balance for a combustion process (10) provides the necessary additional equation for the solution. If w_a lb/hr of air enters a combustion chamber with an enthalpy per lb of i_a and w_f lb/hr of fuel at enthalpy i_f, then

$$q_{\text{from gas}} = (w_a + w_f)(i_g - i_{g,0}) - w_a(i_a - i_{a,0}) - w_f(i_f - i_{f,0}) + w_f(\text{HV})_0 \quad (13.50)$$

where subscript 0 refers to conditions under which the heating value HV was measured and i_g is the enthalpy of the gas at T_g in the furnace.

Example 13.6. Calculate the radiant heat transfer between two large parallel plates at temperatures of 1200 F and 800 F respectively if stagnant CO_2 gas at 1 atm pressure is present in the 3 in. spacing between the plates whose emissivity is 0.6.

From Eq. (13.44), $L = (0.85)\ (4)\ (0.25/2) = 0.425$ ft and $p_c = 1$ atm.

Now assume the gas temperature to be 1000 F; this will be checked later. Then from Fig. 13.20 $\epsilon_c = 0.118$ and $\tau_{1g2} = 0.882$. The gas will be assumed to be grey, so $\alpha_c = \epsilon_c$. (Actually $\alpha_{g1} = 0.114$ and $\alpha_{g2} = 0.116$, which are close to

0.118.) The following sketch and equivalent network reduced from Fig. 13.28 represent the problem.

3"
T_1
1200 F
1660 R
CO_2
1 atm
$\epsilon_1 = \epsilon_2 = 0.60$
T_2
800 F
1260 R

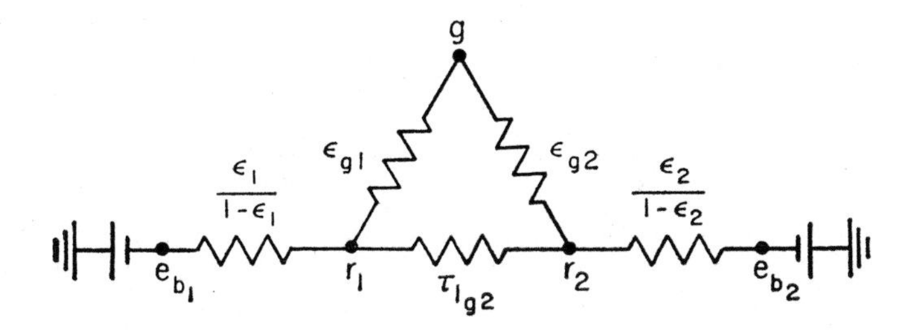

This network is similar to the one in Fig. 13.18, and equations similar to Eqs. (13.31) and (13.23) may be written for this case. Here

$$\frac{1}{\mathfrak{F}_{12}} = \frac{(1-\epsilon_1)}{\epsilon_1} + \frac{1}{F_{12}\tau_{1g2} + \dfrac{1}{(1/\epsilon_{g1}) + (1/\epsilon_{g2})}} + \frac{(1-\epsilon_2)}{\epsilon_2}$$

$$= \frac{0.4}{0.6} + \frac{1}{(1)(0.882) + \dfrac{1}{(1/0.118) + (1/0.118)}} + \frac{0.4}{0.6} = 2.395$$

So

$$\frac{q}{A} = \mathfrak{F}_{12}\sigma(T_1^4 - T_2^4) = \frac{(0.1713)[(16.6)^4 - (12.2)^4]}{2.395} = 3620 \text{ Btu/hr ft}^2$$

Now check the assumption of $T_g = 1000$ F.

$$e_{b_1} = \sigma T_1^4 = (0.1713)(16.6)^4 = 13{,}000, \quad e_{b_2} = 4300$$

$$r_1 = e_{b_1} - \frac{q/A}{\epsilon_1/(1-\epsilon_1)} = 10{,}590$$

$$r_2 = e_{b_2} + \frac{q/A}{\epsilon_2/(1-\epsilon_2)} = 6710$$

Since $\epsilon_{g1} = \epsilon_{g2}$,

$$e_{b_g} = \frac{r_1 + r_2}{2} = 8400$$

or

$$T_g = 100 \sqrt[4]{\frac{8400}{0.1713}} = 1490 \text{ R} = 1030 \text{ F}$$

The problem should be solved again with this as the assumed gas temperature, but since 1030 F is close to 1000 F, no significant improvement in the answer is expected.

13.11 Luminous Flame and Clouds of Particles

Powdered coal or atomized oil flames, soot formed from thermal decomposition of hydrocarbons, and dust particles in flames all radiate as clouds of particles. Powdered coal flames contain particles varying in size from 0.01 in. (250 microns) down, with an average size of around 0.001 in. These particles are essentially opaque to radiation, whereas soot particles, which are in the size range 0.005 to 0.06 microns, act as semitransparent or scattering bodies. Flames of this latter type are called *luminous flames* and obey different optical laws.

The determination of the radiation properties of luminous flames and clouds of particles has not been successfully accomplished. In very large industrial furnaces the emissivity of the combustion space is essentially unity. For smaller size volumes Hottel (6) presents a comparative method for determining flame emissivity.

13.12 Statistical Theory of Thermal Radiation*

The energy distribution function of a batch of electromagnetic radiant energy may be found by the method of quantum statistics. For this analysis electromagnetic radiation is visualized as being a "gas" whose particles are photons instead of molecules. In the statistical method the theory is not concerned with the behavior of individual particles, but with the statistical behavior of a large number of particles.

Let us consider a gas in equilibrium in an isolated enclosure. Its state is completely specified if the position (x, y, z) and the velocity (v_x, v_y, v_z) or equivalently the momentum (p_x, p_y, p_z) of each particle is known. Since the state of a particle is determined by six independent variables, instead of

* The derivation of Bose-Einstein statistics follows in main outline the presentation of F. W. Sears (13).

the usual x, y, z coordinate system in the ordinary space, it is useful to introduce the idea of a six-dimensional *phase space*. Each point in this space, called a *phase point*, represents a state of a particle.

We shall now define a number of statistical concepts associated with the phase space. We shall call a *cell* an element of volume

$$H = dx\, dy\, dz\, dp_x\, dp_y\, dp_z$$

in the phase space, small compared with the dimensions of the enclosure and the range of velocities of the particles, but sufficiently large so as to contain a large number of phase points. Let N_i = the number of phase points in an ith cell. Then

$$N_i = \rho H \tag{13.51}$$

where ρ may be called the local "density" in the phase space.

A complete specification (position and momentum) of each particle within the cell in which its state is represented defines a *microstate* of the gas. A specification of only the number of phase points N_i in each cell defines a *macrostate* of the gas. As an illustration, a bridge hand described as containing 2 spades, 3 hearts, 4 diamonds, and 4 clubs is a macrostate, since only the number of each suit in the hand is specified. When each card is individually identified, we have described a microstate. A fundamental hypothesis of statistical mechanics states all microstates are equally probable. Using again the bridge hand as an illustration, a microstate of completely specified distribution of 2 spades (4, J), 3 hearts (2, 9, K), 4 diamonds (10, J, K, A), and 4 clubs (5, 8, 9, J) is as rare an occurrence as a hand of 13 cards of a single suit.

The statistics in the literature differ essentially in the way of defining and counting the microstates corresponding to a particular macrostate. The earliest known is the Maxwell-Boltzmann statistics which has predicted satisfactorily the behavior of simple gases. When applied to a photon gas, however, it led to serious discrepancies with existing experimental knowledge; the discrepancy led to the discovery of the quantum statistics. In the following we shall derive the pertinent statistics for a photon gas and indicate how it differs from the classical Maxwell-Boltzmann statistics.

The Maxwell-Boltzmann statistics postulates that each molecular state can be represented by a geometrical point in the phase space. In quantum statistics, however, it is recognized that there is a maximum precision to which the position and momentum of a particle can be simultaneously specified (Heisenberg uncertainty principle), and this limit is an element of phase space of volume h^3, where h is the Planck's constant.* We shall call this volume a *sub-cell* (a quantum state). Since $H \gg h^3$, we can imagine superimposed on the subdivision of a phase space into cells of volume H

* $h = 6.624 \times 10^{-27}$ erg-sec; $[h^3] = [\text{erg-sec}]^3 = [\text{length}]^3 \times [\text{momentum}]^3$.

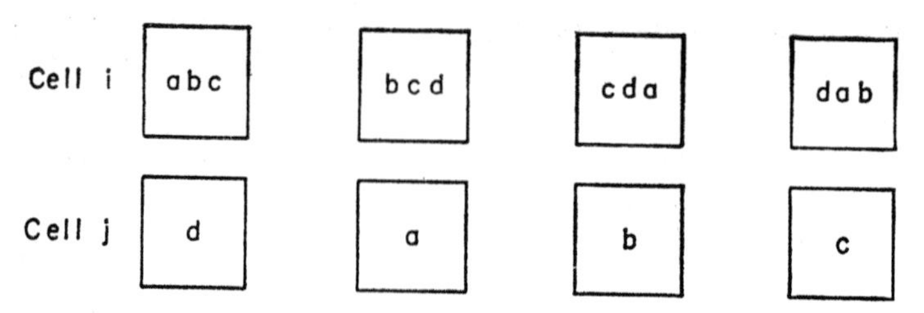

Figure 13.29

a finer subdivision into sub-cells of volume h^3. The number of sub-cells per cell is then given by

$$n = \frac{H}{h^3} \tag{13.52}$$

Consider a phase space divided only into two cells, i and j. A macrostate is defined by specifying the number of phase points in each cell; let us consider a particular macrostate $N_i = 3$, $N_j = 1$. The Boltzmann theory assumes that a molecular state can be identified to any degree of precision so that if molecules a and b exchange cells in the phase space, the microstate is changed. In the example chosen, there are therefore four microstates, Fig. 13.29, corresponding to the particular macrostate.

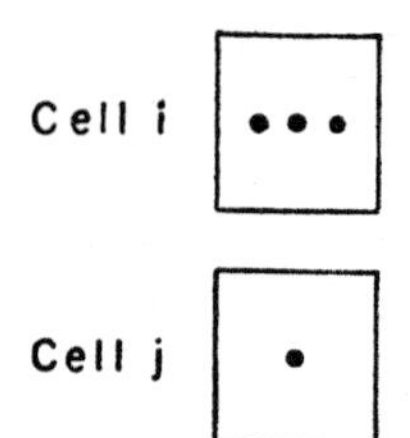

Figure 13.30

In quantum statistics, we do not attempt to tag individual particles. If we replace letters by dots, we find that there is apparently only one microstate possible, Fig. 13.30. However, if we additionally specify four sub-cells per cell, then Fig. 13.31 shows there are 20 possible ways of arranging the three dots in cell i and four ways of arranging the single dot in cell j. For any one arrangement in cell i, we can have any one of the four arrangements in cell j, so that the total number of microstates is

$$W = W_i W_j = 80$$

More generally, if an ith cell contains N_i phase points and n sub-cells, the number of microstates for the ith cell is

$$W_i = \frac{n(n + N_i - 1)!}{n!N_i!}$$

The formula can be readily checked for the case illustrated in Fig. 13.31. The total number of microstates for all cells in a given phase space is

$$W = \text{product } [W_i] = \text{product} \left[\frac{n(n + N_i - 1)!}{n!N_i!}\right] \tag{13.53}$$

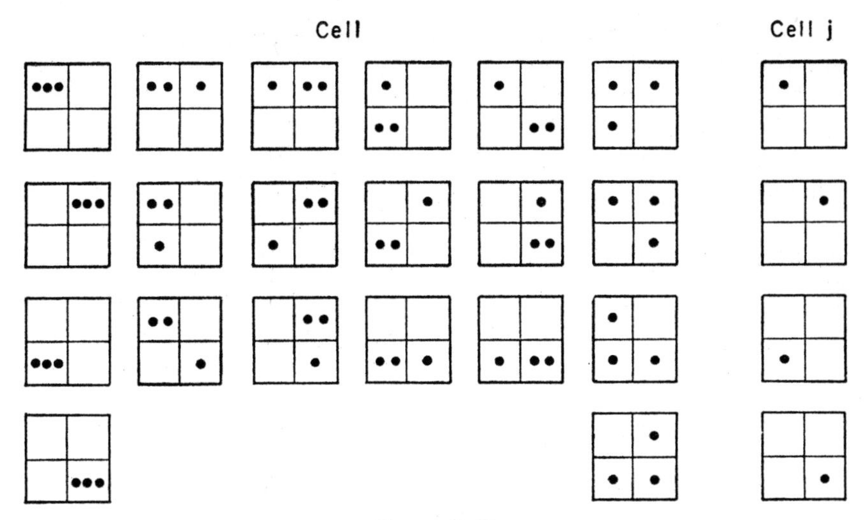

Figure 13.31

W is also known as the *thermodynamic probability of the macrostate* and is related to the entropy S of the system by the following relation:

$$S = \kappa \ln W \tag{13.54}$$

where $\kappa = 1.380 \times 10^{-23}$ joules/C, the Boltzmann constant.

To investigate the distribution of N_i among the cells, we consider that they exist in their most probable or normal distribution and then imagine a shifting of particles among the cells, departing from this normal distribution. This is called a *virtual displacement.* For an isolated system such shifting occurs, keeping the total number of particles and the total energy constant. These constraints are expressed as follows:

$$dN = \sum dN_i = 0 \tag{13.55}$$

$$dU = \sum u_i \, dN_i = 0 \tag{13.56}$$

where u_i represents the energy of a particle when its phase point lies in the ith cell.

The most probable or normal distribution (a thermodynamic equilibrium state) is one for which S or $\ln W$ is a maximum. Letting $d(\ln W) = 0$ and neglecting 1 compared with n and N_i, Eq. (13.53) may be expanded using the Stirling approximation ($\ln M! = M \ln M - M$) as follows:

$$d\,(\ln W) = \sum \left(\ln \frac{n + \bar{N}_i}{\bar{N}_i}\right) dN_i = 0 \tag{13.57}$$

where $\overline{N}_i$ is the number of phase points in an ith cell for the normal distribution.

If each of the displacements dN_i were independent, Eq. (13.57) would be satisfied by setting each of the coefficients equal to zero. However, the dN_i shifts are not independent but are subject to the restraints stated by Eqs. (13.55) and (13.56). These constraints imply that any increase in the population or energy of some cells must be exactly balanced by a corresponding decrease in the population and energy in other cells.

The problem of finding a constrained maximum or minimum function is greatly simplified by the method of undetermined multipliers developed by Lagrange. Equation (13.55) is multiplied by a constant parameter written for convenience as $-\ln B$ and Eq. (13.56) by $-\beta$; the resulting equations are then added to Eq. (13.57) to obtain

$$\sum \left(\ln \frac{n + \bar{N}_i}{\bar{N}_i} - \ln B - \beta u_i \right) dN_i = 0$$

The multipliers are to be determined such that all dN_i's in the new expression are independent. Therefore, we can now equate the coefficients to zero and obtain

$$\ln \frac{n + \bar{N}_i}{\bar{N}_i} - \ln B - \beta u_i = 0$$

or

$$\frac{\bar{N}_i}{n} = \frac{1}{Be^{\beta u_i} - 1} \tag{13.58}$$

Equation (13.58) is the Bose-Einstein statistics. It gives the number of phase points occupying a sub-cell in an ith cell in a normal distribution. We shall now apply the statistics to electromagnetic radiation. Consider an evacuated enclosure with perfectly reflecting walls containing a speck of ideal black matter. The radiant energy in the enclosure can be regarded as a "gas" whose particles are photons possessing energy $u = h\nu$ and momentum $p = h\nu/c$, where c is the speed of light and $\nu = c/\lambda$ is the frequency. Since the photons are not conserved, i.e. they are created or destroyed when they interact with matter, the total number of photons in the enclosure is not constant. Therefore, the constraining condition Eq. (13.56) must be removed or the Lagrangian multiplier $B = 1$. The energy constraint imposed by Eq. (13.57) still applies and may be determined from thermodynamic energy relations to be $\beta = 1/\kappa T$. Equation (13.58) then becomes

$$\frac{\bar{N}_i}{n} = \frac{1}{e^{h\nu/\kappa T} - 1} \tag{13.59}$$

Equation (13.59) gives the number of photons occupying a quantum state in an ith cell. To get the photon density in the enclosure, we must sum Eq. (13.59) over all the cells in the phase space. If the photon distribution in the phase space can be approximated by a continuous function, the summation can be replaced by integration. The quantity n in Eq. (13.59) is by definition the number of quantum states in a cell, or

$$n = \frac{2H}{h^3} = \frac{2}{h^3}\, dx\, dy\, dz\, dp_x\, dp_y\, dp_z$$

where the factor 2, doubling the number of quantum states, is necessary to account for the right- and the left-circular polarizations of the photons or the electromagnetic waves.

Substitution of n into Eq. (13.59) and replacement of N_i by the differential notation d^6N gives, with $h\nu$ replaced by pc,

$$d^6N = \frac{2}{h^3}\frac{1}{e^{pc/\kappa T} - 1}\, dx\, dy\, dz\, dp_x\, dp_y\, dp_z$$

We first integrate this expression over the ordinary space over the limits of the enclosure, which results in simple replacement of $dx\, dy\, dz$ by V, the volume of the enclosure. The resulting expression giving the photon distribution in the momentum space is spherically symmetrical. Consequently, the number of photons in a thin shell of momentum space of radius p and thickness dp is

$$dN_p = \frac{2V}{h^3}\frac{1}{e^{pc/\kappa T} - 1}\, 4\pi p^2\, dp$$

where $p = \sqrt{p_x^2 + p_y^2 + p_z^2}$.

Then in terms of the frequency ν, the number of photons in the frequency range $d\nu$ is given by

$$dN_\nu = \frac{8\pi V}{c^3}\frac{1}{e^{h\nu/\kappa T} - 1}\, \nu^2\, d\nu \qquad (13.60)$$

since $p = h\nu/c$.

Finally, the radiant energy per unit volume or the energy density within a frequency range $d\nu$ is found to be

$$de_\nu = \frac{h\nu\, dN_\nu}{V} = \frac{8\pi h}{c^3}\frac{\nu^3\, d\nu}{e^{h\nu/\kappa T} - 1} \qquad (13.61a)$$

Equation (13.61a) may be written in terms of the wavelength λ by noting that $\nu = c/\lambda$. Then

$$de_v = \frac{8\pi hc\lambda^{-5}\, d\lambda}{e^{ch/\lambda\kappa T} - 1} \qquad (13.61b)$$

Equation (13.61) expresses the radiant energy density in an enclosure as a function of temperature T and the frequency ν or the wavelength λ. The quantity $e_{v,\lambda} = de_v/d\lambda$ from Eq. (13.61b) is called the monochromatic energy density. We shall now show that a simple relation exists between the energy density $e_{v,\lambda}$ of the stream of radiation in an isothermal enclosure and the emissive power $e_{b,\lambda}$ (Eq. 13.5) of a black body.

Suppose an isothermal enclosure contains a black body at the same temperature. The stream of radiation in any one direction in the enclosure is the same as that emitted by the black surface in any direction (since a black body is a perfect absorber, it reflects none of the radiation that falls on it). Energy density in the enclosure, within which radiation travels in all directions, is therefore twice that of hemispherical emission from the black surface by geometrical consideration.

We can relate the energy density (e_v) of black radiation with the emissive power e_b of a black surface. Imagine a point in the space in front of a black surface; the radiation e_v streams equally in all directions (4π steradians) at the velocity of light, c. The intensity of radiation striking the surface is

$$I_b = \frac{e_v c}{4\pi} \tag{13.62}$$

At equilibrium this must equal the intensity of radiation emitted from the black surface, given by Eq. (13.10), or $I_b = e_b/\pi$. Then,

$$e_v = \frac{4e_b}{c} \tag{13.63}$$

Likewise, for monochromatic rays,

$$e_{v,\lambda} = \frac{4e_{b,\lambda}}{c} \tag{13.64}$$

Combining Eqs. (13.61b) and (13.64), we get

$$e_{b,\lambda} = \frac{2\pi h c^2 \lambda^{-5}}{e^{ch/\kappa\lambda T} - 1} \tag{13.6}$$

which is the Planck's equation given in Sec. 13.3.

The total rate of energy emission per unit area is

$$e_b = \int_{\lambda=0}^{\infty} e_{b,\lambda}\, d\lambda = \sigma T^4 \tag{13.7}$$

where $\sigma = 2\pi^5\kappa^4/15c^2h^3 = 0.1713 \times 10^{-8}$ Btu/hr ft^2 R^4 is the Stefan-Boltzmann constant, and the integration is performed with the aid of the definite integral

$$\int_0^{\infty} \frac{x^3\, dx}{e^x - 1} = \frac{\pi^4}{15}$$

It would be interesting to compare the statistical derivation of Eq. (13.7) with the following derivation based upon classical thermodynamics.

Consider the system of photon gas contained in a piston-cylinder enclosure of volume V with perfectly reflecting walls. The electromagnetic radiation or "gas" in the enclosure exerts an average pressure on the walls given by (16)

$$p = \frac{e_v}{3} \tag{13.65}$$

The radiation pressure is usually very small, but it has been experimentally verified.

For a process in which the piston moves, the following equation from the 1st and the 2nd Laws of thermodynamics applies:

$$T\,dS = p\,dV + dU \tag{13.66}$$

The internal energy of the photon gas may be expressed as $U = e_v V$ or

$$dU = e_v\,dV + V\,de_v \tag{13.67}$$

Substitution of Eqs. (13.65) and (13.67) into Eq. (13.66) gives

$$dS = \frac{4}{3}\frac{e_v}{T}\,dV + \frac{V}{T}\,de_v$$

We note that dS is an exact differential, and e_v is a function of temperature only. Therefore,

$$\left[\frac{\partial}{\partial V}\left(\frac{V}{T}\right)\right]_{e_v} = \left[\frac{\partial}{\partial e_v}\left(\frac{4}{3}\frac{e_v}{T}\right)\right]_V$$

$$de_v = \frac{4}{3}\left(de_v - e_v\frac{dT}{T}\right)$$

$$\frac{de_v}{e_v} = 4\frac{dT}{T}$$

$$e_v = CT^4 \tag{13.86}$$

The thermodynamic method leads to an equation of the same form as Eq. (13.7) but does not give any information as to the magnitude of the constant C.

REFERENCES

1. Lummer and Pringsheim, *D. Phys. Ges. Verhandlungen*, **1**, 33 and 215; **2**, 163 (1900).
2. Planck, L., *Ann. Physik*, **4**, 553 (1901).
3. Snyder, N. W., "Review of Thermal Radiation Constants," *Trans. ASME* **76**, 537 (1954).

4. Aschkinass, A., *Ann. Physik,* **17,** (1905).

5. Schmidt, H., and L. Furthmann, *Mitt. Kaiser-Wilhelm-Inst. Eisenforsch. Düsseldorf, Abhandl.,* **109,** 225 (1928).

6. Hottel, H. C., "Radiation," Chap. IV of *Heat Transmission,* by W. H. McAdams, McGraw-Hill, 1954, 3rd Ed.

7. Sieber, W., *Z. Tech. Phys.,* 1941, pp. 130–35.

8. Hamilton, D. C., and W. R. Morgan, "Radiant Interchange Configuration Factors," *Nat. Advisory Comm. Aeronaut. Tech. Note* 2836, Dec. 1952.

9. Oppenheim, A. K., "Radiation Analysis by the Network Method," ASME Paper 54-A-75, Dec. 1954.

10a. Keenan, J. H., *Thermodynamics,* Chaps. XIV and XVI, Wiley, New York, 1941.

10b. Mooney, D. A., *Mechanical Engineering Thermodynamics,* p. 448, Prentice-Hall, Englewood Cliffs, N. J., 1954.

11. Wohlenberg, W. J., *Heat Transfer by Radiation,* Research Bulletin No. 75, Purdue Eng. Exp. Sta. (1941).

12. Koehler, Glenn, *Circuits and Networks,* Macmillan, New York, 1955.

13. Sears, F. W. *Thermodynamics,* Chaps. 14–16, Addison-Wesley, Reading, Mass., 1959.

14. Schmidt, E., and E. Eckert, *Forsch. Gebiete Ingenieurwesen,* **6,** (1935).

15. Sieber, W., *Z. tech. Physik,* **22,** 130–135 (1941).

16. Richtmyer, F. K., and E. H. Kennard, *Introduction to Modern Physics,* McGraw-Hill, New York, 1942, p. 166.

17. Hottell, H. C., and R. B. Egbert, AIChE Trans. **38,** (1942).

PROBLEMS

13.1 A tungsten filament of a lamp is at 4000 F. What percentage of the total radiant energy is in the visible range 0.38μ to 0.76μ ($\mu = 10^{-4}$ cm)?

13.2 Plot $e_{b,\lambda}$ vs. λ for 4000 F abs and 1000 F abs.

13.3 (a) Show that the radiation equation $q/A = \mathfrak{F}\sigma(T_1^4 - T_2^4)$ can be rewritten to define a radiation heat transfer coefficient h_r in the form $q/A = h_r(T_1 - T_2)$ where

$$h_r = \mathfrak{F}\sigma(T_1^2 + T_2^2)(T_1 + T_2) \cong 4\mathfrak{F}\sigma T_{\text{mean}}^3$$

(b) Compare the results of the approximate formulation with the correct equation when $T_1 = 2000$ and $T_2 = 1000$; and when $T_1 = 200$ and $T_2 = 100$.

(c) For cases in which $\mathcal{F} = 1.0$, calculate magnitudes of h_r for the following combinations of temperatures:

T_1, °F	70	200	1000	2000
T_2, °F	60	100	500	1000

13.4 The results of Prob. 13.3(c) suggest the h_r is of the order of magnitude of h associated with natural convection at surfaces in air. Calculate the heat transfer rates associated with natural convection and by radiation from a horizontal wire ($\epsilon = 1.0$) whose surface temperature is 150 F, where the surrounding air and room temperature is 70 F, for wire diameters of 0.001 in., 0.01 in., 0.1 in., and 1 in. The sum of these two quantities is the total rate of heat transferred from the wire surface.

13.5 A thermocouple placed in a gas stream flowing in a duct measures a temperature which is somewhere between the actual gas temperature and the duct wall temperature. For a sufficiently large depth of immersion, conduction effects along the wires (or along a protecting tube, if present) are negligible. At steady state the q/A between thermocouple and gas stream associated with convection equals the q/A between thermocouple and duct wall by radiation, assuming the gas is not a radiating gas.

(a) A bare chromel-alumel thermocouple with wire diameters of 0.01 in. and having surface emissivities of $\epsilon_T = 0.8$ is in a gas flowing in a long 10 in. diameter duct (wall emissivity $\epsilon_w = 0.7$). The gas flow rate produces $h = 20$ Btu/hr ft^2 F. The thermocouple in the gas stream reads $T_T = 1500$ F and a thermocouple peened to the duct wall reads $T_w = 1000$ F. Calculate the actual temperature, T_g, of the gas.

(b) Plot the thermocouple error $(T_g - T_T)$ as a function of thermocouple surface emissivity ϵ_T.

(c) Plot $(T_g - T_T)$ vs. h.

13.6 (a) In Prob. 13.5, how does reducing the thermocouple wire diameter change the thermocouple error? Other conditions, including gas velocity, remain unchanged. Write the equation showing this effect.

(b) The thermocouple error can be substantially reduced by a radiation shield shown in the sketch. Write the governing equations showing how

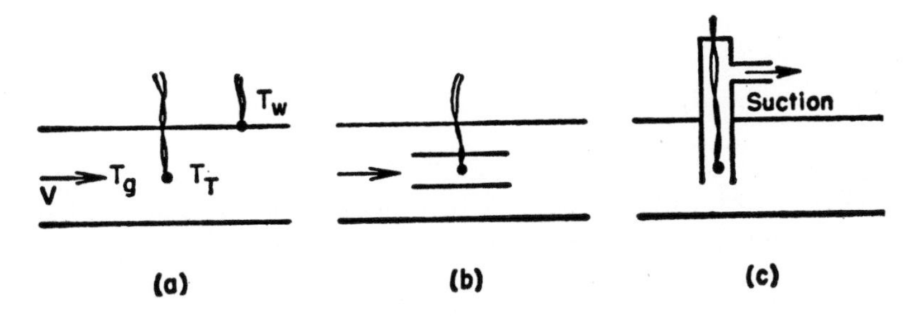

the error is reduced. (W. M. Rohsenow and J. P. Hunsaker, *Trans. ASME*, **69**, 699, Aug. 1947.)

(c) In some cases, simple shielding is inadequate and the thermocouple error is further reduced by increasing significantly the gas velocity by educting gas as shown in the sketch. Write the governing equations showing how gas velocity influences thermocouple error.

13.7 A package of electronic equipment is enclosed in a sheet metal box, a 1-ft cube. If the equipment uses 1500 watts of electrical power, what is the average temperature of the container walls if the wall temperature is assumed uniform all around the package? Also, the emissivity of the walls is $\epsilon = 0.80$ and the room air and surrounding temperature is 70 F.

13.8 Air flows in a long square horizontal duct, 1 ft by 1 ft in cross section, made of sheet metal. The sheet metal is very thin and has a galvanized surface ($\epsilon = 0.15$). A portion of the duct is uninsulated and another portion is covered with $\frac{1}{8}$-in. thick sheet asbestos ($\epsilon = 0.9$, $k = 0.1$ Btu/hr ft F). If the duct wall temperature is 130 F in each section and the air and surrounding temperature is 70 F, compare the rates of heat loss for the two sections of duct.

13.9 Hot air at 1000 F is flowing in a 10 in. diameter duct at a velocity of 100 ft/sec and 14.7 psia. A 5 in. long thermocouple well consisting of a tube (O.D. = 0.5 in., I.D. = 0.4 in.) with a capped end is used to attempt to measure the air temperature. A thermocouple measures the temperature of the end of the tube. The other end of the tube is attached to the duct wall whose inner temperature is 500 F. Assume the thermocouple well tube is also at 500 F at the point of contact at the duct wall. Using an analysis similar to that for "fins," derive the differential equation for determining the temperature distribution along the thermocouple well. State the boundary conditions necessary.

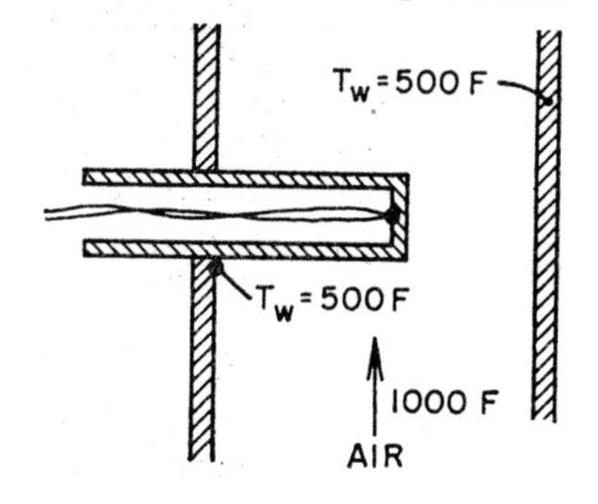

13.10 A large furnace cavity has an inside surface temperature of 1200 F. The 2-ft thick wall is composed of Kaolin insulating brick ($k_m = 0.8$). A hole 6 by 6 in. is open through the furnace wall to the atmosphere which is at 70 F. Calculate the heat loss per ft^2,

(a) through the open hole

(b) through the vertical furnace wall.

13.11 Derive an expression for AF for the disks of radius R_1 and R_2 separated by distance D of Fig. 13.12. Show that

$$AF = (\pi/2)[R_1^2 + R_2^2 + D^2 - \sqrt{(R_1^2 + R_2^2 + D^2} - 4R_1^2R_2^2]$$

13.12 Two flat wooden walls (ϵ = 0.9) are separated by a space in which is a sheet of aluminum foil (ϵ = 0.04). The inner surfaces of the wooden walls are 100 F and 400 F respectively. Neglecting convection and conduction,

(a) what is the heat loss and the temperature of the aluminum foil?

(b) what is the heat loss if the foil is not present?

13.13 A shallow insulated flat pan containing a thin layer of water is placed out of doors at night. The effective black-body temperature of the cloudless night sky is 320 F abs and the air temperature is 60 F.

(a) Will ice form? Justify conclusion by a calculation.

(b) If ice will form, what is the initial rate of formation?

13.14 Air at 1500 F and 1 atm flows at 50 ft/sec in a 12-in. diameter duct (ϵ = 0.9) whose wall temperature is 1000 F. A solid copper (ϵ = 0.4) sphere 0.25 in. in diameter initially at 60 F is inserted into the air stream. Calculate the lengths of time required for the sphere to reach 500 F and 1000 F. What is the steady-state temperature?

13.15 Air at 14.7 psia and 60 F is heated in a steel tube (2.0 in. I.D., 2.2 in. O.D.) at a flow rate of 1000 lb/hr. The tube passes through a brick furnace 24 in. long. The inside dimensions are 6 × 6 × 24 in., the tube passing through the long way. The emissivity of the brick is 0.9 and of the steel is 0.75. The inside furnace wall temperature is 2000 F. Calculate the outlet air temperature and the heat flow rate through the tube walls. (Assume the tube outer wall is at a uniform temperature along its length.)

13.16 Water is evaporated in passing through a grid of 1-in. diameter tubes on 2-in. centers near and parallel with the top of a rectangular furnace. The hot carborundum floor is 2 by 3 ft, the walls are refractory surfaces, and the tube grid is 1.5 ft above the floor. The surface of the floor is maintained at 2500 F and has ϵ = 0.90, and the tube surface will be at 1000 F with ϵ = 0.75. Calculate the rate of heat transfer to the evaporating water in the tubes.

13.17 A steel plate 8 ft by 8 ft by 1 in. is removed from an annealing oven at a uniform temperature of 800 F and hung horizontally in the room at 70 F. Both convection and radiation effects should be considered on top and bottom of the plate. The emissivity is 0.85.

(a) Calculate the initial rate of heat transfer from the plate.

(b) Write an equation for calculating temperature vs. time.

(c) Calculate the time required for the plate to reach 200 F. (Use graphical integration.)

13.18 A gas (60 mol N_2, 25 mol CO_2, 15 mol H_2O) is flowing in steady flow in a long straight duct of triangular cross section (equilateral triangle of 1-ft sides). At one section of the duct the gas velocity is 1000 fps, the temperature is 2000 F, and the pressure is 15 psia. If the wall temperature is 1000 F and its emissivity is 0.8, what is the rate of heat transfer per foot of duct length?

13.19 A long duct 1 ft by 1 ft in cross section carries CO_2 at 1000 F and 1 atm. Three of the sides are no-net-heat-flux refractory surfaces ($\epsilon = 0.8$) and the fourth side is cold surface ($\epsilon = 0.60$) maintained at 250 F. Use $L = 1$ ft to calculate all gas emissivities, absorptivities, and transmissivities. Calculate the rate of heat transfer to the cold surface due to radiation effects.

13.20 A cubical furnace cavity, 10 ft on each edge, has a fuel bed of pure carbon ($HV_0 = 14{,}600$ Btu/lb) at the bottom face. Air (21% O_2 and 79% N_2, by volume) at 60 F is introduced through and over the fuel surface with sufficient excess air to produce products of combustion containing 14% CO_2 by volume. Assume this gas is gray and that its emissivity for all paths is calculated with a mean beam length $L = 0.667 \times 10 = 6.67$ ft (Table 13.1).

(a) Assume the top face is a deep bank of boiler tubes. Essentially its cold surface is maintained at 1000 F abs by boiling water. The four walls are no-net-flux refractory surfaces. All wall surfaces are black, $\epsilon = 1.0$. Calculate and plot the gas temperature for firing rates in the range 10 to 100 lb of fuel burned per square foot of fuel bed area per hour. Assume the gas is at the same temperature as the fuel bed. Also calculate the rate of radiant heat transfer to the cold surface. (*Suggestion*: Assume various gas temperatures and calculate the firing rate.) (Wohlenberg (11))

(b) Repeat the calculations in (a) for the case in which two of the side walls in addition to the top are cold surfaces at 1000 F abs.

(c) Repeat (a) if all five surfaces are cold surfaces at 1000 F abs, e.g., no refractory walls.

(d) Plot gas temperature and furnace heat absorption efficiency vs. firing rate. Furnace heat absorption efficiency is rate of radiant heat transfer to the cold surface divided by the product of heating value times the fuel burning rate.

PART THREE **Mass Transfer**

CHAPTER 14

Introduction to Mass Transfer

In mass transfer systems we encounter a complication not present in our previous studies of momentum and heat transfer in that we are dealing with multi-component systems, and, as we shall see in Section 14.1, there are a number of useful ways of defining their velocities, concentrations, and fluxes. Accordingly, there are a number of ways of stating the rate equations of diffusion; since these are defining equations for the diffusivity, one form is as valid as the next but not necessarily as advantageous when considered in connection with a given application.

For the sake of simplicity, we shall confine our attention to non-reactive, two-component systems and consider only the diffusion flux resulting from a concentration gradient of the diffusing component. We thus ignore the contributions, if any, to the diffusion flux owing to the presence of other driving forces in the system, such as the thermal diffusion, the so-called Soret effect. Likewise, in writing the rate equation of heat conduction in non-isothermal binary systems, the contribution to the heat flux by the concentration gradient, the so-called Dufour effect, is ignored. Both Soret and Dufour effects are examples of coupled phenomena; the general rate equations, when coupling effects are significant, may be derived by a method of irreversible thermodynamics known as the Onsager method.

14.1 Definitions

In Chap. 1, the rate equation for diffusion was given as

$$\left(\frac{w_a}{A}\right)_x = -D\frac{\partial c_a}{\partial x} \tag{14.1}$$

where (w_a/A) is the mass flux and c_a is the mass concentration, mass of component a per unit volume.

Alternate statements of the rate equation appear in the literature. In addition to the mass concentration c_a, the following potentials are used: the mass fraction $c_a^* = c_a/\rho$, also called the *fractional mass concentration*; the molal concentration, n_a, the number of moles of a per unit volume; and the mole fraction, n_a^*, the ratio of the number of moles of a to the total number of moles of the solution or mixture.

In terms of the mass fraction c_a^*, the rate equation is usually written as follows:

$$\left(\frac{w_a}{A}\right)_x = -\rho D\frac{\partial c_a^*}{\partial x} \tag{14.2}$$

Note that the transport coefficient defined by Eq. (14.2) has the dimensions of (length)^2/(time). Coefficients D defined by Eqs. (14.1) and (14.2) are clearly identical when the mass density ρ of the mixture is uniform—that is, in a solid, a liquid solution, or a dilute gaseous mixture. When ρ is not uniform, Eq. (14.2) may be preferred for reasons given in Sec. 15.5.

Equations (14.1) or (14.2) are valid when no temperature or total pressure gradients are present in the system. In a non-isothermal binary mixture, for example, there may be a "coupled" phenomenon of thermal diffusion, that is, diffusion of matter owing to a thermal driving force even in the absence of any concentration gradients. Thermal diffusion is useful in such special applications as isotope separation. A rate equation of mass transfer that includes the thermal diffusion term may be derived by the Onsager method discussed in Chap. 19, or for gases by the kinetic theory. Experiments show, however, that the phenomenon is significant only under extremely larger thermal gradients, not normally encountered in engineering systems.

Let us now consider a mixture of n components which is itself in bulk motion in the x-direction by free or forced convection. It is necessary first to define what we mean by the *bulk motion* of a mixture. Let v_{ix} be the statistical mean velocity of component i in the mixture with respect to stationary coordinate axes. The *bulk velocity* is defined as the velocity of the center of mass of a volume element of the mixture

$$v_x \equiv \frac{1}{\rho}\sum_{i=1}^{n} c_i v_{ix} \tag{14.3a}$$

or for a binary mixture of components a and b,

$$v_x = \frac{1}{\rho}(c_a v_{ax} + c_b v_{bx}) \tag{14.3b}$$

The mass flux of component a can be expressed relative to either a fixed observer or an observer moving with the bulk velocity v_x. The latter viewpoint is associated with the concept of a substantial derivative (D/Dt), defined in Chap. 1. The diffusion flux given by Eq. (14.1) or (14.2) is that seen by the observer moving at v_x. The absolute flux of a seen by the stationary observer, which is given by $c_a v_{ax}$, contains in addition to the diffusion flux $(w_a/A)_x$ a mass flux $c_a v_x$ due to the bulk motion. Therefore

$$c_a v_{ax} = (w_a/A)_x + c_a v_x \tag{14.4}$$

The distinction between the diffusion flux $(w_a/A)_x$ given by Eq. (14.1) or (14.2) and the absolute flux given by Eq. (14.4) should be clearly understood in order to avoid any confusion when deriving the conservation equations. In order to be consistent with our previous treatment of momentum and heat transfer, we use the term *rate equation* for equations of the form (14.1) or (14.2) which express the diffusion flux. Equation (14.4) is useful when we want to know the absolute flux past a fixed section.

The use of either a mass or a molal description of the diffusion process is largely a matter of preference and convenience. We shall discuss both approaches, and use one or the other depending upon the nature of the problem to be analyzed. A parallel molal treatment of the diffusion process would involve rate equations of the following form:

$$\left(\frac{N_a}{A}\right)_x = -D\,\frac{\partial n_a}{\partial x} \tag{14.5}$$

$$\left(\frac{N_a}{A}\right)_x = -\,nD\,\frac{\partial n_a^*}{\partial x} \tag{14.6}$$

where n_a is the molal concentration, n_a^* is the mole fraction, n is the total molal density, and $(N_a/A)_x$ is the molal diffusion flux of a.

A molal bulk velocity in the x-direction may be defined as

$$\mathcal{V}_x \equiv \frac{1}{n}\sum_{i=1}^{n} n_i v_{ix} \tag{14.7a}$$

or for a binary mixture

$$\mathcal{V}_x = \frac{1}{n}(n_a v_{ax} + n_b v_{bx}) \tag{14.7b}$$

The absolute molal flux of a through a plane fixed in space is then given by

$$n_a v_{ax} = \left(\frac{N_a}{A}\right)_x + n_a \mathcal{V}_x \tag{14.8}$$

where $(N_a/A)_x$ is the diffusion flux given by Eq. (14.5) or (14.6). A comparison of Eqs. (14.4) and (14.8) shows that the latter equation cannot be obtained from the former simply by dividing through by the molecular weight M_a since $\mathcal{V}_x$ defined by Eq. (14.7b) is different from v_x defined by Eq. (14.3b).

14.2 Molecular Diffusion in Gases

Theoretical expressions for the diffusion coefficient in gaseous mixtures as a function of the molecular properties of the diffusing gases were derived by Jeans (3), Chapman (4), and Sutherland (5) using the kinetic theory. The form of the equations proposed by these men is similar to Eq. (14.9), differing only in the magnitude of the coefficient. Since the theory is quantitatively rigorous only in the case of the monatomic gas, Gilliland (6) used this form of the equation and evaluated the coefficient empirically based upon published experimental data to obtain the following result:

$$D = 0.0069 \frac{T^{3/2}}{p(V_a^{1/3} + V_b^{1/3})^2} \sqrt{\frac{1}{M_a} + \frac{1}{M_b}} \tag{14.9}$$

TABLE 14.1 ATOMIC VOLUMES

Air	29.9	N_2 in secondary amines	12.0
Antimony	24.2	Oxygen, molecule (O_2)	7.4
Arsenic	30.5	Oxygen coupled to two other elements:	
Bismuth	48.0	In aldehydes and ketones	7.4
Bromine	27.0	In methyl esters	9.1
Carbon	14.8	In ethyl esters	9.9
Chlorine, terminal as in R—Cl	21.6	In higher esters and ethers	11.0
Medial as in R—CHCl—R′	24.6	In acids	12.0
Chromium	27.4	In union with S, P, N	8.3
Fluorine	8.7	Phosphorus	27.0
Germanium	34.5	Silicon	32.0
Hydrogen, molecule (H_2)	14.3	Sulfur	25.6
In compounds	3.7	Tin	42.3
Iodine	37.0	Titanium	35.7
Nitrogen, molecule (N_2)	15.6	Vanadium	32.0
N_2 in primary amines	10.5	Zinc	20.4

For 3-membered ring, deduct 6.0.
For 4-membered ring, deduct 8.5.
For 5-membered ring, deduct 11.5.
For 6-membered ring, deduct 15.0.
For naphthalene ring, deduct 30.0.
For anthracene ring, deduct 47.5.

where D is in ft²/hr, p in atm, T in degrees Rankine, and V_a and V_b are atomic volumes as given in Table 14.1. For compounds, Kopp's law of additive volumes applies. Thus for CO_2, $V = 14.8 + 2(7.4) = 29.6$.

Equation (14.9) shows D to be independent of the ratio of quantities of the gases present. Several experiments attempted to discover an effect of gas composition on D but found only an insignificant effect.

Equation (14.9) should be used only as an approximation where experimental data are not available. Here $D \sim T^{3/2}$, whereas actually in experiments $D \sim T^2$ is usually found. Later advances in theoretical and experimental results have led to more complex correlations of experimental data. Reid and Sherwood (8) and Jost (2) present a more detailed discussion of these results.

Example: Estimate the diffusivity of ethyl alcohol (C_2H_6O) in air at a temperature of 40 F and a pressure of 1 atm.

For air, $V_a = 29.9$ and for C_2H_6O,

$$V_b = 2(14.8) + 6(3.7) + 7.4 = 59.2$$

Then

$$D = 0.0069\frac{(40 + 460)^{3/2}\sqrt{(1/28.95) + (1/46)}}{(1)[29.9^{1/3} + 59.2^{1/3}]^2} = 0.37 \text{ ft}^2/\text{hr}$$

14.3 Molecular Diffusion in Liquids and Solids

Diffusion in liquids occurs at a very much slower rate than in gases. Since the kinetic theory of liquids is not as well developed as that of gases, it is usually assumed as a first approximation that equations of the same general form are applicable to the diffusion of a solute in a solvent as to the diffusion in gases. The rate equation (14.1) is assumed valid for liquids, and an appropriate value of D is determined by experiment for particular cases. Unfortunately, because of experimental difficulties, reliable data on diffusion in liquids are relatively scarce.

Diffusivities for most of the common organic and inorganic materials in the usual solvents such as water, alcohol, and benzene at room temperature lie in the range from 1.0 to 6.0×10^{-5} ft²/hr. The following approximate semi-empirical relation was suggested by Wilke (7) for dilute solutions:

$$D = \frac{4.0 \times 10^{-7} T}{\mu(V^{1/3} - K_1)} \tag{14.10}$$

where D is in ft²/hr, T is in degrees Rankine, μ is the viscosity of the solution in lb_m/ft-hr, V is the atomic volume given by Table 14.1, and K_1 is 2.0, 2.46, and 2.84 for water, methyl alcohol, and benzene respectively.

Diffusion in solids is much slower than in liquids. Diffusion of solid in solid has limited engineering interest, but diffusion of fluids in solids have extensive applications. The mechanism of mass transfer in solids is little understood. Experiments on many solid media, especially fibrous or granular media, show that the mass flux is not proportional to the concentration gradient; in other words, transfer processes in these solids are not strictly diffusional. Nevertheless, rate equation (14.1) is sometimes used, together with an empirically determined effective diffusivity which attempts to take into account the structure of the solid. A typical problem of liquid transfer in a solid, of interest in drying of solids, is given in Chap. 15.

14.4 Concentration Boundary Layer and Mass Transfer Coefficient

Calculation of momentum and heat transfer rates at a solid-fluid interface by the appropriate rate equation requires a knowledge of the velocity

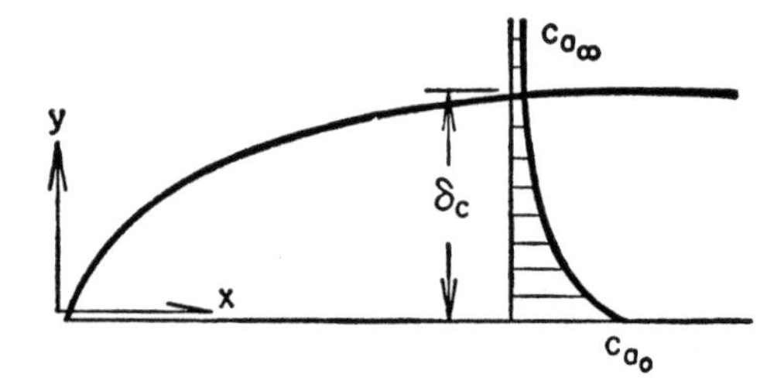

Fig. 14.1. Concentration boundary layer on flat plate.

and the temperature profiles within the boundary layer. Similarly, mass transfer at an interface is determined by the concentration boundary-layer profile. Figure 14.1 shows a fluid mixture, with free-stream velocity and concentration designated respectively by V_∞ and $c_{a\infty}$, flowing over a flat plate; if the plate surface is maintained at a concentration $c_{a0} > c_{a\infty}$, mass a diffuses from the surface into the fluid stream. A concentration boundary layer grows from the leading edge in the same way that a velocity or a thermal boundary layer grows, and the thickness of the concentration boundary layer δ_c may be defined as the distance from the plate where

$$(c_{a0} - c_a) = 0.99(c_{a0} - c_{a\infty})$$

The relative rates of growth of the velocity and the concentration boundary layers are determined by the Schmidt number ($\text{Sc} = \nu/D$) of the fluid (Sec. 1.2d). In addition, if a temperature difference exists between the plate and the free stream, a thermal boundary layer grows concurrently at a relative rate determined by the Prandtl or the Lewis number. When Sc = Pr = Le = 1, all three boundary layers coincide.

By analogy with heat transfer, a mass transfer coefficient may be defined, with reference to Fig. 14.1, as follows*

$$h_D = \frac{(w_a/A)_{y=0}}{c_{a0} - c_{a\infty}} \tag{14.11}$$

The dimensionless mass transfer numbers corresponding to the Nusselt and the Stanton numbers in heat transfer are respectively

$$\frac{h_D L}{D}, \quad \text{Sherwood number (Sh)}, \quad \text{and} \quad \frac{h_D}{V} = \frac{(\text{Sh})}{(\text{Re})(\text{Sc})}$$

where L and V are any characteristic length and velocity of the system.

In many cases of mass transfer across a gas-liquid interface, the resistance to mass transfer in one of the two phases is controlling, requiring calculation of mass transfer coefficient in that phase only. For slightly soluble gases, the controlling resistance is usually in the liquid phase, whereas for highly soluble gases, the controlling resistance is in the gas phase. If, however, the resistance to mass transfer is appreciable in both phases, an "over-all" mass transfer coefficient, analogous in concept to the over-all heat transfer coefficient, may be determined by experiment. This problem is important in absorption processes and is discussed in detail in reference (1).

REFERENCES

1. Sherwood, T. K., and R. L. Pigford, *Absorption and Extraction*, 2nd Ed., McGraw-Hill, New York, 1952.

2. Jost, W., *Diffusion*, Academic, New York, 1952.

* We note that h_D in Eq. (14.11) is defined in terms of the diffusion flux at the surface. Generally, at a solid-fluid or a liquid-gas interface, there is an additional contribution to mass transfer by bulk flow normal to the surface so that the total mass flux with respect to a stationary observer is $c_a v_{ax}$ given by Eq. (14.4). The mass transfer coefficient is defined here in terms of the diffusion flux rather than the total flux, because the coefficient defined in this manner depends on a smaller number of parameters and is simpler to apply. Note also that the diffusion flux might be expected to be approximately proportional to a characteristic concentration difference, as indicated by Eq. (14.11), but the bulk-flow contribution can be relatively independent of any concentration difference. Similarly, when both heat and mass transfer occur, it would be advantageous to retain the definition of the heat transfer coefficient given by Eq. (5.3), which considers only the conduction flux.

3. Jeans, J. H., *Dynamical Theory of Gases*, 3rd Ed., p. 307, Cambridge Univ. Press, New York, 1921.

4. Chapman, W., *Trans. Roy. Soc. (London)*, **A217,** 165 (1917).

5. Sutherland, W., *Phil. Mag.*, **38,** 1 (1894).

6. Gilliland, E. R., *Ind. Eng. Chem.*, **26,** 681 (1934).

7. Wilke, C. R., *Chem. Eng. Progress*, **45,** 218 (1949).

8. Reid, R. C., and T. K. Sherwood, *The Properties of Gases and Liquids*, McGraw-Hill, New York, 1958.

CHAPTER 15

Diffusion in Stationary and Laminar Flow Systems

We begin our study with the simplest case of diffusion of a single component in a non-reactive stationary medium. A mass balance is written for the diffusing component, and typical boundary conditions are indicated. The equation is solved exactly for a number of one-dimensional problems, and the applicability of solutions obtained in Chapter 6 to corresponding mass transfer problems is stressed. The equations of diffusion with superimposed convection are next derived; the different forms of these equations should be carefully noted, particularly with reference to the bulk velocities v_x and $\mathcal{V}_x$ defined in Section 14.1. These equations are applied to diffusion in binary gas mixtures and to problems of forced convection mass transfer in tubes and over flat plates. For very low mass transfer rates, the boundary conditions in convection are analogous to those of corresponding heat transfer problems, and the solutions obtained in the latter may be applied directly to the former, as discussed in Section 15.7. For moderate or high mass transfer rates, the bulk velocity normal to the wall associated with injection or suction at the wall cannot be neglected, as illustrated in connection with the flat plate problem of Section 15.8.

15.1 Equations of Mass Diffusion in Stationary Media

The mass balance for component a diffusing through a control volume Δx, Δy, and Δz in a solid or a stationary fluid medium, $(v_x = v_y = v_z = 0)$ may be obtained as follows (see Fig. 15.1):

Net transfer of a by diffusion in the x-direction:

$$\frac{\partial}{\partial x}\left(\frac{w_a}{A}\right)_x \Delta x\ \Delta y\ \Delta z$$

Similar expressions may be written in the y- and the z-directions.

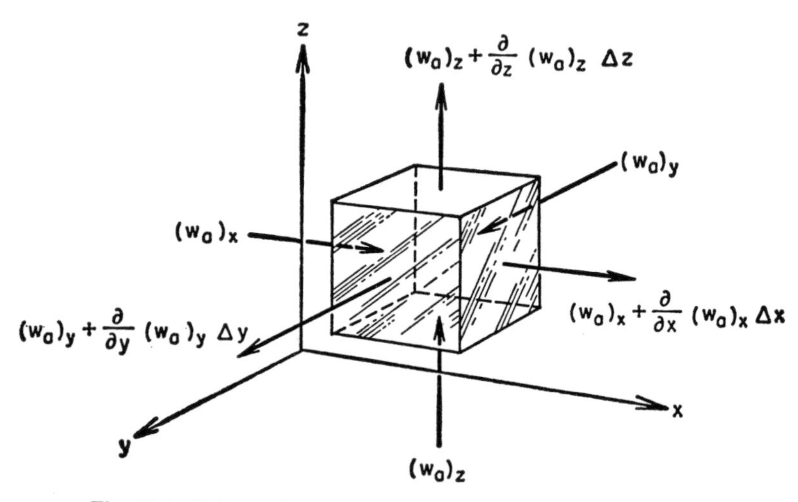

Fig. 15.1. Volume element for deriving mass diffusion equation.

Rate of production or immobilization of a within the volume:

$$Q_a\ \Delta x\ \Delta y\ \Delta z$$

where Q_a is the rate of production per unit volume.

Rate of change of concentration of a within the volume:

$$\frac{\partial c_a}{\partial t}\ \Delta x\ \Delta y\ \Delta z$$

The mass balance, dividing through by $\Delta x\ \Delta y\ \Delta z$, is therefore given by:

$$\frac{\partial c_a}{\partial t} = -\frac{\partial}{\partial x}\left(\frac{w_a}{A}\right)_x - \frac{\partial}{\partial y}\left(\frac{w_a}{A}\right)_y - \frac{\partial}{\partial z}\left(\frac{w_a}{A}\right)_z + Q_a \tag{15.1}$$

If the medium were isotropic and the diffusion obeyed the rate equation (14.1), then Eq. (15.1) becomes

$$\frac{\partial c_a}{\partial t} = \frac{\partial}{\partial x}\left(D\frac{\partial c_a}{\partial x}\right) + \frac{\partial}{\partial y}\left(D\frac{\partial c_a}{\partial y}\right) + \frac{\partial}{\partial z}\left(D\frac{\partial c_a}{\partial z}\right) + Q_a \qquad (15.2)$$

or in vector notation

$$\frac{\partial c_a}{\partial t} = \text{div }(D \text{ grad } c_a) + Q_a$$

If, in addition, D were independent of x, y, z, and c_a, and Q_a were zero,

$$\frac{\partial c_a}{\partial t} = D\left(\frac{\partial^2 c_a}{\partial x^2} + \frac{\partial^2 c_a}{\partial y^2} + \frac{\partial^2 c_a}{\partial z^2}\right) = D\,\nabla^2 c_a \qquad (15.3)$$

A few of the typical boundary conditions in mass transfer are listed in the following:

(a) Specified boundary concentration:

$$\text{at } y = 0, \quad c_a = c_{a0} \qquad (15.4a)$$

(b) Impermeable surface at boundary:

$$\text{at } y = 0, \quad \left(\frac{\partial c_a}{\partial y}\right)_{y=0} = 0 \qquad (15.4b)$$

(c) Specified magnitude of mass flux at boundary:

$$\text{at } y = 0, \quad \left(\frac{\partial c_a}{\partial y}\right)_{y=0} = -\frac{1}{D}\left(\frac{w_a}{A}\right)_{y=0} \qquad (15.4c)$$

(d) The mass flux at a surface is given by Eq. (14.11):

$$\text{at } y = 0, \quad \left(\frac{\partial c_a}{\partial y}\right)_{y=0} = -\frac{h_D}{D}(c_{a0} - c_{a\infty}) \qquad (15.4d)$$

Equations (15.2) and (15.3) are analogous to Eqs. (6.2) and (6.4) in heat conduction, and solutions may be found as in the analogous heat conduction problems. These solutions are extensively discussed by Crank (1). The numerical and analog methods discussed in Chap. 18 are also applicable. We shall examine a few simple cases.

15.2 Steady-State Diffusion through a Plane Membrane

The problem is analogous to that of heat conduction through a slab. We shall assume Eq. (15.3) applies. It simplifies to

$$\frac{d^2c_a}{dx^2} = 0 \tag{15.5}$$

If the boundary conditions were

$$\begin{aligned} x &= 0, \quad c_a = c_{a1} \\ x &= l, \quad c_a = c_{a2} \end{aligned} \tag{15.6}$$

the concentration profile and mass transfer rate would be given respectively by:

$$c_a = (c_{a2} - c_{a1})\frac{x}{l} + c_{a1} \tag{15.7}$$

$$\frac{w_a}{A} = \frac{D}{l}(c_{a1} - c_{a2}) \tag{15.8}$$

as shown in Fig. 15.2.

In certain engineering applications, such as the diffusion of water vapor through a vapor barrier, it is more convenient to express rate of transfer in terms of vapor pressures on the two sides of the membrane rather than in terms of the surface concentrations which are difficult to measure. A new parameter called permeability $\mathcal{P}$ is sometimes used and is defined by:

$$\mathcal{P} \equiv \frac{w_a/A}{(p_{a1} - p_{a2})/l} \tag{15.9}$$

Comparison of Eqs. (15.8) and (15.9) shows that if the relationship (called *sorption isotherms*) between the vapor pressure p_a and the surface concentration of the membrane, c_a, is known, $\mathcal{P}$ may be expressed directly as a function of the diffusion coefficient D.

Composite membranes: Composite membranes in series and in parallel can be treated similar to the thermal or the electrical resistances in series and in parallel (see comment in Sec. 6.3) by defining a diffusional resistance for each membrane:

$$R_D \equiv \frac{c_{a1} - c_{a2}}{w_a} = \frac{l}{AD} \tag{15.10}$$

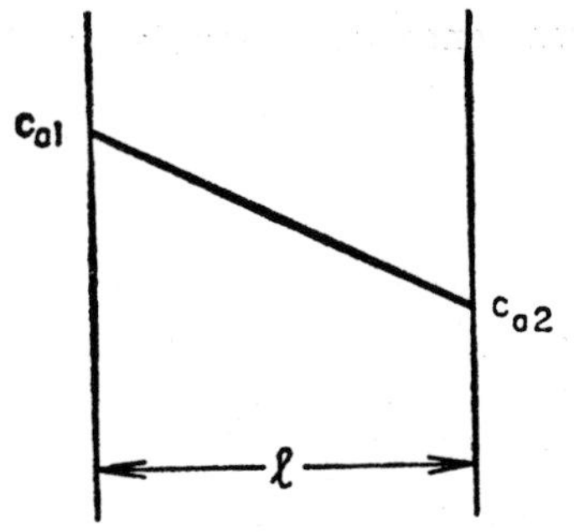

Fig. 15.2. Steady-state concentration distribution in a plane membrane.

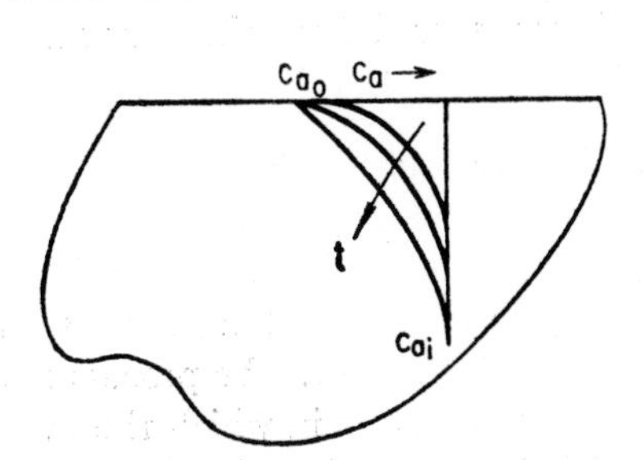

Fig. 15.3. Transient concentration distribution in a semi-infinite medium.

15.3 Transient Diffusion in a Semi-Infinite Medium

Assume Eq. (15.3) applies,

$$\frac{\partial c_a}{\partial t} = D\frac{\partial^2 c_a}{\partial x^2} \tag{15.11}$$

The solution given in Appendix C applies to this case, if the boundary conditions of the problem are as follows (Fig. 15.3):

$$t = 0, \quad c_a = c_{a_i} \quad \text{everywhere}$$

$$t > 0, \quad c_a = 0, \quad \text{at } x = 0$$

With

$$P_i = c_{ai},$$

$$P = c_a,$$

and

$$\mathfrak{D} = D,$$

$$\frac{c_a}{c_{a_i}} = \operatorname{erf}\frac{x}{2\sqrt{Dt}} \tag{15.12}$$

$$\frac{w_a}{A} = -D\left(\frac{\partial c_a}{\partial x}\right)_0 = \frac{-Dc_{a_i}}{\sqrt{\pi Dt}} \tag{15.13}$$

It is easy to verify that Eq. (15.12) satisfies Eq. (15.11) and the specified boundary conditions. The solution involves a single dimensionless parameter $x/2\sqrt{Dt}$, from which we conclude that the rate of penetration of any given concentration varies as $\sqrt{t}$, or conversely, the time required for any point within the medium to attain a given concentration varies as x^2, x being measured from the surface.

15.4 Drying of Solids

Many industrial processes involve the drying of solids—paper pulp, foods, photographic film, etc.—by passing a stream of inert gas over the surface of the solid. The process of drying wet solids has been found to be of two types: one in which the rate of drying is constant and the other in which the rate of drying is decreasing. When the solid is very wet, the initial stages of drying occur while the surface exposed to the inert gas stream remains wet. Then the primary resistance to mass transfer is at the gas-liquid (wet surface) interface. The rate of transfer is governed by the rate of evaporation from the wet surface into the flowing gas stream (see Sec. 15.8 for the laminar flow analysis and Chap. 16 for the evaluation of the mass transfer coefficient in turbulent flow).

When dry spots appear on the surface of the solid, the rate of drying begins to decrease as the internal resistance to diffusion becomes significant. Drying of the solid continues at a decreasing rate until an equilibrium moisture concentration c_e in the solid is reached. The magnitude of c_e depends upon the temperature, pressure, and relative humidity of the inert gas and is determined for each solid empirically. During the falling-rate period, the internal diffusion process of liquid through many non-granular solids may be considered to be represented by an equation similar to Eq. (15.11):

$$\frac{\partial c}{\partial t} = D \frac{\partial^2 c}{\partial x^2} \tag{15.14}$$

if the effects of capillarity and "checking" play a minor role. At the boundary, Eq. (15.4d) may still be considered applicable, but the c_a in Eq. (15.4d) refers to concentration of the vapor a in the gas whereas c in Eq. (15.14) refers to concentration of the liquid in the solid. Since no mass is transferred from the surface when the liquid concentration in the surface $c_0 = c_e$, it is possible to define a new coefficient H_D at the surface such that

$$\frac{w_a}{A} = H_D(c_0 - c_e) = h_D(c_{a0} - c_{a\infty}) \tag{15.15}$$

where c_0 and c_e are respectively the liquid concentration in the solid at the surface and at the equilibrium condition.

These equations are identical in form with those for heat conduction, Sec. 6.13, and hence may be solved by using the charts in that section. For this case

$$Y = \frac{c - c_e}{c_i - c_e}, \quad \frac{Dt}{r_0^2} \text{ replaces } \frac{\alpha t}{r_0^2}$$

$$\frac{H_D r_0}{D} = \frac{\text{surface diffusion resistance}}{\text{internal diffusion resistance}} \text{ replaces } \frac{h r_0}{k}$$

In these equations the concentrations (c) are pounds of liquid per unit volume. If the solid does not contract or expand during drying, then the pound of solid per unit volume remains constant and the concentraton c may be taken as pounds of liquid per pound of dry solid material. Usually, the solid contracts on drying. If the contraction occurs at a rate just sufficient to keep constant the quantity pounds of liquid-plus-solid per unit volume, then c should be interpreted as pounds of liquid per pound of total (solid plus liquid) mass. The former is a "concentration on a dry basis," the latter a "concentration on a wet basis."

Example 15.1: A slab of wood 2 in. thick has a moisture content $c_i = 30$ per cent (based on dry wood) when the falling-rate period begins. The equilibrium moisture content is $c_e = 5$ per cent (based on dry wood) for the conditions of the air existing during the drying. The edges and ends are covered with a moisture-resistant coating. It may be assumed that the surface resistance is negligible; e.g., $(H_D r_0/D) \to \infty$. It is known that for the diffusion of water through this wood, $D = 0.00004$ ft^2/hr.

What is the drying time to reduce the moisture content at the centerline to 10 per cent on a dry basis? It may be assumed that the wood does not shrink during drying.

For these conditions $Y = (10 - 5)/(30 - 5) = 0.20$. Then from Fig. 6.13 with $(H_D r_0/D)$ or $h r_0/k = \infty$, and $x = 0$ for centerline conditions, $X = 0.73$. Since

$$X = \frac{0.00004t}{(\frac{1}{12})^2} = 0.73$$

$t = 127$ hr.

15.5 Equations of Diffusion with Convection

Let us now consider a binary system in bulk motion with velocity $\mathbf{V}$, within which component a diffuses. The convection or bulk velocity $\mathbf{V}$ is defined as in Eq. (14.3).

A mass balance on component a in the mixture flowing through a volume element $\Delta x\ \Delta y\ \Delta z$ fixed in space contains the following quantities:

Rate of change of mass concentration of a within the volume:

$$\frac{\partial c_a}{\partial t}\ \Delta x\ \Delta y\ \Delta z$$

Net transfer of a in the x-direction:

$$-\frac{\partial}{\partial x}(c_a v_{ax})\ \Delta x\ \Delta y\ \Delta z$$

Similar expressions can be written for the y- and the z-directions.

Rate of production or immobilization of a within system:

$$Q_a\ \Delta x\ \Delta y\ \Delta z$$

The conservation equation of component a, dividing through by $\Delta x\ \Delta y\ \Delta z$, is

$$\frac{\partial c_a}{\partial t} = -\frac{\partial}{\partial x}(c_a v_{ax}) - \frac{\partial}{\partial y}(c_a v_{ay}) - \frac{\partial}{\partial z}(c_a v_{az}) + Q_a \tag{15.16}$$

or in vector notation,

$$\frac{\partial c_a}{\partial t} = -\operatorname{div}\ (c_a \mathbf{v}_a) + Q_a \tag{15.16a}$$

Similarly, the conservation equation of component b is

$$\frac{\partial c_b}{\partial t} = -\frac{\partial}{\partial x}(c_b v_{bx}) - \frac{\partial}{\partial y}(c_b v_{by}) - \frac{\partial}{\partial z}(c_b v_{bz}) + Q_b \tag{15.17}$$

If we add Eqs. (15.16) and (15.17), then for the special case in which Q_a and Q_b are zero, the following continuity equation for the mixture results:

$$\frac{\partial (c_a + c_b)}{\partial t} = -\frac{\partial}{\partial x}(c_a v_{ax} + c_b v_{bx}) - \frac{\partial}{\partial y}(c_a v_{ay} + c_b v_{by}) - \frac{\partial}{\partial z}(c_a v_{az} + c_b v_{bz})$$

or

$$\frac{\partial \rho}{\partial t} + \frac{\partial}{\partial x}(\rho v_x) + \frac{\partial}{\partial y}(\rho v_y) + \frac{\partial}{\partial z}(\rho v_z) = 0 \tag{15.18}$$

where $\rho = c_a + c_b$ and v_x, v_y, and v_z are the bulk velocity components defined by Eq. (14.3b).

All the foregoing equations may be written in an alternate form from the viewpoint of an observer moving with the bulk velocity $\mathbf{V}$ (a barycentric description). By definition

$$\frac{D}{Dt} \equiv \frac{\partial}{\partial t} + v_x\frac{\partial}{\partial x} + v_y\frac{\partial}{\partial y} + v_z\frac{\partial}{\partial z}$$

and from Eq. (14.4), $c_a v_{ax} = (w_a/A)_x + c_a v_x$, etc. Substitution of these expressions into Eq. (15.16), with $Q_a = 0$, yields

$$\frac{Dc_a}{Dt} + c_a\left(\frac{\partial v_x}{\partial x} + \frac{\partial v_y}{\partial y} + \frac{\partial v_z}{\partial z}\right) = -\frac{\partial}{\partial x}\left(\frac{w_a}{A}\right)_x - \frac{\partial}{\partial y}\left(\frac{w_a}{A}\right)_y - \frac{\partial}{\partial z}\left(\frac{w_a}{A}\right)_z \qquad (15.19)$$

or

$$\frac{Dc_a}{Dt} + c_a \operatorname{div} \mathbf{V} = -\operatorname{div}\left(\frac{\mathbf{w}_a}{A}\right) \qquad (15.19a)$$

Likewise, Eq. (15.18) may be rewritten using the substantial derivative

$$\frac{D\rho}{Dt} + \rho\left(\frac{\partial v_x}{\partial x} + \frac{\partial v_y}{\partial y} + \frac{\partial v_z}{\partial z}\right) = 0 \qquad (15.20)$$

or

$$\frac{D\rho}{Dt} + \rho \operatorname{div} \mathbf{V} = 0 \qquad (15.20a)$$

Equation (15.19) becomes simpler if instead of using the mass concentration c_a, we use the mass fraction $c_a^* = c_a/\rho$. Since the rules of differentiation are the same for a substantial derivative as for an ordinary derivative,

$$\frac{Dc_a^*}{Dt} = \frac{1}{\rho}\frac{Dc_a}{Dt} - \frac{c_a}{\rho^2}\frac{D\rho}{Dt}$$

which with Eq. (15.20) gives

$$\rho\frac{Dc_a^*}{Dt} = \frac{Dc_a}{Dt} + c_a\left(\frac{\partial v_x}{\partial x} + \frac{\partial v_y}{\partial y} + \frac{\partial v_z}{\partial z}\right)$$

Replacing the left-hand side of Eq. (15.19) by this last result, we get

$$\rho\frac{Dc_a^*}{Dt} = -\frac{\partial}{\partial x}\left(\frac{w_a}{A}\right)_x - \frac{\partial}{\partial y}\left(\frac{w_a}{A}\right)_y - \frac{\partial}{\partial z}\left(\frac{w_a}{A}\right)_z \qquad (15.21)$$

Equations (15.16), (15.19), and (15.21) express the conservation of component a when there is bulk convective motion of the mixture. Comparison of Eqs. (15.19) and (15.21) suggests that in the barycentric description, the conservation equation has the simplest form when c_a^* is the potential.

Substitution of Eq. (14.2) into Eq. (15.21) yields

$$\rho\frac{Dc_a^*}{Dt} = \frac{\partial}{\partial x}\left(\rho D\frac{\partial c_a^*}{\partial x}\right) + \frac{\partial}{\partial y}\left(\rho D\frac{\partial c_a^*}{\partial y}\right) + \frac{\partial}{\partial z}\left(\rho D\frac{\partial c_a^*}{\partial z}\right) \qquad (15.22)$$

For constant total density ρ and constant D, Eq. (15.22) reduces to

$$\frac{Dc_a}{Dt} = D\left(\frac{\partial^2 c_a}{\partial x^2} + \frac{\partial^2 c_a}{\partial y^2} + \frac{\partial^2 c_a}{\partial z^2}\right) \qquad (15.23)$$

Equation (15.23) could have been obtained directly from Eq. (15.19) by noting that for constant total density

$$\operatorname{div} \mathbf{V} = \frac{\partial v_x}{\partial x} + \frac{\partial v_y}{\partial y} + \frac{\partial v_z}{\partial z} = 0$$

We shall also derive the molal conservation equations of diffusion, since frequently it may be more convenient to set up an analysis on a molal rather than a mass basis. If we divide through Eq. (15.16) by the molecular weight M_a, we get

$$\frac{\partial n_a}{\partial t} = -\frac{\partial}{\partial x}(n_a v_{ax}) - \frac{\partial}{\partial y}(n_a v_{ay}) - \frac{\partial}{\partial z}(n_a v_{az}) + Q_a \tag{15.24}$$

Note from Eq. (14.8), $n_a v_{ax} = (N_a/A)_x + n_a \mathcal{V}_x$, etc., where $\mathcal{V}_x$, $\mathcal{V}_y$, $\mathcal{V}_z$ are the molal bulk velocity components defined by Eq. (14.7b). Equation (15.24) may be rewritten from the viewpoint of an observer moving with the molal bulk velocity $\mathcal{V}$ in a form similar to Eq. (15.19) if we defined a *molal substantial derivative* as

$$\frac{\mathfrak{D}}{\mathfrak{D}t} \equiv \frac{\partial}{\partial t} + \mathcal{V}_x \frac{\partial}{\partial x} + \mathcal{V}_y \frac{\partial}{\partial y} + \mathcal{V}_z \frac{\partial}{\partial z}$$

Then, for $Q_a = 0$,

$$\frac{\mathfrak{D}n_a}{\mathfrak{D}t} + n_a\left(\frac{\partial \mathcal{V}_x}{\partial x} + \frac{\partial \mathcal{V}_y}{\partial y} + \frac{\partial \mathcal{V}_z}{\partial z}\right) = -\frac{\partial}{\partial x}\left(\frac{N_a}{A}\right)_x - \frac{\partial}{\partial y}\left(\frac{N_a}{A}\right)_y - \frac{\partial}{\partial z}\left(\frac{N_a}{A}\right)_z \tag{15.25}$$

The molal continuity equation for zero Q_a and Q_b is

$$\frac{\partial n}{\partial t} + \frac{\partial}{\partial x}(n\mathcal{V}_x) + \frac{\partial}{\partial y}(n\mathcal{V}_y) + \frac{\partial}{\partial z}(n\mathcal{V}_z) = 0 \tag{15.26}$$

By the same procedure as was used to obtain Eq. (15.21), we can obtain

$$n\frac{\mathfrak{D}n_a^*}{\mathfrak{D}t} = -\frac{\partial}{\partial x}\left(\frac{N_a}{A}\right)_x - \frac{\partial}{\partial y}\left(\frac{N_a}{A}\right)_y - \frac{\partial}{\partial z}\left(\frac{N_a}{A}\right)_z \tag{15.27}$$

where n is the total molal density, $n_a + n_b$, and $n_a^* \equiv n_a/n$, the mole fraction.

Substitution of Eq. (14.6) into Eq. (15.27) yields

$$n\frac{\mathfrak{D}n_a^*}{\mathfrak{D}t} = \frac{\partial}{\partial x}\left(n\,D\,\frac{\partial n_a^*}{\partial x}\right) + \frac{\partial}{\partial y}\left(n\,D\,\frac{\partial n_a^*}{\partial y}\right) + \frac{\partial}{\partial z}\left(n\,D\,\frac{\partial n_a^*}{\partial z}\right) \tag{15.28}$$

and for constant total number of moles n and constant D,

$$\frac{\mathfrak{D}n_a}{\mathfrak{D}t} = D\left(\frac{\partial^2 n_a}{\partial x^2} + \frac{\partial^2 n_a}{\partial y^2} + \frac{\partial^2 n_a}{\partial z^2}\right) \tag{15.29}$$

Equation (15.29) for constant n and D cannot be obtained from Eq. (15.23) for constant ρ and D by simply dividing the latter equation by M_a, since $\mathfrak{D}/\mathfrak{D}t \neq D/Dt$. This distinction between the two equations should be carefully noted.

15.6 Diffusion in Binary Gas Mixtures

We shall assume for a binary mixture of gases a and b:

1. The Gibbs-Dalton laws hold:

$$p_a + p_b = P$$

2. Each gas is a perfect gas, i.e.,

$$c_a = \frac{p_a M_a}{\mathcal{R}T} \quad \text{or} \quad n_a = \frac{p_a}{\mathcal{R}T}$$

3. The diffusion coefficient D is constant.

4. The process is steady and one-dimensional.

Here p_a and p_b are the partial pressures, P is the total pressure, M_a is the molecular weight of a, and $\mathcal{R}$ is the universal gas constant.

(a) Equimolal Counterdiffusion: $n_a v_{ax} = -n_b v_{bx}$

From Eq. (14.7b), $\mathcal{V}_x = 0$. Since n and D are constant, Eq. (15.29) applies. For the steady, one-dimensional case, with $\mathcal{V}_x = 0$, it reduces to

$$\frac{d^2 n_a}{dx^2} = 0$$

or for a perfect gas

$$\frac{d^2 p_a}{dx^2} = 0 \tag{15.30}$$

For the boundary conditions

$$\text{at } x = 0, \quad p_a = p_{a_1}$$

$$\text{at } x = l, \quad p_a = p_{a_2}$$

Equation (15.30) integrates to

$$p_a = (p_{a_2} - p_{a_1})\frac{x}{l} + p_{a_1} \tag{15.31}$$

Figure 15.4 shows the partial pressure gradient of gas a in a mixture as given by Eq. (15.31). The partial pressure gradient of gas b is equal but of

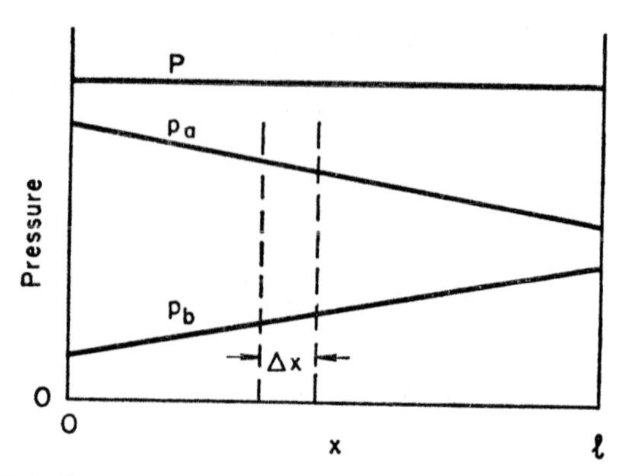

Fig. 15.4. Partial pressure profiles in equimolal counter-diffusion.

opposite sign, since the total pressure is constant by the assumptions of the problem.

The molal flux is given by the rate equation, Eq. (14.5), and for isothermal diffusion in a perfect gas mixture, it may be written in terms of the partial pressure as follows:

$$\frac{N_a}{A} = -\frac{D}{\mathcal{R}T}\frac{\partial p_a}{\partial x} \tag{15.32}$$

or from Eq. (15.31)

$$\frac{N_a}{A} = \frac{D}{\mathcal{R}T}\frac{p_{a_1} - p_{a_2}}{l}$$

(b) Diffusion of Gas a through a Stationary Gas b: $n_b v_{bx} = 0$

Assume the process is at constant total pressure and temperature. Then n = constant, and Eq. (15.29) applies to gas a in the following reduced form:

$$\mathcal{V}_x \frac{dn_a}{dx} = D\frac{d^2 n_a}{dx^2} \tag{15.33}$$

The molal bulk velocity $\mathcal{V}_x$ may be determined from the condition that gas b must be stationary; from Eq. (14.8),

$$n_b v_{bx} = 0 = \left(\frac{N_b}{A}\right)_x + n_b \mathcal{V}_x$$

or

$$\mathcal{V}_x = \frac{D}{n_b}\frac{dn_b}{dx} \tag{15.34}$$

Substitution of Eq. (15.34) into Eq. (15.33) gives

$$\frac{1}{n_b}\frac{dn_b}{dx}\frac{dn_a}{dx} = \frac{d^2n_a}{dx^2} \tag{15.35}$$

For perfect gases, $n_a = p_a/\mathfrak{R}T$ and $n_b = p_b/\mathfrak{R}T$, and for constant total pressure, $dp_a/dx = -dp_b/dx$. Therefore, Eq. (15.35) may be rewritten in terms of p_b as follows:

$$\frac{1}{p_b}\left(\frac{dp_b}{dx}\right)^2 = \frac{d^2p_b}{dx^2} \tag{15.36}$$

Equation (15.36) in equivalent form is

$$\frac{d}{dx}\left(\frac{1}{p_b}\frac{dp_b}{dx}\right) = 0$$

which integrates directly to

$$\ln p_b = A_1x + A_2 \tag{15.37}$$

where A_1 and A_2 are constants of integration.

For the boundary conditions: $x = 0$, $p_b = p_{b_1}$ and $x = l$, $p_b = p_{b_2}$, Eq. (15.37) becomes

$$\ln\left(\frac{p_b}{p_{b_1}}\right) = \left[\ln\left(\frac{p_{b_2}}{p_{b_1}}\right)\right]\frac{x}{l} \tag{15.38a}$$

or

$$\ln\left(\frac{P - p_a}{P - p_{a_1}}\right) = \left[\ln\left(\frac{P - p_{a_2}}{P - p_{a_1}}\right)\right]\frac{x}{l} \tag{15.38b}$$

These solutions are shown in Fig. 15.5. In order to satisfy the condition of constant total pressure, a gradient of p_a in the mixture must be accompanied by a gradient of p_b. Consequently, diffusion of a gives rise to a tendency for b to diffuse in the opposite direction. In order that gas b be stationary, this tendency must be exactly compensated by a bulk motion of the mixture in the direction of diffusion of gas a. The mass flux of a for an observer moving with the molal bulk velocity $\mathcal{V}_x$ is given by the rate equation (14.5). For this observer, gas b is apparently diffusing in the opposite direction at a rate given by the same equation. However, in many engineering applications, we may be interested in knowing the absolute flux of a with respect to stationary coordinate axes. The absolute molal flux is given by the sum of contributions from bulk motion and diffusion, Eq. (14.8):

$$\left(\frac{N_a}{A}\right)_{\text{abs}} = n_a v_{ax} = n_a\mathcal{V}_x - D\frac{dn_a}{dx} \tag{15.39}$$

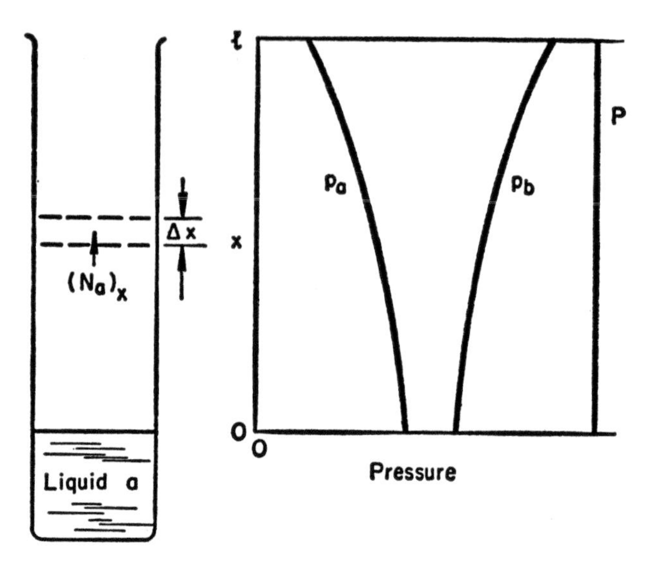

Fig. 15.5. Partial pressure profiles for diffusion of gas a in a stationary gas b.

Substituting $\mathcal{V}_x$ from Eq. (15.34) into Eq. (15.39) and rewriting the resultant expression in terms of the partial pressure p_a, we get:

$$\left(\frac{N_a}{A}\right)_{\text{abs}} = -\frac{D}{\mathcal{R}T}\left(\frac{P}{P - p_a}\right)\frac{dp_a}{dx} \tag{15.40}$$

Equation (15.40) is known as *Stefan's equation.* Since the partial pressure profiles are known from Eq. (15.38), then by differentiation

$$\frac{dp_a}{dx} = -\left(\frac{P - p_a}{l}\right)\ln\left(\frac{P - p_{a_2}}{P - p_{a_1}}\right)$$

and the absolute flux of a may be evaluated from Eq. (15.40) as follows:

$$\left(\frac{N_a}{A}\right)_{\text{abs}} = \frac{DP}{\mathcal{R}Tl}\ln\left(\frac{P - p_{a_2}}{P - p_{a_1}}\right) = \frac{DP}{\mathcal{R}Tl}\frac{p_{a_1} - p_{a_2}}{p_{bm}} \tag{15.41}$$

where $p_{bm} \equiv (p_{b_2} - p_{b_1})/\ln(p_{b_2}/p_{b_1})$ is the logarithmic mean of p_{b_1} and p_{b_2}.

Typical applications of Eq. (15.38) and (15.41) are evaporation and sublimation processes, which usually involve diffusion of the new vapor phase (gas a) in stationary gas (gas b). One of the principal methods of measuring the diffusion coefficients (D) in binary gaseous mixtures is to place one of the two components, in liquid state, in a small glass tube (Fig. 15.5), filling it to near the top. The second gas is passed over the top of the tube, and the rate of diffusion is measured by the rate of fall of liquid level in the tube. The coefficient D is then given by Eq. (15.41).

Example 15.2: Calculate the rate of burning of a pulverized carbon particle, $d_0 = 0.1$ in. diameter, in an atmosphere of pure oxygen at 1800 F_{abs} and 1 atm pressure, assuming a very large blanketing layer of CO_2 has formed around the particle. At the carbon surface $p_{CO_2} = 1$ atm, $p_{O_2} = 0$, and at a very large radius, $p_{CO_2} = 0$ and $p_{O_2} = 1$ atm.

In Eq. (14.9), $V_{CO_2} = 29.6$, $V_{O_2} = 14.8$, $M_{CO_2} = 44$, $M_{O_2} = 32$, $T = 1800\ F_{abs}$, and $p = 1.0$; so $D = 4.0$ ft²/hr. Since $C + O_2 \rightarrow CO_2$, this is a case of equimolal counterdiffusion. In spherical coordinates, Eq. (15.32) becomes, with a representing CO_2,

$$\frac{N_a}{4\pi r^2} = -\frac{D}{\mathcal{R}T}\frac{dp_a}{dr}$$

Separating variables and integrating between $r = r_0$ and ∞, this becomes

$$p_{a_0} - p_{a_\infty} = \frac{N_a \mathcal{R}T}{4\pi D}\left(\frac{1}{r_0} - \frac{1}{\infty}\right)$$

Since $p_{a_\infty} = 0$,

$$N_{CO_2} = \frac{4\pi D p_{a_0} r_0}{\mathcal{R}T} = \frac{4\pi(4.0)(1.0)(0.05/12)(14.7)(144)}{(1545)(1800)}$$

$$= 16 \times 10^{-5} \text{ lb/moles/hr}$$

Since one mole of carbon burns to one mole of CO_2, this is also the molal consumption of carbon, or $16 \times 10^{-5} \times 12 = 19.2 \times 10^{-4}$ lb/hr of carbon.

15.7 Forced Convection with Mass Transfer in a Tube—Low Mass Transfer Rates

The problem may be the evaporation of liquid from the wetted wall of a tube into air in laminar flow; or, the tube itself may be made of a substance which dissolves and diffuses in the fluid flowing through it (Fig. 15.6).

Let us assume the fluid entering the tube has an initial concentration c_{a_1}, and the concentration at the tube wall is maintaned constant at c_{a_0}. We shall assume the concentration of a is sufficiently dilute so that the

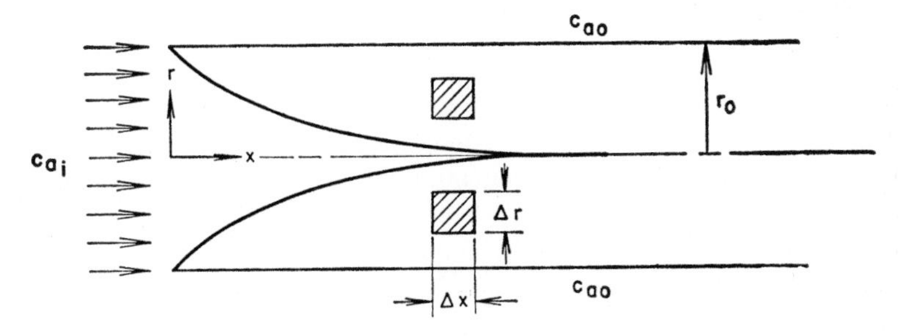

Fig. 15.6. Mass transfer in a tube.

total density ρ is approximately constant. We wish to find the concentration profile within the tube at some distance from the inlet where the velocity and concentration profiles are fully developed.

Mass balance applied to component a in an annular control volume in the entrance region for the steady-state case and for constant total density and diffusivity D yields

$$v_x \frac{\partial c_a}{\partial x} + v_r \frac{\partial c_a}{\partial r} = D\left[\frac{1}{r}\frac{\partial}{\partial r}\left(r\frac{\partial c_a}{\partial r}\right) + \frac{\partial^2 c_a}{\partial x^2}\right] \tag{15.42}$$

The terms on the left represent net efflux of a associated with fluid motion in the x- and r-directions, and the terms on the right represent transfer of a by diffusion resulting from concentration gradients in the x- and r-directions. Alternately, Eq. (15.42) can be obtained by direct reduction of Eq. (15.23) written in cylindrical coordinates.

In the fully developed region, $v_r \cong 0$. If in addition, diffusion in the axial direction can be neglected, Eq. (15.42) simplifies to

$$v_x \frac{\partial c_a}{\partial x} = D\left[\frac{1}{r}\frac{\partial}{\partial r}\left(r\frac{\partial c_a}{\partial r}\right)\right] \tag{15.43}$$

Equation (15.43) is of the same form as Eqs. (7.18) and (7.70); hence, the solutions of Sections 7.3 and 7.7 are equally valid here for the comparable boundary conditions of uniform concentration at the wall, c_{a_0}, and uniform mass flux $(w_a/A)_0$ at the wall. For example, for a fully developed (parabolic) velocity distribution in a circular tube with uniform c_{a_0} at the wall, $h_D(2r_0)/D = 3.66$ (from Table 7.1). In the entrance region of tubes, Figs. 7.19 and 7.20 apply with Nu_D replaced by $h_D(2r_0)/D$ and Pr replaced by Sc. It should be emphasized that these comparable solutions are valid only when v_{r_0} at the wall is very small, namely for low rates of mass transfer. This is not a severe limitation in many engineering applications involving mass transfer.

15.8 Forced Convection with Mass Transfer over a Flat Plate—Laminar Boundary Layer

Mass transfer cooling has been studied as a means of protecting a surface from a hot gas stream, for example, the walls of combustion chambers or surfaces of high-speed vehicles. It may be accomplished by injecting a foreign gas through the wall into the boundary-layer fluid (there is no mass transfer if the injected gas is the same as the boundary-layer fluid), by a continuously supplied liquid film which evaporates, or by constructing the walls with a solid substance which sublimes into the hot boundary-layer fluid. In all cases, simultaneously with the transfer of momentum and heat, there is mass transfer normal to the wall.

(*a*) *Exact Solution*

Consider steady, two-dimensional flow of an incompressible fluid over a porous flat plate, initially at uniform velocity, concentration, and temperature; gas a diffuses upward through the plate, as shown in Fig. 15.7. For constant properties of the fluid mixture (properties of the diffusing

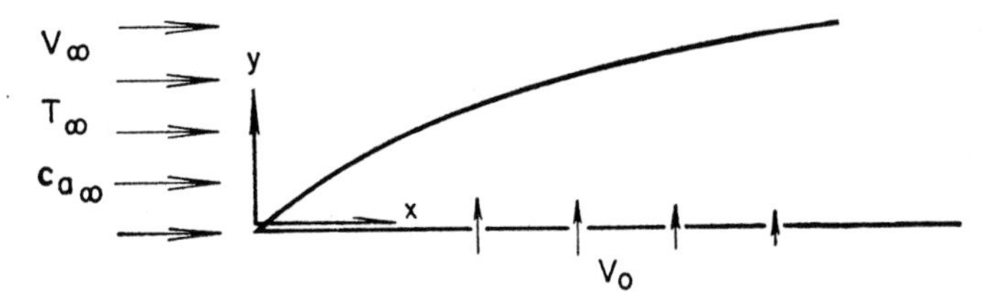

Fig. 15.7. Mass transfer on a flat plate.

gas are identical with the properties of the free-stream fluid), the continuity equation of the mixture and the simplified boundary-layer equations of momentum, energy, and mass may be written as follows:

$$\frac{\partial v_x}{\partial x} + \frac{\partial v_y}{\partial y} = 0 \tag{15.44}$$

$$v_x \frac{\partial v_x}{\partial x} + v_y \frac{\partial v_x}{\partial y} = \nu \frac{\partial^2 v_x}{\partial y^2} \tag{15.45}$$

$$v_x \frac{\partial T}{\partial x} + v_y \frac{\partial T}{\partial y} = \alpha \frac{\partial^2 T}{\partial y^2} \tag{15.46}$$

$$v_x \frac{\partial c_a}{\partial x} + v_y \frac{\partial c_a}{\partial y} = D \frac{\partial^2 c_a}{\partial y^2} \tag{15.47}$$

Equations (15.45) and (15.46) were obtained in Chap. 7 for zero mass transfer. The assumption that these same equations are valid in the presence of mass transfer implies that any additional momentum and energy fluxes associated with mass transfer are negligible. The more general conservation equations of momentum and energy in multicomponent systems are derived in reference (2). Equation (15.47) is a simplification of Eq. (15.23) for the steady, two-dimensional case with the assumption that the diffusion flux in the x-direction is negligibly small compared with that in the y-direction.

A typical set of boundary conditions that may be specified are:

$$y = 0, \quad v_x = 0, \quad v_y = v_0, \quad T = T_0, \quad c_a = c_{a0}$$

$$y = \infty, \quad v_x = V_\infty, \quad T = T_\infty, \quad c_a = c_{a\infty}$$

The condition $v_y = v_0$ accounts for the bulk motion that generally accompanies diffusion from a wall [Sec. 15.6(b)].

Each of the equations above can again be reduced to an ordinary differential equation by similarity transformation:

$$\frac{d^3f}{d\eta^3} + \frac{1}{2} f \frac{d^2f}{d\eta^2} = 0 \tag{15.48}$$

$$\frac{d^2\theta}{d\eta^2} + \frac{1}{2}\left(\frac{\nu}{\alpha}\right) f \frac{d\theta}{d\eta} = 0 \tag{15.49}$$

$$\frac{d^2\phi}{d\eta^2} + \frac{1}{2}\left(\frac{\nu}{D}\right) f \frac{d\phi}{d\eta} = 0 \tag{15.50}$$

where $\qquad \eta \equiv \dfrac{y}{2\sqrt{x}}\sqrt{\dfrac{V_\infty}{\nu}}, \qquad \theta \equiv \dfrac{T - T_0}{T_\infty - T_0}, \qquad \phi \equiv \dfrac{c_a - c_{a0}}{c_{a\infty} - c_{a0}}$

and f is given by Eq. (3.22), $f(\eta) \equiv \psi/\sqrt{V_\infty \nu x}$.

Boundary conditions in terms of the new variables are

$$\eta = 0, \quad f' = 0, \quad \theta = 0, \quad \phi = 0, \quad f = \frac{2v_0}{V_\infty}\sqrt{\text{Re}_x} = \text{constant}$$

$$\eta = \infty, \quad f' = 1, \quad \theta = 1, \quad \phi = 1$$

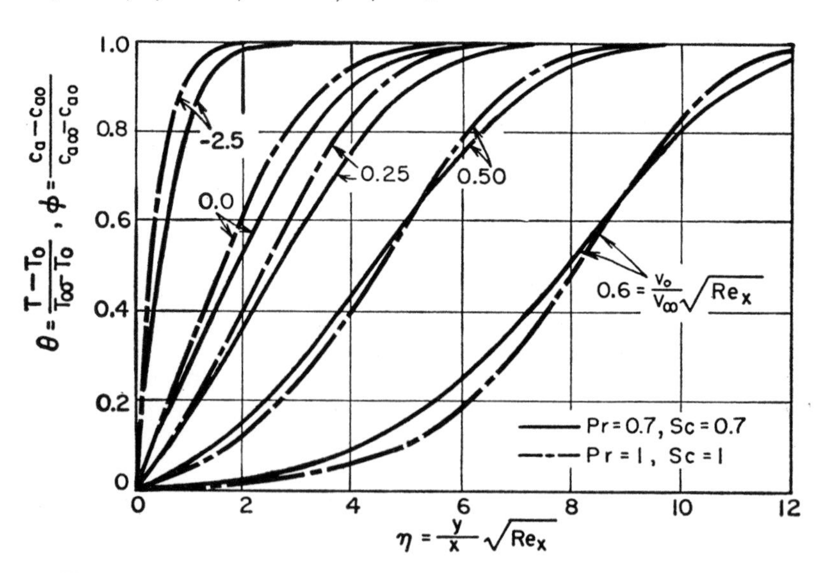

Fig. 15.8. Temperature and concentration profiles in laminar boundary layer on a flat plate for Pr = Sc = 0.7 and Pr = Sc = 1. Curves for Pr = Sc = 1 also represent velocity profiles [Hartnett and Eckert (3)].

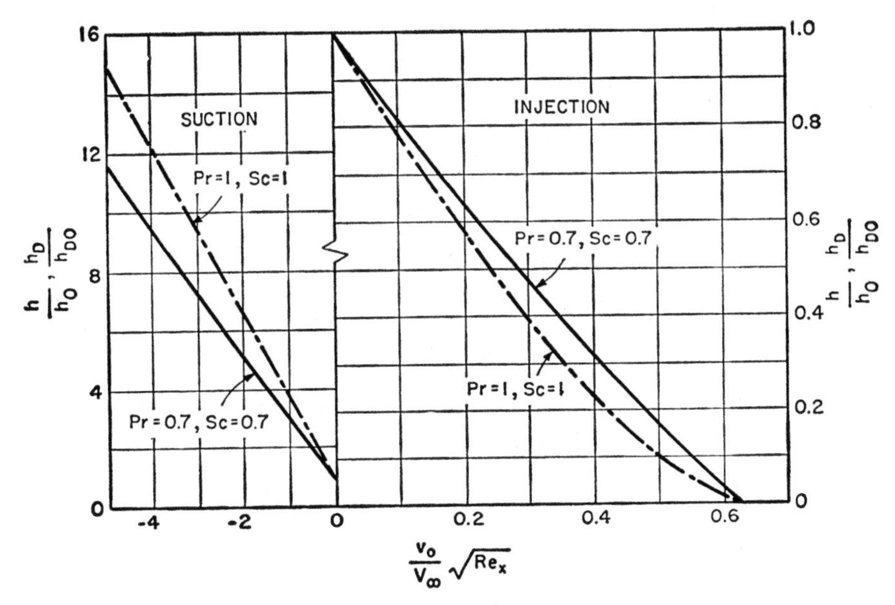

Fig. 15.9. Heat-transfer and mass-transfer coefficients for laminar flow over a flat plate. Subscript zero indicates respective coefficients for essentially zero bulk flow normal to wall [Hartnett and Eckert (3)].

We note that the boundary conditions on f' (dimensionless velocity), θ (dimensionless temperature), and ϕ (dimensionless concentration) are identical. Therefore, when Pr = Sc = 1, solutions to the differential equations (15.48), (15.49), and (15.50) must be the same; in other words, the velocity, temperature, and concentration profiles within the boundary layer must coincide. These results taken from reference (3) are shown in Fig. 15.8. The dimensionless temperature and concentration profiles are also shown for Pr = Sc = 0.7; the velocity profile remains unchanged.

These profiles show a strong dependence on the parameter f, or equivalently v_0. Mass transfer towards the plate, negative v_0, gives steeper profiles, whereas mass transfer away from the plate, positive v_0, gives flatter profiles. This latter effect can be used to protect a surface from hot gas streams. Figure 15.9 shows the local heat and mass transfer coefficients plotted against the parameter f for Pr = Sc = 0.7 and Pr = Sc = 1.0.

The condition f = constant along the wall requires that v_0 at the wall vary as $1/\sqrt{x}$.. Results of calculations (5) for local velocity distribution and wall shear stress for the case of uniform v_0 along the wall agree very closely with those of Fig. 15.9 for the same magnitude of v_0.

The foregoing results apply only at low velocities since compressibility and frictional effects were not taken into account in the analysis. Figure 15.10 reported by Eckert (4) gives the variation of the recovery factor,

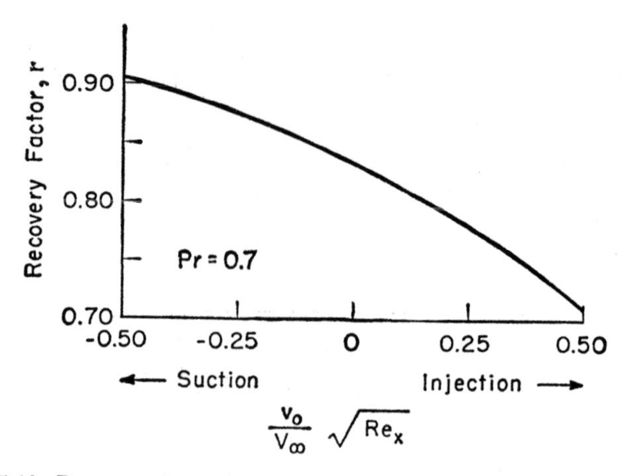

Fig. 15.10. Recovery factor for laminar flow of air over a flat plate with injection and suction of air through the plate [Eckert (4)].

Eq. (11.5), in high-velocity laminar flow of air (Pr = 0.7) over a flat plate as a function of the injection parameter. With the adiabatic wall temperature known, the heat flux at the wall may be calculated by using these results on heat transfer coefficients, Fig. 15.9.

(b) *Approximate Integral Method*

To simplify the discussion, we shall assume that the properties of the fluid mixture remain approximately constant and that the injection velocity v_0 is uniform along the plate. Consider the control volume of unity depth, shown in Fig. 15.11. Continuity requires

$$\frac{d}{dx}\left[\int_0^\delta \rho v_x \, dy\right] \Delta x = \rho_\infty v_1 \, \Delta x + \rho_0 v_0 \, \Delta x$$

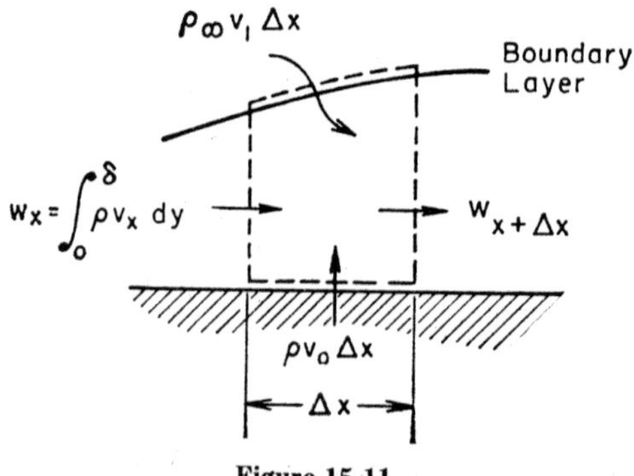

Figure 15.11

or for constant density $\rho_\infty \cong \rho_0 \cong \rho$

$$v_1 = \frac{d}{dx}\left[\int_0^\delta v_x \, dy\right] - v_0 \tag{15.51}$$

The momentum integral equation may be obtained by a procedure similar to the one outlined in Sec. 3.4; using the continuity equation (15.51), the following equation results:

$$\frac{d}{dx}\left[\int_0^\delta (V_\infty - v_x)v_x \, dy\right] = \nu \left(\frac{\partial v_x}{\partial y}\right)_{y=0} + v_0 V_\infty \tag{15.52}$$

Likewise, the energy and the mass transfer integral equations may be derived as follows:

$$\frac{d}{dx}\left[\int_0^{\delta_T} (T_\infty - T)v_x \, dy\right] = \alpha \left(\frac{\partial T}{\partial y}\right)_{y=0} + (T_\infty - T_0)v_0 \tag{15.53}$$

$$\frac{d}{dx}\left[\int_0^{\delta_c} (c_{a\infty} - c_a)v_x \, dy\right] = D \left(\frac{\partial c_a}{\partial y}\right)_{y=0} + (c_{a\infty} - c_{a0})v_0 \tag{15.54}$$

Note that Eqs. (15.52) and (15.53) reduce, respectively, to Eqs. (3.37) and (7.44) when $v_0 \approx 0$. These integral equations may be solved by selecting suitable velocity, temperature, and concentration profiles that match the boundary conditions, as outlined in earlier chapters.

REFERENCES

1. Crank, J., *Mathematics of Diffusion*, Clarendon Press, Oxford, 1956.
2. Merk, H. J., *App. Sci. Res.*, A, **8,** 73 (1958).
3. Harnett, J. P., and E. R. G. Eckert, *Trans ASME*, **79,** 247 (1957).
4. Eckert, E. R. G., *Heat Transfer Symposium*, University of Michigan Press, 1953.
5. Sparrow, E. M., and E. R. G. Eckert, *Journ. Aero Space Science*, **28,** (1961).

PROBLEMS

15.1 Calculate the diffusivity of water vapor in air at 1 atmosphere total pressure and at a temperature of 80 F using the Gilliland equation (14.9). Calculate the corresponding Schmidt number using the value of the kinematic viscosity for pure air.

15.2 Estimate the diffusivity of ammonia (NH_3) present in dilute amount in water at 54 F. Compare your result with the value from the International Critical Tables listed in Table E.6.

15.3 Consider steady, one-dimensional diffusion through a plane membrane. If the diffusivity D varies with concentration, $D = D_0(1 + Ac_a)$, where D_0 and A are constants. Simplify and solve the differential equation, Eq. (15.2), and show the profiles for positive and negative values of A.

15.4 A stream of high-velocity air at 80 F and 60% relative humidity is used to dry a slab of clay 1 in. thick; the four thin edges of the slab are coated to prevent evaporation and the two large faces are exposed to the air. The equilibrium moisture content of the clay in contact with the air stream is 2% on a wet basis (pounds of water per pound of wet clay). The diffusivity is 0.0002 ft²/hr. At the beginning of the falling-rate period the average moisture content is 15% on a wet basis.

(a) Calculate and plot the average moisture content in pounds of water per pound of dry clay at the end of 5, 10, and 20 hours of drying. Neglect the diffusional resistance at the surface of the clay.

(b) The same slab of clay is dried in a stream of the same air but at a lower velocity so that the surface resistance equals the internal resistance to diffusion. Calculate the average moisture on a dry basis at the end of 5, 10, and 20 hours.

15.5 The clay of Prob. 15.4 is fashioned into a solid sphere 9 in. in diameter. Neglecting the surface resistance, calculate and plot the average water content on a dry basis at the end of 5, 10, and 40 hours of drying.

For zero surface resistance, the values of X and Y in terms of the space-mean concentration are as follows for the sphere.

X	0.01	0.05	0.10	0.40	1.0
Y	0.69	0.40	0.23	0.012	0

15.6 Gases A and B are initially separated by a partition and are at the same pressure and temperature. When the partition is removed, there is interdiffusion of the gases. Write the differential equation for the concentration distribution of gas A, and the necessary number of boundary conditions (do not solve the equation). Write down a statement of the analogous heat transfer problem. (See illustration for Prob. 1.6.)

15.7 One method of measuring diffusion coefficients of vapors is to measure rate of evaporation of a liquid in narrow tubes. In one such experiment, a glass tube 1 cm in diameter was filled with water at 68 F to within 4 cm of the top. Dry air at 68 F and 14.7 psia was blown across the top of the tube. At the end of 24 hours of steady-state operation, the level of the water dropped 0.1 cm. Calculate the diffusivity of the air-vapor system at 68 F.

15.8 A long, porous cylinder of radius r_i is concentric inside another cylinder of r_0. A salt solution diffuses through the porous cylinder into the liquid in the annulus. At the outer cylinder the salt is absorbed. If the concentration of salt at r_i is c_i and at r_0 is c_0, and the liquid in the annulus is stagnant, derive the expression giving variation of salt concentration with radius at steady state. If $c_i = 0.65$ lb$_m$/ft³ and $c_0 = 0.05$ lb$_m$/ft³, calculate the rate of transfer of salt. Assume $D = 0.00005$ ft²/hr.

15.9 Nitrogen diffuses steadily through a stagnant layer of air which is 0.2 in. thick. The concentration of N_2 is 0.005 lb/ft^3 at one face and zero at the other. The total pressure is 1 atm and the temperature is 70 F.

(a) Calculate the time required for 1 lb of N_2 to diffuse across 1 sq ft of this air film.

(b) For the same conditions as above, but with a stagnant layer of water 0.2 in. thick instead of air, calculate the time required for 1 lb of N_2 to diffuse through 1 sq ft of the water film.

15.10 A pulverized coal particle burns in pure oxygen at 2200 F. The process is limited by diffusion of the oxygen counterflow to the CO_2 formed at the particle surface. Assume the coal is pure carbon and has an initial diameter of 0.004 in. If $D = 4.0$ ft^2/hr, calculate the time required for 90% of the carbon to burn away. What is the final diameter of the coal particle? Plot particle diameter vs. time.

15.11 Write the differential equation for neutron concentration in a long cylindrical element of fissionable material. Assume the rate of production of neutrons is proportional to neutron concentration and the movement of neutrons in the fissionable material obeys Fick's diffusion equation. State the conditions for the analogous heat conduction problem.

15.12 As in the case of heat conduction, Sec. 6.7, derive an expression for the critical radius for maximum diffusion rate from a cylindrical rod.

15.13 Dry air at 1500 F, 14.7 psia flows at 100 ft/sec over a porous flat plate. Superheated steam at 500 F is forced through the porous wall at a rate of 0.93 lb/hr ft^2. Calculate the rate of heat transfer from the air at a position 1 ft from the leading edge of the plate.

15.14 Derive the appropriate integral equation of mass transfer from an ablative plate (surface vaporizes and material diffuses into the hot gas stream) under forced convection. Sketch your system and show all the mass fluxes. Clearly state any assumptions you make in the derivation.

CHAPTER 16

Mass Transfer in Turbulent Flow and Experimental Results

Problems associated with the transfer processes in turbulent flow were discussed in earlier chapters. In this chapter the analogy between the transfer of momentum and heat is extended to the transfer of mass; only a brief outline is given, since much of the material in Sec. 16.1 and 16.2 parallels that of Chapter 8.

Experimental results are presented for low mass transfer rates in isothermal binary systems. The mass transfer correlations for such systems may be compared directly with the corresponding heat transfer correlations given in Chap. 8. Mass transfer data for free convection are scarce and are not discussed. In non-isothermal binary systems, simultaneous heat and mass transfer occur, one of the most familiar examples being the wet-bulb thermometer, Sec. 16.4.

16.1 Mass Transfer in Turbulent Flow

When mass transfer occurs in a turbulent flow field, the concentration of the diffusing substance fluctuates both in time and in space. The instantaneous mass concentration at a point in the flow field may be expressed as

$$c_a = \bar{c}_a + c_a' \tag{16.1}$$

where the barred quantity represents the mean component independent of time and the primed quantity the high frequency fluctuating component.

Consider mass transfer at the surface of a flat plate or in the entrance region between parallel plates. In turbulent, as well as in laminar, flow the equation of the concentration boundary layer in two dimensions for assumed constant total density is given by

$$\frac{\partial c_a}{\partial t} + v_x \frac{\partial c_a}{\partial x} + v_y \frac{\partial c_a}{\partial y} = \frac{\partial}{\partial y}\left(D \frac{\partial c_a}{\partial y}\right) + \frac{\partial}{\partial x}\left(D \frac{\partial c_a}{\partial x}\right) \tag{16.2a}$$

where in turbulent flow v_x and v_y are the instantaneous bulk velocities defined by Eqs. (8.1) and 14.3), and c_a is the instantaneous mass concentration defined by Eq. (16.1). Equation (16.2a) in its alternate form is

$$\frac{\partial c_a}{\partial t} + \frac{\partial}{\partial x}(v_x c_a) + \frac{\partial}{\partial y}(v_y c_a) = \frac{\partial}{\partial y}\left(D \frac{\partial c_a}{\partial y}\right) + \frac{\partial}{\partial x}\left(D \frac{\partial c_a}{\partial x}\right) \tag{16.2b}$$

If we substitute Eqs. (8.1) and (16.1) in Eq. (16.2b), and time-average the resulting equation term by term, then by an identical procedure as in Sec. 8.1 for the turbulent momentum and energy transfers, we get the following equation:

$$\frac{\partial \bar{c}_a}{\partial t} + \bar{v}_x \frac{\partial \bar{c}_a}{\partial x} + \bar{v}_y \frac{\partial \bar{c}_a}{\partial y} = \frac{\partial}{\partial y}\left[D \frac{\partial \bar{c}_a}{\partial y} - \overline{v_y' c_a'}\right] + \frac{\partial}{\partial x}\left[D \frac{\partial \bar{c}_a}{\partial x} - \overline{v_x' c_a'}\right] \tag{16.3}$$

Confining our attention to mass transfer in the y-direction, an eddy mass diffusivity may be defined such that

$$\overline{v_y' c_a'} = -\epsilon_D \frac{\partial \bar{c}_a}{\partial y} \tag{16.4}$$

A rate equation of turbulent mass transfer may in turn be postulated as follows:

$$\left(\frac{w_a}{A}\right)_{\text{app}} = \left(\frac{w_a}{A}\right)_{\text{molec}} + \left(\frac{w_a}{A}\right)_{\text{eddy}} = -(D + \epsilon_D) \frac{\partial \bar{c}_a}{\partial y} \tag{16.5}$$

For the mean steady-state and fully developed flow, neglecting axial diffusion, $\partial[\ ]/\partial x$; then Eq. (16.3) becomes

$$\bar{v}_x \frac{\partial \bar{c}_a}{\partial x} = -\frac{\partial}{\partial y}\left(\frac{w_a}{A}\right)_{\text{app}} \tag{16.6}$$

Equation (16.6) is clearly similar to Eq. (8.16). Integration of Eq. (16.6) between the indicated limits gives the following expression for the distribution of $(w_a/A)_{\text{app}}$ in the region of fully developed flow between parallel plates, $2y_0$ apart, with $y = 0$ at the wall:

$$\frac{(w_a/A)_{\text{app}}}{(w_a/A)_0} = M_D\left(1 - \frac{y}{y_0}\right) \tag{16.7}$$

where

$$M_D = \frac{\dfrac{1}{y_0 - y}\displaystyle\int_y^{y_0} \bar{v}_x \frac{\partial \bar{c}_a}{\partial x}\,dy}{\dfrac{1}{y_0}\displaystyle\int_0^{y_0} \bar{v}_x \frac{\partial \bar{c}_a}{\partial x}\,dy}$$

For a wetted wall tube, with uniform $(w_a/A)_0$ along the tube, the distribution of $(w_a/A)_{\text{app}}$ is given by Fig. 8.2 where we need only replace (q/A) by (w_a/A) and the Prandtl number by the corresponding Schmidt number.

Again, since theory is incomplete regarding turbulence and no experimental information exists on concentration fluctuations in a turbulent flow field, we must rely upon some assumed analogy between the turbulent transfer of mass and the turbulent transfer of momentum or energy in order to relate mass transfer rates to known friction factor or heat transfer data. Since the latter information has generally been collected for zero mass transfer, we must also assume that the data are not significantly altered by mass transfer. The validity of results obtained under this assumption is thus limited to dilute concentrations of the diffusing component.

Proceeding as in Sec. 8.4, if we combine Eqs. (16.5) and (16.7), assuming $M_D = 1$ and noting that for a circular tube we need only replace y_0 by r_0, and that $(w_a/A)_0$ is the flux from wall *into* fluid, we get

$$\left(\frac{w_a}{A}\right)_0\left(1 - \frac{y}{r_0}\right) = -(D + \epsilon_D)\,\frac{\partial c_a}{\partial y} \tag{16.8}$$

If the momentum transfer and the heat transfer are assumed unaffected by mass transfer, we can write as before

$$\frac{\tau_0}{\rho}\left(1 - \frac{y}{r_0}\right) = (\nu + \epsilon_m)\,\frac{\partial v}{\partial y} \tag{8.21}$$

$$\frac{(q/A)_0}{\rho c_p}\left(1 - \frac{y}{r_0}\right) = -(\alpha + \epsilon_h)\,\frac{\partial T}{\partial y} \tag{8.22}$$

16.2 Analogy Results

The Reynolds result is obtained if it is assumed that flow is turbulent throughout the tube cross section, so that $\nu \ll \epsilon_m$ and $D \ll \epsilon_D$ everywhere. Dividing Eq. (8.21) by Eq. (16.8) and integrating between the wall and the location where both the velocity and the concentration have their mean values V and c_m, we get

$$\frac{\tau_0}{\rho\left(\dfrac{w_a}{A}\right)_0} = \frac{1}{E_D}\frac{V}{(c_0 - c_m)} \tag{16.9}$$

where $E_D \equiv \epsilon_D/\epsilon_m$. Or, introducing the definitions of f and h_D and assuming $E_D = 1$,

$$\frac{h_D}{V} = \frac{f}{2} \tag{16.10}$$

Equation (16.10) agrees well with experimental results provided $Sc \cong 1.0$. For $E = E_D = 1.0$, combine Eqs. (8.24) and (16.10) to eliminate $f/2$. Then

$$\frac{h}{\rho c_p} = h_D \tag{16.11}$$

The Karman result is obtained by an analysis similar to the one outlined in Sec. 8.6 using the universal velocity distribution of Fig. 4.14 but neglecting D compared with ϵ_D and ν compared with ϵ_m in the turbulent core $y^+ > 30$ (Fig. 4.14); Sherwood (6) performed this analysis and arrived at the following result:

$$\frac{h_D}{V} = \frac{f/2}{1 + \sqrt{f/2}[5\,(\mathrm{Sc} - 1) + 5\ln(1 + 5\,\mathrm{Sc}) - 5\ln 6]} \tag{16.12}$$

This equation is identical with the Kármán result (5) for heat transfer if h_D/V is replaced by $h/c_p\rho V$ and Sc by Pr. Agreement of Eq. (16.12) with data on vaporization of liquids in wetted-wall columns is quite good.

16.3 Experimental Results

At low mass transfer rates at a surface, a heat transfer analog with comparable boundary conditions exists for most mass diffusional problems. This implies that many of the empirical and semi-empirical correlations presented in Chap. 8 may be applied to corresponding mass transfer systems by a simple change of notation, such as

$$T \rightarrow c_a$$

$$\mathrm{Pr} \rightarrow \mathrm{Sc}$$

$$\mathrm{Nu} \rightarrow \mathrm{Sh}$$

where $\mathrm{Sh} \equiv h_D d/D$ is the *Sherwood number*. The flow parameters such as the Reynolds number will be the same. We shall explore and verify a number of these correlations.

At high mass transfer rates the analogy is disturbed by the existence of a finite bulk flow normal to the surface. To obtain the total flux, some knowledge about the bulk flow is needed.

Consider now the case in which the bulk flow is of the type that can be approximated by the diffusion of one gas in a stationary second gas, as in the case of a liquid on a wetted-wall evaporating in an air stream. We want to obtain an approximate relationship between the diffusion flux and the absolute flux of component a. Equation (15.41) applies in the following form:

$$\left(\frac{w_a}{A}\right)_{\text{abs}} = \frac{D}{l}\frac{P}{p_{pm}}(c_{a_0} - c_{a\infty}) \tag{16.13a}$$

Then for this case, if we define $h_{D_{\text{abs}}}$ as follows:

$$\left(\frac{w_a}{A}\right)_{\text{abs}} = h_{D_{\text{abs}}}(c_{a_0} - c_{a\infty}) \tag{16.13b}$$

there results

$$h_{D_{\text{abs}}} = \frac{D}{l}\frac{P}{p_{bm}} \tag{16.13c}$$

where l is an effective "film" thickness for diffusion, Fig. 16.1.

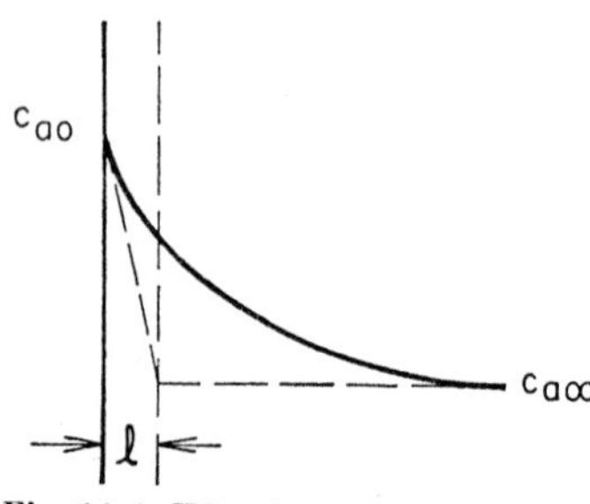

Fig. 16.1. Effective "film" thickness for mass diffusion.

The diffusive flow of component a that is superimposed on the bulk motion is

$$\frac{w_a}{A} = -D\frac{dc_a}{dx} = D\frac{c_{a_0} - c_{a\infty}}{l} \tag{16.14a}$$

and from the definition of h_D, Eq. (14.11),

$$\frac{w_a}{A} = h_D(c_{a_0} - c_{a\infty}) \tag{16.14b}$$

Then

$$h_D = D/l \tag{16.14c}$$

Comparing Eq. (16.13c) with (16.14c) we observe the following relation:

$$h_{D_{\text{abs}}} = h_D(P/p_{bm}) \tag{16.15}$$

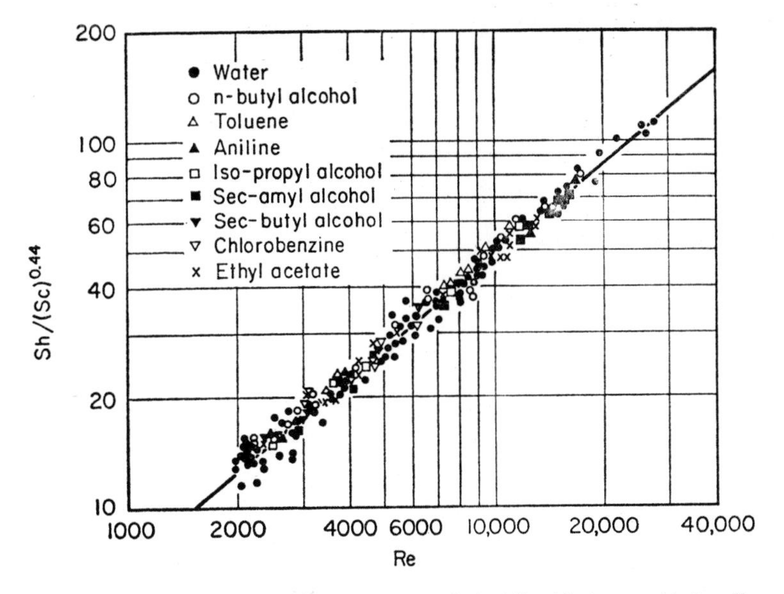

Fig. 16.2. Correlation of data on evaporation of liquids in a wetted-wall column. Solid line represents Eq. (16.16), [Gilliland (1)].

for evaporation from a wetted wall. This is an approximation relating h_D with $h_{D_{\text{abs}}}$ for this case. In other systems the relation between h_D and $h_{D_{\text{abs}}}$ may be quite different from this.

Gilliland (1) compiled experimental data on vaporization of different liquids into air in wetted-wall columns with air and liquid in parallel and counterflow and at approximately the same temperature. He suggested that these data can be correlated, Fig. 16.2, successfully by

$$\frac{h_{D_{\text{abs}}}d}{D}\frac{p_{bm}}{P} = 0.023\left(\frac{Gd}{\mu}\right)^{0.83}\left(\frac{\mu}{\rho D}\right)^{0.44} \tag{16.16a}$$

or

$$\frac{h_D d}{D} = 0.023\left(\frac{Gd}{\mu}\right)^{0.83}\left(\frac{\mu}{\rho D}\right)^{0.44} \tag{16.16b}$$

The Reynolds number in the correlation is based on the velocity of the gas relative to the wall, not to the liquid surface. Equation (16.16) is recommended for Reynolds numbers in the range of 2000 to 35,000 and Schmidt numbers in the range of 0.6 to 2.5. Equation (16.16) may be compared with the McAdams equation for heat transfer for turbulent flow in tubes, Eq. (8.34).

The analogy among momentum, heat, and mass transfer in forced-convection systems may also be stated empirically by using the Chilton-

Colburn j factors defined by the following equations:

$$j_H = \frac{h}{\rho c_p V}(\mathrm{Pr})^{2/3} \tag{16.17a}$$

$$j_D = \frac{h_D}{V}(\mathrm{Sc})^{2/3} \tag{16.17b}$$

For fully developed flow in a round tube, the analogy may be stated in terms of the j factors and the friction factor f as follows:

$$j_H = j_D = \frac{f}{2} \tag{16.17c}$$

where f may be found from the Moody chart, Fig. 4.1, for both laminar and turbulent flows.

In the case of a flat plate, the j factors are analogous to the skin friction coefficient and for average values are given by Eq. (7.41a) for laminar flow, and Eq. (8.47) for turbulent flow. Therefore,

laminar:
$$j_H = j_D = \frac{C_f}{2} = \frac{0.664}{\sqrt{\mathrm{Re}_L}} \tag{16.18a}$$

turbulent:
$$j_H = j_D = \frac{C_f}{2} = \frac{0.037}{(\mathrm{Re}_L)^{0.2}} \tag{16.18b}$$

Figure 16.3 shows fair agreement with experimental data for mass transfer data on flat plates. It should be noted that the data represent only a limited range of Schmidt numbers from 0.6 for the vaporization of water to approximately 3 for the vaporization of several organic liquids into air.

For flow across bluff bodies such as spheres and cylinders, the drag coefficients based on the total drag are found to be much greater than the j factors, although $j_H = j_D$ still holds. However, if we consider only that part of the total drag due to skin friction alone, obtained by subtracting the measured form drag from the measured total drag, satisfactory analogy is noted between the j factors and the skin friction coefficient, as shown in Fig. 16.4 for a circular cylinder. An explanation is that heat conduction and mass diffusion at the wall are directly analogous to the shear stress, but not to the normal stress or pressure. Therefore, when pressure drop associated with form drag is significant, as in the case of bluff bodies, departure of total drag coefficient from j factors might be expected. The mass transfer data in Fig. 16.4 were obtained for the evaporation of water into air from a cylinder, and for the absorption of water vapor by solid cylinders of caustic. The line j_H vs. Re_d is representative of heat transfer data on single cylinders placed transverse to air flow, and the line $C_f/2$ vs. Re_d represents measured skin friction data.

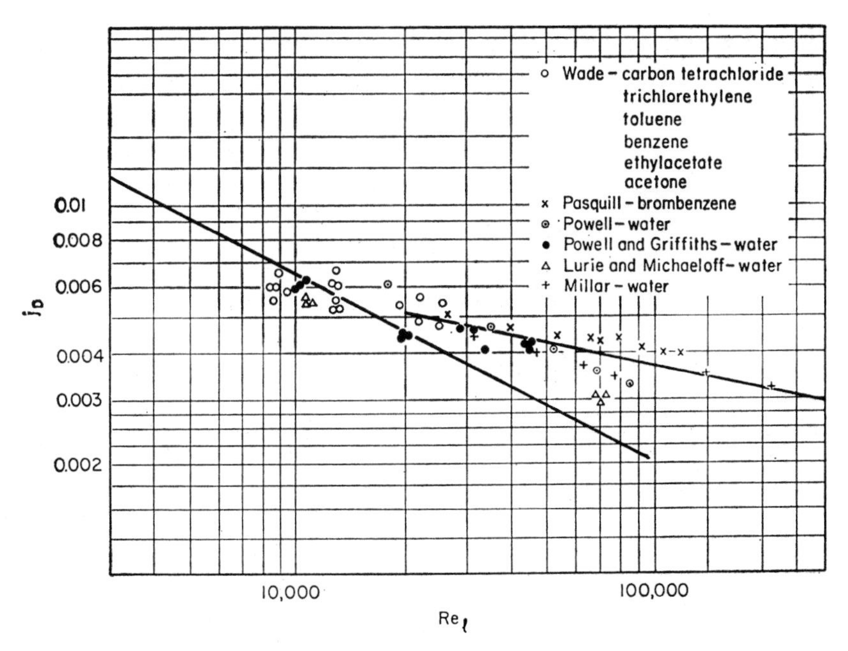

Fig. 16.3. Correlation of data on evaporation from flat plates. Solid lines represent Eqs. (16.18a) and (16.18b), [Sherwood and Pigford (2)].

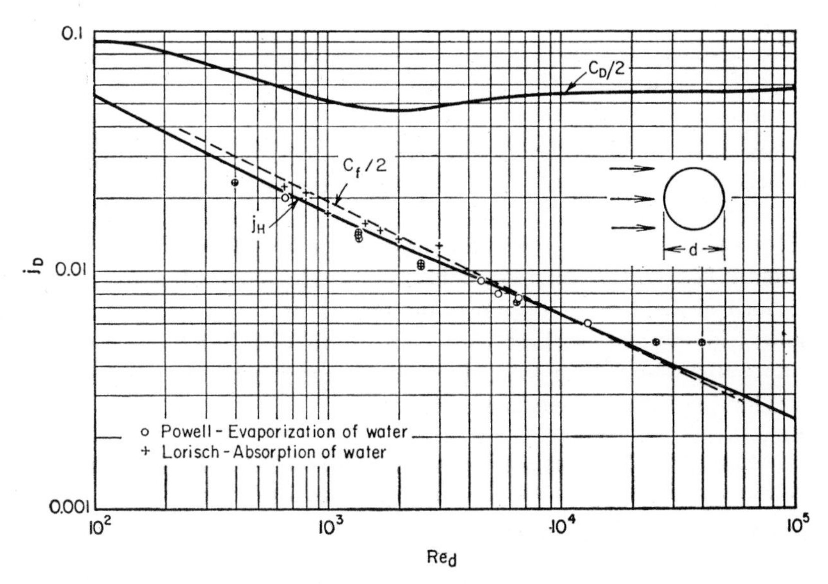

Fig. 16.4. Mass transfer data compared with skin friction, total drag, and heat transfer data on a circular cylinder [Sherwood and Pigford (7)].

Experimental results for mass transfer to or from a single sphere as, for example, a liquid drop or a solid sphere, under forced convection conditions were correlated by Froessling (3) by the following semi-empirical equation:

$$\text{Sh} = 2.0[1 + 0.276 \text{Re}_d^{1/2} \text{Sc}^{1/3}] \tag{16.19}$$

It is to be noted that as the flow velocity approaches zero, Sh approaches in the limit the theoretical value of 2 for mass transfer to or from a liquid drop in still air. This asymptote is identical with the heat conduction result for a sphere in an infinite stationary medium for which Nu = 2.

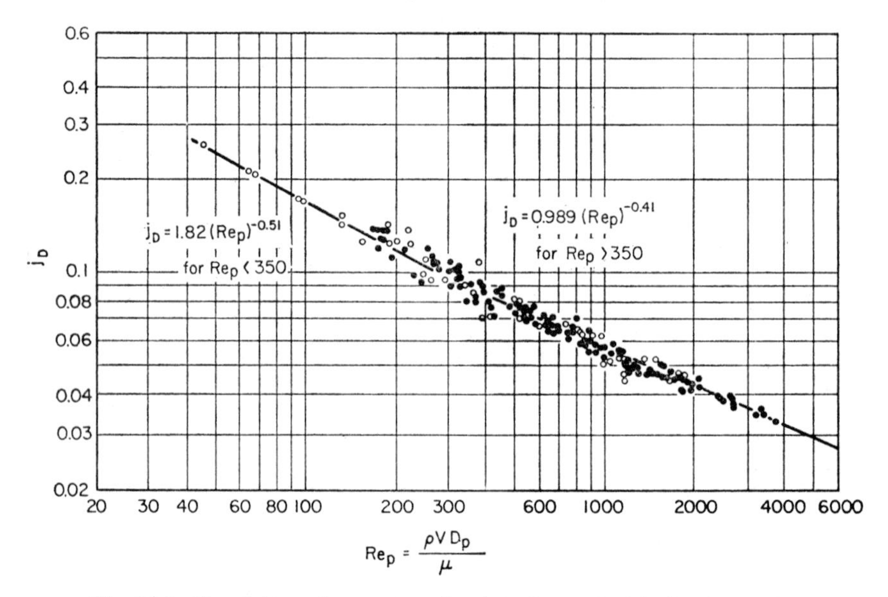

Fig. 16.5. Correlation of mass transfer data for packed beds of granular solids with through air circulation [Wilke and Hougen (4)].

Mass transfer data for packed beds of granular solids may be correlated in terms of j_D and a modified Reynolds number Re_p as shown in Fig. 16.5(4); experimental data were obtained on drying of small wet spheres and cylinders in random packing with through-air circulation. D_p represents the diameter of spheres; for cylinders, D_p represents the diameter of a sphere having the same surface area.

Example 16.1: Calculate the mass transfer coefficient h_D of water vapor in air in turbulent flow at 200 ft/sec at 1 atm and 70 F over (a) a flat plate 1 ft long; (b) a sphere 2 in. in diameter. Assume concentration of vapor in air is sufficiently dilute so that $p_{bm}/P \approx 1$.

Air at 1 atm and 70 F,

$$\rho = 0.074 \text{ lb/ft}^3, \quad \nu = 16.88 \times 10^{-5} \text{ ft}^2/\text{sec}$$

Also from Appendix E,

$$D = 23.6 \times 10^{-5} \text{ ft}^2/\text{sec}$$

Using these values,

$$\text{Sc} = \nu/D = 0.715$$

(a)

$$\text{Re}_L = \frac{(200)(1)}{16.88 \times 10^{-5}} = 1{,}185{,}000 \text{ turbulent flow}$$

From Eq. (16.18b),

$$j_D = \frac{0.037}{(\text{Re}_L)^{0.2}} = 0.00226$$

From Eq. (16.17b),

$$h_D = \frac{(0.00226)(200)}{(0.715)^{2/3}} = 0.565 \text{ lb/ft}^2 \text{ sec (lb/ft}^3)$$

(b)

$$\text{Re}_d = \frac{(200)(1/6)}{16.88 \times 10^{-5}} = 197{,}000$$

From Eq. (16.19),

$$h_D = 0.31 \text{ lb/ft}^2 \text{ sec (lb/ft}^3)$$

Example 16.2: Estimate the value of h_D for the absorption of NH_3 by the wet surface of a cylinder placed in a turbulent air stream flowing across the cylinder at 15 ft/sec. No data on mass transfer exist for this process, but heat transfer tests with the same geometry and air velocity show $h = 10$ Btu/hr ft² F.

For air, Pr = 0.74, $\rho = 0.075$ lb/ft³, and $c_p = 0.24$ Btu/lb F. For dilute NH^3-air mixtures, $p_{bm} = P$ and Sc = 0.61.

Assume $j_D = j_H$. Then from Eqs. (16.17a) and (16.17b)

$$h_D = \frac{h}{\rho c_p}\left(\frac{\text{Pr}}{\text{Sc}}\right)^{2/3} = \frac{10}{(0.075)(0.24)}\left(\frac{0.74}{0.61}\right)^{2/3} = 642 \frac{\text{lb}}{\text{ft}^2 \text{ hr (lb/ft}^3)}$$

16.4 Simultaneous Heat and Mass Transfer

The simultaneous transfer of heat and mass occurs in nearly all cases involving mass transfer between phases such as occur in cooling towers, dryers, dehumidifying equipment, and gas absorption equipment. In some cases, the heat transfer may be negligible; in others it may dominate the design consideration.

As an illustration of this combined process, consider the operation of the ordinary wet-bulb thermometer, Fig. 16.6. If air con-

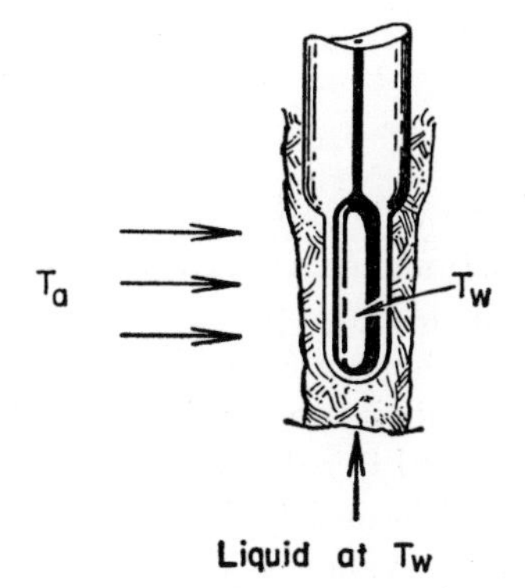

Fig. 16.6. Wet-bulb thermometer.

taining some moisture at a partial pressure $p_{w,a}$ is passed across the wet wick of a wet-bulb thermometer, the temperature T_w assumed by the thermometer will be sufficient to produce a heat transfer rate just large enough to supply the latent heat of the evaporating liquid, provided the thermometer exchanges heat only with the air and provided the liquid added to the wick along the stem is at T_w. This may be written as

$$\frac{q}{A} = (h + h_r)(T_a - T_w) = \frac{w_{\text{evap}}}{A} h_{fg,w} \tag{16.20}$$

where h is associated with the convection process, and h_r with radiation. The rate of evaporation is governed by the mass transfer process which may be expressed as

$$\frac{w_{\text{evap}}}{A} = \frac{h_D}{R_w T}(p_{ww} - p_{w,a}) \tag{16.21}$$

where p_{ww} refers to the partial pressure of the water vapor in saturated air at T_w and R_w is the gas constant of the water vapor.

From Eqs. (16.17a) and (16.17b) with $j_H = j_D$

$$\frac{h}{h_D} = \rho c_p \left(\frac{\text{Sc}}{\text{Pr}}\right)^{2/3} \tag{16.22}$$

The specific humidity (lb water vapor/lb dry air) ω for low partial pressures of water vapor may be written as follows (8):

$$\omega = \frac{M_w}{M_a} \frac{p_w}{P - p_w} \approx \frac{M_w}{M_a} \frac{p_w}{P} \tag{16.23}$$

Combining Eqs. (16.20) through (16.23)

$$\frac{\omega_w - \omega_a}{T_a - T_w} = \frac{c_p}{h_{fg,w}} \left(\frac{\text{Sc}}{\text{Pr}}\right)^{2/3} \left(1 + \frac{h_r}{h}\right) \tag{16.24}$$

The wet- and dry-bulb temperatures, T_w and T_a, are measured, and ω_w is determined from Eq. (16.23) since p_w depends only on T_w. The term (h_r/h) may be estimated from the radiation and convection correlations in Chaps. 8 and 13 but is usually small compared with unity. The other quantities on the right are known *properties* of the fluids. Then Eq. (16.24) is used to calculate ω_a and Eq. (16.23) to calculate $p_{w,a}$.

The adiabatic saturation temperature, T_{ad} of a mixture of air and water vapor, is the temperature the stream attains in becoming saturated with water vapor by being brought in contact with liquid water, also at T_{ad}, while no heat is exchanged with the surroundings (adiabatic). Then the enthalpy change of the two fluids (the stream and the liquid evaporated) is zero.

$$c_p(T_a - T_{\text{ad}}) = h_{fg,w}(\omega_{\text{ad}} - \omega_a) \tag{16.25}$$

or

$$\frac{(\omega_{\text{ad}} - \omega_a)}{(T_a - T_{\text{ad}})} = \frac{c_p}{h_{fg,w}} \tag{16.26}$$

Equations (16.24) and (16.26) show that in general $T_{\text{ad}} \neq T_w$ for a particular gas stream having temperature T_a and humidity ω_a. However, for air-water vapor systems Sc/Pr is slightly less than unity while $(1 + h_r/h)$ is slightly greater than unity making $T_{\text{ad}} \cong T_w$. This is not true for other gas-vapor systems. Equation (16.24) also shows that the wet-bulb temperature is relatively insensitive to changes in air velocity since air velocity affects only h and h_r/h is small for most operating conditions.

Example 16.3: The wet- and dry-bulb temperatures in moist air at a total pressure 14.7 psia are 65 F and 80 F, respectively. For the existing air velocity across the thermometer, it was estimated by heat transfer data and by Eq. (8.50) that $h = 4$ Btu/hr ft^2 F and $h_D = 260$ lb/ft^2 hr (lb/ft^3). Heat transfer by radiation may be neglected. For this air-water vapor mixture Pr = 0.74 and Sc = 0.60. Calculate the specific humidity of the moist air.

At 65 F, $p_{ww} = 0.6222$ psia from Steam Tables. The saturated specific humidity is calculated by Eq. (16.23):

$$\omega_a = \frac{18}{28.95}\,\frac{0.6222}{14.078} = 0.0275 \text{ lb } H_2O/\text{lb dry air}$$

Other properties are $c_p = 0.24$ and $h_{fg,w} = 1057$ Btu/lb. With these quantities Eq. (16.24) becomes

$$\frac{0.0275 - \omega_a}{80 - 65} = \frac{0.24}{1057}\left(\frac{0.74}{0.60}\right)^{2/3}$$

or $\omega_a = 0.0236$ lb H_2O/lb dry air and from Eq. (16.23), $p_{wa} = 0.538$ psia.

REFERENCES

1. Gilliland, E. R., and T. K. Sherwood, *Ind. Eng. Chem.*, **26,** 516 (1934).

2. Sherwood, T. K., and R. L. Pigford, *Absorption and Extraction*, McGraw-Hill, New York, 1952, p. 66.

3. Froessling, N., *Gerlands Beitr. Geophys.*, **32,** 170 (1938).

4. Wilke, C. R., and O. A. Hougen, *Trans. AIChE*, **41,** 445 (1945).

5. Von Kármán, T., *Trans. ASME*, **61,** 705 (1939).

6. Sherwood, T. K., *Trans. AIChE*, **36,** 817 (1940).

7. Sherwood and Pigford, *op. cit.*, p. 70.

8. Keenan, J. H., *Thermodynamics*, Chap. 13, Wiley, New York, 1941.

9. Chilton, T. H., and A. P. Colburn, *Ind. Eng. Chem.*, **26,** 1183 (1934).

PROBLEMS

16.1 Starting with Eq. (16.2a), perform the operations outlined in Sec. 16.1 to obtain Eq. (16.3).

16.2 Air flows at 10 ft/sec at 1 atm and 60 F in a wetted-wall tube of 2-in. I.D. Compare the mass transfer coefficient of water vapor in air calculated by the following methods ($D = 0.9$ ft²/hr):

(a) the Kármán analogy, Eq. (16.12)

(b) the Gilliland equation, Eq. (16.16)

(c) the Chilton-Colburn j-factors, Eq. (16.17)

16.3 A 1-ft square plate of solid sodium chloride is immersed parallel to a stream of water flowing at 4 ft/sec. The temperature of the water and plate is maintained constant at 60 F. If transition from laminar to turbulent flow occurs at $\text{Re}_x = 10^5$, plot the variation of the local mass transfer coefficient along the plate.

Calculate the rate at which the plate is losing salt based on an average h_D estimated from your plot. Use the value of D given in Appendix E.6.

16.4 A layer of liquid water 0.05 in. thick is evaporating in dry air at 85 F and 1 atm total pressure. The water is maintained at 70 F, and molecular diffusion is assumed to take place through an effective gas film of a thickness of 0.15 in. Calculate the time required for the water to evaporate completely. ($D = 0.9$ ft²/hr.)

16.5 Determine the humidity of the exit air stream flowing through a wetted-wall tube of 1.00-in. diameter and 30-in long, if the entering air is at 14.7 psia, 80 F, and has 0.003 lb H_2O/lb air and is flowing at a rate of 20 lb/hr. The water at the surface is maintained at 75 F by heat exchange at the outer wall of the tube.

16.6 (a) Mass transfer from a single sphere in a fluid stream may be calculated from Eq. (16.19), which gives the average h_D over the surface of the sphere. Note that $\text{Sh} \to 2$ as $V_\infty \to 0$. Show that $\text{Sh} = 2$ for radial diffusion from the surface of a sphere into still air.

(b) Estimate the rate at which a $\frac{1}{2}$-inch sphere of naphthalene will sublime when placed in an air stream flowing at 40 ft/sec. Air and naphthalene are both at 60 F. For the diffusion of naphthalene vapor in air, assume $D = 0.2$ ft²/hr.

16.7 Small water droplets $\frac{1}{8}$ in. in diameter fall through dry air at 100 F. Find the distance a drop must fall in order to reduce its volume by 50%. Assume the drop falls steadily at its terminal velocity (calculated from Fig. 4.17) and its surface temperature is 60 F. Use $D = 0.95$ ft²/hr and evaluate all other properties at the average temperature of 80 F.

16.8 Dry air at 70 F and 14.7 psia flows past a very porous wet plate (which is waterlogged) at 100 fps. Calculate the rate at which water evaporates from this plate if sufficient heat is supplied to maintain the plate temperature at 70 F. (Take a 1-ft-wide plate, 6 in. long in the flow direction.)

16.9 The wick of a wet-bulb thermometer is wet with benzene. The velocity of dry air passing across the thermometer is so large that radiation to the wick may be neglected. The observed wet-bulb temperature is −2.6 F and the dry-bulb temperature is 129 F. Barometric pressure is 14.7 psia and the surface coefficient of heat transfer is found from other data to be 10 Btu/hr ft² F. For benzene at −2.6 F,

$$p_{ww} = 0.388 \text{ psia}, \qquad h_{fg} = 200 \text{ Btu/lb}$$

$$\text{Pr} = 0.71, \qquad \text{Sc} = 1.71$$

From the data, calculate the mass transfer coefficient h_D.

16.10 A wet-bulb thermometer covered with a wick 0.5 in. in diameter and 10 in. long is inserted in a completely dry air stream which flows across the thermometer with a mass velocity of 10,000 lb/hr ft² at a pressure of 1 atm. At steady-state conditions the thermometer reads 100 F. The correlation for heat transfer coefficient at the surface of the wick is given approximately by Eq. (8.51). Radiation effects may be neglected. Calculate the temperature of the air.

16.11 In Prob. 16.7, calculate the surface temperature of the drop using the method of Sec. 16.4, and compare your answer with the value of 60 F given in that problem. Is the given value high or low?

16.12 The following data [reference: Sherwood and Woertz, *Ind. Eng. Chem.*, **31,** 1034 (1939)] were obtained on water vapor concentrations in a turbulent stream of CO_2 flowing at a Reynolds number of 102,000 in a duct 5.06 cm wide. Water vapor was transferred at a constant rate from one vertical wall covered with water to the opposite wall covered by a film of strong $CaCl_2$ solution

Position, cm from water wall	*Partial pressure H_2O, mm Hg*
0 (H_2O)	21.12
0.27	17.14
0.43	17.00
1.08	16.58
1.72	16.08
2.37	15.74
2.69	15.49
3.34	15.12
4.63	14.27
4.79	13.90
5.06 ($CaCl_2$)	9.54

Temperature of experiment, 23 C.

The rate of transfer of water vapor was 7.14 g of liquid water per min for a total cross-sectional area of 12,630 cm^2.

(a) Prepare a graph of partial pressure of H_2O vs. position in the duct, and calculate the value of the eddy diffusivity for the main central portion in which the gradient is essentially a straight line. Express the result as cm^2/sec.

(b) If the molecular diffusivity of the system of CO_2 and water vapor is 0.20 cm^2/sec at 23 C, calculate the ratio of the eddy diffusivity to the molecular diffusivity.

PART FOUR Special Topics

CHAPTER 17

Dimensional Analysis

Dimensional analysis is a method of deducing logical groupings of the variables involved in a process. It is applied to systems which are geometrically similar, e.g., all comparable lengths in two systems have identical ratios. In two circular tubes, for example,

$$K_L = \frac{d_1}{d_2} = \frac{L_1}{L_2} = \frac{x_1}{x_2} = \frac{r_1}{r_2} = \cdots$$

where x_1 and x_2 are comparable points along the length and r_1 and r_2 are comparable radii. This also implies $x_1/d_1 = x_2/d_2$ or $L_1/d_1 = L_2/d_2$, etc.; hence, only one dimension is necessary to establish the size of a system.

We wish to investigate the conditions under which there exists:

(1) kinematic similarity, the ratio of comparable velocities at comparable points being identical, e.g.,

$$\frac{v_{x1}}{v_{x2}} = \frac{v_{y1}}{v_{y2}} = \frac{V_1}{V_2}$$

(2) dynamic similarity, the ratio of forces identical at comparable points, e.g.,

$$\frac{(\text{inertia force})_1}{(\text{inertia force})_2} = \frac{(\text{viscous force})_1}{(\text{viscous force})_2}$$

(3) thermal similarity, the ratio of temperature differences identical at comparable points, e.g.,

$$\frac{(T_a - T_{\text{ref}})_1}{(T_a - T_{\text{ref}})_2} = \frac{(T_b - T_{\text{ref}})_1}{(T_b - T_{\text{ref}})_2}$$

where T_a and T_b are the temperatures at locations a and b in each system, and T_{ref} is an arbitrary identifiable reference temperature* at comparable points in the two systems. Similar conditions of similarity exist for mass concentration, electric charge, etc.

The above-mentioned conditions of similarity when extended to the boundaries of the systems imply similarity of boundary conditions.

17.1 Dimensions

In engineering, we represent physical concepts by symbols or dimensions, as for instance length by L and velocity by V. Through experience, we have learned that we can select a certain number of dimensions as fundamental, and express all other dimensions in terms of products of powers of these fundamental dimensions. In momentum transfer, generally only three fundamental dimensions suffice to express all other quantities, the customary choice being either mass M, length L, and time Θ, or force F, length L, and time Θ. The selection of either mass or force as a fundamental dimension has been argued in terms of which is the more basic, but it is really immaterial, since they are directly related by Newton's second law. This confusion, which should not exist, is somewhat compounded by haphazard use of the unit "pound" to represent both the pound-force F and the pound-mass M. In heat transfer, additional fundamental dimensions such as temperature T, and sometimes heat H, are usually specified. The dimension of heat may be eliminated since it can be expressed from the First Law of thermodynamics as the dimensions of work, (FL). Occasionally, in problems involving ideal gases, it may even be convenient to eliminate temperature as a fundamental dimension, since it is related to pressure and volume by the equation of state. By now, it should be clear that the word "fundamental" as used here is quite arbitrary. The choice of a particular set of dimensions as fundamental in a given class of problems is dictated solely by convenience and practice.

The dimensions of commonly used quantities in heat transfer analysis are listed in Table 17.1, with reference to two alternative sets of fundamental dimensions, $ML\Theta T$ and $ML\Theta TH$.

* Examples of suitable reference temperatures are as follows: for flow in a tube T_{ref} may be taken as the uniform temperature of the fluid upstream of the heated section, or if the wall temperature is uniform this may be used as T_{ref}.

TABLE 17.1

Quantity	Symbol	Dimension	
Mechanical Quantities		$ML\Theta$	$FL\Theta$
Diameter, Length	d, L	L	L
Area	A	L^2	L^2
Velocity	V	L/Θ	L/Θ
Acceleration	a, g	L/Θ^2	L/Θ^2
Mass	M	M	$F\Theta^2/L$
Density	ρ	M/L^3	$F\Theta^2/L^4$
Specific weight	γ	$M/L^2\Theta^2$	F/L^3
Viscosity	μ	$M/L\Theta$	$F\Theta/L^2$
Kinematic viscosity	ν	L^2/Θ	L^2/Θ
Mass flow rate	w	M/Θ	$F\Theta/L$
Force	F	ML/Θ^2	F
Pressure, shear	p, τ	$M/L\Theta^2$	F/L^2
Surface tension	σ	M/Θ^2	F/L
Heat and Mass Transfer Quantities		$ML\Theta T$	$ML\Theta TH$
Temperature	T	T	T
Heat transfer coefficient	h, U	M/Θ^3T	$H/\Theta L^2T$
Thermal conductivity	k	ML/Θ^3T	$H/\Theta LT$
Thermal diffusivity	α	L^2/Θ	L^2/Θ
Specific heat	c	L^2/Θ^2T	H/MT
Coefficient of expansion	β	$1/T$	$1/T$
Heat transfer rate	q	ML^2/Θ^3	H/Θ
Entropy	s	ML^2/Θ^2T	H/T
Diffusivity	D	L^2/Θ	L^2/Θ
Mass concentration of a	c_a	M/L^3	M/L^3
Mass transfer coefficient	h_D	L/Θ	L/Θ
Electrical Quantities		$RIl\Theta$	$Ql\Theta$
Voltage	e	IR	lQ/Θ^2
Resistance	R	R	l/Θ
Current	i	I	Q/Θ
Charge	Q	$I\Theta$	Q
Capacitance	C	Θ/R	Θ^2/l
Inductance	l	l	l

The methods of dimensional analysis are founded upon the principle of dimensional homogeneity, which states that all equations describing the behavior of physical systems must be dimensionally consistent, i.e., each term with reference to a given set of fundamental dimensions must have the same dimensions. When the equations governing a process are known and solvable, dimensional analysis suggests logical grouping of quantities for presenting the results. When the mathematical equations governing certain processes are unknown or too complex, dimensional analysis lays the foundation of an efficient experimental program for obtaining the results, by reducing the number of variables requiring investigation and by indicating a possible form of the semi-empirical correlations that may be formulated. It should be borne in mind, however, that dimensional analysis by itself cannot provide quantitative answers, and thus cannot be a substitute for the exact or the approximate mathematical solutions. It is nevertheless an important tool to learn to use, especially in instances when mathematical anslysis is impractical or when some rapid, qualitative answers are needed. Two general methods, both corollaries of the principle of dimensional homogeneity and both variously known as dimensional analysis, will be discussed in some detail: (a) rearrangement of differential equation and (b) pi theorem.

17.2 Rearrangement of Differential Equations

If the differential equations and boundary conditions can be written to describe any process adequately, they may be put into dimensionless form either by inspection or by the method outlined in the following examples.

Flow in Entrance Region between Parallel Plates

The differential equation applicable to this problem in the steady state is Eq. (3.3):

$$v_x \frac{\partial v_x}{\partial x} + v_y \frac{\partial v_x}{\partial y} = -\frac{1}{\rho}\frac{\partial p}{\partial x} + \nu \frac{\partial^2 v_x}{\partial y^2} \tag{3.3}$$

and a boundary condition is

$$(\tau_{yx})_0 = -\mu \left(\frac{\partial v_x}{\partial y}\right)_{y=0} \tag{17.1}$$

When two systems are geometrically similar, all dimensions have the same ratios,

$$K_L = \frac{L_1}{L_2} = \frac{r_{0_1}}{r_{0_2}} = \frac{d_1}{d_2} = \frac{x_1}{x_2} = \frac{y_1}{y_2} \tag{17.2a}$$

where x_1 and x_2 and also y_1 and y_2 are corresponding points in the two systems. The differential equation and boundary conditions require certain relations between velocities, dp/dx, and fluid properties at these corresponding points. For kinematic and dynamic similarity we define the following ratios:

$$K_\rho = \frac{\rho_1}{\rho_2}, \quad K_\nu = \frac{\nu_1}{\nu_2}, \quad K_{dp/dx} = \frac{(dp/dx)_1}{(dp/dx)_2}$$

$$K_V = \frac{v_{x1}}{v_{x2}} = \frac{v_{y1}}{v_{y2}} = \frac{V_1}{V_2} \tag{17.2b}$$

where K_L, K_ρ, etc. are all constant ratios and V is the average velocity in the x-direction.

To investigate the conditions which must be satisfied by these ratios we write Eqs. (3.3) and (17.1) for system 1 and replace v_{x_1}, x_1, v_{y_1}, y_1, etc. by their equivalents from Eq. (17.2) with the following result:

$$\frac{K_V^2}{K_L}\left(v_{x2}\frac{\partial v_{x2}}{\partial x_2} + v_{y2}\frac{\partial v_{x2}}{\partial y_2}\right) = -\frac{K_{dp/dx}}{K_\rho}\frac{1}{\rho_2}\frac{\partial p_2}{\partial x_2} + \frac{K_\nu K_V}{K_L^2}\nu_2\frac{\partial^2 v_{x2}}{\partial y_2^2} \tag{17.3}$$

Note that $\partial^2(K_V v_{x_2})/\partial(K_L y_2)^2 = (K_V/K_L^2)\,\partial^2 v_{x_2}/\partial y_2^2$.

It is apparent that if Eq. (3.2) applies equally well to systems 1 and 2, then

$$\frac{K_\nu K_V/K_L^2}{K_V^2/K_L} = 1$$

$$\frac{K_{dp/dx}/K_\rho}{K_V^2/K_L} = 1 \tag{17.4}$$

Substitution of Eq. (17.2) into the first equation of (17.4) gives the following dimensionless group:

$$\frac{K_V K_L}{K_\nu} = 1 \quad \text{or} \quad \frac{V_1 d_1}{\nu_1} = \frac{V_2 d_2}{\nu_2} = \frac{Vd}{\nu}, \quad \text{Reynolds number} \tag{17.5}$$

Similarly, for the second group

$$\frac{K_{dp/dx}K_L}{K_\rho K_V^2} = 1 \quad \text{or} \quad \frac{(dp/dx)d}{\rho V^2}, \quad \text{dimensionless} \tag{17.6}$$

This dimensionless quantity is defined as $2f$, where f is the friction factor, Eq. (4.3). This analysis leads to f and Re as being the significant dimensionless groups.

Likewise, following the procedure above for boundary condition, Eq. (17.1),

$$K_{\tau_0}\tau_{0_2} = -\frac{K_\mu K_V}{K_L}\mu_2\left(\frac{\partial v_{x_2}}{\partial y_2}\right)_0 \tag{17.7}$$

then

$$\frac{K_{\tau_0}K_L}{K_\mu K_V} = 1 \quad \text{or} \quad \frac{\tau_0 d}{\mu V}, \quad \text{dimensionless} \tag{17.8a}$$

This is not a commonly used dimensionless group, but when divided by the

Reynolds number Vd/ν it becomes

$$\frac{\tau_0}{\rho V^2}, \quad \text{dimensionless} \tag{17.8b}$$

which is equivalent to $2f$ from Eq. (4.9) where f is the friction factor.

The analysis above suggests that for flow between parallel plates, a case for which Eq. (3.2) applies, pressure-drop results may be correlated as f vs. Vd/ν where d is the plate spacing. If L were allowed to vary keeping d constant we might expect f to be a function both of Vd/ν and of L/d.

One-Dimensional Transient Heat Conduction

Consider the transient conduction problem of Sec. 6.13 with the differential equation and boundary conditions given by:

$$\frac{\partial\theta}{\partial t} = \alpha \frac{\partial^2\theta}{\partial x^2} \tag{17.9}$$

$$\left(\frac{\partial\theta}{\partial x}\right)_{x=r_0} = -\frac{h}{k}(\theta)_{x=r_0} \tag{17.10}$$

where $\theta \equiv T - T_f$. Here the reference temperature in the system is taken as T_f, the constant temperature of the fluid. When two systems are similar, the magnitudes of variables in system 1 relate to those in system 2 in the following ratios:

$$K_L = \frac{x_1}{x_2} = \frac{r_{0_1}}{r_{0_2}}$$

$$K_c = \frac{c_1}{c_2}, \quad K_\rho = \frac{\rho_1}{\rho_2}, \quad K_k = \frac{k_1}{k_2}, \quad K_h = \frac{h_1}{h_2} \tag{17.11}$$

$$K_T = \frac{\theta_1}{\theta_2}, \quad K_t = \frac{t_1}{t_2}, \quad K_\alpha = \frac{\alpha_1}{\alpha_2}$$

To investigate the conditions which must be satisfied by these ratios, we write Eqs. (17.9) and (17.10) for system 1, then replace T_1, t_1, c_1, etc. by their equivalents from Eq. (17.11) with the result:

$$\frac{K_T}{K_L^2}\frac{\partial^2\theta_2}{\partial x_2^2} = \frac{K_T}{K_t K_\alpha}\frac{1}{\alpha_2}\frac{\partial\theta_2}{\partial t_2} \tag{17.12}$$

$$\frac{K_T}{K_L}\left(\frac{\partial\theta_2}{\partial x_2}\right)_{x_2=r_{0_2}} = -\frac{K_h K_T}{K_k}\frac{h_2}{k_2}\theta_2 \tag{17.13}$$

It is apparent that if Eqs. (17.9) and (17.10) apply equally well to system 2, then

$$\frac{K_T}{K_L^2} = \frac{K_T}{K_t K_\alpha} \tag{17.14}$$

$$\frac{K_T}{K_L} = \frac{K_h K_T}{K_k} \tag{17.15}$$

Substitution from Eq. (17.11) gives the following dimensionless groups

$$\frac{r_{01}^2}{\alpha_1 t_1} = \frac{r_{02}^2}{\alpha_2 t_2} = \frac{r_0^2}{\alpha t} \tag{17.16}$$

and

$$\frac{h_1 r_{01}}{k_1} = \frac{h_2 r_{02}}{k_2} = \frac{h r_0}{k} \tag{17.17}$$

Since

$$\frac{\partial^n}{\partial m^n}\left(\frac{T - T_f}{T_i - T_f}\right) = \frac{1}{(T_i - T_f)} \frac{\partial^n T}{\partial m^n}$$

the obvious choice of nondimensional temperature and space groups is

$$\frac{x}{r_0} \quad \text{and} \quad Y = \frac{T - T_f}{T_i - T_f}$$

Then Eqs. (17.9) and (17.10) may be rearranged in terms of the four dimensionless groups as follows:

$$\frac{\partial^2 Y}{\partial (x/r_0)^2} = \frac{\partial Y}{\partial (\alpha t/r_0^2)} \tag{17.18}$$

$$\left(\frac{\partial Y}{\partial (x/r_0)}\right)_{x=r_0} = \left(\frac{h r_0}{k}\right)(1 - Y_{x=r_0}) \tag{17.19}$$

Often in the simpler cases the equations can be put into nondimensional form by inspection without taking the steps outlined in Eqs. (17.12) through (17.15).

Application to Three-Dimensional Equations

Consider the third equation of Eq. (3.60), the special form of the Navier-Stokes equation for incompressible fluids and constant properties. Take the body force Z as the gravitational force ρg. Then, the terms of this z-direction equation are as listed in the second line of Table 17.2. Following the dimensional analysis procedures given earlier in this section the coefficients of the various terms are shown in the third line.

TABLE 17.2 DIMENSIONLESS GROUPS OBTAINABLE FROM MOMENTUM EQUATION EQ. (3.60) (REF 6)

	Unsteady term	Inertia terms	Static-pressure force	Viscous forces	Gravity force
	$\frac{\partial v_z}{\partial t}$	$v_x \frac{\partial v_z}{\partial x} + v_y \frac{\partial v_z}{\partial y} + v_z \frac{\partial v_z}{\partial z}$	$-\frac{1}{\rho} \frac{\partial p}{\partial z}$	$\nu \left(\frac{\partial^2 v_z}{\partial x^2} + \frac{\partial^2 v_z}{\partial y^2} + \frac{\partial^2 v_z}{\partial z^2} \right)$	g
	$\frac{K_V}{K_t}$	$\frac{K_V^2}{K_L}$	$\frac{K_{\partial p/\partial z}}{K_\rho}$	$\frac{K_\nu K_V}{K_L^2}$	K_g
$\frac{\text{Inertia}}{\text{Viscous}}$		←	$\left\{ \frac{VL}{\nu} \right\}$ Reynolds number	→	
$\frac{\text{Inertia}}{\text{Gravity}}$		←	$\left\{ \frac{V^2}{Lg} \right\}$ Froude number		→
$\frac{\text{Static Pressure}}{\text{Inertia}}$		← $\left\{ \frac{L(\partial p/\partial z)}{\rho V^2} \right\}$ Euler number	→		
$\frac{\text{Gravity}}{\text{Unsteady Term}}$	←		$\left\{ \frac{gt}{V} \right\}$		→

TABLE 17.3 DIMENSIONLESS GROUPS OBTAINABLE FROM ENERGY EQUATION, EQ. (7.84) (REF 6)

	Unsteady term	Convection terms	Conduction terms	Viscous terms
	$\frac{\partial \theta}{\partial t}$	$v_x \frac{\partial \theta}{\partial x} + v_y \frac{\partial \theta}{\partial y} + v_z \frac{\partial \theta}{\partial z}$	$\alpha\left(\frac{\partial^2 \theta}{\partial x^2} + \frac{\partial^2 \theta}{\partial y^2} + \frac{\partial^2 \theta}{\partial z^2}\right)$	$\frac{\mu}{\rho c} \quad \Phi$
	$\frac{K_T}{K_t}$	$\frac{K_V K_T}{K_L}$	$\frac{K_\alpha K_T}{K_L^2}$	$\frac{K_\mu}{K_\rho K_c} \quad \frac{K_V^2}{K_L^2}$
$\frac{\text{Conduction}}{\text{Unsteady term}}$	←	$\left\{\frac{\alpha t}{L^2}\right\}$ Fourier number	→	
$\frac{\text{Convection}}{\text{Conduction}}$		← $\left\{\frac{VL}{\alpha}\right\}$ Peclet number = Re·Pr	→	
$\frac{\text{Viscous}}{\text{Conduction}}$			← $\left\{\frac{\mu V^2}{k\theta}\right\}$	→

TABLE 17.4 DIMENSIONLESS GROUPS OBTAINABLE FROM MASS TRANSFER EQUATION, EQ. (15.22)

	Unsteady term	Convection terms	Diffusion terms
	$\frac{\partial c_a^*}{\partial t}$	$v_x \frac{\partial c_a^*}{\partial x} + v_y \frac{\partial c_a^*}{\partial y} + v_z \frac{\partial c_a^*}{\partial z}$	$D\left(\frac{\partial^2 c_a^*}{\partial x^2} + \frac{\partial^2 c_a^*}{\partial y^2} + \frac{\partial^2 c_a^*}{\partial z^2}\right)$
	$\frac{K_c}{K_t}$	$\frac{K_V K_c}{K_L}$	$\frac{K_D K_c}{K_L^2}$
Diffusion / Unsteady term	←	$\left\{\frac{Dt}{L^2}\right\}$	→
Convection / Diffusion		← $\left\{\frac{VL}{D}\right\}$ Re Sc	→

The operations shown in Eq. (17.4) are equivalent to expressing the ratios of viscous to inertia forces. Similarly, in Table 17.2, the results of taking ratios of the various terms of the equation are shown in the remainder of the table. Only a few of the possible combinations of dimensionless groups are shown.

In a similar way the energy equation (7.84) and the mass transfer equation (15.22) may be treated as in Tables 17.3 and 17.4, showing here only a few of the possible groups. More extensive tables are presented by Klinkenberg and Mooy (6).

A dimensional analysis of a problem is performed by writing all of the pertinent governing equations and boundary-condition equations in dimensionless form, following a procedure such as the one outlined previously in this section. Each equation provides a number of dimensionless groups equal to one less than the number of different kinds of terms in the equation.

17.3 Pi Theorem

When the governing equations of a problem are unknown, an alternative approach in the application of dimensional analysis is necessary. At the very start, it is necessary to know, or more typically to guess, the independent variables that determine the behavior of a particular dependent variable of interest. These can usually be found by logic or intuition developed from previous experiences with problems of a similar nature, but there is no way to insure that all essential quantities have been included. Rayleigh (4) first used this method and the rules of algebra to combine the many variables of a problem into dimensionless groups. A simpler and perhaps more systematic way of determining the groups was suggested by Buckingham (3) and has come to be known as the *pi theorem*, a formal statement of which follows:

> If there are n_1 physical quantities of importance and n_2 fundamental dimensions, then there exist a maximum number (n_{max}) of these quantities which in themselves cannot form a dimensionless group, where $n_{max} \leq n_2$. Then by combining one each of the remaining quantities with the n_{max} quantities, there can be formed $n\pi$ dimensionless groups where $n\pi = n_1 - n_{max}$. The dimensionless groups thus formed are called "π terms" and identified by the symbols $\pi_1, \pi_2, \cdots, \pi_{n_\pi}$.

In the application of the pi theorem, the following rules are helpful:

(a) Compile a tentative list of all significant variables. Suppose in a certain problem, $n_1 = 7$. The physical equation can be expressed symbolically as

$$f(Q_1, Q_2, Q_3, \cdots, Q_7) = 0$$

where Q's represent the physical quantities. If all these quantities can be expressed in terms of four fundamental dimensions, say L, M, Θ, T, then $n_2 = 4$.

(b) Next select n_{max} physical quantities called primary Q's, which by themselves cannot form a dimensionless group. In almost all practical cases, $n_{max} = n_2$, so a useful rule is to select the primary Q's so that each contains one of the fundamental dimensions at least once. Obviously, there is no unique set of primary Q's, and experience will dictate the best choice.

(c) Form each π term by expressing ratio of the remaining Q's to the product of powers of the primary Q's. In the case discussed, this procedure yields three π terms, and the original symbolic equation relating seven variables may now be replaced by

$$\phi(\pi_1, \pi_2, \pi_3) = 0$$

We shall illustrate the method with a few representative problems.

Pipe flow. We shall assume the pressure drop per unit length, $\Delta p/L$, in a pipe is a function of the fluid properties ρ and μ, the pipe diameter d, the velocity of flow V, and the pipe surface roughness e. To the question why the surface tension or even the color of the fluid is not a significant variable, there is no firm answer short of direct experimental proof. We make the best guesses possible and then proceed. The physical equation for this case is

$$\frac{\Delta p}{L} = \phi(\rho, \mu, d, V, e) \tag{17.20}$$

In terms of the fundamental dimensions M, L, and Θ, the corresponding dimensional formula of each of the quantities above is found from Table

17.1. We arbitrarily select as primary quantities V, d, and ρ, which clearly contain among them the three fundamental dimensions. The π groups may now be formed as follows:

$$\pi_1 = \frac{\Delta p/L}{\rho^a V^b d^c}, \qquad \pi_2 = \frac{\mu}{\rho^a V^b d^c}, \qquad \pi_3 = \frac{e}{\rho^a V^b d^c}$$

The exponents a, b, c are to be determined for each group to make it dimensionless. Let us write π_1 in terms of its dimensions:

$$|\pi_1| = \left|\frac{\Delta p/L}{\rho^a V^b d^c}\right| = \frac{ML^{-2}\Theta^{-2}}{(ML^{-3})^a(L\Theta^{-1})^b(L)^c}$$

Now collect the exponents of each dimension and equate to zero

$$M: \qquad 1 - a = 0$$

$$L: \qquad -2 + 3a - b - c = 0$$

$$\Theta: \qquad -2 + b = 0$$

Solving these three equations simultaneously leads to $a = 1$, $b = 2$, and $c = -1$; or

$$\pi_1 = \frac{\Delta p}{(L/D)\rho V^2}$$

Similarly,

$$|\pi_2| = \left|\frac{\mu}{\rho^a V^b d^c}\right| = \frac{ML^{-1}\Theta^{-1}}{(ML^{-3})^a(L\Theta^{-1})^b(L)^c}$$

which leads to $a = 1$, $b = 1$, and $c = 1$, so that

$$\pi_2 = \frac{\mu}{\rho V d} = \frac{1}{\text{Re}}$$

By inspection,

$$\pi_3 = \frac{e}{d}$$

Some algebraic operations were necessary to obtain π_1 and π_2, but π_3 is obvious by inspection. With skill and experience, considerable saving in time is possible. The original equation can be expressed as a relation involving the three π terms,

$$2f = \frac{\Delta p}{(L/d)\rho V^2} = \psi\left(\text{Re}, \frac{e}{d}\right) \tag{17.21}$$

introducing the definition of the friction factor, Eq. (4.3).

The choice of a different set of primary quantities will lead to different but equally valid π groups. For instance, the choice of ρ, μ, d gives, by the same method as before,

$$\pi_1' = \frac{(\Delta p)\rho d^3}{L\mu^2}, \qquad \pi_2' = \frac{\rho V d}{\mu}, \qquad \pi_3' = \frac{e}{d}$$

It is seen that the new dimensionless pressure-drop group is a combination of the groups found previously, $\pi_1' = (\pi_1)/(\pi_2^2)$. This new group is less familiar and useful, but certainly no less valid.

Heat transfer in forced convection. Assume the heat transfer coefficient in fully developed forced convection in a tube is a function of the following variables:

$$h = \phi(V, \rho, \mu, c_p, k, d) \tag{17.22}$$

All these quantities can be expressed in terms of the four fundamental dimensions M, L, Θ, and T. We select as primary quantities V, d, ρ, and k, and proceed as before in forming the π terms:

$$|\pi_1| = \left|\frac{h}{V^a d^b \rho^c k^d}\right| = \frac{M\Theta^{-3}T^{-1}}{(L\Theta^{-1})^a(L)^b(ML^{-3})^c(ML\Theta^{-3}T^{-1})^d}$$

Equating the exponents of each dimension in the numerator and the denominator,

$$\begin{aligned} M&: & 1 &= c + d \\ L&: & 0 &= a + b - 3c + d \\ \Theta&: & -3 &= -a - 3d \\ T&: & -1 &= -d \end{aligned}$$

and solving these simultaneous equations, we get $a = 0$, $b = -1$, $c = 0$, and $d = 1$. Thus

$$\pi_1 = \frac{hd}{k} = \mathrm{Nu}$$

Similarly, we can get

$$\pi_2 = \frac{\mu}{\rho V d} = \frac{1}{\mathrm{Re}}$$

$$\pi_3 = \frac{c_p \mu}{k} = \mathrm{Pr}$$

It is seen that heat transfer data in forced convection can be correlated in terms of these three dimensionless groups:

$$\mathrm{Nu} = \psi(\mathrm{Re}, \mathrm{Pr}) \tag{17.23}$$

Alternate groups such as the Stanton number (St) can be formed by choosing a different set of primary quantities. Note that $\text{St} = \text{Nu}/(\text{Re}\cdot\text{Pr})$.

Heat transfer in free convection. In free convection over a heated vertical plate, motion is caused by a buoyancy force. The velocity V is no longer an independent variable, and in its place, we might expect the following physical quantities would be significant—the gravitational acceleration g, the temperature difference ΔT between the undisturbed fluid and the heated surface, and the coefficient of volume expansion β of the fluid. Also, a characteristic length might be the length L of the plate. Experience and perhaps intuition indicate to us that the gravity g acts on a control volume of the heated fluid only in conjunction with the property β, suggesting the two quantities might be combined as a single independent variable $(g\beta)$. The heat transfer coefficient can be expressed as

$$h = \phi(\rho, \mu, c_p, k, L, g\beta, \Delta T) \tag{17.24}$$

In terms of the fundamental dimensions M, L, Θ, T, and H, we might select as primary quantities ρ, L, μ, ΔT, and k; we can form three π groups by an identical procedure as before:

$$\pi_1 = \frac{hL}{k} = \text{Nu}$$

$$\pi_2 = \frac{c_p\mu}{k} = \text{Pr}$$

$$\pi_3 = \frac{L^3\rho^2 g\beta(\Delta T)}{\mu^2} = \text{Gr}$$

Dimensional analysis thus tells us that the heat transfer data in free convection might be correlated by three dimensionless parameters instead of the original eight independent variables:

$$\text{Nu} = \psi(\text{Gr}, \text{Pr}) \tag{17.25}$$

Note that the π terms obtained by dimensional analysis may be combined in any product of powers which may seem desirable so long as the number of distinct groups remains the same; this means we can replace, for instance, Gr by the Rayleigh number, $\text{Ra} = \text{Gr} \cdot \text{Pr}$, or $\text{Nu} = \psi_2(\text{Ra}, \text{Pr})$. Table 17.5 lists some dimensionless groups in heat and mass transfer.

17.4 Application of Dimensional Analysis to Model Design

Qualitative results of dimensional analysis may be combined with carefully designed experiments to provide quantitative information. Typical examples of such a procedure are the more recent experimental results

TABLE 17.5 SOME DIMENSIONLESS GROUPS

Group	Symbol	Name
$\Delta p/\rho V^2$	Eu	Euler number
$\alpha t/r_0$	Fo	Fourier number
$(L/d)(k/Vd\rho c_p)$	Gz [= (L/d)/RePr]	Graetz number
$g\beta(\Delta T)L^3\rho^2/\mu^2$	Gr	Grashof number
λ/L	Kn	Knudsen number
α/D	Le	Lewis number
V/V_{sound}	Ma	Mach number
hL/k, hd/k	Nu	Nusselt number
$Vd\rho c_p/k$	Pe (= Re Pr)	Peclet number
$c_p\mu/k$	Pr	Prandtl number
$g\beta(\Delta T)L^3\rho^2 c_p/\mu k$	Ra (= Gr Pr)	Rayleigh number
$\rho VD/u$, $\rho VL/\mu$	Re	Reynolds number
$\mu/\rho D$	Sc	Schmidt number
$h_D d/D$	Sh	Sherwood number
h/c_pG	St (= Nu/Re Pr)	Stanton number

leading to the McAdams and the Colburn correlations for heat transfer in tubes. It is not always possible or desirable, however, to conduct experiments on an actual system (prototype), and perhaps the cost of building an exact duplicate for test may be prohibitive. One solution is to study the behavior of a scaled-down or a scaled-up model of the prototype. Dimensional analysis tells us that if a model were scaled and a different fluid used, and the test conditions were maintained so that all the significant π terms had identical values as in the prototype operation, accurate prediction on the performance of the prototype should be possible from measurements made on the model. In practice, complete similarity of π terms between model and prototype is not always easy or possible to attain, especially if the π terms are numerous.

Faced with this dilemma, an engineer typically compromises by satisfying only those π terms more directly pertinent to his investigation, and allows for the effect of the remaining terms by some empirical methods. These methods are discussed in detail in various literature on model testing (5) (7).

REFERENCES

1. Stokes, G. G., *Trans. Cambr. Phil. Soc.*, **8** (1845); reprinted *Mathematics and Physics Papers*, pp. 95–104, Cambridge, London, 1905.

2. Nusselt, W., *Z. ver. deut. Ing.*, **53,** 1750, 1808 (1909); *Mitt. Forsch.*, **89,** 1 (1910); *Gesundh. Ing.*, **38,** 477, 490 (1915).

3. Buckingham, E., *Phys. Rev.*, **4,** 345 (1914); also, *Trans. ASME*, **35,** 262 (1915).
4. Rayleigh, Lord *Nature*, **95,** 66 (1915); *Phil. Mag.*, **34,** 59 (1892).
5. Bridgeman, P. W., *Dimensional Analysis*, Yale Univ. Press, New Haven, 1931.
6. Klinkenberg, A., and H. H. Mooy, *Chem. Eng. Prog.*, **44,** 17 (1948).
7. Langhaar, H. L., *Dimensional Analysis and Theory of Models*, Wiley, New York, 1951.

PROBLEMS

1. A brick has dimensions 2 by 4 by 6 in. What are the dimensions of three geometrically similar bricks each having at least one side with a dimension of 7 in.?

2. State the similarity condition for (a) mass diffusion in a dilute binary system, and (b) electrical conduction in a stationary conducting solid body.

3. Write Eq. (3.3) in terms of dimensionless groups only.

4. Using the differential equations discussed in Sec. 7.4, show by dimensional analysis that the velocity distribution for free convection between parallel plates can be represented by $v_x^* = f(\mathrm{Gr}, \mathrm{Pr}, y^*)$.

5. Using the differential equations of Sec. 7.5, [Eqs. (7.32) and (3.18) and the conditions at the boundary], show that the dimensionless groups for representing the heat transfer performance of a flat plate are $\mathrm{Nu}_L = f(\mathrm{Re}_L, \mathrm{Pr})$.

6. Using Eqs. (7.32) and (7.53), show by dimensional analysis that heat transfer data from a vertical plate with natural convection may be correlated by $\mathrm{Nu} = f(\mathrm{Gr}, \mathrm{Pr})$.

7. For the combined heat and mass transfer problem of Sec. 15.8, show by dimensional analysis of the differential equations that

$$\frac{c_a - c_{a0}}{c_{a\infty} - c_{a0}} = f\left(\frac{y}{x}, \mathrm{Re}_x, \mathrm{Pr}, \mathrm{Sc}, \frac{v_0}{V_\infty}\right)$$

8. An infinite flat plate of half-thickness r_0 is initially at a uniform temperature T_i; suddenly an electric current begins to flow, producing a heat source of W Btu/hr ft^3. The plate is cooled on each side by a fluid of T_i with a surface heat transfer coefficient of h. The differential equation and surface boundary conditions are

$$\frac{\partial^2 T}{\partial x^2} = \frac{1}{\alpha}\frac{\partial T}{\partial t} - \frac{W}{k}$$

$$\left(\frac{\partial T}{\partial x}\right)_{x=r_0} = \frac{h}{k}\,(T_{x=r_0} - T_i)$$

Determine by both dimensional analysis methods discussed that the significant dimensionless groups are:

$$\frac{k(T - T_i)}{Wr_0^2}, \quad \frac{\alpha t}{r_0^2}, \quad \frac{hr_0}{k}, \quad \frac{x}{r_0}$$

when it is assumed that $(T - T_i) = \phi(h, k, \rho, c, r_0, x, t, W)$.

9. Show for the problem discussed in Sec. 6.12 that the appropriate dimensionless groups are

$$\frac{T_f - T_2}{T_f - T_1}, \quad \frac{tAh}{V\rho c}$$

10. Using the pi theorem, show that experimental data for mass diffusion at the wall of a pipe with air flowing in turbulent flow should be correlated by Sh $= f(\text{Re}, \text{Sc})$; see Sec. 16.3.

CHAPTER 18

Numerical and Analog Methods

When analytical solutions of differential equations are difficult, numerical or analog methods often provide the needed answers. In this chapter, we shall concentrate our attention primarily on numerical and analog solutions by finite-difference methods, extremely useful tools in heat transfer for solving problems of steady- and unsteady-state diffusion of heat and mass in bodies of complex shapes and unusual boundary conditions. The essence of finite-difference methods consists of replacing the pertinent differential equations by finite-difference equations; in other words, the derivatives in the former are replaced by finite-difference ratios in the latter. Physically, this is tantamount to replacing a continuous system by a network of finite elements, called a *lumped-parameter network*.

In the applications that follow, we shall illustrate the transformation of the differential equations into finite-difference equations, but we shall also show that it is actually simpler in many instances to derive the difference equations by direct application of pertinent conservation laws to each finite element. Relatively simple problems are chosen so that existing mathematical formulations and results can be utilized when necessary to interpret the finite-difference procedures.

The discussion of finite-difference methods may logically be divided into steady-state and unsteady-state problems. Steady-state problems are characterized by boundary conditions being specified over the entire domain of the solutions, which are sometimes called "jury" solutions (2) because they must satisfy a jury composed of the boundary conditions. In unsteady-state problems, solutions are not bounded in time in that initial conditions are prescribed, but conditions at a later time are determined by progress of the solutions themselves; they are sometimes called "marching" solutions.

18.1 Finite-Difference Solutions of Steady-State Problems

(a) *Thin Straight Fin*

Let us consider the fin problem of Sec. 6.9. The differential equation is given by Eq. (6.39):

$$\frac{d^2\theta}{dx^2} - \frac{hP}{kS}\theta = 0 \tag{6.39}$$

where $\theta = T - T_a$. Imagine the fin to be divided into a number of equal-width (Δx) elements, as shown in Fig. 18.1. In this example assume the

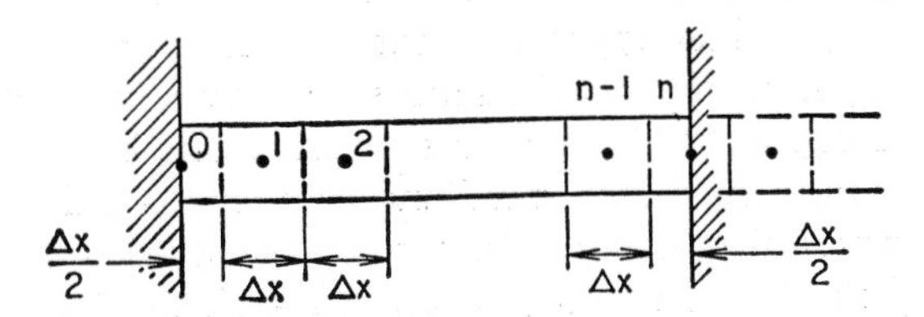

Fig. 18.1. Thin fin subdivided into finite difference elements.

temperature at point 0 to be T_0, the temperature of the surrounding atmosphere, T_a, to be known, and that the outer end of the fin is insulated. Taylor series expansions of function θ about point 2, for example, give the values of θ at neighboring points $1(x - \Delta x)$ and $3\ (x + \Delta x)$:

$$\theta_1 = \theta_2 - (\Delta x)\left(\frac{d\theta}{dx}\right)_2 + \frac{(\Delta x)^2}{2!}\left(\frac{d^2\theta}{dx^2}\right)_2 - \cdots \tag{18.1}$$

$$\theta_3 = \theta_2 + (\Delta x)\left(\frac{d\theta}{dx}\right)_2 + \frac{(\Delta x)^2}{2!}\left(\frac{d^2\theta}{dx^2}\right)_2 + \cdots \tag{18.2}$$

Or, adding the series (18.1) and (18.2) we obtain:

$$\frac{d^2\theta}{dx^2} = \frac{\theta_3 - 2\theta_2 + \theta_1}{(\Delta x)^2} \tag{18.3}$$

where terms of the order of $(\Delta x)^4$ and higher are neglected. Substituting Eq. (18.3) into Eq. (6.39) and letting $\Gamma \equiv (hP/Sk)(\Delta x)^2$, we get the following finite-difference equation:

$$\theta_3 - (2 + \Gamma)\theta_2 + \theta_1 = 0 \tag{18.4}$$

We note that the equation is algebraic and can be solved readily for θ_2 in terms of θ of the neighboring points 1 and 3. The same equation can be obtained by deriving the differential equation directly in finite-difference form. An energy balance applied to element 2 requires for the steady state that the net heat transferred to the element be zero. This results in the equation

$$Sk\frac{T_1 - T_2}{\Delta x} - Sk\frac{T_2 - T_3}{\Delta x} - hP(\Delta x)(T_2 - T_a) = 0 \tag{18.5}$$

or

$$T_1 - (\Gamma + 2)T_2 + T_3 + \Gamma T_a = 0$$

This equation is identical with Eq. (18.4), since $\theta \equiv T - T_a$. In arriving at Eq. (18.5) it was assumed that dT/dx at the interface between elements 1 and 2 may be approximated by $(T_1 - T_2)/(\Delta x)$ with a similar approximation at the interface between elements 2 and 3. It was also assumed that $\int hP(T - T_a)\,dx$ over the surface of element 2 could be approximated by $hP(\Delta x)(T_2 - T_a)$. These approximations become better as Δx is assumed smaller, e.g., a greater number of elements for a given fin length.

Application of Eq. (18.5) to each element results in a set of algebraic equations that must be solved simultaneously for the unknown temperatures. Two numerical procedures for accomplishing this will be discussed: *relaxation* and *iteration*. In both methods, we evaluate Γ at each element, guess initially at the unknown temperatures in the interior based on the known boundary conditions, and then adjust these temperatures so that condition (18.5) is satisfied everywhere.

In the relaxation procedure, we recognize that unless our initial guesses are exactly correct, and most likely they will not be, Eq. (18.5) will not hold at all points and a residual defined as for instance at point 2 by

$$R_2 = T_1 - (\Gamma + 2)T_2 + T_3 + \Gamma T_a \tag{18.6a}$$

will exist at all points.

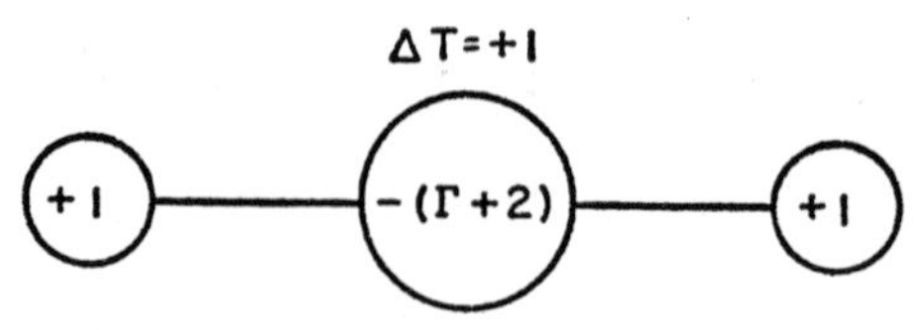

Fig. 18.2. One-dimensional relaxation pattern for fin element.

If the end of the fin is insulated, divide the fin into elements in such a way that a half-element exists at the end. Then imagine the existence of a mirror image of the fin, shown dotted in Fig. 18.1. At point n the residual is

$$R_n = T_{n-1} - \left(\frac{\Gamma}{2} + 1\right)T_n + \frac{\Gamma}{2}T_a \tag{18.6b}$$

After the residuals at each point have been calculated and found not to be zero, we must adjust or "relax" our original estimation of the value of T at each point so that the residuals are progressively reduced. Let us examine what happens if the guessed value of temperature at point 2 were to be increased by one unit. Inspection of Eq. (18.6a) shows that only the residuals at points 1, 2, and 3 are affected. From Eq. (18.6a) the residuals at point 2 would be changed by an amount, $-(\Gamma + 2)$, and at the points 1 and 3 would each be changed by an amount, $+1$. A diagram of this rule, called a *relaxation pattern*, can be made as shown in Fig. 18.2. This indicates that a change in guessed temperature of $+1$ at a point causes a change of $-(\Gamma + 2)$ in the residual at the point and a change of $+1$ in the residuals at each of the adjacent points. The only exception for this case is at point n. Figure 18.3 shows the relaxation pattern for this point. Here a unit change in temperature at n causes a change of $-(\frac{1}{2}\Gamma + 1)$ in the residual at n and a change of $+1$ in the residual at $(n - 1)$.

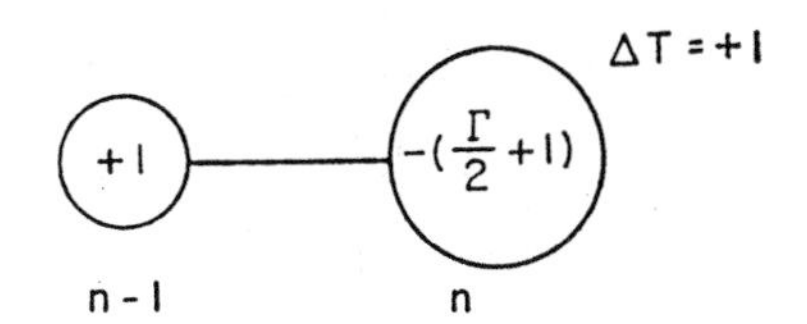

Fig. 18.3. One-dimensional relaxation pattern for element at tip of fin.

Example 18.1: Imagine a fin to be divided into equal elements with half-width elements at points 0 and n, as in Fig. 18.1 with $n = 5$. The tip of the fin is insulated, $T_0 = 100$ F, $T_a = 0$ F, and $\Gamma = 1$. The procedure for applying the relaxation method outlined above is shown in Table 18.1. In line 0 the initial guessed values are listed along with the residuals as calculated by Eq. (18.6). Efforts are now directed toward reducing these residuals to zero at each point.

The largest initial residual occurs at point 2. If T_2 is reduced by 5 F, $\Delta T_2 = 5$, then the new residuals at points 1, 2, and 3 would be as shown in line 1. This same procedure is applied to the point at which the largest residual exists until all residuals have been reduced to small values. The numbers in the temperature columns of Table 18.1 represent magnitudes for ΔT except in line 0. In the residual columns the numbers represent running totals.

Down to line 8, integer changes in T were employed; at this point all residuals are $\leq |1|$, and there are approximately an equal number of positive and negative residuals, e.g., algebraic sum of the residuals $\cong 0$. It is advisable to check the arithmetic at this point by totalling the initial temperature guess

TABLE 18.1 RELAXATION SOLUTION OF EXAMPLE 18.1 $T_0 = 100$, $T_a = 0$, $\Gamma = 1$

Line	T_1	R_1	T_2	R_2	T_3	R_3	T_4	R_4	T_5	R_5
0	40	0	20	−8	12	−7	9	−7	8	−3
1		−5	−5	+7		−12				
2				+2	−5	+3		−12		
3						−1	−4	0		−7
4								−4	−4	−1
5	−2	+1		0						
6						−3	−2	+2		−3
7								0	−2	0
8				−1	−1	0		−1		
9	38	+1	15	−1	6	0	3	−1	2	0

and the subsequent changes resulting in the values shown in line 9. With these new assumed temperature values a new set of residuals may be computed, and any errors which may have appeared earlier in the table may be forgotten because line 9 is correct in itself. Further changes will probably involve fractional changes in temperature. The temperature distribution as listed in line 9 is probably precise to ±1 F.

By continuing this process with fractional degree changes, we can reduce the residuals to as many decimal places as time permits. Even if the residuals are reduced to zero exactly, the resulting temperature distribution is not exact because of the approximations made in deriving Eq. (18.5). These approximations, and hence the results, are better for smaller values of Δx. However, finer grids entail more time and effort to reach a solution, and an intelligent compromise must be made by the calculator.

The solution in Table 18.1 was obtained using constant Γ. A solution could have been obtained with equal ease if Γ were not constant, i.e., if k, h, P, or S varied along the fin length. In contrast, we have seen how varying properties greatly complicate analytical solutions.

We shall now discuss an iterative procedure. Equation (18.5), solved for T_2, becomes

$$T_2 = \frac{T_1 + T_3 + \Gamma T_a}{\Gamma + 2} \tag{18.7}$$

which, for Example 18.1 with $T_a = 0$ and $\Gamma = 1$, becomes $T_2 = (T_1 + T_3)/3$.

As before, we first guess the temperatures at the interior points in the fin. Iteration involves successive application of Eq. (18.7) from point to point. Ideally, the solution is accomplished when two successive iterations at each point give identical magnitudes. Table 18.2 shows the results of applying iteration to the example of Table 18.1.

Because of slow convergence, the iteration procedure is usually less efficient than the relaxation procedure for "hand" calculation, but since iteration computations follow a fixed routine, they are more readily usable on automatic computing machines.

TABLE 18.2 ITERATION SOLUTION OF EXAMPLE 18.1 $T_0 = 100$, $T_a = 0$, $\Gamma = 1$

Line	T_1	T_2	T_3	T_4	T_5
0	40	20	12	9	8
1	40	17.3	9.7	6.7	6
2	39.1	16.6	8	5.2	4.5
3	38.9	15.7	7.3	4.2	3.5
4	38.5	15.4	6.6	3.6	2.8
5	38.4	15.0	6.3	3.1	2.4

(b) *Thick Fins or Walls*

In problems involving, for example, thick fins and walls of thick-walled furnaces, it is necessary to consider temperature changes in more than one direction. Consider as an example a thick fin shown in Fig. 18.4 in which temperature gradients in the x- and y-directions are considered. Temperature gradients in the z-direction are assumed to be zero, and the z-direction thickness is taken as unity.

Divide the fin into elements of dimension Δx, Δy. The horizontal rows are labelled A, B, C, $\cdots$, and the vertical columns are labelled 0, 1, 2, 3, $\cdots$. Each element is then located by a letter and a number. We shall proceed to solve this problem by relaxation procedures.

A residual is defined at each point by a procedure similar to that of Eqs. (18.5) and (18.6). For a typical interior point $D3$, an energy balance in finite-difference form may be written as

$$k_y\,\Delta x\left(\frac{T_{C3}-T_{D3}}{\Delta y}+\frac{T_{E3}-T_{D3}}{\Delta y}\right)+k_x\,\Delta y\left(\frac{T_{D2}-T_{D3}}{\Delta x}+\frac{T_{D4}-T_{D3}}{\Delta x}\right)=0$$

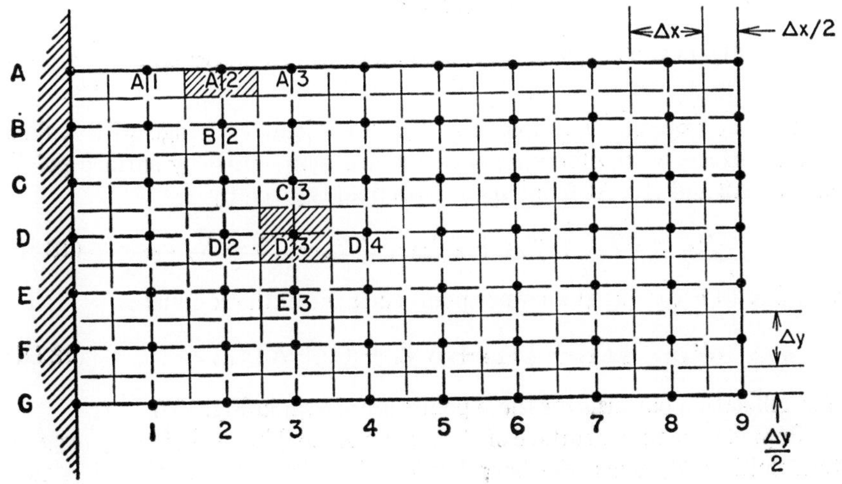

Fig. 18.4. Thick fin subdivided into finite difference elements.

or a residual which should be zero may be defined as

$$R_{D3} = \frac{k_y}{k_x}\frac{\Delta x}{\Delta y}(T_{C3} + T_{E3}) + \frac{\Delta y}{\Delta x}(T_{D2} + T_{D4}) - 2\left(\frac{k_y}{k_x}\frac{\Delta x}{\Delta y} + \frac{\Delta y}{\Delta x}\right)T_{D3} \tag{18.8a}$$

In most cases involving a homogeneous material $k_x = k_y$. Also, it will be convenient to take $\Delta x = \Delta y$. Then Eq. (18.8a) reduces to

$$R_{D3} = T_{D2} + T_{E3} + T_{D4} + T_{C3} - 4T_{D3} \tag{18.8b}$$

The relaxation pattern at point $D3$ is given by Fig. 18.5.

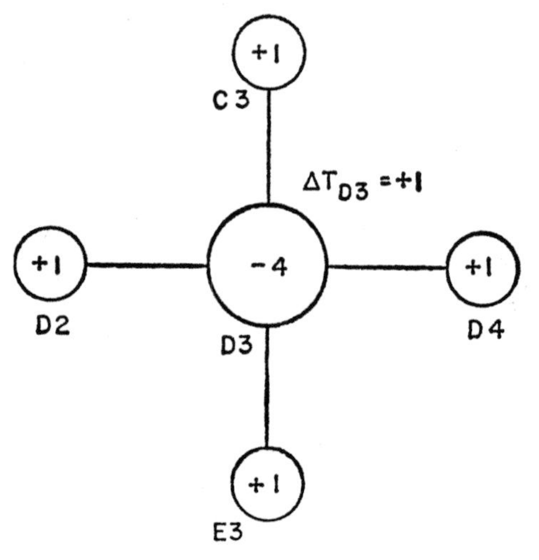

Fig. 18.5. Two-dimensional relaxation pattern for interior element.

For the example shown in Fig. 18.4 with heat exchange with a fluid at the surfaces of the fin, special points are represented by surface points such as $A2$ and $A9$. At point A_2, the residual is given by

$$R_{A2} = \tfrac{1}{2}T_{A1} + T_{B2} + \tfrac{1}{2}T_{A3} + NT_a - (2 + N)T_{A2} \tag{18.8c}$$

where $N \equiv h(\Delta y)/k$; at a corner point such as $A9$ the residual is

$$R_{A9} = \tfrac{1}{2}T_{A8} + \tfrac{1}{2}T_{B9} + NT_a - (1 + N)T_{A9} \tag{18.8d}$$

The relaxation patterns for these points are shown in Fig. 18.6.

Suppose the upper surface of the fin is insulated instead of being exposed to a fluid at T_a. Then the boundary condition at point $A2$ requires that $\partial T/\partial y$ at the surface be zero or symmetry exists at the surface. Then, as

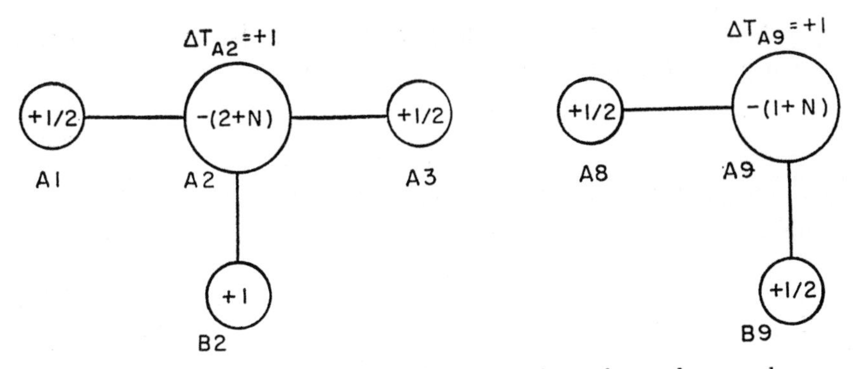

Fig. 18.6. Two-dimensional relaxation pattern for surface and corner elements of fin.

shown in Fig. 18.7, the residual at $A2$ may be obtained from Eq. (18.18b) with the result

$$R_{A2} = T_{A1} + 2T_{B2} + T_{A3} - 4T_{A2} \tag{18.8e}$$

The procedure for solving these two-dimensional, steady-state problems is identical with that shown in Table 18.1. A column listing magnitudes of temperature and of residual is needed for each point in the network. An initial temperature distribution is assumed, and at each point the residual is calculated with the proper residual relation from Eq. (18.8). Then the residuals are "relaxed" as before by using the proper relaxation pattern applicable at each point. When the residuals have been reduced to near zero with as many positive values as there are negative values, the resulting temperature distribution is an approximate solution to the problem.

(c) *Calculation of Heat Transfer Rate*

The result of the finite-difference solution is a temperature distribution throughout a lumped-parameter network. The heat transfer rate across

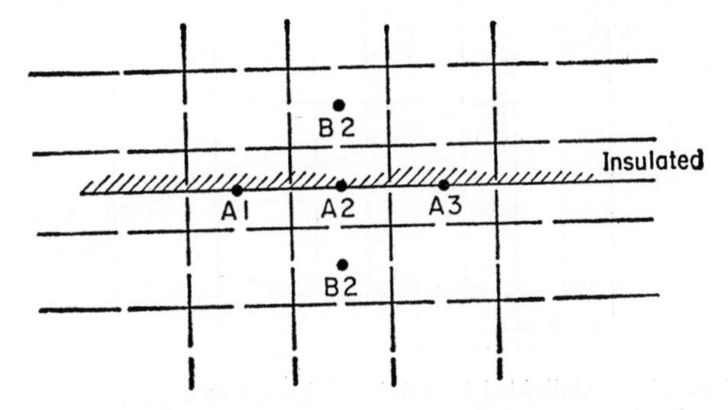

Fig. 18.7. Element at an insulated surface.

one of the faces of an element of the fin in Fig. 18.4—for example, the face between points $D3$ and $D4$—is

$$\delta q = \Delta y k_x \frac{T_{D3} - T_{D4}}{\Delta x} \tag{18.9a}$$

for a unit depth in the z-direction. For points along the lower edge—for example, $G3$,

$$\delta q = \Delta x k_y \frac{T_{F3} - T_{G3}}{\Delta y} \tag{18.9b}$$

also

$$\delta q = \Delta x h (T_{G3} - T_a) \tag{18.9c}$$

Equation (18.9b) represents the heat transfer rate across the face between points $F3$ and $G3$, and Eq. (18.9c) the rate across the face at the surface point $G3$. As Δx and Δy are made smaller, these two heat transfer quantities become equal. As an approximation either equation may be applied to each element at the surface. The sum of these values of δq along the surface represents the total heat transfer rate crossing the surface of the fin.

18.2 Finite-Difference Solutions of Unsteady-State Problems

We would expect a logical extension of the finite-difference methods to apply to unsteady-state problems, with time entering as an additional independent variable. A relaxation procedure requires that the boundary conditions be completely specified. In the unsteady-state "marching" solutions, the boundary in time is open, but a relaxation procedure can still be used if, for example, the ultimate steady-state solution can be treated as the prescribed boundary at some large time. More commonly, iteration

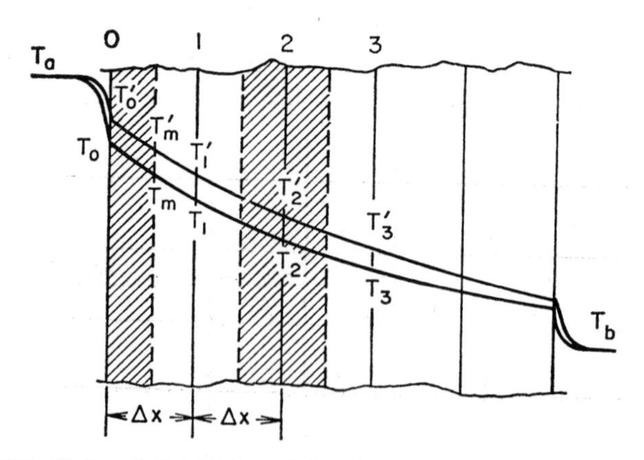

Fig. 18.8. Plate subdivided into finite difference elements, showing temperature distributions corresponding to a "marching" solution.

procedures are used to solve unsteady-state problems. However, seemingly correct procedures can result in solution instability, as we shall see.

Infinite Flat Plate

The plate of Fig. 18.8 is assumed to have no temperature gradients in the y- and z-directions. The governing differential equation for this case is given in Chap. 6:

$$\frac{\partial T}{\partial t} = \alpha \frac{\partial^2 T}{\partial x^2} \tag{6.53}$$

In changing the differential equation into finite-difference form, it is helpful to interpret the transformation with the aid of a lumped-parameter

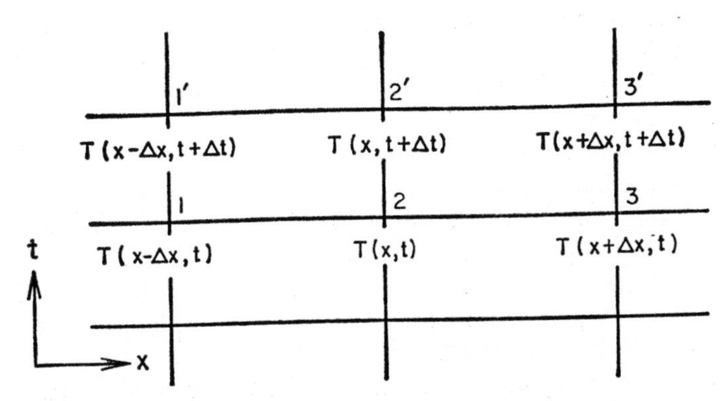

Fig. 18.9. Lumped-parameter network in space and time.

network in space x and time t, Fig. 18.9. Taylor series expansions of function T about point 2 are given below:

$$T_3 = T_2 + \Delta x\left(\frac{\partial T}{\partial x}\right)_2 + \frac{(\Delta x)^2}{2!}\left(\frac{\partial^2 T}{\partial x^2}\right)_2 + \cdots \tag{18.10}$$

$$T_1 = T_2 - \Delta x\left(\frac{\partial T}{\partial x}\right)_2 + \frac{(\Delta x)^2}{2!}\left(\frac{\partial^2 T}{\partial x^2}\right) - \cdots \tag{18.11}$$

$$T_{2'} = T_2 + \Delta t\left(\frac{\partial T}{\partial t}\right)_2 + \frac{(\Delta t)^2}{2!}\left(\frac{\partial^2 T}{\partial t^2}\right)_2 + \cdots \tag{18.12}$$

Equations (18.10) and (18.11) give

$$\left(\frac{\partial^2 T}{\partial x^2}\right)_2 = \frac{T_1 - 2T_2 + T_3}{(\Delta x)^2} \tag{18.13}$$

Equation (18.12) gives

$$\left(\frac{\partial T}{\partial t}\right)_2 = \frac{T_{2'} - T_2}{\Delta t} \tag{18.14}$$

neglecting terms of order $(\overline{\Delta x})^4$ and $(\overline{\Delta t})^2$ and higher. When we substitute Eqs. (18.13) and (18.14) into Eq. (6.53) and let $M = (\Delta x)^2/\alpha(\Delta t)$, we get the following finite-difference equation in place of the original differential equation:

$$T_{2'} = \frac{T_1 + (M - 2)T_2 + T_3}{M} \tag{18.15}$$

The terms in Eq. (18.15) are rearranged so that the future temperature $T_{2'}$ is expressed *explicitly* in terms of its present temperature T_2 and the present temperatures at the neighboring points, T_1 and T_3. In accordance with Eq. (18.15), then, the unsteady-state solutions can be marched out from prescribed initial conditions by successive application of the equation from point to point.

Before proceeding with a numerical example, we shall rederive the difference equation, Eq. (18.15), by direct energy balance on a finite element. Let the two curves shown in Fig. 18.8 represent the temperature distribution in the plate at times t and $t+\Delta t$. It will be assumed that the time interval Δt is so small that: (i) the temperature gradients $(T_1 - T_2)/\Delta x$ and $(T_2 - T_3)/\Delta x$ at time t may be used during the interval Δt as the temperature gradients at the dotted faces between faces 1 and 2 and faces 2 and 3; (ii) the temperature at point 2 is not influenced by temperature changes to the left of face 1 and to the right of face 3; and (iii) the internal energy change for the shaded element 2 may be computed using the temperature at position 2.

With these assumptions, an energy balance for the shaded element Δx in width is

$$kA\frac{(T_1 - T_2)}{\Delta x} - kA\frac{(T_2 - T_3)}{\Delta x} = \frac{\rho c A(\Delta x)}{\Delta t}(T_{2'} - T_2)$$

or

$$T_{2'} = \frac{T_1 + (M - 2)T_2 + T_3}{M} \tag{18.16}$$

Equation (18.16) is identical with Eq. (18.15). The quantity M that appears in Eq. (18.15) or (18.16) is a dimensionless modulus relating the selected magnitudes of Δx to Δt. It has an important bearing on the stability of the solution, as discussed in reference (4). We shall simply accept here the criterion that the solution behaves satisfactorily if M is taken large enough, which for the interior points is $M \geq 2$. Too small a value of M will be indicated by a pronounced oscillation or divergence of the solution. Equation (18.16) for particular values of M becomes

$$M = 2, \qquad T_{2'} = \frac{T_1 + T_3}{2}$$

$$M = 3, \qquad T_{2'} = \frac{T_1 + T_2 + T_3}{3}$$

$$M = 4, \qquad T_{2'} = \frac{T_1 + 2T_2 + T_3}{4}$$

This finite-difference method with $M = 2$ is the basis of a graphical procedure attributed to Schmidt and Binder (1). The modification employing values of M other than 2 was developed by Dusinberre (5), (6) and Nessi and Nissole (7).

If the boundary temperature T_0 is prescribed, then Eq. (18.16) may be applied successively at all points to obtain the temperature distribution at successive time intervals, Δt. However, if the surface is exposed to a fluid, the energy balance on the surface half-element is

$$hA(T_a - T_0) - \frac{kA(T_0 - T_1)}{\Delta x} = A\left(\frac{\Delta x}{2}\right)\frac{\rho c}{\Delta t}(T_{1/4'} - T_{1/4})$$

For this half-element, point $\frac{1}{4}$ (midway between 0 and m) is so close to point 0 that little accuracy is lost in assuming $(T_{1/4'} - T_{1/4}) \cong (T_{0'} - T_0)$; then this last equation becomes

$$T_{0'} = \frac{2NT_a + [M - (2N + 2)]T_0 + 2T_1}{M} \tag{18.17}$$

where $N \equiv h(\Delta x)/k$.

For convergence of the solution, Fowler (4) showed that $M \geq \sqrt{N^2 + 1} + 1$. A more stringent requirement is $M \geq (2N + 2)$ which is necessary to prevent oscillatory results, as illustrated by the following example.

Example 18.2: The slab of Fig. 18.8 is initially at a uniform temperature $T = 0$, and the face at 0 is suddenly exposed to a fluid which is maintained at a constant temperature $T_a = 100$ F while T_b is maintained at 0 F. Assume that the values of h, k, and Δx are such that $N = 1.0$ and $n = 5$. Then Fowler's convergence rule for the surface points requires $M \geq 2.414$. For the interior points $M \geq 2$. Select $M = 3$. Then for interior points

$$T_{2'} = \tfrac{1}{3}(T_1 + T_2 + T_3)$$

and at the surfaces

$$T_{0'} = \tfrac{1}{3}(2T_a - T_0 + 2T_1)$$

and

$$T_{5'} = \tfrac{1}{3}(2T_b - T_5 + 2T_4)$$

TABLE 18.3 SLAB PROBLEM OF FIG. 18.8

$T_a = 100, \quad N = 1, \quad M = 3, \quad T_b = 0$

$n_t \diagdown n_x$	0	1	2	3	4	5
0	0	0	0	0	0	0
1	66.7	0	0	0	0	0
2	44.4	22.2	0	0	0	0
3	66.7	22.2	7.4	0	0	0
4	59.3	32.1	9.9	2.5	0	0
5	68.4	33.4	14.8	4.1	0.8	0
6	66.1	38.8	17.4	6.6	1.6	0.5
7	70.5	40.8	20.9	8.5	2.9	0.9
8	70.4	44.0	23.4	10.8	4.1	1.6
9	72.5	46.0	26.0	12.8	5.5	2.2
⋮	⋮	⋮	⋮	⋮	⋮	⋮
∞	85.7	71.5	57.2	42.9	28.6	14.3

The results of applying these equations successively are shown in Table 18.3. Here the time interval between each successive line is

$$\Delta t = \frac{(\Delta x)^2}{\alpha M}$$

where $M = 3$, n_t is the number of time intervals elapsed, and n_x the number of space intervals from one wall.

Inspection of the results in Table 18.3 shows a damped oscillatory performance of the temperature T_0 at the surface. This results from the fact that with $M = 3$ and $N = 1$ the coefficient of the T_0 term in Eq. (18.17) is negative. To avoid these oscillatory results, make $M \geq (2N + 2)$ which makes the coefficient of T_0 greater than or equal to zero; for the example in Table 18.3, $M \geq 4$ will eliminate oscillations. The requirements for convergence and for no oscillation at the interior points, from Eq. (18.16), is still $M \geq 2$.

In neglecting the higher-order terms, Eqs. (18.13) and (18.14) represent the approximations to $\partial^2 T/\partial x^2$ at position 2 and $\partial T/\partial t$ at a point midway between 2 and 2′ in Fig. 18.9. At point 2′ we may write

$$\left(\frac{\partial^2 T}{\partial x^2}\right)_{2'} = \frac{T_{1'} - 2T_{2'} + T_{3'}}{(\Delta x)^2} \tag{18.18}$$

To get $\partial^2 T/\partial x^2$ at the point midway between points 2 and 2′, take the average of the expressions (18.13) and (18.18). Then

$$\left(\frac{\partial^2 T}{\partial x^2}\right)_{x,t+(\Delta t/2)} = \frac{T_1 + T_3 + T_{1'} + T_{3'} - 2T_2 - 2T_{2'}}{2(\Delta x)^2} \tag{18.19}$$

Combining Eq. (18.19) with Eq. (18.14) for $(\partial T/\partial t)_{t+(\Delta t/2)}$ and Eq. (6.53) results in the following comptability equation:

$$[T_{1'} + T_{3'} - (2M + 2)T_{2'}]_{t+\Delta t} + [T_1 + T_3 + (2M + 2)T_2]_t = 0 \tag{18.20}$$

This equation involves temperatures at time $t + \Delta t$ which are unknown, as well as temperatures at time t which are known; therefore, the equation cannot be solved explicitly for the unknown temperatures at $t + \Delta t$. Instead, application of Eq. (18.20) from point to point along a row results in a set of simultaneous equations that give *implicitly* the unknown temperatures. The solution may still be marched out, but at each time step in the marching process we must solve a set of simultaneous equations. This may be accomplished by relaxation or iteration at *each* time interval. For the relaxation solution, we define a residual at point $2'$ from Eq. (18.20) as follows:

$$R_{2'} = [T_{1'} + T_{3'} - (2M + 2)T_{2'}] + [T_1 + T_3 + (2M + 2)T_2] \tag{18.21}$$

This leads to the relaxation pattern shown in Fig. 18.10.

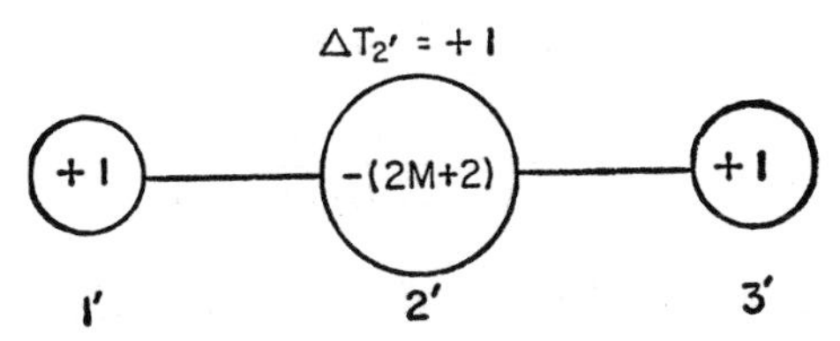

Fig. 18.10. Relaxation pattern at a given instant in the "marching" solution.

For the iteration process we solve Eq. (18.20) for $T_{2'}$ as follows:

$$T_{2'} = \frac{T_{1'} + T_{3'} + T_1 + T_3 + (2M + 2)T_2}{2M + 2} \tag{18.22}$$

This equation applied at each position is solved successively either "by hand" or on an automatic computing machine.

The use of the simple explicit marching solution, Eqs. (18.16) and (18.17), requires that $M \geq 2$ or $M \geq (2N + 2)$ to obtain stable results. Sometimes this limitation requires many thousands of steps to traverse only a few seconds in time. In these cases the seemingly more tedious solutions using either Eq. (18.21) or Eq. (18.22) do, in fact, involve less work since M may be taken as close to zero as we please, thus providing larger time intervals. The relaxation or iteration procedure at each step is indeed tedious, but this is offset by having to calculate fewer steps to cover the same time period. A similar procedure could employ Eq. (18.37).

In the discussion of the numerical methods, our choice of examples was limited to heat conduction. However, since all these methods are in essence special procedures for the approximate solution of certain types of differential equations, they can be applied identically to problems in mass transfer (by simple change of symbols), fluid flow, electricity, elasticity, etc., so long as the pertinent differential equations have the proper mathematical structure. Also, in limiting our discussion to an outline of the underlying principles, we omitted many important details—largely in the nature of variations of the basic procedures. Readers may refer for these details to (3), (6).

18.3 Analogs

Two systems are said to be *analogous* when the mathematical equations governing their performance are identical in form. These equations may be algebraic or differential. For an example, let us take the equations governing one-dimensional transient conduction of heat and electric charge:

$$\frac{\partial T}{\partial t} = \alpha \frac{\partial^2 T}{\partial x^2} \tag{6.53}$$

$$\frac{\partial e}{\partial t} = \frac{1}{R_l C_l} \frac{\partial^2 e}{\partial x^2} \tag{18.23}$$

In the second equation, e is the electrical potential, R_l is the resistance per unit length, and C_l is the capacitance per unit length of the conducting path. We immediately recognize that there is complete analogy between the temperature field and the electrical potential field if α and $1/R_l C_l$ vary in the same manner with either x or the potentials (T and e). Analogs do not give general analytical solutions; they can only provide answers to specific problems for which they are set up. They do offer important advantages, however, in experimental engineering where an analog might be used if the original system proved impractical for measurements because of size, time, or cost. In this respect, the analog methods are far more flexible and powerful than the method of model-testing of dimensional analysis, since analogs need not and usually do not physically resemble their prototypes. Transfer processes comprise an important area of analog applications; in this section, we shall examine a number of fluid-flow and electrical analogs of heat and mass transfer systems.

The conversion of a system to its analog involves two essential steps. The first is a qualitative step consisting of duplicating functionally all elements of the original system in the analog; this means the analog elements should be capable of reproducing faithfully all significant characteristics of the prototype. There are no general, infallible sets of rules to accomplish this, in contrast with the method of dimensional analysis in which the equality of π-terms is sufficient assurance of correctness of a model. Consider unsteady-state heat conduction in a furnace wall, with the temperature distribution in the wall governed by Eq. (6.53). Accurate measurement of heat transfer rates presents rather formidable instrumentation difficulties. Since electrical quantities such as the current and voltage can be measured quite easily, electrical analogs of thermal systems have found wide use. In the electrical system, we might expect the electrical resistance to duplicate the reciprocal thermal conductance of the wall and the capacitance to duplicate the heat capacity. A distributed resistor is readily available, but a distributed electrical capacitor is difficult and costly to build. For the present, we shall assume an analog is available with a distributed capacitance, as, for example, a cable over a ground, Fig. 18.11. A more realistic analog containing a finite number of capacitors will be discussed later in connection with network analogs.

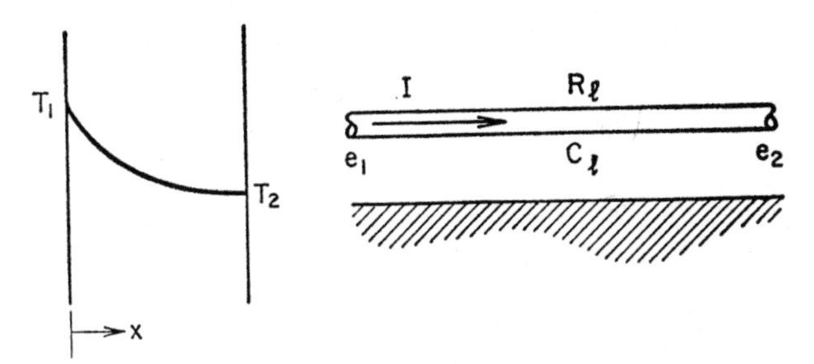

Fig. 18.11. Electrical analog of transient heat transfer in a wall. Electric cable with capacitance-to-ground.

The next consideration is quantitative. Complete analogy requires that the corresponding elements in the two systems have the same numerical values. In practice, this one-to-one numerical correspondence is often not feasible, and analog quantities must be scaled in relation to their corresponding thermal quantities. We shall illustrate this with a numerical example. The cable in Fig. 18.11 obeys Eq. (18.23) and complete analogy requires $\alpha \sim 1/R_lC_l$. Let $T_1 = 1000$ F, $T_2 = 100$ F, and $\alpha = 0.01$ ft²/hr (a typical value for a brick wall), and let us assume an interest in the temperature distribution in the interval 0 to 3 hours. In Eq. (6.53), let us introduce the scaling factors K_t for the time, $K_e = T/e$, $K_{RC} = (1/\alpha)/R_lC_l$, and K_x for the geometry:

$$\frac{\partial(K_e e)}{\partial(K_t t)} = \frac{1}{K_{RC}} \frac{1}{R_l C_l} \frac{\partial^2(K_e e)}{\partial(K_x x)^2} \tag{18.24}$$

or

$$\frac{K_e}{K_t}\frac{\partial e}{\partial t} = \frac{K_e}{K_{RC}K_x^2}\frac{1}{R_lC_l}\frac{\partial^2 e}{\partial x^2}$$

Obviously all the K's cannot be independent. In fact, from inspection,

$$K_t = K_{RC}K_x^2$$

Let us choose a convenient time and length scale, say $K_t = 50$ and $K_x = 1$. This means 3 hours thermal time is scaled down to 3.6 minutes electrical time. Then, $K_{RC} = K_t/K_x^2 = 50$. Correspondingly,

$$R_l C_l = \frac{1/\alpha}{K_{RC}} = \frac{100}{50} = 2 \text{ ohm-farad/ft}^2$$

If we choose a capacitance of 500 μfarad/ft, we need a resistance of 4000 ohm/ft in the analog. Note that K_e is independent and may be selected at any magnitude, e.g., 1v = 100 F.

(a) Field Analogs

An important class of analogs exists in two-dimensional steady-state systems. The differential equations for the steady-state heat conduction and mass diffusion in an isotropic plane were derived in earlier chapters, and they are

$$\frac{\partial^2 T}{\partial x^2} + \frac{\partial^2 T}{\partial y^2} = 0 \tag{18.25}$$

$$\frac{\partial^2 c_a}{\partial x^2} + \frac{\partial^2 c_a}{\partial y^2} = 0 \tag{18.26}$$

During steady flow of current in a homogeneous conducting plane, the electrical potential field satisfies

$$\frac{\partial^2 e}{\partial x^2} + \frac{\partial^2 e}{\partial y^2} = 0 \tag{18.27}$$

In the thermal, diffusional, and electrical systems above, the flux lines are everywhere normal to their respective equipotential lines, forming an orthogonal net. All these two-dimensional, steady-state systems are analogous in that they satisfy the Laplace equation, and any one may be chosen as the analog of the others.

Electrical field analog. Heat conduction (or mass diffusion) is simulated by current flow in a homogeneous, electrically conducting medium which may be liquid, such as an electrolyte contained in a shallow tank, or a solid, such as an electrically conducting sheet. The tank or sheet is prepared in the same configuration as the prototype thermal (or diffusion) system; a constant temperature boundary in the latter is simulated in the electrical analog by a highly conducting boundary electrode, and an adiabatic boundary by a dielectric boundary material. Figure 18.12 shows the

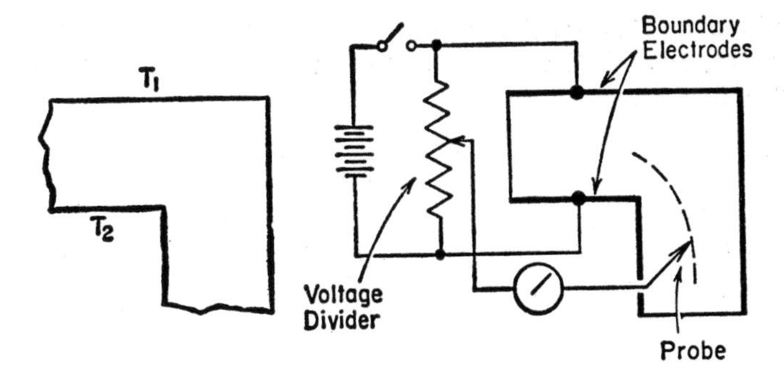

Fig. 18.12. Electrical field analog of heat transfer through corner of thick-walled duct.

electrical analog of the temperature field in a thick corner of a duct wall. The equipotential lines may be traced with a stylus or a probe, each line corresponding to a selected voltage level on the voltage divider. The procedure is straightforward, but to attain a reasonable accuracy, precautions must be taken against such sources of errors as impurities and stray voltages in the electrolyte, polarization of the electrodes, nonuniformity of resistance in the conducting sheets—to name a few.

Fluid-flow field analog. The two-dimensional, steady, irrotational flow of an ideal fluid is analogous to the flow of current or heat in a homogeneous plane. The condition of continuity for the flow system is, from Eq. (1.12),

$$\frac{\partial v_x}{\partial x} + \frac{\partial v_y}{\partial y} = 0 \tag{18.28}$$

The condition of irrotational flow* in a plane is given by

$$\frac{\partial v_x}{\partial y} - \frac{\partial v_y}{\partial x} = 0 \tag{18.29}$$

* Rotation ω at point $P(x, y)$ may be defined as the average angular velocity of any two mutually perpendicular elements (Δx and Δy) at P. Thus from Fig. 18.13,

$$\omega = \frac{1}{2}\left(\frac{\partial v_y}{\partial x} - \frac{\partial v_x}{\partial y}\right)$$

Irrotational flow exists when

$$\frac{\partial v_y}{\partial x} = \frac{\partial v_x}{\partial y}$$

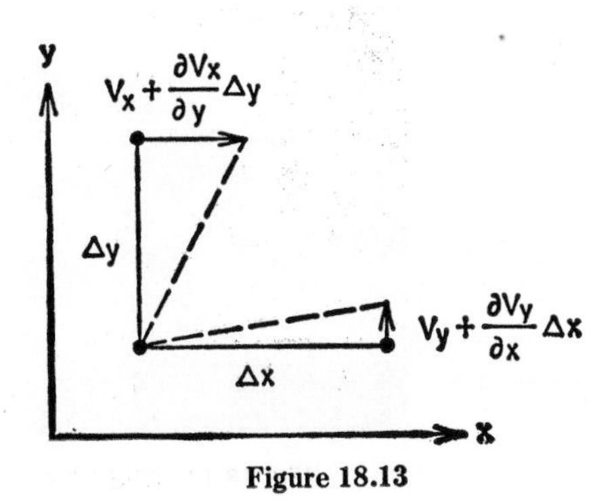

Figure 18.13

The continuity condition provides the basis for the definition of a stream function $\psi(x, y)$ such that the velocity components v_x and v_y everywhere in the flow field are given by

$$v_x = \frac{\partial \psi}{\partial y} \qquad v_y = -\frac{\partial \psi}{\partial x} \tag{18.30}$$

Note that $\psi(x, y)$ satisfies Eq. (18.28) by definition. Since flow is irrotational, $\psi(x, y)$ must also satisfy Eq. (18.29) or

$$\frac{\partial^2 \psi}{\partial x^2} + \frac{\partial^2 \psi}{\partial y^2} = 0 \tag{18.31}$$

Likewise, the irrotationality condition implies existence of a velocity potential function $\phi(x, y)$ such that

$$v_x = \frac{\partial \phi}{\partial x} \qquad v_y = \frac{\partial \phi}{\partial y} \tag{18.32}$$

Since $\phi(x, y)$ must satisfy continuity, Eq. (18.28), we get

$$\frac{\partial^2 \phi}{\partial x^2} + \frac{\partial^2 \phi}{\partial y^2} = 0 \tag{18.33}$$

In accordance with Eq. (18.30), the velocity vector is tangent to ψ, and in accordance with Eq. (18.32), the velocity vector is normal to ϕ. Lines of constant ψ and constant ϕ on the xy plane thus form an orthogonal net. A fluid-flow field analog is based on the similarity of Eq. (18.33) to Eqs. (18.25), (18.26), and (18.27). However, an important distinction should be noted; although T, c_a, and e are the driving potentials for their respective fluxes, ϕ is not the driving potential for the fluid flow.

If a flow field were constructed with the same geometry and boundary conditions as a system on which, for instance, heat transfer information is desired, we could equate the velocity potential lines (constant ϕ) in the

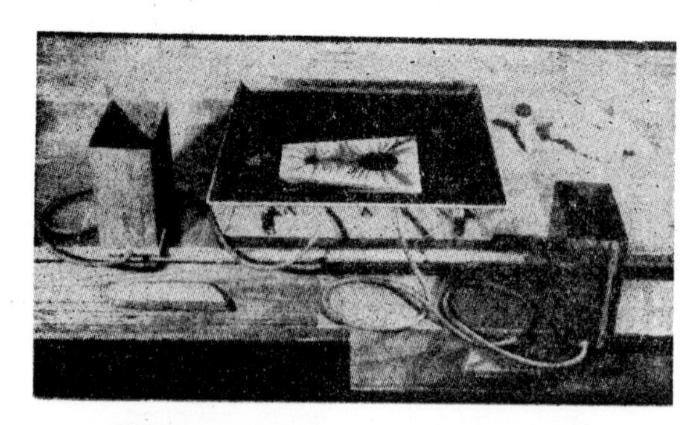

Fig. 18.14. Fluid flow analog [Moore (11)].

former to the isothermal lines in the latter and the streamlines (constant ψ) to lines of constant heat flow (adiabatic lines). If, then, the streamlines were rendered visible by some techniques of flow visualization, the resulting flow pattern could provide useful information about the thermal field. Hele-Shaw *et al.* (10) were among the first to build a successful fluid-flow analog, and Moore (11) developed the "fluid mapping" technique to a high degree of perfection. Figure 18.14 shows a typical fluid map obtained by Moore.

(*b*) *Network Analogs*

We shall confine our study of network analogs to the electrical analogs, which are the most important because of their flexibility and ease of measurement. For steady-state problems, only resistors are needed. On the other hand, unsteady-state problems may be represented by a resistance network or a resistance-capacitance network. Circuits with inductances have also been suggested.

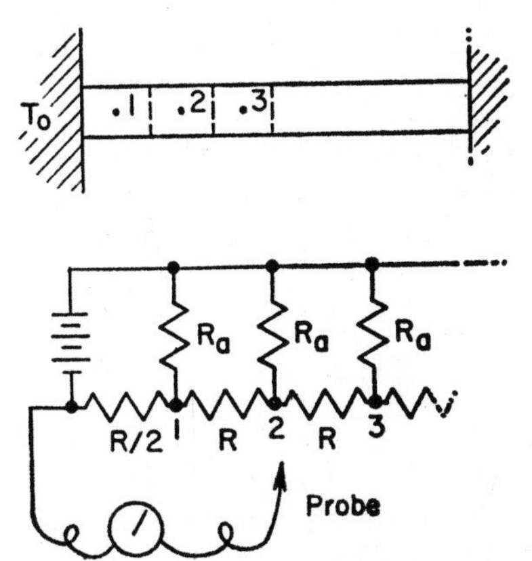

Fig. 18.15. Lumped electrical resistance analog of steady state heat transfer in a fin.

Steady-state network analog. Consider the fin problem of Sec. 18.1. It is shown there that lumping the system, as in Fig. 18.15, is tantamount to replacing the differential equation (6.39) by a set of approximate algebraic equations of the form of Eq. (18.4) or (18.5). At point 2, for instance, the equation is

$$T_1 - (\Gamma + 2)T_2 + T_3 + \Gamma T_a = 0 \tag{18.5}$$

where $\Gamma \equiv hP(\Delta x)^2/Sk$.

We shall now construct an analogous electrical network to the lumped system. This means the analogy will be based on the similarity of the equations on the finite-difference (algebraic) level rather than on the differential equation level. Consider the resistance network shown in Fig. 18.15. When the switch is closed, current balance at a junction (Kirchhoff's law) as, for example, at junction 2 yields

$$\frac{e_1 - e_2}{R} + \frac{e_3 - e_2}{R} + \frac{e_a - e_2}{R_a} = 0 \tag{18.34}$$

or

$$e_1 - \left(\frac{R}{R_a} + 2\right)e_2 + e_3 + \frac{R}{R_a}e_a = 0$$

Let us choose the series resistance R proportional to the reciprocal of the thermal conductance of element Δx,

$$K_R = \frac{\Delta x / kS}{R}$$

and the parallel resistances R_a proportional to the reciprocal of the surface conductance of element Δx,

$$K_{R_a} = \frac{1/hP\ \Delta x}{R_a}$$

where K_R and K_{R_a} are scaling factors. Also, let $K_e = T/e$. There is complete analogy between the temperature T in Eq. (18.5) and the potential e in Eq. (18.34) if $\Gamma = R/R_a$. From the definitions of K_R and K_{R_a},

$$\frac{K_R}{K_{R_a}} = \frac{hP(\Delta x)^2}{kS}\frac{R_a}{R} = \Gamma\frac{R_a}{R}$$

Therefore, the scaling factors for the resistors must be chosen so that $K_R = K_{R_a}$. The scaling factor $K_e = T/e$ is clearly independent and may be chosen at any convenient value. We shall illustrate the method with a numerical example.

Consider Example 18.1 of Sec. 18.1, $T_0 = 100$ F, $T_a = 0$ F, and $\Gamma = 1$. Let us further specify the thermal conductance $kS/\Delta x = 2$ and the surface conductance $hP\ \Delta x = 2$. If we select $K_R = K_{R_a} = \frac{1}{2}$, then $R = R_a = 1$ ohm. If we select $K_e = 10$, then 1 v = 10 F. The electrical analog can now be set up. At point 2, for instance, of the thermal and the electrical systems,

$$T_0 - T_2 = K_e(e_0 - e_2)$$

or

$$T_2 = T_0 - K_e(e_0 - e_2) = 100 - 10(e_0 - e_2)$$

In the earlier numerical solution, we had to solve a set of simultaneous finite-difference equations (compatibility condition) by a relaxation (Table 18.1) or an iteration (Table 18.2) procedure to find T_2. With the analog, the compatibility condition is physically satisfied, and we simply measure the potential difference $(e_0 - e_2)$ with a probe.

The heat transfer rate at any point along the fin may be found by measuring the current at the corresponding point in the electrical analog, if the current scaling factor, $K_I = q/I$, is known. The rate of heat conduction within the fin element 1–2 is given by

$$q = \frac{kS}{\Delta x}(T_1 - T_2)$$

Introducing the scaling factors, we get

$$K_I I = \frac{K_e(e_1 - e_2)}{K_R R}$$

or

$$K_I = \frac{K_e}{K_R} = 20 \text{ Btu/hr/amp}$$

K_I may also be found from the equation of heat transfer at the surface of element 2, $q = hP\,\Delta x(T_2 - T_a)$; the same procedure as above gives $K_I = K_e/K_{R_a} = 20$.

Transient network analog. Let us consider the unsteady-state heat conduction in a wall, governed by Eq. (6.53). An analogous equation for the current flow in an R-C circuit is Eq. (18.23). An electrical analog based on these two partial differential equations was previously described; since it requires a distributed capacitance, it has little practical value. Let us instead lump the thermal system—we have a choice of lumping it with respect to x or t alone, or both x and t. We might consider lumping Eq. (6.53) spacewise, but not timewise. Equation (6.53) combined with Eq. (18.13) gives at a point n:

$$\frac{\partial T_n}{\partial t} = \alpha \frac{T_{n-1} - 2T_n + T_{n+1}}{(\Delta x)^2} \tag{18.35}$$

The corresponding equation for the R-C circuit, from Eq. (18.23), is

$$\frac{\partial e_n}{\partial t} = \frac{1}{R_l C_l} \frac{e_{n-1} - 2e_n + e_{n+1}}{(\Delta x)^2} \tag{18.36a}$$

or

$$\frac{\partial e_n}{\partial t} = \frac{1}{RC}(e_{n-1} - 2e_n + e_{n+1}) \tag{18.36b}$$

The distributed resistance and capacitance per unit length in Eq. (18.36a) are replaced by lumped resistance and capacitance in Eq. (18.36b). Comparison of Eqs. (18.35) and (18.36b) shows that the analogy is complete between T_n and e_n if

$$\frac{1}{RC} \sim \frac{\alpha}{(\Delta x)^2}$$

In addition,

$$R \sim \frac{\Delta x}{kS} \quad \text{and } C \sim S(\Delta x)\rho c$$

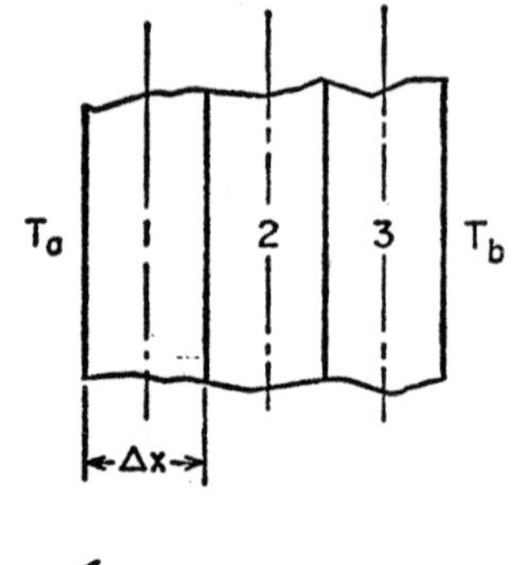

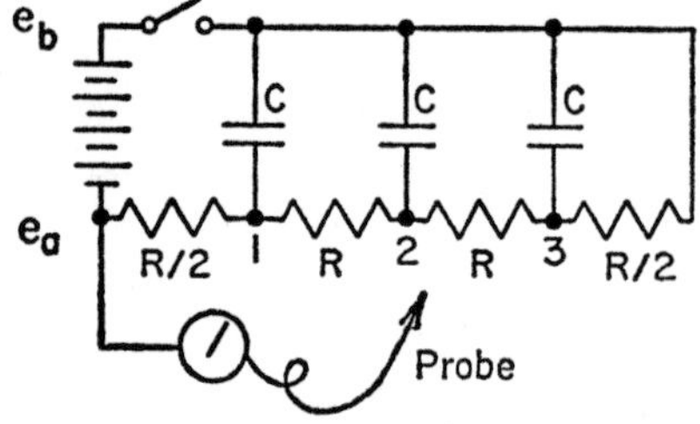

Fig. 18.16. Lumped electrical resistance-capacitance analog of transient heat transfer in a wall.

The scaling of these quantities follows the same general procedure outlined in previous examples.

Figure 18.16 shows the *R-C* network analog of the infinite wall, subdivided into three elements for simplicity. At time $t = 0$, the switch is closed, which simulates raising the temperature of one side of the slab to T_a. The time rate of change of e at points 1, 2, and 3 may be recorded on an oscilloscope or strip-chart recorder.

If Eq. (6.53) were lumped with respect to both x and t, and if the time derivatives were replaced by the finite-difference ratio taken at the *end* of the time interval Δt, we would get at point x and time t the following algebraic approximation to the original differential equation:

$$\frac{T_{x,t+\Delta t} - T_{x,t}}{\Delta t} = \alpha\left[\frac{T_{x-\Delta x,t+\Delta t} - 2T_{x,t+\Delta t} + T_{x+\Delta x,t+\Delta t}}{(\Delta x)^2}\right] \tag{18.37}$$

The solutions based on Eq. (18.37) are stable for all values of the parameter $M = (\Delta x)^2/\alpha\,\Delta t$. An electrical analog of Eq. (18.37) using only resistors

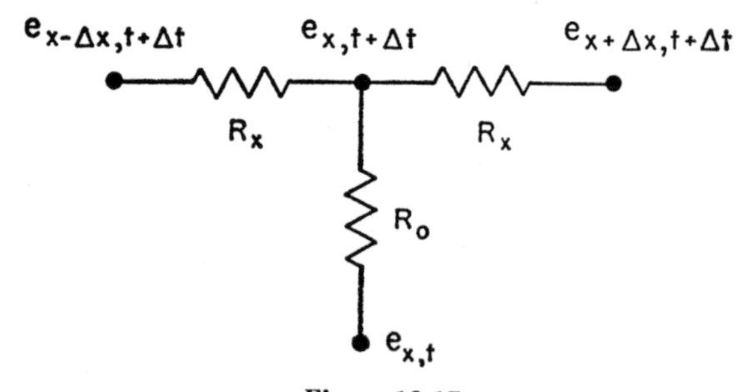

Figure 18.17

has been suggested by Leibmann (12). Consider a network junction shown in Fig. 18.17. A current balance at the junction yields

$$(e_{x,t+\Delta t} - e_{x,t})\frac{R_x}{R_0} = e_{x-\Delta x,t+\Delta t} - 2e_{x,t+\Delta t} + e_{x+\Delta x,t+\Delta t} \tag{18.38}$$

Equations (18.37) and (18.38) are analogous if $R_0/R_x \sim \alpha\,\Delta t/(\Delta x)^2$. Thus, an electrical analog consisting only of resistances R_0 and R_x can be built to march out transient temperature distribution expressed by Eq. (18.37). After each time step, the potential $e_{x,t}$ must be reset to its new value $e_{x,t+\Delta t}$. The process can be laborious. Special circuit arrangements allowing greater working speeds are described by Liebmann (12).

18.4 Flux Plot

A simple and rapid procedure for obtaining an orthogonal net of constant flow and constant potential lines in two-dimensional, steady-state systems will be described. A "heat flow" net may be drawn simply by filling in the region of interest with a net of equally spaced adiabatic and isothermal lines, carrying out the construction by trial-and-error steps until all lines intersect at right angles. Accuracy of construction may be improved by sketching in the diagonals which should also form an orthogonal net. The method is useful for systems which have complex geometry but simple isothermal or adiabatic boundaries; otherwise, the numerical or analog methods may be preferred.

Let us take the example of steady-state heat conduction through the walls of a hollow rectangular duct, Fig. 18.18. Assume the inner wall is at 200 F, the outer wall at 0 F, and k is uniform. On a scale drawing of the system, sketch in by eye a network of curvilinear squares. If L^b is the distance between a pair of adiabatic lines, L^a the distance between a pair of isotherms, then $L_a = L_b$. Heat transfer through the element shown in Fig. 18.18 is given for unit depth by

$$q = -kL^b \frac{T_3 - T_2}{L^a} = k(T_2 - T_3)$$

If N_a is the number of lanes formed by the adiabatic lines, and N_b the

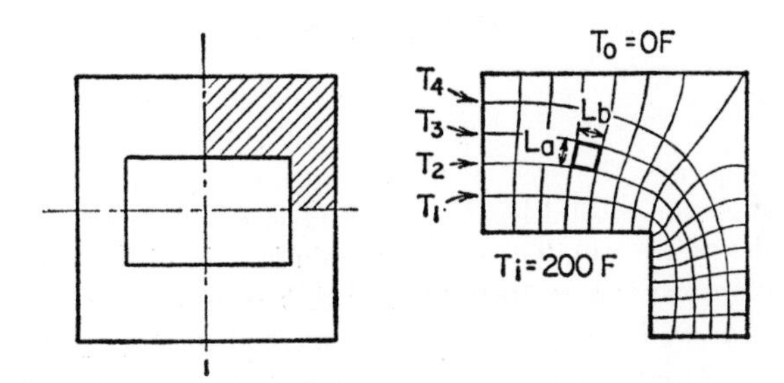

Fig. 18.18. Flux plot for heat transfer through corner of thick-walled duct.

number of equal temperature increments in the segment of interest, then,

$$T_2 - T_3 = \frac{(T_i - T_0)}{N_b}$$

Since the heat transfer rate in each of the lanes is the same,

$$q_{\text{total}} = k\frac{N_a}{N_b}(T_i - T_0) \tag{18.39}$$

For the example shown in Fig. 18.18, if $k = 2$,

$$q_{\text{total}} = 2(\tfrac{15}{5})(200) = 1200 \text{ Btu/hr.}$$

REFERENCES

1a. Schmidt, E., *Beitr. tech. Mech. u. tech. Phys.*, A. Föppl, Festschrift, 179 (1924).

1b. Binder, L., Aussere Wärmeleitung und Erwärmung Elect. Masch., Munich (1911).

2. Richardson, L. F., *Trans. Roy. Soc. (London)*, **A226,** 229 (1927).

3. Crandall, S. H., *Engineering Analysis*, McGraw-Hill, New York, 1956.

4. Fowler, M., *Quart. of Appl. Math.*, **3,** 4 (1946).

5. Dusinberre, G. M., *Trans. ASME*, **65,** 703 (1945).

6. Dusinberre, G. M., *Heat Transfer Calculations by Finite Differences*, International, Scranton, Pa., 1961.

7. Nessi, A., and L. Nissole, *Méthode Graphiques pour l'Étude des Installations de Chauffage*, Dunod, Paris, 1929.

8. Southwell, R. V., *Relaxation Methods in Engineering Science*, Oxford Univ. Press, New York, 1940.

9. Emmons, H. W., *Quart. of Appl. Math.*, **2,** 3 (1944).

10. Hele-Shaw, H. S., A. Hay, and P. H. Powell, *Jour. Inst. Elec. Engrs.*, **34,** 21 (1904–1905).

11. Moore, A. D., *Jour. Appl. Phys.*, **20**:8, 790 (1949); *Trans. AIEE*, **69**:2, 1615 (1950).

12. Liebmann, G., ASME Paper 55-SA-15 (1955).

PROBLEMS

18.1 Assume the fin in Fig. 18.1 has an uninsulated tip. Find the residual and the relaxation pattern for the half-width element n.

18.2 (a) Determine the steady-state temperature distribution in the fin of Fig. 18.1 whose total length is 5 in. The cross section of the fin is rectangular, $\frac{1}{8}$ by $\frac{1}{4}$ in. The end of the fin is uninsulated and the following conditions exist: $h = 10$ Btu/hr ft² F, $k = 16$ Btu/hr ft F, $T_0 = 200$ F and $T_a = 60$ F. Solve by relaxation method with $\Delta x = 1$ in.

(b) Calculate the rate of heat transfer at the base of the fin and the fin efficiency. (See Chap. 6.)

18.3 Solve Prob. 2 by the iteration method.

18.4 Determine the temperature distribution in a steel turbine disk of dimensions shown, considering temperature gradients in the radial direction only. The gas temperature at the tip end is 1200 F, and along the side of the disk the gas temperature is 700 F. The turbine disks are welded together at the hubs; therefore, the sides at the $1\frac{1}{2}$-in. thickness may be considered to be insulated. The rotor assembly is hollow and in the 2-in. diameter hole the air temperature is 200 F. For the disk $k = 26$ Btu/hr ft F. At the tip end $h = 50$ Btu/hr ft² F;

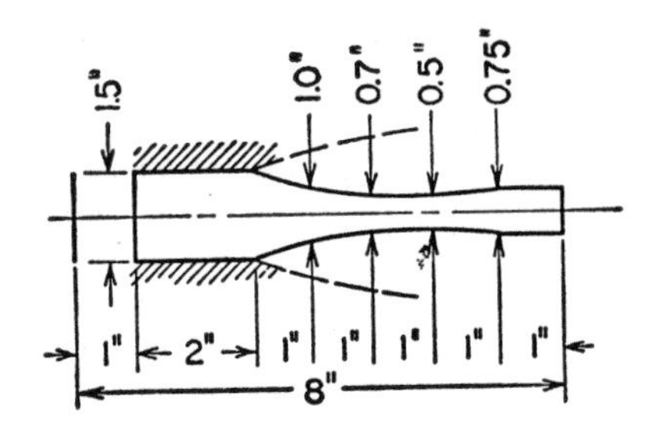

along the sides assume h varies linearly with the radius from a value of 50 at $r = 8$ in. to a value of 18.75 at $r = 3$ in.; and in the hole $h_i = 5$ Btu/hr ft² F.

18.5 Derive Eq. (18.8b) by Taylor series expansions such as those in Eqs. (18.1) and (18.2).

18.6 Modify Eqs. (18.5) and (18.8b) for cases in which a distributed heat source W_i (Btu/hr ft³) exists, e.g., resulting from an electric current, nuclear reaction, etc.

18.7 Solve the example of Sec. 18.2, Table 18.3, with $M = 4$ instead of 3.

18.8 A refractory wall consists of a layer of firebrick 9 in. thick in perfect thermal contact with a layer of red building brick 6 in. thick. As a result of previous heating and cooling operations the following temperature distribution was set up in the wall:

9": $k = 0.60$, $\rho = 115$, $c = 0.18$
6": $k = 0.40$, $\rho = 129$, $c = 0.24$

x	T	x	T	x	T
0 in.	200 F	5	640	10	205
1	425	6	540	11	160
2	670	7	440	12	130
3	738	8	350	13	110
4	720	9	285	14	100
				15	95

At this instant a radiant heater is started resulting in a constant and uniform $(q/A)_0 = 100$ Btu/hr ft² transferred to the wall at $x = 0$. A heat transfer coefficient $h = 1.2$ Btu/hr ft² F exists at the surface at $x = L$ which is exposed to a fluid whose temperature is maintained constant at 0 F.

In the interest of accuracy of the solution, take no elemental Δx greater than 3 in. thick but $\Delta x = 3$ in. is permissible. Determine the temperature distribution at the end of approximately 4.5 hr using the smallest permissible value of M.

18.9 Solve Prob. 2 for the transient temperature distributions if the fin is initially uniformly at 60 F and the boundary conditions stated in Prob. 2 are suddenly imposed.

18.10 Solve Prob. 4 for the transient temperature distribution if the rotor disk is initially uniformly at 200 F and the boundary conditions as stated in Prob. 4 are suddenly imposed.

18.11 High-pressure steam lines have been observed to have fine random cracks on the inner surface. Since the pipes are much thicker than the low-pressure steam lines where these cracks are not observed, it is suspected that the cracks are a result of "thermal shock"; that is, high-temperature steam is admitted to the cold pipe producing thermal stresses.

Consider a bare steel pipe of I.D. 1 in. and O.D. 2 in. in a room at 70 F. Initially, it is in temperature equilibrium with the room; then the steam valve is opened and superheated steam at 800 psia and 1000 F flows with a mass velocity of 5 lb/ft^2 sec through the pipe. Calculate estimated temperature distributions at various time intervals after the steam is admitted.

18.12 The example of Table 18.3 may be solved by using the Binder-Schmidt graphical method, using $M = 2$. In this procedure, the future temperature at a point is given by the arithmetic mean of the present temperatures at the two adjacent points. Such averaging can be performed graphically by simply connecting alternate temperature points by straight lines.

(a) If, as in Table 18.3, the fluid temperature T_a is given instead of the surface temperature, the graphical method can still be applied by extending the wall a fictitious distance d such that T_a is the surface temperature of this extended wall. Show that $d = k/h$. Note that in the construction of the temperature lines, the distance d is not subdivided.

(b) In the example of Table 18.3, if after three time intervals T_a were suddenly reduced to zero, find the temperature distribution after the next three time periods.

18.13 The energy equation representing the conditions in the thermal entrance region for flow between parallel flat plates is

$$v_x \frac{\partial T}{\partial x} = \alpha \frac{\partial^2 T}{\partial y^2}$$

Assuming the velocity distribution is fully developed (parabolic) on approaching the heated section, set up a finite difference procedure for finding the temperature distribution after the fluid enters the heated section whose wall temperature is uniform.

18.14 For heat transfer associated with flow over a flat plate (Sec. 7.5) with uniform wall temperature, assume that $f(\eta)$ is known from Eq. (3.28) or Fig. 3.5. Set up the solution for temperature distribution, Eq. (7.34), in finite difference form—both iteration and relaxation forms.

18.15 Draw an electrical resistance network analog of the turbine disk of Prob. 18.4.

18.16 Design an electrical $R - C$ network analog for the refractory wall of problem 18.8. Use the same wall subdivisions as in the finite difference solution.

18.17 A soap film suspended with uniform edge tension on a closed boundary is dilated with a uniform pressure on one side of its surface. The differential equation of the film surface $z(x, y)$ in static-equilibrium under the external pressure p and the tensile forces within the film is (for small deflections):

$$\frac{\partial^2 z}{\partial x^2} + \frac{\partial^2 z}{\partial y^2} + \frac{p}{\sigma} = 0$$

where σ is the constant surface tension of the film. We note that this equation is analogous to the two-dimensional heat conduction equation with a distributed heat source.

(a) Derive the above equation by writing the condition of static equilibrium on a differential element of the film surface.

(b) Show how T may be related to z, and how the boundary conditions may be duplicated in a soap film analog. What are the advantages and disadvantages of this analog?

18.18 A freely suspended soap film under zero dilation pressure satisfies the Laplacian equation (see Prob. 18.17). Draw the shape of a film analog of steady-state heat conduction in a hollow cylinder with

$$r_i = 4 \text{ in.,} \qquad T_i = 800 \text{ F}$$

$$r_0 = 8 \text{ in.,} \qquad T_0 = 100 \text{ F}$$

18.19 The retaining wall shown is constructed of concrete. Assume that the flat surface is at 120 F and the fluted surface is uniformly at 60 F. By drawing a flux plot, estimate the rate of heat transfer through a repeating section.

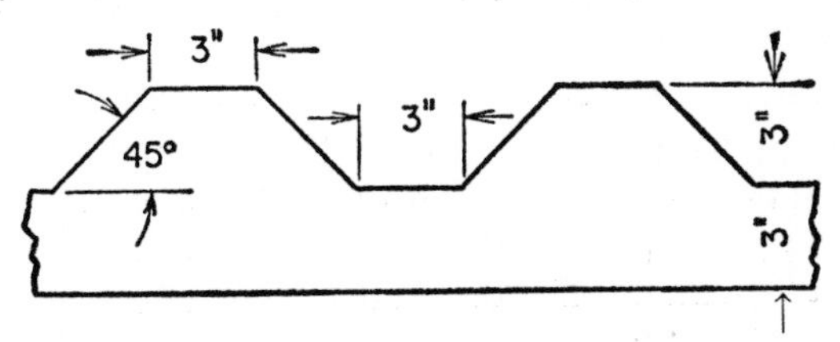

18.20 (a) A 4-in. O.D. pipe is buried with its centerline 2 ft below the surface of the ground. If the pipe surface is at a temperature of 190 F and the surface of the ground is at 40 F, estimate the rate of heat loss per unit length of the pipe using the method of the flux plot. Assume the mean thermal conductivity of the soil is 0.3 Btu/hr ft F.

(b) If the isotherms and the adiabatic lines in part (a) were reversed, devise a physical system that the flux plot will represent.

CHAPTER 19

Rate Equations by Methods of Irreversible Thermodynamics

When a number of transfer processes occur simultaneously in a system, the pertinent rate equations may be complicated by the presence of coupling or mutual interference among these processes. In Section 19.1, this phenomenon is explained with an example of heat conduction in an anisotropic plane; note that in this instance coupling is evidenced by fluxes of the same type (both are energy fluxes). In Section 19.2, a rigorous method for deriving the general rate equations, known as the Onsager method, is outlined. The method is applied in Section 19.3 to the thermoelectric phenomenon; by an identical procedure, coupled heat and mass transfer processes giving rise to the Soret (thermal diffusion) and Dufour effects may be analyzed (4).

19.1 Heat Conduction in Anisotropic Plane

The rate equations of Chap. 1 were formulated on empirical evidence. Implicit in their formulation is a fundamental assumption that the direction of transfer is everywhere normal to the equipotential lines. Many materials

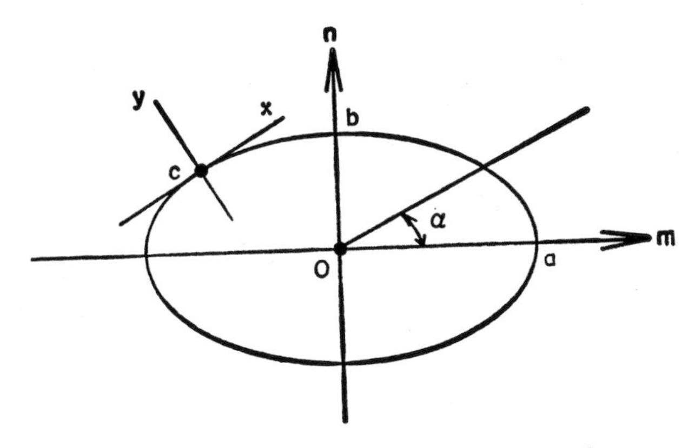

Fig. 19.1. Heat conduction in an anisotropic plane with a heat source at the origin O.

of engineering importance are nonisotropic, their transport coefficients such as thermal conductivity and diffusivity being directionally sensitive. In such a material, if we consider an arbitrary plane cut in the medium, we find that the maximum and the minimum values of the coefficients occur along some preferred axes known as the *principal axes*. Let us consider heat conduction in an anisotropic plane with a heat source at the origin of the coordinate system (8). An isotherm in that plane takes the shape of an ellipse, with the major and minor axes of the ellipse, axes m and n in Fig. 19.1, coinciding with the principal axes of conductivities. Along these principal axes, such as at points a and b, the direction of heat conduction is normal to an isotherm. At any point c not on the principal axes, the heat flux has a component parallel to the isotherm. We shall derive the correct rate equation for this case, and use the result to introduce a general method of finding rate equations of multiple transfer processes.

Let x and y be the axes tangential and normal respectively to an isotherm at c, and let α be the angle between the axes x and m. From Fig. 19.1,

$$\left(\frac{q}{A}\right)_x = \left(\frac{q}{A}\right)_m \cos\alpha + \left(\frac{q}{A}\right)_n \sin\alpha = -k_m \cos\alpha \frac{\partial T}{\partial m} - k_n \sin\alpha \frac{\partial T}{\partial n} \tag{19.1}$$

$$\left(\frac{q}{A}\right)_y = -\left(\frac{q}{A}\right)_m \sin\alpha + \left(\frac{q}{A}\right)_n \cos\alpha = k_m \sin\alpha \frac{\partial T}{\partial m} - k_n \cos\alpha \frac{\partial T}{\partial n} \tag{19.2}$$

Noting that

$$\frac{\partial T}{\partial m} = \frac{\partial T}{\partial x}\frac{\partial x}{\partial m} + \frac{\partial T}{\partial y}\frac{\partial y}{\partial m}$$

and

$$\frac{\partial T}{\partial n} = \frac{\partial T}{\partial x}\frac{\partial x}{\partial n} + \frac{\partial T}{\partial y}\frac{\partial y}{\partial n}$$

and that

$$\frac{\partial y}{\partial m} = -\sin\alpha, \quad \frac{\partial y}{\partial n} = \cos\alpha, \quad \frac{\partial x}{\partial m} = \cos\alpha, \quad \frac{\partial x}{\partial n} = \sin\alpha$$

we can rewrite Eqs. (19.1) and (19.2) as follows:

$$\left(\frac{q}{A}\right)_x = -(k_m \cos^2\alpha + k_n \sin^2\alpha)\frac{\partial T}{\partial x} + (k_m - k_n)\sin\alpha\cos\alpha\frac{\partial T}{\partial y} \tag{19.3a}$$

$$\left(\frac{q}{A}\right)_y = (k_m - k_n)\sin\alpha\cos\alpha\frac{\partial T}{\partial x} - (k_m \sin^2\alpha + k_n \cos^2\alpha)\frac{\partial T}{\partial y} \tag{19.4a}$$

If x and y coincide with the principal axes m and $n(\alpha = 0)$, Eqs. (19.3a) and (19.4a) reduce by inspection to

$$\left(\frac{q}{A}\right)_x = \left(\frac{q}{A}\right)_m = -k_m\frac{\partial T}{\partial m}$$

$$\left(\frac{q}{A}\right)_y = \left(\frac{q}{A}\right)_n = -k_n\frac{\partial T}{\partial n}$$

The results are in accord with our earlier statement regarding heat transfer along the principal axes. Also, if the medium were isotropic, so that $k_m = k_n = k_x = k_y = k$, Eq. (19.3a) and (19.4a) would reduce to the familiar simple form,

$$\left(\frac{q}{A}\right)_x = -k\frac{\partial T}{\partial x}$$

$$\left(\frac{q}{A}\right)_y = -k\frac{\partial T}{\partial y}$$

Equations (19.3a) and (19.4a) are the rate equations for heat conduction in an anisotropic plane. They are of the form

$$\left(\frac{q}{A}\right)_x = -k_{xx}\frac{\partial T}{\partial x} - k_{xy}\frac{\partial T}{\partial y} \tag{19.3b}$$

$$\left(\frac{q}{A}\right)_y = -k_{yx}\frac{\partial T}{\partial x} - k_{yy}\frac{\partial T}{\partial y} \tag{19.4b}$$

where
$$k_{xx} = k_m \cos^2 \alpha + k_n \sin^2 \alpha$$
$$k_{xy} = (k_n - k_m) \sin \alpha \cos \alpha$$
$$k_{yx} = (k_n - k_m) \sin \alpha \cos \alpha$$
$$k_{yy} = k_m \sin^2 \alpha + k_n \cos^2 \alpha$$

Two significant facts concerning these equations may be noted:

1. Equation (19.3) states that a temperature gradient in the y-direction can cause heat conduction in the x-direction, or heat conduction in the x-direction can establish a temperature gradient in the y-direction. The same is true of Eq. (19.4). Fluxes $(q/A)_x$ and $(q/A)_y$ are said to be *coupled* because one cannot exist without the other in an anisotropic plane.
2. The transport coefficients responsible for the coupling, namely the cross coefficients k_{xy} and k_{yx}, are equal.

An identical analysis can be performed for the mass diffusion in an anisotropic plane, such as in crystals or polymer films. The example of heat conduction in an anisotropic plane illustrates coupling fluxes of the same type (both are energy fluxes). Coupling may also occur among dissimilar fluxes, such as between an energy flux and a mass flux (thermal diffusion) and between an energy flux and an electrical flux (thermoelectricity). Experience has shown, however, that coupling is not indiscriminate, but occurs most commonly among transfer fluxes of the same tensorial rank.* Thus, coupling is possible among mass, energy, and electrical fluxes (all are first-order tensors), but has not been observed, for example, between an energy flux and a momentum flux (a second-order tensor). Experience has also revealed that the equality of cross coefficients exists only for certain choices of fluxes and potentials. These choices may not always be obvious. A rigorous method of defining them is outlined in the following section.

Example 19.1: A special anisotropic graphite has conductivity values, $k_m = 100$ and $k_n = 2$ Btu-ft/hr ft^2 F, along the principal axes m and n. A slab of the material 0.2 in. thick is exposed on one side to 1000 F and the other to 100 F. If the isothermal surfaces make an angle $\alpha = 30$ deg with the axis m as shown in the figure, find the magnitudes of heat flux along and normal to the isothermal surfaces.

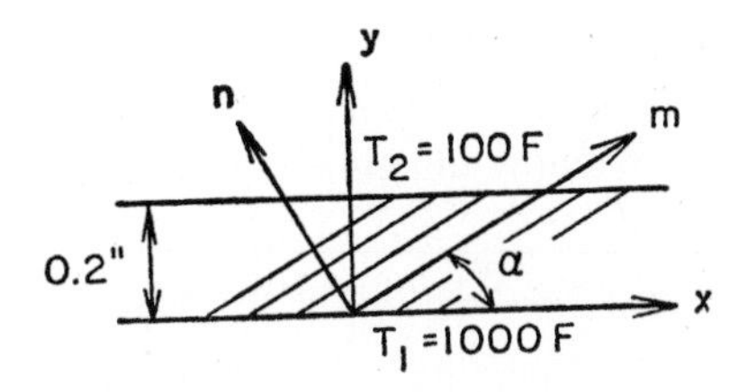

* A scalar quantity is a zeroth-order tensor, a vector is a first-order tensor, and a stress tensor is a second-order tensor. A more general statement for an isotropic system is that coupling can occur between tensors whose difference in rank is even, Curtiss (9).

Solution: Let the axes x and y be the directions parallel and normal respectively to the isothermal surfaces.

In Eqs. (19.3b) and (19.4b), $\partial T/\partial x = 0$. Therefore

$$\left(\frac{q}{A}\right)_x = -k_{xy}\frac{\partial T}{\partial y}$$

$$\left(\frac{q}{A}\right)_y = -k_{yy}\frac{\partial T}{\partial y}$$

$$k_{xy} = (k_n - k_m)\sin\alpha\cos\alpha = (98)(0.5)(0.866) = 42.4$$

$$k_{yy} = k_m\sin^2\alpha + k_n\cos^2\alpha = (100)(0.25) + (2)(0.75) = 26.5$$

Finally, $$\left(\frac{q}{A}\right)_x = (42.2)\frac{(900)}{0.0167} = -22.8 \times 10^5 \text{ Btu/hr ft}^2$$

$$\left(\frac{q}{A}\right)_y = (26.5)\frac{(900)}{0.0167} = 14.3 \times 10^5 \text{ Btu/hr ft}^2$$

Note that the heat flux in the isothermal plane is larger than that in the normal direction. Note also that if the material were placed in a thermal conductivity apparatus in the position shown in the figure, the measured conductivity would be $k_{yy} = 26.5$ Btu-ft/hr ft² F.

19.2 The Onsager Method

Let us re-examine the rate equations given in Chap. 1. We note that they are all of the form

$$J = LX \tag{19.5}$$

where J represents a flux,

X represents a potential gradient or a "driving force,"

L is the corresponding transport coefficient.

Their formulation was entirely empirical and intuitive, and they served to define the transport coefficients μ, k, and D.

We now imagine a situation involving two simultaneous processes. Let J_1, J_2 and X_1, X_2 represent respectively fluxes and forces describing the processes. We make the following two assumptions:

1. Each flux is dependent on all driving forces acting on the system. This is certainly less restrictive than the assumption that each flux depends only on a particular force.
2. The dependence is linear. This is identical with the assumption made in the formulation of the simple rate equations and is a reasonable approximation for states not too far disturbed from equilibrium.

The rate equations for the two simultaneous processes may therefore be expressed as follows:

$$J_1 = L_{11}X_1 + L_{12}X_2$$
$$J_2 = L_{21}X_1 + L_{22}X_2 \tag{19.6}$$

These equations as yet have no physical meaning, since the fluxes and forces have not been identified. Let us suppose J_1 is an energy flux and J_2 a mass flux. We might intuitively expect X_1 to be a temperature gradient and X_2 a concentration gradient (or any one of a number of driving forces for mass transfer, such as partial pressure, activity, etc.). We note that mass transfer can result from a temperature gradient, and energy transfer from a concentration gradient; our intuition is confused. The difficulty of interpreting the four transport coefficients on the basis of empirical data or intuition becomes obvious. When more than two processes occur simultaneously, the difficulty multiplies. Clearly, there is need for a comprehensive macroscopic theory that would lead to formulation of rate equations of any degree of complexity on a logical basis. A powerful tool in this connection is a method based on the irreversible thermodynamics, called the *Onsager method*.

Since every diffusional process with a finite transfer rate is irreversible, it may be quantitatively identified in terms of a net entropy increase of the universe, i.e., a diffusional transfer process generates entropy. In the Onsager method, the amount of entropy generated is shown to be given by the product of a flux and a force, as the following example illustrates.

Consider steady-state, one-dimensional heat conduction in a solid M of cross-sectional area A, and consider an element Δx of the conductor. Replace adjacent sections of the conductor by large equilibrium regions, Fig. 9.2, each region being in equilibrium with the side of the conductor

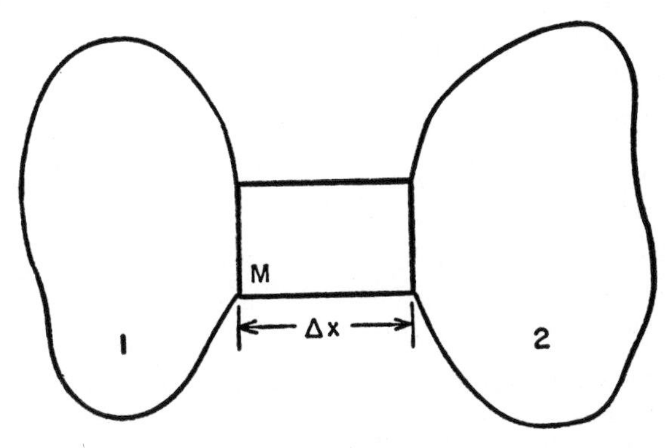

Fig. 19.2. One-dimensional heat conduction in a solid.

with which it is in contact.* Then the heat flux J_u crossing the boundaries of the element Δx may be expressed in terms of the energy U of the equilibrium regions as follows:

$$J_{u1} = -\frac{1}{A}\frac{dU_1}{dt}, \qquad J_{u2} = \frac{1}{A}\frac{dU_2}{dt}$$

Applying the first law of thermodynamics to the closed system $1 + 2 + M$, we get

$$\frac{dU_1}{dt} + \frac{dU_2}{dt} = 0 \quad \text{or} \quad J_{u1} = J_{u2} = J_u \tag{19.7}$$

The rate of entropy generation may be written as

$$\dot{S} = \dot{S}_1 + \dot{S}_2 \tag{19.8}$$

since $\dot{S}_M$ is zero for the assumed steady state of the conductor. For an equilibrium region of constant volume, $dS = dU/T$, or

$$\begin{aligned} \dot{S}_1 &= -A\tau J_{u1} \\ \dot{S}_2 &= A(\tau + \Delta\tau) J_{u2} \end{aligned} \tag{19.9}$$

where τ† and $\tau + \Delta\tau$ are the reciprocal temperatures of the equilibrium regions 1 and 2 respectively. Combining Eqs. (19.7), (19.8), and (19.9), we have

$$\dot{S} = -A\tau J_{u1} + A(\tau + \Delta\tau) J_{u2} = A J_u \, \Delta\tau$$

or

$$\Phi = J_u \frac{d\tau}{dx} \tag{19.10}$$

where Φ expresses the rate of irreversibility or the rate of entropy production per unit volume $(A\,\Delta x)$ of the conductor and is given by the product of a flux, J_u, and a driving force, $X = d\tau/dx$. The quantity Φ can also be interpreted in terms of "entropy fluxes" crossing the boundaries of the element Δx, entropy flux being defined by

$$J_{s_1} \equiv -\frac{1}{A}\frac{dS_1}{dt}$$

Then

$$\Phi = \frac{dJ_s}{dx} = J_u \frac{d\tau}{dx}$$

which is identical with Eq. (19.10).

* The phenomenological treatment of Keenan (1) is followed throughout. The analysis of thermoelectric phenomena in Sec. 9.3 is taken from Hatsopoulos and Keenan (2).

† In this chapter, the symbol τ will be used to represent the reciprocal temperature T^{-1}. Throughout the rest of the book, τ represents the shear stress.

When there are n simultaneous processes, an expression for the entropy production Φ can be found by a generalization of the method leading to Eq. (19.10), and the result can be expressed as

$$\Phi = J_1X_1 + J_2X_2 + \cdots + J_nX_n \tag{19.11}$$

where the flux and force with identical subscripts are called *conjugate*.

The rate equations among the n fluxes and forces may be written with the assumption of linear and general dependence of each flux on all forces:

$$J_1 = L_{11}X_1 + L_{12}X_2 \cdots + L_{1n}X_n$$

$$J_2 = L_{21}X_1 + L_{22}X_2 \cdots + L_{2n}X_n$$

$$J_n = L_{n1}X_1 + L_{n2}X_2 \cdots + L_{nn}X_n$$

Onsager (3) showed that when the fluxes and forces in these rate equations satisfy Eq. (19.11), the following *reciprocal relations* involving the coupling coefficients are valid:

$$L_{12} = L_{21}, \qquad L_{13} = L_{31}, \cdots, L_{ik} = L_{ki}$$

Onsager proved these relations by statistical consideration of microscopic fluctuations about an equilibrium state. In the phenomenological treatment, we simply accept the reciprocal relations as an axiom not deducible from the laws of classical thermodynamics (1).

We shall also note without proof that the Onsager method offers considerable flexibility in the choice of the conjugate pairs. The magnitude of the total entropy generated in a given system is, of course, uniquely determined, but the quantity Φ can be split into a sum of products in many alternate ways. In other words, the method places no restriction on what fluxes and forces we select so long as they generate the necessary amount of entropy. It would be desirable naturally to select conjugate pairs such that the transfer coefficients they define are the most useful or the easiest to determine. We shall illustrate the method with a familiar example.

19.3 Thermoelectric Phenomena

Consider an element Δx of a metallic conductor M of cross-sectional area A through which energy and electric current flow in steady state. The first step in an analysis is to find by the Onsager method an expression for the total entropy generated by the two processes. Replace adjacent sections of the element by large equilibrium regions 1 and 2, as shown in Fig. 19.3 (2). In each region, U is the energy, τ is the reciprocal temperature, n is the

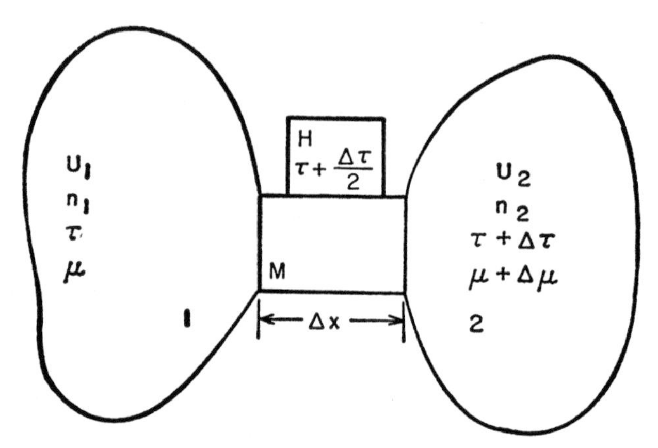

Fig. 19.3. One-dimensional coupled flows of energy and electric charge in a homogeneous conductor (2).

number of unit positive electric charges, and μ is the electrochemical potential.*

The energy flux J_u and the electrical flux J_n crossing the boundaries of the element Δx may be expressed in terms of the properties of the equilibrium regions by

$$J_{u1} = -\frac{1}{A}\frac{dU_1}{dt}, \quad J_{u2} = \frac{1}{A}\frac{dU_2}{dt}$$

$$J_{n1} = -\frac{1}{A}\frac{dn_1}{dt}, \quad J_{n2} = \frac{1}{A}\frac{dn_2}{dt}$$

Let the conductor M exchange heat Q_H with a reservoir H at a temperature $(\tau + \Delta\tau/2)^{-1}$. When M is at steady state, the First Law of thermodynamics applied to the closed system $1 + 2 + M$ yields

$$\frac{dQ_H}{dt} = \frac{dU_1}{dt} + \frac{dU_2}{dt} = A(J_{u2} - J_{u1}) = A\,\Delta J_u \tag{19.12}$$

* The *electrochemical potential* in a solid conductor is defined as the change of the Helmholtz free energy ψ per unit negative charge introduced into the conductor at constant temperature and volume:

$$\mu = \left(\frac{\partial\psi}{\partial n}\right)_{T,V}$$

where $\psi \equiv U - TS$.

It can be shown that the difference in electrochemical potential between two regions of a conductor equals the work necessary to move a unit charge from one region into another, provided the process is executed at constant temperature and volume for each of the regions. In an isothermal system, the electrochemical potential difference is identical with the familiar electrical potential difference between two points.

From the conservation of electric charges:

$$J_{n2} - J_{n1} = \Delta J_n = 0 \tag{19.13}$$

The rate of entropy generation may be written as

$$\dot{S} = \dot{S}_1 + \dot{S}_2 + \dot{S}_H \tag{19.14}$$

omitting $\dot{S}_M$ which is zero for the assumed steady state of the conductor.

The quantities in Eq. (19.14) may be found by applying the Gibbs equation (7), $dS = \tau\, dU - \mu\tau\, dn$, to the equilibrium regions 1 and 2,

$$\dot{S}_1 = -A\tau J_{u1} + A\mu\tau J_{n1}$$

$$\dot{S}_2 = A(\tau + \Delta\tau) J_{u2} - A(\mu + \Delta\mu)(\tau + \Delta\tau) J_{n2}$$

and to the heat reservoir H,

$$\dot{S}_H = -\left(\tau + \frac{\Delta\tau}{2}\right)\dot{Q}_H$$

Substituting these equations in Eq. (19.14) and using the results of Eqs. (19.12) and (19.13), we get the following expression for the rate of entropy generation:

$$\dot{S} = A J_u \Delta\tau - A J_n \Delta(\mu\tau)$$

to first-order small quantities. The rate of entropy generation per unit volume of the conductor is then

$$\Phi = J_u \frac{d\tau}{dx} - J_n \frac{d(\mu\tau)}{dx} \tag{19.15}$$

From inspection of Eq. (19.15), we select the conjugate fluxes and forces of the two processes; the rate equations may now be written as

$$J_u = L_{11}\frac{d\tau}{dx} - L_{12}\frac{d(\mu\tau)}{dx} \tag{19.16}$$

$$J_n = L_{21}\frac{d\tau}{dx} - L_{22}\frac{d(\mu\tau)}{dx} \tag{19.17}$$

Onsager's reciprocal relation specifies $L_{12} = L_{21}$. To determine the remaining three coefficients, we need three experimental relations. Two are quite obvious, namely,

1. *The Fourier equation for pure heat conduction.* In terms of the symbols used in this chapter, it may be written

$$(J_u)_{J_n=0} = -k\frac{dT}{dx} = \frac{k}{\tau^2}\frac{d\tau}{dx}$$

or, by eliminating $d(\mu\tau)/dx$ from Eqs. (19.16) and (19.17) and setting $J_n = 0$, we get

$$L_{11} - \frac{L_{12}^2}{L_{22}} = \frac{k}{\tau^2} \tag{19.18}$$

2. *The Ohm's law for an isothermal conductor*, or alternatively, the isothermal Joule effect. The Ohm's law is strictly applicable to an isothermal, homogeneous conductor:

$$\left(J_n = -\frac{1}{r}\frac{de}{dx}\right)_{T=\text{const}}$$

where r is the electrical resistance per unit length of the conductor and e is the electrical potential.

In the case of an isothermal homogeneous conductor, the change in the electrochemical potential μ from point to point is identical with the change in the electrical potential e. Therefore, we may write

$$\left(\frac{de}{dx}\right)_{T=\text{const}} = \frac{d\mu}{dx}$$

and from Eq. (19.17) we get

$$L_{22} = \frac{1}{r\tau} \tag{19.19}$$

Instead of Ohm's law, we could use the isothermal Joulean heating effect to obtain an alternative expression to Eq. (19.19).

In order to determine completely the rate equations, (19.16) and (19.17), we need one additional empirical relation. In other words, we must postulate one of the coupled effects and then verify by experiment a relation that expresses this effect.

For the thermoelectric effects, we already know the answer, but let us proceed as if the answer were not known. From Eq. (19.17), taking for example the case $J_n = 0$, we get

$$\left(\frac{d(\mu\tau)/dx}{d\tau/dx}\right)_{J_n=0} = \frac{L_{21}}{L_{22}}$$

or

$$\left[\tau\left(\frac{d\mu/dx}{d\tau/dx}\right) + \mu = \frac{L_{21}}{L_{22}}\right]_{J_n=0} \tag{19.20}$$

The ratio $(d\mu/dx)/(d\tau/dx)$ in Eq. (19.20) states that when the coupling coefficient $L_{21} \neq 0$, an electrical potential gradient may be induced by a temperature gradient even though the current flow is zero. We verify this observation in the laboratory and decide that it would be useful to define a new transport property h. Then with Eqs. (19.16) and (19.17)

$$h \equiv \tau\left(\frac{d\mu/dx}{d\tau/dx}\right)_{J_n=0} = \left(\frac{J_u}{J_n}\right)_{T=\text{const}} - \mu \tag{19.21}$$

Combining Eqs. (19.20) and (19.21), we get

$$\frac{L_{21}}{L_{22}} = h + \mu \tag{19.22}$$

Equations (19.18), (19.19), and (19.22) can now be solved for the unknown coefficients. Upon substitution into Eqs. (19.16) and (19.17), we get

$$J_u = \left[\frac{k}{\tau^2} + \frac{(h+\mu)^2}{r\tau}\right]\frac{d\tau}{dx} - \left(\frac{h+\mu}{r\tau}\right)\frac{d(\mu\tau)}{dx} \tag{19.23}$$

$$J_n = \left(\frac{h+\mu}{r\tau}\right)\frac{d\tau}{dx} - \left(\frac{1}{r\tau}\right)\frac{d(\mu\tau)}{dx} \tag{19.24}$$

Equations (19.23) and (19.24) are the complete rate equations for the simultaneous transfer of energy and electric charge in a conductor. The rate equation of electrical conduction, Eq. (19.24), may be readily simplified to the following form in terms of the gradients of temperature T and potential μ:

$$J_n = -\frac{1}{r}\left(\frac{h}{T}\frac{dT}{dx} + \frac{d\mu}{dx}\right) \tag{19.25}$$

Equation (19.25) has the following interpretation: In a nonisothermal conductor, current flows not only because of an electrochemical potential gradient $d\mu/dx$ but also because of a temperature potential gradient, the magnitude of the latter depending upon a temperature function h defined by Eq. (19.21).

Example 19.2: *The thermocouple problem.* Consider a thermocouple circuit consisting of conductors of dissimilar metals A and B, and a voltmeter V as shown in Fig. 19.4. Junctions 1 and 2 of metal A to metal B are kept at temperatures T_1 and T_2 respectively, while junctions a and b of metal B with the voltmeter are maintained at T_0 by means of a heat reservoir H.

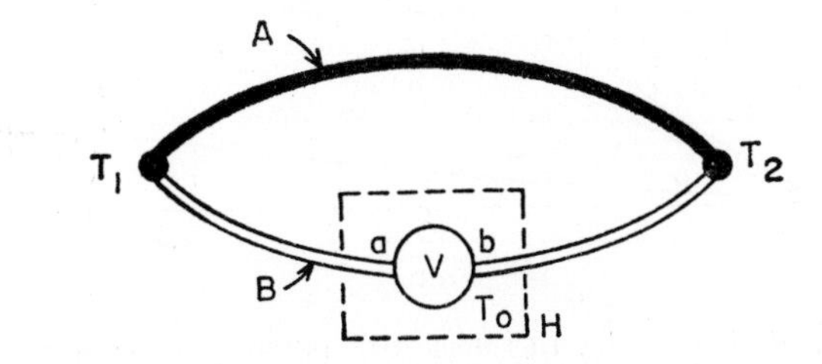

Fig. 19.4. Thermocouple.

The electrical potential difference between a and b measured by the voltmeter is in this case the same as the electrochemical potential difference, because a and b are identical metals at identical temperatures. Therefore,

$$\begin{aligned}
\Delta e = \mu_b - \mu_a &= \int_a^b d\mu \\
&= \int_a^b \frac{d\mu}{dx} \cdot \frac{dx}{d\tau}\, d\tau \\
&= \left(\int_{\tau_0}^{\tau_1} \frac{d\mu}{dx}\frac{dx}{d\tau}\, d\tau\right)_B + \left(\int_{\tau_1}^{\tau_2} \frac{d\mu}{dx}\frac{dx}{d\tau}\, d\tau\right)_A + \left(\int_{\tau_2}^{\tau_0} \frac{d\mu}{dx}\frac{dx}{d\tau}\, d\tau\right)_B \\
&= \left(\int_{\tau_1}^{\tau_2} \frac{d\mu}{dx}\frac{dx}{d\tau}\, d\tau\right)_A - \left(\int_{\tau_1}^{\tau_2} \frac{d\mu}{dx}\frac{dx}{d\tau}\, d\tau\right)_B \qquad (19.26)
\end{aligned}$$

Upon substituting $d\mu/dx$ from Eq. (19.25) into Eq. (19.26), and noting that $J_{nA} = -J_{nB}$, we get after rearrangements,

$$\Delta e = \int_{\tau_1}^{\tau_2} (h_A - h_B)\frac{d\tau}{\tau} - J_n \int_{\tau_1}^{\tau_2} (r_A + r_B)\frac{dx}{d\tau}\, d\tau$$

The voltmeter reading at zero current (potentiometer reading) is therefore given by

$$(\Delta e)_{J_n=0} = \int_{\tau_1}^{\tau_2} (h_A - h_B)\frac{d\tau}{\tau} = \int_{T_1}^{T_2} (h_B - h_A)\frac{dT}{T} \qquad (19.27)$$

This potential difference developed by the thermocouple in the absence of current flow is called the *Seebeck emf*. Equation (19.27) shows it is a function only of the temperatures of the junctions and the difference of the properties $(h_B - h_A)$ of the materials. Other thermoelectric quantities, namely the Thomson coefficient and the Peltier coefficient, may be similarly derived (2).

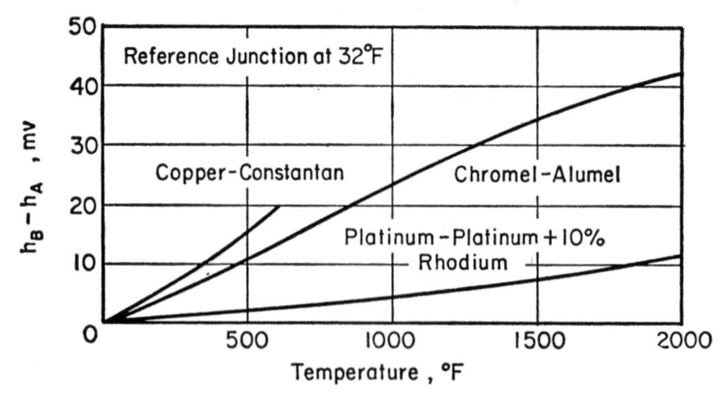

Fig. 19.5. Variation of the thermoelectric property difference $(h_B–h_A)$ with temperature.

Figure 19.5 is a plot of $(h_B - h_A)$ vs. temperature for several thermocouples. The values of $h_B - h_A = T\, de/dT \approx T\, \Delta e/\Delta T$ were taken from standard tables (6).

REFERENCES

1. Keenan, J. H., "Concepts Employed in the Thermodynamics of Coupled Irreversible Flows," *Appl. Mech. Rev.*, **49** (Feb. 1955).
2. Hatsopoulos, G. N., and J. H. Keenan, "Analysis of Thermoelectric Effects by Methods of Irreversible Thermodynamics," ASME Paper 58 A–1.
3. Onsager, L., "Reciprocal Relations in Irrev. Processes," Part I, *Phys. Rev.*, **37,** 405 (1931); Part II, *Phys. Rev.*, **38,** 2265 (1931).
4. De Groot, S. R., *Thermodynamics of Irreversible Processes*, Interscience Publishers, New York, 1952.
5. Denbigh, K., *Thermodynamics of the Steady State*, Wiley, New York, 1951.
6. *Temperature—Its Measurement and Control in Science and Industry*, Reinhold Publishing Corp., New York, 1941.
7. Gibbs, J. W., *Collected Works*, **1,** 63, Longmans Green, New York, 1931.
8. Carslaw, H. S. and J. C. Jaeger, *Conduction of Heat in Solids*, Oxford U. Press, New York, 1959, 2nd Ed., pp. 38–49.
9. Curtiss, C. F., "Thermodynamics of Irreversible Processes," Sec. J, *High Speed Aerodynamics and Jet Propulsion*, **I,** Princeton U. Press, Princeton, N. J., 1955, p. 791.

CHAPTER 20

Kinetic Theory of Transport Properties

In this chapter, we shall derive the transport properties of gases from elementary considerations of the kinetic theory and briefly review the results of the more rigorous theories.

According to the molecular hypothesis, the primary building blocks of nature are molecules—the smallest entities that retain the chemical identity of a substance—and the various states of matter differ essentially in the arrangement and the state of motion of these molecules. It might be reasonable to expect then that the macroscopic behavior of matter is a consequence of, and may be predicted from, the microscopic behavior of the individual molecules. The thermodynamic properties of matter such as the pressure and the specific heat are defined for a system in a state of thermodynamic equilibrium, and may be derived by the methods of statistical thermodynamics. Transfer processes, on the other hand, occur in systems that may be in a steady state, but certainly are not at thermodynamic equilibrium. A rigorous treatment of the transfer processes requires the solution of an integro-differential equation derived by Boltzmann. However, for the limited purpose of explaining the significant characteristics of the transfer properties, an extremely simple treatment based on the concept of the mean free path may be adequate, as discussed by Sears (1). A hypothetical "ideal" gas possessing the following characteristics may be postulated:

1. The molecules are hard spheres resembling billiard balls, having diameter d and mass m.

2. The molecules exert no force on one another except when they collide.

3. The collisions are perfectly elastic and obey the classical conservation laws of momentum and energy.

4. The molecules are uniformly distributed throughout the gas. They are in a state of continuous motion and are separated by distances, large compared with their diameter d.

5. All directions of molecular velocities are equally probable. The speed (magnitude of velocity) of a molecule can have any value between zero and infinity.

20.1 Mean Free Path

A significant parameter that governs the mechanism of transfer in gases is the free path, defined as the distance travelled by a molecule between two successive collisions. A molecule travelling a free path of a certain distance is in effect transferring momentum, energy, or mass over that distance. For our hypothetical gas of billiard ball molecules, this distance is well defined. The first step in our study is to find the mean free path and the distribution of free paths of the molecules.

Imagine a collection of molecules of diameter d colliding at random. We single out a molecule as it travels in straight paths from one collision to the next; its speed and direction of motion change with each collision. We wish to find the average distance the molecule travels between collisions. Imagine that at a given instant all molecules but the one in question are frozen in position and this molecule moves with the average speed $\bar{V}$. At the instant of a collision, the center-to-center distance of the two molecules is d. This distance would be the same if the moving molecule had a radius d and the stationary molecules were reduced to geometrical points. The collision cross section or the target area σ of the moving molecule is then

$$\sigma = \pi d^2$$

In time t the moving molecule sweeps out a cylindrical volume of length $\bar{V}t$ and cross section σ. Any molecule whose center is in this cylinder will be struck by the moving molecule. The number of collisions in time t is given by

$$\sigma n \bar{V} t$$

where n is the number of molecules present in a unit volume, assumed to be uniformly distributed in space.

The average distance between collisions, or the mean free path λ, is given by the ratio of the distance covered in time t over the number of collisions in this time:

$$\lambda = \frac{\bar{V}t}{\sigma n \bar{V}t} = \frac{1}{\sigma n} \tag{20.1}$$

The collision frequency Z of a molecule is defined as the number of collisions per unit time experienced by the molecule, or

$$Z = \sigma n \bar{V} = \frac{\bar{V}}{\lambda} \tag{20.2}$$

The results above were obtained assuming all molecules except one were stationary. Clausius calculated the mean free path assuming all molecules moved at the average speed $\bar{V}$ and obtained

$$\lambda = \frac{0.750}{\sigma n}$$

If the molecules are assumed to possess a Maxwellian speed distribution (Fig. 20.1) characteristic of a gas that is in equilibrium, the average speed

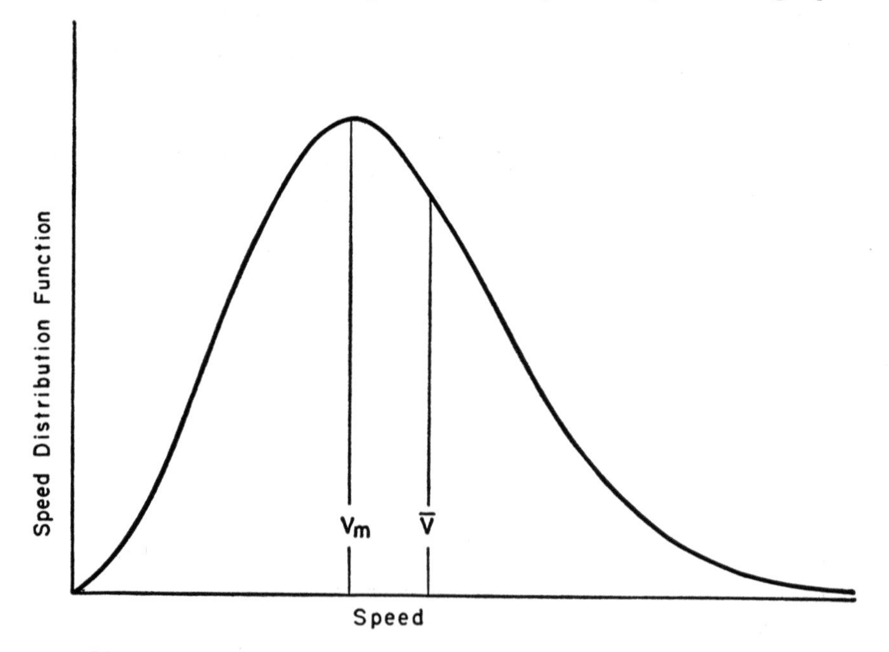

Fig. 20.1. Maxwellian speed distribution. V_m is the most probable speed and $\bar{V}$ is the average speed.

$\bar{V}$ and the most probable speed V_m are (1):

$$\bar{V} = \sqrt{\frac{8\kappa T}{\pi m}} \tag{20.3a}$$

$$V_m = \sqrt{\frac{2\kappa T}{m}} \tag{20.3b}$$

where $\kappa = 1.380 \times 10^{-23}$ joules/°C, the Boltzmann constant. For this case,

$$\lambda = \frac{0.707}{\sigma n} \tag{20.4}$$

It is seen that the more rigorous assumptions have the end result of simply changing the magnitude of the numerical coefficient in the expression for the mean free path.

To find the distribution of free paths, consider a large group of molecules N_0 originating their free paths at a given instant. Some experience collision after travelling a very short distance; others travel very far before experiencing a collision. Let N be the number of these N_0 molecules which have not experienced a collision after travelling a distance x (Fig. 20.2). We assume the number dN, which will experience a collision in travelling from x to $x + dx$, is proportional to N and the distance dx, or

$$dN = -P_c N\, dx \tag{20.5}$$

The quantity P_c is a proportionality constant expressing the collision probability over the distance dx, with the negative sign introduced to indicate decreasing N with increasing x. Integration of Eq. (20.5) and writing $N = N_0$ when $x = 0$ gives

$$N = N_0 e^{-P_c x} \tag{20.6}$$

By the definition of the mean free path we can write

$$\lambda = \frac{1}{N_0} \int_0^{N_0} x\, dN$$

where dN is the number of molecules that have a particular free path of length x. Substitution of x from Eq. (20.6) and integration of the resulting expression give

$$\lambda = \frac{1}{P_c}$$

Equation (20.6) may now be written

$$N = N_0 e^{-x/\lambda} \tag{20.7}$$

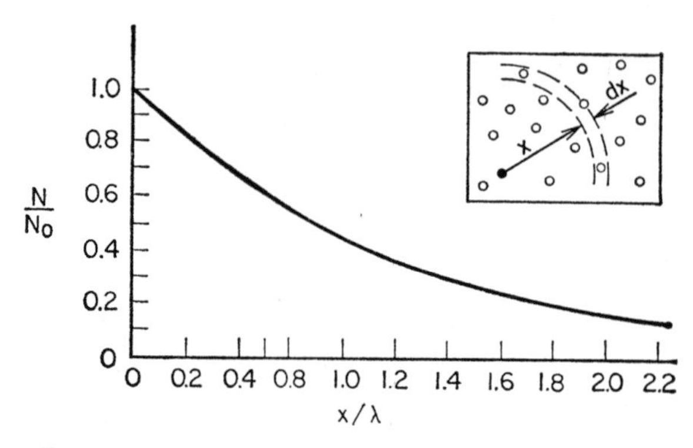

Fig. 20.2. Graph of the survival equation, Eq. (20.7).

Equation (20.7) may be called the survival equation. It gives the number N out of the original number N_0 having free paths longer than x. This number decreases exponentially as shown in Fig. 20.2.

20.2 Transport Properties

Consider Fig. 20.3, identical with Fig. 1.4 in Chap. 1. We shall denote the momentum flux, the energy flux, and the mass flux of a diffusing species in a dilute mixture by a common symbol (p/A) and the property transferred by the common symbol P. Consider an imaginary plane $y = y_1$ which is being crossed by molecules in either direction. If the condition is imposed that there is no bulk motion of the gas in the y-direction, the molecules cross the y_1 plane from both sides with equal frequency. The molecular flux in either direction can then be found as follows.

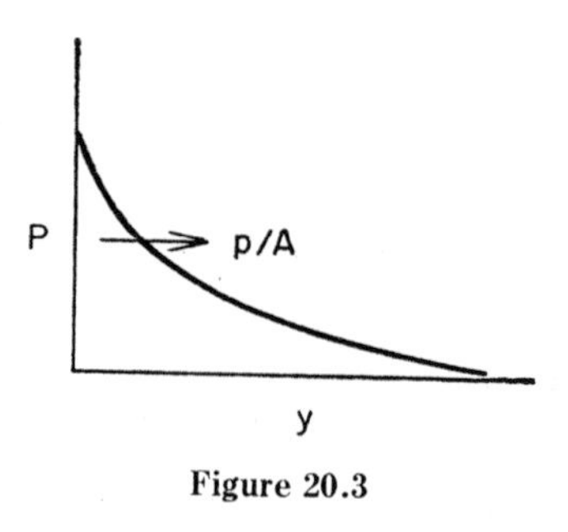

Figure 20.3

Consider a volume element dV at a distance r from the y_1 plane, Fig. 20.4. Note that the volume element is infinitesimal compared with the physical dimensions of the system, but contains a sufficiently large number of molecules so that its properties are negligibly influenced by individual molecules crossing its boundaries. The total number of molecules in dV is $n\,dV$. The total number of collisions within dV in time dt is given by $\frac{1}{2}Zn\,dV\,dt$ where Z is the collision frequency of a molecule, n is the molecular density, and the factor $\frac{1}{2}$ is necessary so that each collision is not counted twice since two molecules are involved in each collision. At each collision, two new free paths originate, so that the total number of molecules, or new free paths, originating in dV is $Zn\,dV\,dt$. If we assume that these

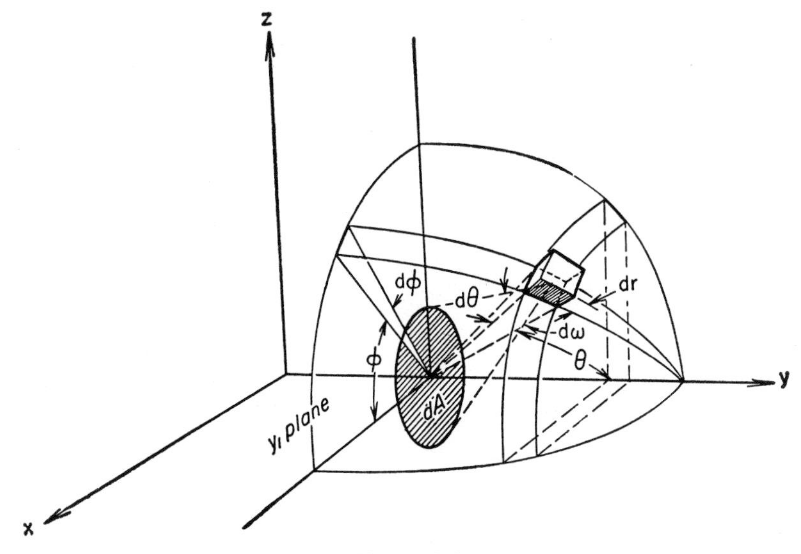

Figure 20.4

molecules are uniformly distributed in direction throughout the solid angle 4π, then the number headed toward an element of area dA in the y_1-plane is

$$\frac{d\omega}{4\pi} Zn\, dV\, dt$$

where $d\omega$ is the solid angle subtended at the center of dV by dA. All these molecules do not reach dA. The number that leave dV and reach dA without further collisions along the way may be found by multiplying the last expression by N/N_0 from the survival equation, Eq. (20.7):

$$\left(\frac{d\omega}{4\pi} Zn\, dV\, dt\right) e^{-r/\lambda}$$

Now, $d\omega = dA \cos\theta/r^2$ and $dV = r^2 \sin\theta\, d\theta\, d\phi\, dr$. The total number of molecules from the right side crossing dA in time dt is

$$\begin{aligned} N_{\text{total}} &= \int_{\theta=0}^{\pi/2} \int_{\phi=0}^{2\pi} \int_{r=0}^{\infty} \frac{dA \cos\theta}{4\pi} Zn \sin\theta\, e^{-r/\lambda}\, d\theta\, d\phi\, dr\, dt \\ &= \tfrac{1}{4} Zn\lambda\, dA\, dt \end{aligned} \tag{20.8}$$

Note that the integral over r is extended to infinity, justified by the fact that the dimensions of the physical system are very much larger than the molecular free path.

These molecules crossing the y_1-plane may be pictured as carrying properties characteristic of an average distance $\bar{y}$ to the left and to the right of the plane at which they made their last collisions. To find $\bar{y}$, multiply each molecule crossing from dV by its distance $r \cos \theta$ from the y_1-plane, integrate over θ, ϕ, and r, and divide the resulting expression by N_{total} crossing the plane:

$$\bar{y} = \frac{1}{N_{\text{total}}}\left[\int_{\theta=0}^{\pi/2}\int_{\phi=0}^{2\pi}\int_{r=0}^{\infty}\left(\frac{dA \cos \theta}{4\pi} Zn \sin \theta e^{-r/\lambda}\right)(r \cos \theta)\, d\theta\, d\phi\, dr\, dt\right]$$

$$= \frac{\frac{1}{6}Zn\lambda^2\, dA\, dt}{\frac{1}{4}Zn\lambda\, dA\, dt} = \tfrac{2}{3}\lambda \tag{20.9}$$

The fact that $\bar{y}$ is directly related to λ is not surprising. It may be concluded from Eq. (20.9) that the molecules crossing the y_1-plane from either direction carry property values characteristic of the planes two-thirds of the mean free path distance from the plane. Usually, we are interested in the transfer processes occurring in systems only slightly removed from equilibrium. In such a system, the property gradients are small over a distance of the order of a mean free path. If the property P were assigned a value P_1 at the y_1-plane, then at a distance $\bar{y}$ to the right of the plane

$$P_{\bar{y}} = P_1 + \tfrac{2}{3}\lambda \frac{dP}{dy}$$

and at a distance $\bar{y}$ to the left of the plane

$$P_{-\bar{y}} = P_1 - \tfrac{2}{3}\lambda \frac{dP}{dy}$$

From Eq. (20.8), the number of molecules crossing the y_1-plane per unit area dA and unit time dt is $\frac{1}{4}Zn\lambda$. Then the net flux of property P crossing the plane in the direction shown in Fig. 20.3 is given by the difference between the above quantities, multiplied by this molecular flux:

$$\frac{p}{A} = (\tfrac{1}{4}\, Zn\lambda)\left(-\frac{4}{3}\lambda \frac{dP}{dy}\right)$$

$$= -\tfrac{1}{3}Zn\lambda^2 \frac{dP}{dy} \tag{20.10}$$

$$= -\tfrac{1}{3}n\bar{V}\lambda \frac{dP}{dy}$$

Viscosity. In Chap. 1, the viscosity of a fluid was defined by the phenomenological rate equation, Eq. (1.1). In Eq. (20.10), we note that if P were the x-direction momentum of the molecules of mass m, $P = mv_x$, then $p/A = \tau_{yx}$ and we can write

$$\tau_{yx} = -\tfrac{1}{3}\, n\bar{V}\lambda \frac{d(mv_x)}{dy} = -\tfrac{1}{3}\, nm\bar{V}\lambda \frac{dv_x}{dy} \tag{20.11}$$

Comparison of Eq. (20.11) with Eq. (1.1) yields the following expression for the viscosity:

$$\mu = \tfrac{1}{3}nm\bar{V}\lambda = \tfrac{1}{3}\rho\bar{V}\lambda \tag{20.12}$$

where $\rho = nm$, the density of the gas.

For a gas with a Maxwellian velocity distribution, (20.12) combined with $\sigma = \pi d^2$ and (20.3) and (20.4) gives

$$\mu = \frac{2}{3\pi^{3/2}} \frac{(m\kappa T)^{1/2}}{d^2} \tag{20.13}$$

A significant conclusion from the kinetic theory is that the viscosity of a gas is independent of the pressure or the density and depends only on the temperature. This conclusion is in good qualitative agreement with experimental data. For instance, μ is approximately independent of the pressure up to about 10 atmospheres. Typical values of μ, λ, and d for some gases are listed (11) in Table 20.1. Note that values of λ and d given in the Table should be regarded only as equivalent elastic-sphere mean free paths and diameters. They serve as convenient concepts which should not be interpreted as signifying actual existence of such molecules.

TABLE 20.1

Gas	Molecular weight M	Viscosity μ dyne sec/cm	Mean free path λ cm	Molecular diameter d cm
H_2..........	2.016	871×10^{-7}	11.77×10^{-6}	2.74×10^{-8}
Helium......	4.002	1943	18.62	2.18
CH_4.........	16.03	1077	5.16	4.14
NH_3.........	17.03	970	4.51	4.43
Neon........	20.18	3095	13.22	2.59
O_2..........	32.00	2003	6.79	3.61
Air..........	28.96	1796	6.40	3.72

All values at 15 C.

Thermal conductivity. The thermal conductivity was defined phenomenologically by Eq. (1.2). It is possible to derive an expression for the thermal conductivity of a gas by repeating the argument leading up to Eq. (20.10). In a nonisothermal system, the molecules moving from the warmer region into the colder region carry with them more energy than those moving in the opposite direction, resulting in a net transfer of energy.

In Eq. (20.10), letting $P = \bar{e}$ where $\bar{e}$ is the average thermal energy of a molecule,

$$\frac{q}{A} = -\tfrac{1}{3}\, n\bar{V}\lambda\, \frac{d\bar{e}}{dy} \tag{20.14}$$

Comparison of Eq. (20.14) with Eq. (1.2), noting that

$$d\bar{e}/dy = (d\bar{e}/dT)\,(dT/dy),$$

gives

$$k = \tfrac{1}{3}\, n\bar{V}\lambda\, \frac{d\bar{e}}{dT}$$

Now, when a gas is heated at constant volume, all of the energy supplied goes into increasing the energy of the molecules, so that $n(d\bar{e}/dT) = \rho c_v$, where ρ is the density and c_v is the specific heat per unit mass at constant volume. Also, in the case of the hypothetical gas under consideration, $c_v = \frac{3}{2}\kappa/m$. Therefore,

$$k = \tfrac{1}{3}\rho\bar{V}\lambda c_v = \tfrac{1}{2}n\bar{V}\lambda\kappa \tag{20.15}$$

For a gas with a Maxwellian velocity distribution, Eq. (20.15) combines with Eqs. (20.3) and (20.4) to yield

$$k = \frac{1}{\pi^{3/2}\, d^2}\left(\frac{\kappa^3 T}{m}\right)^{1/2} \tag{20.16}$$

TABLE 20.2

Gas	Viscosity, μ dyne-sec/cm	Thermal conductivity, k cal/cm sec deg	Specific heat, c_v cal/gm deg	c_p/c_v	$k/\mu c_v$
Helium	1875×10^{-7}	0.344×10^{-3}	0.753	1.66	2.44
Neon	2986	0.1104	0.150	1.64	2.47
Argon	2100	0.0387	0.0763	1.67	2.42
H_2	840	0.416	2.40	1.41	2.06
N_2	1664	0.0566	0.178	1.406	1.91
O_2	1918	0.0573	0.156	1.395	1.92
H_2O at 100 C	1215	0.0551	0.366	1.32	1.24
CO_2	1377	0.0340	0.151	1.31	1.64
NH_3	915	0.0514	0.401	1.32	1.40
CH_4	1027	0.0718	0.400	1.31	1.75
C_2H_4	948	0.0404	0.282	1.25	1.51
C_2H_6	854	0.0428	0.325	1.23	1.54

All values at 0 C except in the case of H_2O.

The simple theory thus predicts that the thermal conductivity, like the viscosity, is independent of the pressure and increases with increasing temperature.

Dividing Eq. (20.12) by Eq. (20.15), we get

$$\frac{k}{\mu c_v} = 1 \tag{20.17}$$

This result is significant because it involves only quantities that are directly measurable and thus provides a check on the validity of the simple theory. Actual measured values of the group $k/\mu c_v$ for a number of monatomic and polyatomic gases are given (11) in Table 20.2. It is seen that the result given by Eq. (20.17) agrees with the data only as regards order of magnitude, the measured values ranging between 1.5 and 2.5.

Mass diffusivity. In a gaseous mixture, diffusion results from random molecular motion whenever there is a concentration gradient of any molecular specie. For simplicity, we shall consider a binary mixture in which the total pressure and temperature are uniform. Equation (20.10) was derived for a single gas; in order for this result to apply to a binary mixture we shall assume that the molecules of a single specie diffuse into others of the same specie (so-called self-diffusion). Of course, if all the molecules were exactly alike, there would be no way experimentally to identify the diffusion process. However, the diffusion of molecules that are isotopes of the same element and the diffusion of tracer molecules are practical examples of the self-diffusion process.

Let n_1 denote the molal concentration of the diffusing molecules. The property that is transferred by one of these molecules in mass transfer is the molecule itself, so that $P = n_1/n$ and $(p/A) = N_1/A$. Then Eq. (20.10) may be written

$$\frac{N_1}{A} = -\tfrac{1}{3}\, n\bar{V}\lambda \frac{d(n_1/n)}{dy} \tag{20.18a}$$

or for the case of uniform total n,

$$\frac{N_1}{A} = -\tfrac{1}{3}\, \bar{V}\lambda \frac{dn_1}{dy} \tag{20.18b}$$

Comparison of Eq. (20.18b) with Eq. (14.5) gives

$$D = \tfrac{1}{3}\bar{V}\lambda \tag{20.19}$$

The coefficient D given by Eq. (20.19) is called the *coefficient of self-diffusion.*

If we divide Eq. (20.12) by (20.19), we obtain

$$\frac{\mu}{\rho D} = 1$$

Measured values of this group for the diffusion of isotopic tracer molecules yield values between 1.3 and 1.5, indicating better quantitative agreement of theory with data than in the case of heat conduction.

Assuming that the results of kinetic theory for a single gas are valid for this mixture of identical molecules, then if the distribution of velocity in the mixture is Maxwellian, Eq. (20.19) combined with Eqs. (20.3) and (20.4) yields

$$D = \frac{2}{3\pi^{3/2}d^2n}\left(\frac{\kappa T}{m}\right)^{1/2} \tag{20.20}$$

20.3 Summary of Theoretical and Semi-empirical Equations of Transport Properties

Viscosity. The simple model considered earlier gave the viscosity of a Maxwellian gas as

$$\mu = \frac{2}{3\pi^{3/2}} \frac{(m\kappa T)^{1/2}}{d^2} \tag{20.13}$$

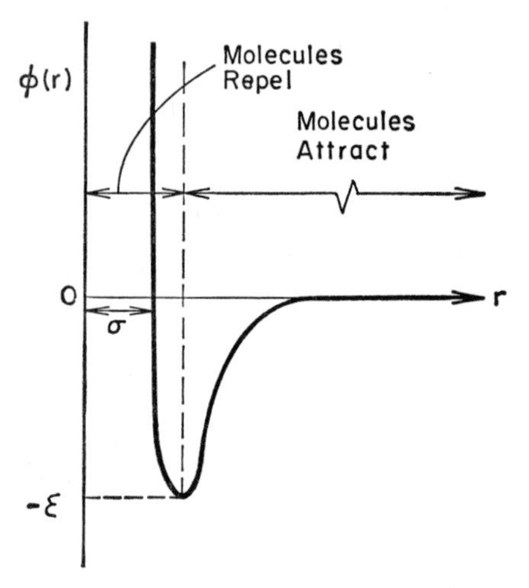

Fig. 20.5. Lennard-Jones potential function.

The result gives a qualitatively correct picture, but is quantitatively inaccurate when applied to real gases. The modern theories of transport properties give expressions for the properties in terms of the forces of attraction and repulsion between molecules instead of the hard-elastic-sphere model. These theories are reviewed in reference (5). A fairly successful semi-empirical function for such forces is the Lennard-Jones potential:

$$\phi(r) = 4\epsilon\left[\left(\frac{\sigma}{r}\right)^{12} - \left(\frac{\sigma}{r}\right)^{6}\right] \tag{20.21}$$

where $\phi(r)$ is the potential energy of interaction between molecules, r is the molecular separation distance, and σ and ϵ are constants characteristic of the particular molecular specie. Equation (20.21) is shown plotted in Fig. 20.5, and values of σ and ϵ for some gases are listed in Table 20.3 (5).

TABLE 20.3 LENNARD-JONES CONSTANTS OF SOME GASES, CALCULATED FROM VISCOSITY DATA

Gas	Formula	ϵ/κ, °K	σ, Å
Air		97	3.617
Argon	A	124	3.418
Benzene	C_6H_6	440	5.270
Carbon dioxide	CO_2	190	3.996
Carbon tetrachloride	CCl_4	327	5.881
Helium	He	10.22	2.576
Hydrogen	H_2	33.3	2.968
Methane	CH_4	136.5	3.822
Mercury	Hg	851	2.898
Nitrogen	N_2	91.5	3.681
Oxygen	O_2	113	3.433
n-Pentane	n-C_5H_{12}	345	5.769

Source: Hirschfelder, Curtiss, and Bird (5).

The Chapman-Enskog theory (2) using the Lennard-Jones potential yields the following equation for the viscosity of a gas at low pressures or densities:

$$\mu = 2.6693 \times 10^{-5} \frac{\sqrt{MT}}{\sigma^2 \Omega_\mu} \tag{20.22}$$

Here μ is in poise, T is in °K, and σ is the collision diameter in Å. The quantity Ω_μ is the so-called *collision integral* of the Chapman-Enskog theory; it has the value of unity for the hypothetical rigid-sphere molecule of diameter σ. Values of Ω_μ are listed in Table 20.4 (5) as a function of a dimensionless temperature parameter $\kappa T/\epsilon$. The viscosity calculated by this method agrees reasonably well with existing experimental data on monatomic and polyatomic gases over the temperature range 100 K to 1500 K. The values predicted by Eq. (20.22) are less reliable for polar molecules such as water vapor and highly elongated molecules such as n-heptane.

The use of Eq. (20.22) to calculate μ requires knowing σ and ϵ/κ, which in turn are generally calculated from viscosity measurements. A modified form of Eq. (20.22) suggested by Bromley and Wilke (6) from semi-empirical considerations may also be used:

$$\mu = 3.33 \times 10^{-5} \frac{(MT_c)^{1/2} f_1(\kappa T/\epsilon)}{v_c^{2/3}} \tag{20.23}$$

where
μ is in poises
T_c is the critical temperature, °K
v_c is the critical volume, cm^3/g mole
$f_1(\kappa T/\epsilon)$ is an empirical function tabulated in Table (20.4).

TABLE 20.4 VALUES OF Ω-INTEGRAL FOR VISCOSITY, THERMAL CONDUCTIVITY AND DIFFUSION COEFFICIENT AND OF THE VISCOSITY FUNCTION $f_1(\kappa T/\epsilon)$, BASED ON THE LENNARD-JONES POTENTIAL

$\kappa T/\epsilon$	Ω_μ or Ω_k	Ω_D	$f_1(\kappa T/\epsilon)$
0.3	2.785	2.662	0.1969
0.4	2.492	2.318	0.2540
0.5	2.257	2.066	0.3134
0.6	2.065	1.877	0.3751
0.7	1.908	1.729	0.4384
0.8	1.780	1.612	0.5025
0.9	1.675	1.517	0.5666
1.0	1.587	1.439	0.6302
2.0	1.175	1.075	1.2048
4.0	0.9700	0.8836	2.0719
6.0	0.8963	0.8124	2.751
8.0	0.8538	0.7712	3.337
10	0.8242	0.7424	3.866
20	0.7432	0.6640	6.063
40	0.6718	0.5960	9.488
60	0.6335	0.5596	12.324
80	0.6076	0.5352	14.839
100	0.5882	0.5130	17.137
200	0.5320	0.4644	26.80
400	0.4811	0.4170	41.90

Source: Hirschfelder, Curtiss, and Bird (5), and Bromley and Wilke (6).

To calculate the viscosity of a binary mixture, Wilke (7) suggests the following approximate equation:

$$\mu_{\text{mix}} = \frac{\mu_1}{1 + (x_2/x_1)\phi_{12}} + \frac{\mu_2}{1 + (x_1/x_2)\phi_{21}} \tag{20.24}$$

where x_1, x_2 are mole fractions of the components in the mixture and

$$\phi_{12} = \frac{[1 + (\mu_1/\mu_2)^{1/2}(M_2/M_1)^{1/4}]^2}{2\sqrt{2}(1 + M_1/M_2)^{1/2}}$$

$$\phi_{21} = \frac{[1 + (\mu_2/\mu_1)^{1/2}(M_1/M_2)^{1/4}]^2}{2\sqrt{2}(1 + M_2/M_1)^{1/2}}$$

The temperature dependence of viscosity predicted by the simple theory, Eq. (20.13), is too weak. Many semi-empirical expressions are known for predicting the variation of viscosity with temperature; a typical one is the Sutherland (3) equation,

$$\mu = \frac{a\sqrt{T}}{1 + S/T} \tag{20.25}$$

where a is an empirical constant and S is known as the Sutherland constant.

Thermal conductivity. The simple theory gave for the thermal conductivity of a Maxwellian gas

$$k = \frac{1}{\pi^{3/2} d^2}\left(\frac{\kappa^3 T}{m}\right)^{1/2} \tag{20.16}$$

Again, the result is only qualitatively accurate. The Chapman-Enskog theory (2) applied to monatomic gases with no internal degrees of freedom, using the Lennard-Jones potential, gives

$$k = \frac{1.9891 \times 10^{-4}}{\sigma^2 \Omega_k} \sqrt{\frac{T}{M}} \tag{20.26}$$

where k is in cal/sec cm C, T is in °K, σ is in Å, and Ω_k-integral is identical with the Ω_μ-integral listed in Table 20-4.

The relation between k and μ for monatomic gases, according to the Chapman-Enskog theory, is

$$\frac{k}{\mu c_v} = \frac{5}{2}$$

as compared with the value of unity given by the simple theory, Eq. (20.17). This value agrees well with measured values for monatomic gases given in Table 20.2.

In order to extend the rigorous theory to polyatomic molecules, the rotational and the vibrational energies of the molecules must be considered in addition to the kinetic energy of translation. Proper allowance for the distribution of energy among the several modes must be made. A relatively simple expression for k of polyatomic gases was developed by Eucken (8) semi-empirically:

$$k_{\text{poly}} = k_{\text{mon}}\left(\frac{4}{15}\frac{M}{R}c_v + \frac{3}{5}\right) \tag{20.27a}$$

The quantity within brackets is the correction applied by Eucken to account for the internal degrees of freedom in polyatomic molecules. In terms of the Prandtl number, the expression above becomes

$$\text{Pr} = \frac{\mu c_p}{k} = \frac{\gamma}{\frac{9}{4}\gamma - \frac{5}{4}} \tag{20.27b}$$

where $\gamma = c_p/c_v$.

For air, Eq. (20.27b) gives Pr = 0.736. The accuracy is less satisfactory in the case of more complex gases. An improved empirical expression developed by Bromley (9) is discussed in reference (4).

The thermal conductivity of a gas mixture may be calculated by a procedure analogous to that for calculating the viscosity of a gas mixture, but the accuracy attainable is only fair. An extremely simple rule for calculating k of a mixture of nonpolar gases is due to Brokaw (10):

$$k_m = \tfrac{1}{2}(k_{sm} + k_{rm}) \tag{20.28}$$

where

$$k_{sm} \equiv x_1 k_1 + x_2 k_2 + \cdots$$

$$\frac{1}{k_{rm}} \equiv \frac{x_1}{k_1} + \frac{x_2}{k_2} + \cdots.$$

The Sutherland equation for the effect of temperature on k is similar to Eq. (20.25):

$$k = \frac{b\sqrt{T}}{1 + S/T}$$

where b is an empirical constant different in magnitude from a in Eq. (20.25).

Diffusion coefficient. The simple theory gave the coefficient of self-diffusion as

$$D_{11} = \frac{2}{3\pi^{3/2}\, d^2 n}\left(\frac{\kappa T}{m}\right)^{1/2} \tag{20.20}$$

This equation applies to a binary mixture of almost identical gases. For dissimilar gases, the result for rigid-sphere molecules undergoing elastic collisions can be expressed in the following form using the ideal-gas equation of state:

$$D_{12} = \frac{bT^{3/2}\left(\dfrac{1}{M_1} + \dfrac{1}{M_2}\right)^{1/2}}{pd^2} \tag{20.29}$$

where b is a constant, d is the center-to-center distance of two molecules upon impact, and p is the absolute pressure.

The Chapman-Enskog theory, using the Lennard-Jones potential and the ideal-gas equation of state, yields

$$D_{12} = \frac{1.858 \times 10^{-3}\, T^{3/2}\left(\dfrac{1}{M_1} + \dfrac{1}{M_2}\right)^{1/2}}{p\sigma_{12}^2 \Omega_D} \tag{20.30}$$

where D_{12} is in cm²/sec, T is in °K, p is in atm, $\sigma_{12} = \frac{1}{2}(\sigma_1 + \sigma_2)$ is in Angstrom units, and values of Ω_D-integral are listed in Table 20.4.

The foregoing equations are the basis for the Gilliland equation (14.9) given in Chap. 14, in which the constant b was determined empirically from hundreds of experimental data points. Other methods of estimating the diffusion coefficients are reviewed in detail by Reid and Sherwood (4).

REFERENCES

1. Sears, F. W., *Thermodynamics*, Chaps. 11–13, Addison-Wesley, 1959.

2. Chapman, S., and T. G. Cowling, *Mathematical Theory of Non-Uniform Gases*, 2nd Ed., Cambridge University Press, London, 1951.

3. Sutherland, W., *Phil. Mag.*, **36,** 507 (1893).

4. Reid, R. C., and T. K. Sherwood, *The Properties of Gases and Liquids*, McGraw-Hill, 1958.

5. Hirschfelder, J. O., C. F. Curtiss, and R. B. Bird, *Molecular Theory of Gases and Liquids*, Wiley, 1954.

6. Bromley, L. A., and C. R. Wilke, *Ind. Eng. Chem.*, **43,** 1641 (1951).

7. Wilke, C. R., *J. Chem. Phys.* **18,** 517 (1950).

8. Eucken, A., *Forsch. Gebiete Ingenieurw.*, **11B,** 6 (1940).

9. Bromley, L. A., *Univ. of Calif. Radiation Lab Report*, UCRL-1852, Berkeley, Calif., June 1952.

10. Brokaw, R. S., *Ind. Eng. Chem.*, **47,** 2398 (1955).

11. Kennard, E. H., *Kinetic Theory of Gases*, McGraw-Hill, 1938.

APPENDIX A

Two-Dimensional Steady State Diffusion

SOLUTION OF $\partial^2 P/\partial x^2 + \partial^2 P/\partial y^2 = 0$

Consider the following boundary conditions for $P(x, y)$:

$$P(0, y) = 0, \quad P(L, y) = 0, \quad P(x, \infty) = 0, \quad P(x, 0) = P_0, \quad \text{uniform}$$

Assume the solution of the form

$$P = X \cdot Y \tag{A.1}$$

where X is a function of x alone and Y a function of y alone. This substituted into the differential equation yields

$$\frac{1}{Y}\frac{d^2Y}{dy^2} = -\frac{1}{X}\frac{d^2X}{dx^2} \tag{A.2}$$

Each side of this equation is a constant, because the left side is a function of y alone and the right side of x alone. Let this constant be a positive quantity, λ^2. Then the solution of the original partial differential equation is reduced to the solution of the following two ordinary differential equations:

$$\frac{d^2X}{dx^2} + \lambda^2 X = 0$$

$$\frac{d^2Y}{dy^2} - \lambda^2 Y = 0 \tag{A.3}$$

The solution of the first could be sines and cosines, and of the second, exponentials; hence, with Eq. (A.1)

$$P = (K_1 \sin \lambda x + K_2 \cos \lambda x)(K_3 e^{\lambda y} + K_4 e^{-\lambda y}) \tag{A.4}$$

This shows why λ^2 was taken as positive. If Eq. (A.2) had been set equal to $-\lambda^2$, the solution would have been sines and cosines in y and exponentials in x. The boundary conditions would not be satisfied by this choice. If the constant were taken to be zero, the solutions would be $(ax + b)(cy + d)$ which again cannot satisfy these boundary conditions.

From boundary condition $P(0, y) = 0$, $K_2 = 0$ since $\cos 0 = 1$ and since the exponential terms in y cannot both be zero over all of y. From $P(x, \infty) = 0$, $K_3 = 0$ since $e^{\infty} = \infty$. Equation (A.4) then reduces to

$$P = Ke^{-\lambda y} \sin \lambda x \tag{A.5}$$

With $P(L, y) = 0$, this reduces to $\sin \lambda L = 0$ since $e^{-\lambda y}$ and K are not zero. This requires $\lambda_n = n\pi/L$ where $n = 0, 1, 2, 3, \cdots$. Then because the differential equation (A.5) is linear, these solutions may be added as follows:

$$P = \sum_{n=0}^{\infty} K_n e^{-n\pi y/L} \sin \frac{n\pi x}{L} \tag{A.6}$$

The equation $\sin \lambda_n L = 0$ is called the *eigenfunction* and λ_n, the *eigenvalues*.

For boundary conditions $P(x, 0) = P_0$, Eq. (A.6) becomes

$$P_0 = \sum_{n=0}^{\infty} K_n \sin \frac{n\pi x}{L} \tag{A.7}$$

To determine K_n, multiply both sides of this equation by $\sin m\pi x/L$ where m is one particular integer value of n; then integrate between $x = 0$ and L.

$$P_0 \int_{x/L=0}^{1} \sin \frac{m\pi x}{L}\, d\left(\frac{x}{L}\right) = \int_{x/L=0}^{1} \sum_{n=0}^{\infty} K_n \sin \frac{n\pi x}{L} \sin \frac{m\pi x}{L}\, d\left(\frac{x}{L}\right)$$

The integral on the left is $2/n\pi$ where $n = 1, 3, 5, \cdots$, odd. The integral on the right is zero when $m \neq n$ and when $m = n$ is $K_n/2$; therefore

$$K_n = \frac{4P_0}{n\pi}, \quad n \text{ odd} \tag{A.8}$$

The final solution is

$$\frac{P}{P_0} = \sum_{\substack{n=1 \\ n,\ \text{odd}}}^{\infty} \frac{4}{n\pi} e^{-n\pi y/L} \sin \frac{n\pi x}{L} \tag{A.9}$$

The solution obtained here applies to diffusion process in general, e.g. heat conduction in a semi-infinite slab, Sec. 6.10, mass transfer such as vapor through a stagnant gas or even diffusion of an electron "gas" in an electrically conducting medium, provided that the boundary conditions are satisfied.

APPENDIX B

One-Dimensional Transient Diffusion: Slab

SOLUTION OF $\partial^2 P/\partial x^2 = (1/\mathfrak{D})\, \partial P/\partial t$

Consider the following boundary and initial conditions for $P(x, t)$:

$$\text{(a)}\ P(x, 0) = P_i, \qquad \text{(b)}\ \frac{\partial P(0, t)}{\partial x} = 0, \qquad \text{(c)}\ \frac{\partial P(r_0, t)}{\partial x} = -NP(r_0, t)$$

where N is a constant.

Following the procedure of Appendix A, we assume a product solution of the form $P = X(x) \cdot \Theta(t)$. Then the differential equation becomes

$$\frac{1}{X}\frac{d^2X}{dx^2} = \frac{1}{\mathfrak{D}\Theta}\frac{d\Theta}{dt} = -\lambda^2, \quad \text{a constant}$$

or

$$\frac{d^2X}{dx^2} + \lambda^2 X = 0 \tag{B.1}$$

$$\frac{d\Theta}{dt} + \mathfrak{D}\lambda^2\Theta = 0 \tag{B.2}$$

The product of the solutions of these two equations is

$$P = \exp(-\lambda^2 \mathfrak{D} t)(K_1 \cos \lambda x + K_2 \sin \lambda x) \tag{B.3}$$

Also

$$\frac{\partial P}{\partial x} = \exp(-\lambda^2 \mathfrak{D} t)(K_2 \lambda \cos \lambda x - K_1 \lambda \sin \lambda x) \tag{B.4}$$

To satisfy boundary condition (b), $\partial P(0, t)/\partial x = 0$, K_2 from Eq. (B.4) must be zero. Then with $K_2 = 0$, substitute Eq. (B.3) and (B.4) into boundary condition (c) with the following result:

$$-\exp(-\lambda^2 \mathfrak{D} t) K \lambda \sin \lambda r_0 = -N \exp(-\lambda^2 \mathfrak{D} t) K \cos \lambda r_0$$

or simplifying,

$$\cot \lambda_n r_0 = \frac{1}{N r_0} (\lambda_n r_0) \tag{B.5}$$

where subscript n on λ indicates the infinite number of roots.

Equation (B.5) is best solved graphically as in Fig. B.1.

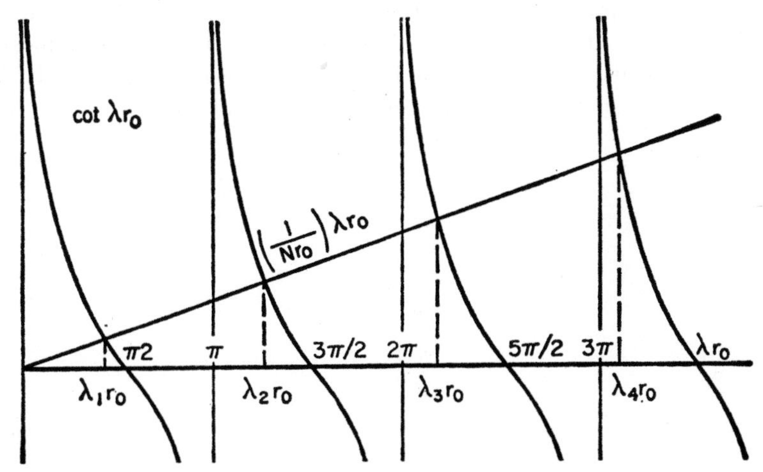

Fig. B.1. Graphical solution of Eq. (B.5).

Since the differential equation is linear, this infinite number of solutions may be added. Then from Eq. (B.3), with $K_2 = 0$ and $\lambda = \lambda_n$ determined from Eq. (B.5),

$$P = \sum_{n=1}^{\infty} K_n \exp(-\lambda_n^2 \mathfrak{D} t) \cos \lambda_n x \tag{B.6}$$

Substitute this into $P(x, 0) = P_i$, multiply both sides of the resulting equation by $\cos \lambda_m x \, dx$, and integrate between $x = 0$ and r_0. Then

$$P_i \int_0^{r_0} \cos \lambda_m x \, dx = \int_0^{r_0} \sum_{n=1}^{\infty} K_n \cos \lambda_n x \cos \lambda_m x \, dx$$

When $m \neq n$, the integral on the right is zero and when $m = n$ it is

$$K_n\left[\frac{r_0}{2} + \frac{1}{2\lambda_n} \sin \lambda_n r_0 \cos \lambda_n r_0\right]$$

The integral on the right is $(1/\lambda_n) \sin \lambda_n r_0$; so

$$K_n = \frac{2P_i \sin \lambda_n r_0}{\lambda_n r_0 + \sin \lambda_n r_0 \cos \lambda_n r_0} \tag{B.7}$$

The final solution is then

$$\frac{P}{P_i} = 2 \sum_{n=1}^{\infty} \left(\frac{\sin \lambda_n r_0}{\lambda_n r_0 + \sin \lambda_n r_0 \cos \lambda_n r_0}\right) \exp\,(-\lambda_n^2 \mathfrak{D} t) \cos \lambda_n x \tag{B.8}$$

This solution applies to diffusion processes in general for the particular boundary conditions. For heat and mass transfer, its application is discussed in Secs. 6.13 and 15.4. It also applies to momentum transfer if $N = \infty$ or $P(r_0, t) = 0$. This corresponds to two infinite parallel plates in a stagnant fluid. If the plates are suddenly placed into motion in, say, the y-direction, the solution modified for $P(r_0, t) = 0$ determines the transient velocity distribution in the fluid between the plates. If N is finite, this corresponds to the case of slip-flow at the wall.

APPENDIX C

One-Dimensional Transient Diffusion: Infinite Bodies

SOLUTION OF $\partial^2 P/\partial x^2 = (1/\mathfrak{D})\, \partial P/\partial t$

Consider the following boundary conditions for $P(x, t)$:

$$P(x, 0) = f(x), \qquad P(\infty, t) \neq \infty, \quad \text{and} \quad P(-\infty, t) \neq \infty$$

If m is a particular value of x and $f(m)$ the corresponding value of $f(x)$, a solution of the differential equation is

$$P(x, t) = \frac{f(m)}{2\sqrt{\pi \mathfrak{D} t}} \exp\left(-\frac{(x-m)^2}{4\mathfrak{D}t}\right) \tag{C.1}$$

Since m may take on any magnitude and since the differential equation is linear, the following addition of an infinity of solutions is itself a solution:

$$P(x, t) = \frac{1}{2\sqrt{\pi \mathfrak{D} t}} \int_{m=-\infty}^{\infty} f(m) \exp\left[-\frac{(x-m)^2}{4\mathfrak{D}t}\right] dm \tag{C.2}$$

Equation (C.2) expresses P as a function of x and t. Here m is merely a variable of integration.

By introducing a new variable β in place of m,

$$\beta \equiv \frac{(m - x)}{2\sqrt{\mathfrak{D}t}} \tag{C.3}$$

Equation (C.2) can be written as follows:

$$P = \frac{1}{\sqrt{\pi}} \int_{\beta=-\infty}^{\infty} f[x + 2\sqrt{\mathfrak{D}t}\,\beta]e^{-\beta^2}\,d\beta \tag{C.4}$$

which is an alternate form of the solution. This reduces to $P = f(x)$ at $t = 0$ since

$$\frac{1}{\sqrt{\pi}} \int_{-\infty}^{\infty} e^{-\beta^2}\,d\beta = \text{unity}$$

Example C.1:

$$f(x) = 0, \qquad x < a$$

$$f(x) = P_i, \text{ constant}, \qquad a \leq x \leq b$$

$$f(x) = 0, \qquad x > b$$

Then Eq. (C.2) becomes

$$\frac{P}{P_i} = \frac{1}{2\sqrt{\pi \mathfrak{D}t}} \int_{m=a}^{b} \exp\left[-\frac{(x - m)^2}{4\mathfrak{D}t}\right] dm \tag{C.5}$$

since the integral is zero in the region where $f(x) = 0$. Equation (C.5) may also be written in the following form:

$$\frac{P}{P_i} = \frac{1}{\sqrt{\pi}} \int_{\beta=(a-x)/2\sqrt{\mathfrak{D}t}}^{(b-x)/2\sqrt{\mathfrak{D}t}} e^{-\beta^2}\,d\beta \tag{C.6}$$

or

$$\frac{P}{P_i} = \frac{1}{2}\left[\operatorname{erf}\left(\frac{b - x}{2\sqrt{\mathfrak{D}t}}\right) - \operatorname{erf}\left(\frac{a - x}{2\sqrt{\mathfrak{D}t}}\right)\right] \tag{C.7}$$

where

$$\operatorname{erf} w \equiv \frac{2}{\sqrt{\pi}} \int_{\beta=0}^{w} e^{-\beta^2}\,d\beta \tag{C.8}$$

TABLE C.1 VALUES OF GAUSS' ERROR INTEGRAL

w	erf w	w	erf w	w	erf w
0.00	0.0000	0.4	0.4284	1.3	0.9340
0.01	0.0113	0.5	0.5205	1.4	0.9523
0.02	0.0226	0.6	0.6039	1.5	0.9661
0.04	0.0451	0.7	0.6778	1.6	0.9763
0.06	0.0676	0.8	0.7421	1.8	0.9891
0.08	0.0901	0.9	0.7969	2.0	0.9953
0.10	0.1125	1.0	0.8427	2.2	0.9981
0.20	0.2227	1.1	0.8802	2.5	0.9996
0.30	0.3286	1.2	0.9103	3.0	1.0000

Example C.2: Semi-infinite body

$$P(x, 0) = P_i, \qquad x > 0$$

$$P(0, t) = 0$$

This is identical to writing Eq. (C.2) or (C.4) with $f(x)$ having odd symmetry about $x = 0$ or

$$P(x, 0) = P_i, \qquad x > 0$$

$$P(x, 0) = -P_i, \qquad x < 0$$

Then Eq. (C.2) may be written as

$$P(x, t) = \frac{1}{2\sqrt{\pi \mathfrak{D} t}} \left\{ \int_{m=-\infty}^{0} -P_i \exp\left[-\frac{(x - m)^2}{4\mathfrak{D}t} \right] dm + \int_{m=0}^{\infty} P_i \exp\left[-\frac{(x - m)^2}{4\mathfrak{D}t} \right] dm \right\} \qquad \text{(C.9)}$$

In the first integral change m to $-m$ and dm to $-dm$ and integrate from $+\infty$ to 0; then change the sign and invert the limits of integration to get

$$\frac{P}{P_i} = \frac{1}{\sqrt{\pi}} \int_{\beta=-x/2\sqrt{\mathfrak{D}t}}^{\infty} e^{-\beta^2}\, d\beta + \int_{\gamma=\infty}^{+x/2\sqrt{\mathfrak{D}t}} e^{-\gamma^2}\, d\gamma$$

$$= \frac{1}{\sqrt{\pi}} \int_{-x/2\sqrt{\mathfrak{D}t}}^{+x/2\sqrt{\mathfrak{D}t}} e^{-\phi^2}\, d\phi = \operatorname{erf}\left(\frac{x}{2\sqrt{\mathfrak{D}t}} \right) \qquad \text{(C.10)}$$

where $\beta \equiv (m - x)/2\sqrt{\mathfrak{D}t}$ and $\gamma = (m + x)/2\sqrt{\mathfrak{D}t}$

These solutions are equally applicable to heat transfer, Sec. 6.15, mass transfer, Sec. 15.3, and momentum transfer, Sec. 3.6.

APPENDIX D

Summary of Vector Notation

Any vector **A** can be represented uniquely by

$$\mathbf{A} = A_x\mathbf{i} + A_y\mathbf{j} + A_z\mathbf{k} \qquad \text{(D.1)}$$

where A_x, A_y, A_z are the scalar components of the vector and **i**, **j**, and **k** are the unit vectors in the x-, y-, and z-directions. For example, the velocity vector **A** may be represented by $\mathbf{V} = v_x\mathbf{i} + v_y\mathbf{j} + v_z\mathbf{k}$.

The *scalar* or *dot product* of two vectors **A** and **B** is the scalar quantity that represents the product of the length of one of the vectors by the projected length of the other vector upon the first.

$$\mathbf{A}\cdot\mathbf{B} = AB \cos\,(A, B) \qquad \text{(D.2a)}$$

In terms of the Cartesian components of the two vectors

$$\mathbf{A}\cdot\mathbf{B} = A_xB_x + A_yB_y + A_zB_z \qquad \text{(D.2b)}$$

The *vector* or *cross product* of two vectors **A** and **B** is a vector **C** which is normal to the plane of the vectors **A** and **B** and so directed that the vectors **A, B, C** form a right-handed system. The vector product may be conveniently expressed in terms of the Cartesian components of the vectors **A** and **B** in determinant form:

$$\mathbf{A} \times \mathbf{B} = \begin{vmatrix} \mathbf{i} & \mathbf{j} & \mathbf{k} \\ A_x & A_y & A_z \\ B_x & B_y & B_z \end{vmatrix} \tag{D.3}$$

The *gradient* of a scalar function ϕ is

$$\text{grad } \phi = \nabla\phi = \mathbf{i}\frac{\partial\phi}{\partial x} + \mathbf{j}\frac{\partial\phi}{\partial y} + \mathbf{k}\frac{\partial\phi}{\partial z} \tag{D.4}$$

where the differential operator "del" is defined by

$$\nabla \equiv \mathbf{i}\frac{\partial}{\partial x} + \mathbf{j}\frac{\partial}{\partial y} + \mathbf{k}\frac{\partial}{\partial z} \tag{D.5}$$

The *divergence* and *curl* of a vector A are given respectively by

$$\text{div } \mathbf{A} = \nabla \cdot \mathbf{A} = \frac{\partial A_x}{\partial x} + \frac{\partial A_y}{\partial y} + \frac{\partial A_z}{\partial z} \tag{D.6}$$

$$\text{curl } \mathbf{A} = \nabla \times \mathbf{A} = \begin{vmatrix} \mathbf{i} & \mathbf{j} & \mathbf{k} \\ \frac{\partial}{\partial x} & \frac{\partial}{\partial y} & \frac{\partial}{\partial z} \\ A_x & A_y & A_z \end{vmatrix} \tag{D.7}$$

The divergence of grad ϕ is a second-order differential quantity that occurs frequently in the study of transfer processes and can be expressed as follows:

$$\text{div (grad } \phi) = \nabla \cdot \nabla\phi = \nabla^2\phi = \frac{\partial^2\phi}{\partial x^2} + \frac{\partial^2\phi}{\partial y^2} + \frac{\partial^2\phi}{\partial z^2} \tag{D.8}$$

where the Laplacian operator ∇^2 is defined by

$$\nabla^2 \equiv \frac{\partial^2}{\partial x^2} + \frac{\partial^2}{\partial y^2} + \frac{\partial^2}{\partial z^2} \tag{D.9}$$

Other second-order differential quantities can be expressed as follows:

$$\text{div curl } \mathbf{A} = \nabla \cdot \nabla \times \mathbf{A} = 0 \tag{D.10}$$

$$\text{curl grad } \phi = \nabla \times \nabla\phi = 0 \tag{D.11}$$

$$\text{curl curl } \mathbf{A} = \nabla \times \nabla \times \mathbf{A} = \text{grad div } \mathbf{A} - \nabla^2\mathbf{A} \tag{D.12}$$

Note that we can perform the gradient of a scalar function and the divergence and curl of a vector function, but the operator ∇^2 can be applied both to a scalar quantity such as a temperature and to a vector quantity such as a velocity. Equation (D.12) may be taken as the defining equation for the Laplacian of a vector quantity.

The substantial derivative D/Dt introduced in Chap. 1 may be operated on a scalar function ϕ as well as a vector function $\mathbf{A}$. Its definition in the vector notation follows:

$$\frac{D\phi}{Dt} = \frac{\partial \phi}{\partial t} + \mathbf{V} \cdot \nabla \phi \tag{D.13}$$

$$\frac{D\mathbf{A}}{Dt} = \frac{\partial \mathbf{A}}{\partial t} + (\mathbf{V} \cdot \nabla)\mathbf{A} \tag{D.14}$$

The *Stokes theorem* states that in a surface S bounded by a closed curve L, if $\mathbf{A}$ is a continuous vector function on that surface, the line integral of $\mathbf{A}$ equals the surface integral of curl $\mathbf{A}$:

$$\oint_L \mathbf{A} \cdot d\mathbf{L} = \iint_S (\nabla \times \mathbf{A}) \cdot \mathbf{n}\, dS \tag{D.15}$$

where $\mathbf{n}$ is the unit normal vector at any point on the surface.

The *Gauss theorem* states that in a volume V bounded by a surface S, if $\mathbf{A}$ is a continuous vector function in that volume, the surface integral of $\mathbf{A}$ equals the volume integral of div $\mathbf{A}$:

$$\iint_S \mathbf{A} \cdot \mathbf{n}\, dS = \iiint_V (\nabla \cdot \mathbf{A})\, dV \tag{D.16}$$

To illustrate an application of the Gauss theorem, the continuity equation, Eq. (1.11a), may be derived as follows. Consider a control volume V bounded by a surface S. The continuity equation is written by equating the rate of accumulation of mass inside V to the net mass flux entering across the surface S. The rate of accumulation of mass in the volume V is

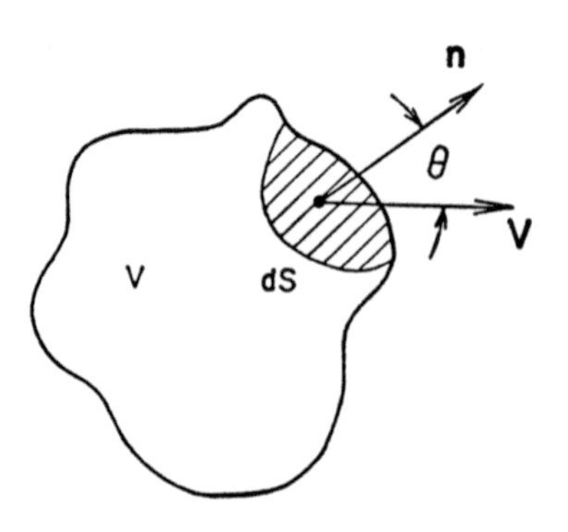

Figure D.1

$$\frac{\partial}{\partial t} \iiint_V \rho\, dV$$

The net rate of outflow across the surface S is

$$\iint_S \rho \mathbf{V} \cdot \mathbf{n}\, dS$$

The conservation of mass requires

$$\frac{\partial}{\partial t}\iiint_V \rho \, dV + \iint_S \rho\mathbf{V}\cdot\mathbf{n} \, dS = 0 \tag{D.17}$$

By the Gauss theorem

$$\iint_S \rho\mathbf{V}\cdot\mathbf{n} \, dS = \iiint_V \text{div } (\rho\mathbf{V}) \, dV$$

Therefore, Eq. (D.17) becomes

$$\frac{\partial}{\partial t}\iiint_V \rho \, dV + \iiint_V \text{div } (\rho\mathbf{V}) \, dV = 0 \tag{D.18}$$

This equation is valid no matter how small we take the control volume; we can, therefore, drop the integral sign in Eq. (D.18) and write:

$$\frac{\partial \rho}{\partial t} + \text{div } (\rho\mathbf{V}) = 0$$

APPENDIX E

Thermophysical Properties

TABLE E.1 PROPERTY VALUES OF SOME SOLIDS

(1) Metals and Alloys	ρ (lb_m/ft^3) (68 F)	c_p (Btu/lb_m F) (68 F)	k (Btu/hr ft F) (68 F)	(212 F)	(1112 F)	α (ft^2/hr) (68 F)
Aluminum, pure	169	0.214	118	119		3.665
Brass (70%, Cu, 30% Zn)	532	0.092	64	74		1.322
Constantin (60% Cu, 40% Ni)	557	0.098	13	13		0.237
Copper, pure	559	0.0915	223	219	204	4.353
Iron, pure	493	0.108	42	39	23	0.785
cast (C ~ 4%)	454	0.10	30			0.666
wrought (C < 0.5%)	490	0.11	34	33	21	0.634
Lead, pure	710	0.031	20	19.3		0.924
Magnesium, pure	109	0.242	99	97		3.762
Molybdenum	638	0.060	71	68	61	2.074
Nickel, pure (99.9%)	556	0.1065	52	48		0.882
impure (99.2%)	556	0.106	40	37	32	0.677
Silver, pure	657	0.056	242	240		6.601
Steel, mild, 1% C	487	0.113	25	25	19	0.452
Stainless steel (18 Cr, 8 Ni)	488	0.11	9.4	10	13	0.172
Tin, pure	456	0.054	37	34		1.505
Tungsten	1208	0.032	94	87	65	2.430
Zinc, pure	446	0.092	64.8	63		1.591

TABLE E.1—Continued

(2) Nonmetals	T (°F)	ρ (lb_m/ft^3)	c_p (Btu/lb_m F)	k (Btu/hr ft F)	α (ft^2/hr)
Aerogel, silica	100	5.3	0.205	0.013	0.012
Asbestos	32	36	0.25	0.087	0.010
	800	36		0.130	
Brick, common	68	100	0.20	0.20–0.10	0.01–0.02
fire clay	1472	145	0.23	0.79	0.024
Bakelite	68	79.5	0.38	0.134	0.0044
Concrete	68	119–144	0.21	0.47–0.81	0.019–0.027
Corkboard	100	10	0.4	0.025	0.006
Diatomaceous earth, powdered	100	14	0.21	0.030	0.01
Fiber insulating board	100	14.3		0.024	
Glass, window	68	162	0.16	0.51	0.020
Glass wool, fine	100	1.5		0.031	
packed	100	6.0		0.022	
Ice	32	57	0.46	1.28	0.048
Magnesia, 85%	100	17		0.039	
Marble	68	156–169	0.193	1.6	0.054
Paper				0.075	
Rock wool	100	12		0.023	
Rubber, hard	32	74.8	0.48	0.087	0.0024
Wood, oak, ⊥ to grain	70	51	0.57	0.12	0.004
Wood, oak, ‖ to grain	70	51	0.57	0.23	0.0069

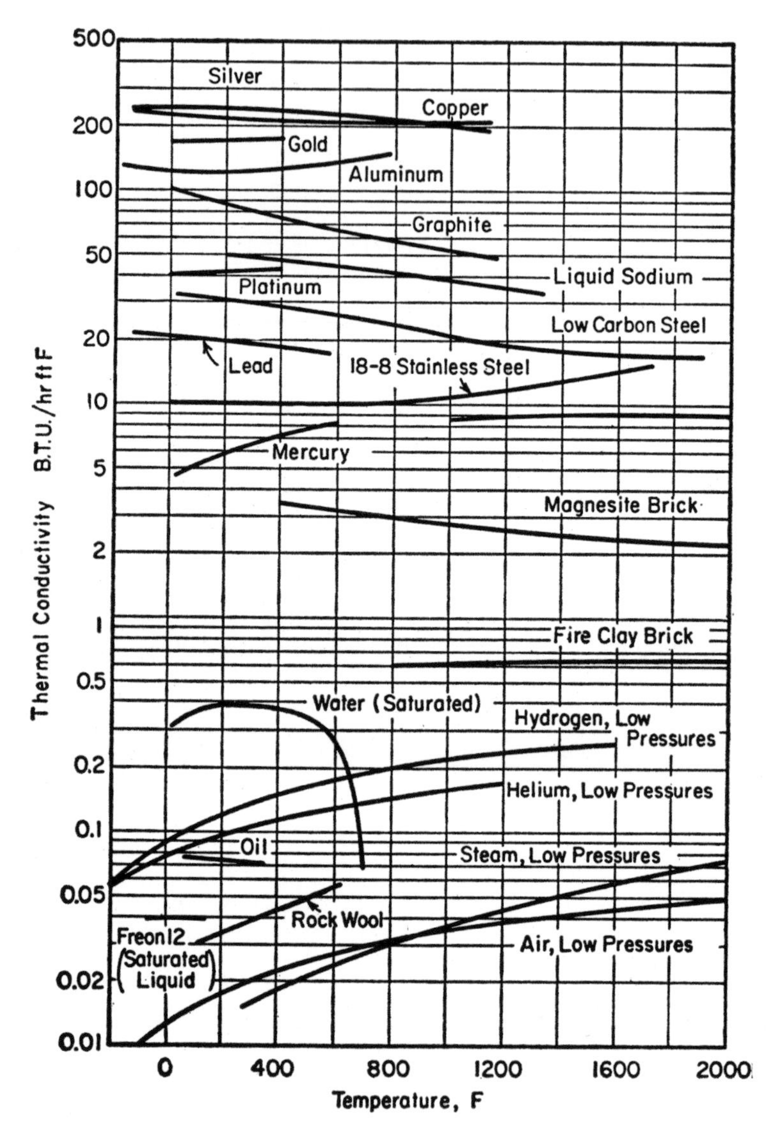

500
200
100
50
20
10
5
2
1
0.5
0.2
0.1
0.05
0.02
0.01
Thermal Conductivity B.T.U./hr ft F
0
400
800
1200
1600
2000
Temperature, F
Silver
Copper
Gold
Aluminum
Graphite
Liquid Sodium
Platinum
Low Carbon Steel
Lead
18-8 Stainless Steel
Mercury
Magnesite Brick
Fire Clay Brick
Water (Saturated)
Hydrogen, Low Pressures
Helium, Low Pressures
Oil
Steam, Low Pressures
Rock Wool
Freon 12 (Saturated Liquid)
Air, Low Pressures

Figure E.1

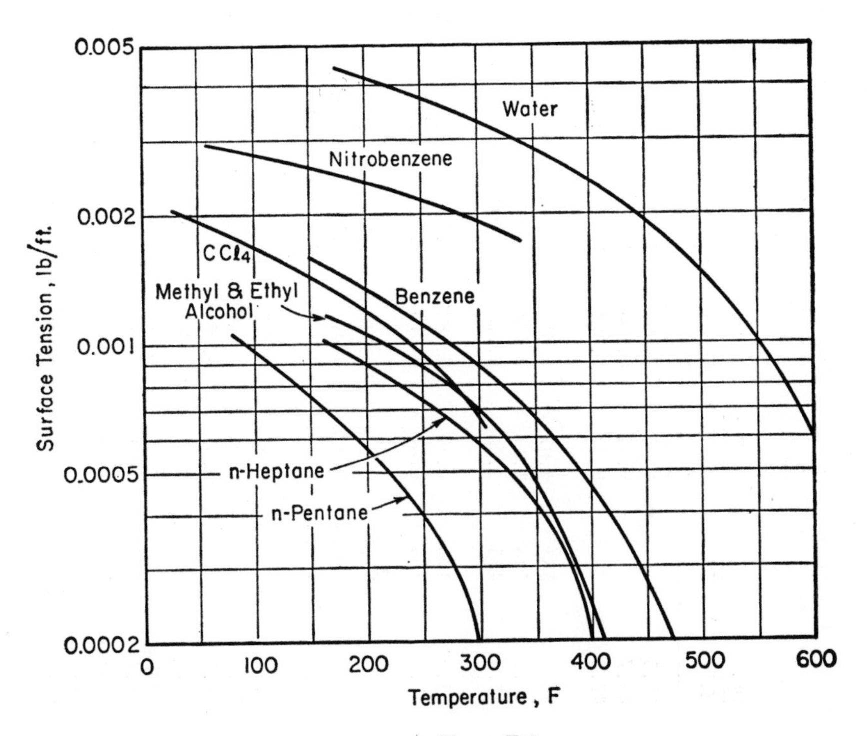

0.005
0.002
0.001
0.0005
0.0002
Surface Tension, lb/ft.
Water
Nitrobenzene
CCl_4
Methyl & Ethyl Alcohol
Benzene
n-Heptane
n-Pentane
0
100
200
300
400
500
600
Temperature, F

Figure E.2

TABLE E.2 PROPERTY VALUES OF SOME LIQUIDS AT SATURATION STATE

(1) Liquids	T °F	p psia	ρ $\frac{lb_m}{ft^3}$	h_{fg} $\frac{Btu}{lb}$	c_p $\frac{Btu}{lb_m\ F}$	k $\frac{Btu}{hr\ ft\ F}$	μ $\frac{lb_m}{hr\ ft}$	ν $\frac{ft^2}{hr}$	$\alpha \times 10^3$ $\frac{ft^2}{hr}$	Pr	$\beta \times 10^3$ $\frac{1}{°R}$
n-Butyl alcohol	60		50.5		0.55	0.097	8.15	0.16	3.49	46.6	
	100		49.7		0.61	0.096	4.65	0.094	3.16	29.5	0.45
	200		47.2		0.77	0.094	1.39	0.029	2.58	11.3	
Ammonia	−20	18.4	42.4	584	1.07	0.317	0.634	0.0150	6.94	2.15	
	32	62.2	40.0	543	1.11	0.312	0.580	0.0145	7.03	2.05	1.2
	120	286	35.2	455	1.22	0.275	0.469	0.0128	6.40	1.99	
Benzene	177	14.7	51.0	170	0.46	0.083	0.76	0.015	3.54	4.2	0.6
	256	50	47.5	155	0.51	0.076	0.51	0.011	3.14	3.4	
Freon-12 (CCl_2F_2)	−20	15.3	93.0	71.8	0.214	0.040	0.90	0.0097	2.01	4.8	1.03
	32	44.8	87.2	66.6	0.223	0.042	0.72	0.0083	2.16	3.8	1.72
	120	171.8	75.9	54.0	0.244	0.039	0.56	0.0074	2.12	3.5	
Light oil	60		57.0		0.43	0.077	210	3.68	3.14	1170	0.38
	100		56.0		0.46	0.076	55	0.98	2.95	340	0.39
	200		54.0		0.51	0.074	9.0	0.17	2.69	62	0.42
	300		51.8		0.54	0.073	3.0	0.058	2.62	22	0.45
Water	32	0.089	62.4	1076	1.01	0.319	4.33	0.070	5.07	13.4	−0.04
	60	0.26	62.3	1060	1.00	0.340	2.71	0.044	5.47	7.9	0.08
	100	0.95	62.0	1037	1.00	0.364	1.65	0.027	5.88	4.5	0.2
	200	11.5	60.1	978	1.00	0.394	0.74	0.012	6.55	1.9	0.4
	500	681	49.0	714	1.19	0.349	0.26	0.0052	5.99	0.87	1.0

(2) Liquid Metals	T	ρ	c_p	k	μ	ν	α	Pr	$\beta \times 10^4$
Bismuth	600	625	0.0345	9.5	3.88	0.0062	0.44	0.014	0.65
	1000	608	0.0369	9.0	2.68	0.0044	0.40	0.011	0.70
	1400	591	0.0393	9.0	1.95	0.0033	0.39	0.0084	
Lead	700	658	0.038	9.3	5.86	0.0089	0.37	0.024	
	1000	646	0.037	8.9	4.07	0.0063	0.37	0.017	
	1300	633		8.6					
Mercury	50	847	0.033	4.7	3.90	0.0046	0.17	0.027	1.0
	200	834	0.033	6.0	2.92	0.0035	0.22	0.016	
	600	802	0.032	8.1	2.08	0.0026	0.31	0.0084	
Potassium	300	50.4	0.19	26.0	0.907	0.018	2.7	0.0066	
	800	46.3	0.18	22.8	0.440	0.0095	2.7	0.0035	
	1300	42.1	0.18	19.1	0.328	0.0078	2.5	0.0031	
Sodium	200	58.0	0.33	49.8	1.68	0.029	2.6	0.011	1.5
	700	53.7	0.31	41.8	0.70	0.013	2.5	0.0050	
	1300	48.6	0.30	34.5	0.44	0.0091	2.4	0.0038	

TABLE E.3 PROPERTY VALUES OF SOME GASES AT ATMOSPHERIC PRESSURE

Gases	T (°F)	ρ ($\frac{lb_m}{ft^3}$)	c_p ($\frac{Btu}{lb_m F}$)	k ($\frac{Btu}{hr\,ft\,F}$)	μ ($\frac{lb_m}{hr\,ft}$)	ν ($\frac{ft^2}{hr}$)	α ($\frac{ft^2}{hr}$)	Pr	$g\beta/\nu^2$ ($\frac{1}{ft^3 F}$)
Dry air	0	0.0863	0.239	0.013	0.040	0.457	0.633	0.722	4.2×10^6
	60	0.0763	0.240	0.015	0.043	0.56	0.799	0.712	2.5
	100	0.0709	0.241	0.016	0.046	0.648	0.919	0.706	1.8
	200	0.0601	0.242	0.018	0.052	0.864	1.25	0.694	0.86
	500	0.0413	0.248	0.025	0.068	1.63	2.40	0.680	0.16
	800	0.0315	0.257	0.030	0.081	2.56	3.74	0.685	
	1500	0.0202	0.277	0.041	0.106	5.22	7.37	0.709	
Carbon dioxide	0	0.13	0.184	0.0076	0.0316	0.241	0.313	0.77	15.8×10^6
	100	0.11	0.203	0.010	0.0378	0.352	0.455	0.77	6.1
	500	0.063	0.247	0.012	0.060	0.957	1.27	0.75	0.48
	2000	0.025	0.309	0.050	0.12	4.81	6.55	0.74	7.3×10^3
Helium	−200	0.021	1.24	0.054	0.030	1.44	2.05	0.70	770×10^3
	0	0.012	1.24	0.078	0.044	3.68	5.29	0.70	70
	200	0.0083	1.24	0.098	0.056	6.71	9.49	0.71	15
	1000	0.0038	1.24	0.16	0.093	24.4	34.0	0.72	0.48
Hydrogen	0	0.0060	3.39	0.094	0.0194	3.20	4.62	0.70	87×10^3
	200	0.0042	3.44	0.122	0.0249	5.94	8.45	0.69	18
	2000	0.0011	3.76	0.307	0.059	51.8	74.2	0.72	0.060
Steam	212	0.0372	0.45	0.0145	0.0314	0.842	0.870	0.96	0.88×10^6
	500	0.0258	0.47	0.0228	0.0455	1.76	1.88	0.94	0.14
	1000	0.0169	0.51	0.0388	0.0691	4.07	4.50	0.91	1.7×10^4
	2000	0.0100	0.60	0.076	0.109	10.9	12.7	0.86	0.14

TABLE E.4 PROPERTIES OF THE STANDARD ATMOSPHERE

Altitude (ft)	Temperature (°F)	Absolute pressure (lb_f/ft^2)	Density (lb_m/ft^3)	Speed of sound (ft/sec)
0	59	2116	0.0765	1117
5,000	41	1761	.0660	1098
10,000	23	1455	.0566	1078
15,000	6	1194	.0482	1058
20,000	−12	972	.0408	1037
25,000	−30	785	.0343	1017
30,000	−48	628	.0288	995
35,000	−66	498	.0238	973
40,000	−68	392	.0189	971
45,000	−68	308	.0148	971
50,000	−68	242	.0117	971
60,000	−68	151	.0072	971
70,000	−68	94	.0045	971
80,000	−68	58	.0028	971
90,000	−68	36	.0017	971
100,000	−68	22	.0011	971
150,000	114	3	9.8×10^{-5}	1174
200,000	159	0.7	2.0×10^{-5}	1220
250,000	−8	0.1	4.8×10^{-6}	1042
500,000	450	10^{-4}	3.1×10^{-9}	...

TABLE E.5 NORMAL TOTAL EMISSIVITY OF SOME SURFACES*

Surface	Temperature (°F)†	Emissivity
A. Metals		
Aluminum:		
highly polished plate, pure	440–1070	0.04–0.06
oxidized at 1110 F	390–1110	0.11–0.19
commercial sheet	212	0.09
Brass:		
polished	100–600	0.10
oxidized at 1110 F	390–1110	0.61–0.59
Chromium, polished	100–2000	0.08–0.36
Copper:		
polished	212	0.05
plate, heated at 1110 F	390–1110	0.57
Gold, pure, highly polished	440–1160	0.02–0.03

* Selected from larger table compiled by Hottel (7).

† When temperatures and emissivities appear in pairs separated by dashes, linear interpolation is permissible.

TABLE E.5 NORMAL TOTAL EMISSIVITY OF SOME SURFACES*—Continued

Surface	Temperature (°F)†	Emissivity
Iron and Steel (excluding stainless):		
iron, polished	800–1880	0.14–0.38
cast iron, polished	392	0.21
oxidized at 1110 F	390–1110	0.64–0.78
wrought iron, polished	100–480	0.28
dull oxidized	70–680	0.94
iron plate, rusted	67	0.69
steel, polished	212	0.066
oxidized at 1110 F	390–1110	0.79
rolled sheet steel	70	0.66
steel plate, rough	100–700	0.94–0.97
Lead, gray oxidized	75	0.28
Mercury	32–212	0.09–0.12
Molybdenum filament	1340–4700	0.10–0.20
Nickel:		
polished	212	0.07
plate, oxidized at 1110 F	390–1110	0.37–0.48
Platinum:		
polished plate, pure	440–1160	0.05–0.10
wire	440–2510	0.07–0.18
Silver, pure, polished	440–1160	0.02–0.03
Stainless steel:		
polished	212	0.074
type 310, oxidized from furnace service	420–980	0.90–0.97
Tin, bright	122	0.06
Tungsten, filament, aged	80–6000	0.03–0.35
Zinc:		
commercial pure, polished	440–620	0.05
galvanized sheet	212	0.21
B. Nonmetals		
Asbestos	100–700	0.93–0.94
Brick		
red, rough	70	0.93
fire clay	1832	0.75
Carbon:		
filament	1900–2560	0.53
lampblack, rough deposit	212–932	0.84–0.78
Glass (Pyrex, lead, soda)	500–1000	0.95–0.85
Marble, light gray, polished	72	0.93
Paints, lacquers, and varnishes:		
white enamel	73	0.91
flat black lacquer	100–200	0.96–0.98
aluminum paints	212	0.27–0.67
oil paints, 16 colors	212	0.92–0.96
Porcelain, glazed	72	0.92
Quartz, opaque	570–1540	0.92–0.68
Water	32–212	0.95–0.96
Wood, oak, planed	70	0.90

*† See footnotes on page 523.

TABLE E.6

A. VALUES OF D AND Sc FOR SOME GASES IN AIR AT 32 F AND 1 ATM†

Gas	D (ft^2/hr)	Sc
Acetone	0.32*	1.60
Ammonia	0.84	0.61
Benzene	0.30	1.71
Butane	0.29*	1.77
n-Butyl alcohol	0.27	1.88
Carbon dioxide	0.53	0.96
Carbon tetrachloride	0.24*	2.13
Chlorine	0.36*	1.42
Ethane	0.42*	1.22
Ethyl alcohol	0.40	1.30
Hydrogen	2.37	0.22
Methane	0.61*	0.84
Methyl alcohol	0.51	1.00
Naphthalene	0.20	2.57
Nitrogen	0.52*	0.98
n-Octane	0.20	2.62
Oxygen	0.69	0.74
Pentane	0.26*	1.97
Toluene	0.27	1.86
Water	0.85	0.60

B. VALUES OF DIFFUSION COEFFICIENTS OF SOME MATERIALS IN LIQUIDS†

Solute	Solvent	T (°F)	Concentration (lb moles/ft^3)	D (ft^2/hr $\times 10^5$)
H_2	Water	61	very dilute	18.2
N_2	Water	72	very dilute	7.9
CO_2	Water	68	very dilute	7.0
Cl_2	Water	54	0.0062	5.4
Br_2	Water	54	0.0062	3.5
NH_3	Water	54	0.062	6.2
NaCl	Water	64	0.0062	4.8
			0.025	4.6
			0.062	4.8
			0.187	5.3
Glycerol	Water	59	very dilute	2.7
CO_2	Ethyl alcohol	63	very dilute	12.4
NaI	Methyl alcohol	57	0.0062	3.9

† The values of D are obtained from the International Critical Tables (1) where available; otherwise, D is found from Eq. (14.9) and these values are indicated by an asterisk. In computing Sc, the value of ν for pure air is used, 0.512 ft^2/hr.

TABLE E.7 CONVERSION FACTORS

Geometry:	1 cm = 0.394 in. = 0.0328 ft
	1 gal = 0.134 ft^3
Mass:	1 gm = 2.205 × 10^{-3} lb_m
	1 slug = 32.174 lb_m
Density:	1 gm/cm^3 = 62.43 lb_m/ft^3
Force:	1 dyne = 1 gm cm/sec^2 = 2.248 × 10^{-6} lb_f
Pressure:	1 $dyne/cm^2$ = 0.0021 lb_f/ft^2
	1 in. Hg = 70.73 lb_f/ft^2
	1 in. H_2O = 5.20 lb_f/ft^2
Viscosity:	1 centipoise = 2.42 lb_m/hr ft
	1 lb_f-hr/ft^2 = 4.17 × 10^8 lb_m/hr ft
Energy:	1 ft lb_f = 0.001285 Btu
	1 kw-hr = 3413 Btu
	1 hp = 2544 Btu
	1 cal = 4.187 joule = 3.97 × 10^{-3} Btu
Energy flux:	1 cal/sec cm^2 = 13,272 Btu/hr ft^2
	1 $watt/cm^2$ = 3171 Btu/hr ft^2
Specific heat:	1 cal/gm C = 1 Btu/lb_m F
Thermal conductivity:	1 cal/sec cm C = 241.9 Btu/hr ft F

References for Physical Properties

1. *International Critical Tables*, McGraw-Hill, 1929.
2. Keenan, J. H., and J. Kaye, *Gas Tables*, Wiley, 1945.
3. Keenan, J. H., and F. G. Keyes, *Thermodynamic Properties of Steam*, Wiley, 1936.
4. Marks, L. S., *Mechanical Engineers' Handbook*, 5th Ed., McGraw-Hill, 1951.
5. Perry, J. H., *Chemical Engineers' Handbook*, 3rd Ed., McGraw-Hill, 1950.
6. *Liquid Metals Handbook*, 2nd Ed., U.S. Government Printing Office, Washington, D. C., 1952.
7. McAdams, W. H., *Heat Transmission*, 3rd Ed., McGraw-Hill, 1954. Table of Normal Total Emissivity compiled by H. C. Hottel.
8. Sherwood, T. K., and R. L. Pigford, *Absorption and Extraction*, McGraw-Hill, 1952.
9. *Handbook of Chemistry and Physics*, 40th Ed., Chemical Rubber Publishing Co., Cleveland, 1959.
10. Hilsenrath, J., et al., *Tables of Thermal Properties of Gases*, Nat. Bur. of Standards, Circ. 564, 1955.
11. Wilkes, G. B., *Heat Insulation*, Wiley, 1950.
12. Warfield, C. N., *Tentative Tables for the Properties of the Upper Atmosphere*, NACA TN1200, 1947.

APPENDIX F

Review of Property Relations of Equilibrium Gas Mixtures

In order to determine the state of a thermodynamic system consisting of a mixture of perfect gases, the masses m_i of the component gases must be specified in addition to the usual state variables of a single-component system. If the given gas mixture is inert, then all the m_i's are constant and the mixture itself can be treated as a single, perfect gas. An example is dry air. In a chemically reacting gas mixture, on the other hand, the composition of the mixture usually varies with both the pressure and the temperature, and the mixture does not behave like a perfect gas. In this brief review of the property relations, we shall confine our attention to inert mixtures and to a particularly simple but important case of varying-composition mixture, namely gas-vapor mixtures.

Mixtures of Inert Ideal Gases

The following equations relate the properties of the mixture to the properties of its constituents. The mixture properties are indicated by quantities without subscripts.

The mass of the mixture equals the sum of the masses of its constituents:

$$m = \sum m_i$$

The pressure of the mixture equals the sum of the partial pressures of its constituents:

$$P = \sum p_i$$

where the partial pressure p_i is defined as the pressure of component i if it alone occupied the volume of the mixture at the temperature of the mixture.

The volume of the mixture equals the sum of the partial volumes of its constituents:

$$V = \sum V_i$$

where the partial volume V_i is defined as the volume that the component i would occupy if it were separated out at the pressure and the temperature of the mixture.

A simple relation exists between the partial pressure, the partial volume, and the composition of the mixture:

$$\frac{p_i}{p} = \frac{V_i}{V} = x_i$$

where x_i is the ratio of the number of moles of component i to the total number of moles of the mixture, namely the mole fraction.

In addition, we can write for the mixture:

Internal energy:

$$mu = \sum m_i u_i$$

Enthalpy:

$$mi = \sum m_i i_i$$

Entropy:

$$ms = \sum m_i s_i$$

Specific heats:

$$mc_v = \sum m_i c_{vi}, \qquad mc_p = \sum m_i c_{pi}$$

Gas constant:

$$mR = \sum m_i R_i$$

Note that the ratio of specific heats ($\gamma = c_p/c_v$) does not follow this additive rule. The rules for computing viscosity and thermal conductivity of mixtures are discussed in Chap. 20.

All the foregoing relations may be written on a molal basis by replacing the mass of the mixture m by the total number of moles n, and the mass of each component m_i by its corresponding number of moles n_i. These relations apply approximately to mixtures of real gases at low pressures.

Gas-Vapor Mixtures

In most applications, vapor can be treated as a perfect gas. However, changes in the mixture composition generally occur when the vapor component undergoes phase changes. A familiar example is the atmospheric air which may be regarded as a mixture of dry air (treated as a single component) and water vapor. The following definitions have been found useful in the study of such mixtures.

The specific humidity is the ratio of the mass of vapor to the mass of dry gas in the mixture. Then, assuming each component can be represented as a perfect gas, $m_i = M_i p_i V/\mathcal{R}T$, the specific humidity may be written as follows:

$$\omega = \frac{m_v}{m_g} = \frac{M_v}{M_g} \frac{p_v}{P - p_v}$$

For air and water vapor mixtures $M_v/M_g = 18/29.95 = 0.622$. Note that ω is not the mass fraction of vapor in the mixture.

The relative humidity ϕ is the ratio of mass of vapor in the gas to the mass of vapor when the gas is saturated. Again, assuming the perfect gas equation applies

$$\phi = \frac{m_v}{m_{v,\,\text{sat}}} = \frac{p_v}{p_{v,\,\text{sat}}}$$

The dew point temperature is the saturation temperature of the vapor corresponding to its partial pressure in the mixture. It is the temperature at which condensation begins when the mixture is cooled at constant pressure.

Changes in the equilibrium properties of gas-vapor mixtures, as during drying or humidification processes, may be calculated by considering each component separately and using the appropriate tables or equations of state. An alternate and simpler procedure is provided by the psychrometric chart available for mixtures of air and water vapor.

Index

I

J

K

L

BOILING HEAT TRANSFER

Methods of Augmentation

1) free convection
2) incipient boiling
must get above boiling Temp
Bubbles come from cavities - nucleation site
which trap vapor. This is necessary for boiling.

The peak of the curve is the burnout. After this is film boiling.
The film prevents q/A so the curve slumps after the peak. Result of a multitude of bubbles coming together.

In 1 the q/A increases due to more & more activated cavities and more stirring from the bubbles.

AUGMENTATION

Techniques
- surface promoters
- Displaced "
- Vortex flow
- additives
- capillary devices
- surface vibration
- fluid vibration
- EHD or BHD (electric fields)
- Injection or suction

Changing the surface to a rougher type increases q/A because of increased cavities in nucleate boiling. No change in film boiling because fluid is not in contact with wall